T0230665

TRANSMISSION LINES, MATCHING, AND CROSSTALK

TRANSMISSION LINES, MATCHING, AND CROSSTALK

Kenneth L. Kaiser

Kettering University
Flint, Michigan

A CRC title, part of the Taylor & Francis imprint, a member of the Taylor & Francis Group, the academic division of T&F Informa plc.

Published in 2006 by
CRC Press
Taylor & Francis Group
6000 Broken Sound Parkway NW, Suite 300
Boca Raton, FL 33487-2742

International Standard Book Number-10: 0-8493-6362-4 (Hardcover)
International Standard Book Number-13: 978-0-8493-6362-7 (Hardcover)
Library of Congress Card Number 2005050489

Library of Congress Cataloging-in-Publication Data

Kaiser, Kenneth L.
Transmission lines, matching, and crosstalk / by Kenneth L. Kaiser.
p. cm.
Includes bibliographical references and index.
ISBN 0-8493-6362-4 (alk. paper)
1. Telecommunication lines. 2. Impedance matching. 3. Crosstalk. I. Title.

TK5103.15.K35 2005
621.319--dc22 2005050489

Taylor & Francis Group
is the Academic Division of T&F Informa plc.

Visit the Taylor & Francis Web site at
http://www.taylorandfrancis.com

and the CRC Press Web site at
http://www.crcpress.com

Preface

Over time, the topic of transmission lines (and electromagnetics) has been given less coverage in electrical engineering courses. As many working engineers are fully aware, this is not because the topic is unimportant or irrelevant but because the number of other important topics in electrical engineering has blossomed. Transmission lines, matching techniques for lines, and crosstalk between lines have, in the age of the ubiquitous cellphone, once again become vital topics. In many applications, strange, wondrous, and unexpected results can sometimes be explained if transmission line theory is understood.

The transmission line and related topics contained in this book were selected from my previously published reference and textbook published by CRC Press entitled, *Electromagnetic Compatibility Handbook*. I hope this "spinoff" book, being only a small fraction of the *Handbook*, can be more conveniently used and more easily owned by individuals interested mainly in these topics. One of the main purposes of this particular book is to demystify transmission lines, matching, and crosstalk and to help explain the source and limitations of the approximations, guidelines, models, and rules-of-thumb seen in this field. For further reference and personal edification, many of the examples contained in this book were written by me to document the answers to questions I had about the subject matter. Although the chapters in this book are fairly self contained, it is assumed that the reader has a rudimentary background in electromagnetics, including vector analysis.

I have tried to be diligent in crediting all of the sources that were used in solving a problem, generating tabled results, or understanding an unfamiliar or a confusing concept, and I apologize in advance for any oversights. (If, by chance, you locate some material in this book not appropriately referenced, which is possible considering the book's length and the number of editing iterations, please e-mail me. I will include the addition on my web site and in any future editions of this book.)

So as not to burden the reader with frequent citation interruptions, in most cases the references are grouped together at the end of each problem statement and located between brackets: [ ]. The references are listed by last name roughly from most used to least used. (The date of publication is also given when necessary to avoid confusion with other authors with the same surname or different publications by the same author.) In some cases, as with many of the tables, specific references are provided for particular equations or results. When possible, I have tried to use original sources.

The program Mathcad was used to generate most of the plots and solve many of the equations. The major reason this program was selected is that it is easily understood even with little or no prior experience with Mathcad. Although it is somewhat unorthodox, entire Mathcad programs have been provided in many cases. This allows all of the variable assignments, assumptions, equations, and possible mistakes to be clearly seen. I used the program VISIO, sometimes with an embedded Mathcad output, to generate all of the figures.

Much time was devoted to crafting the tables in this book. A brief comment concerning the accuracy of the many expressions contained in this book, including in the tables, is appropriate. Not in this book are the many derivations and checks. Although most of the equations in this book were personally derived, obviously, not every equation in this book has been personally obtained from basic principles. Extensive use has been made of the years of hard work of myriad scholars. It can be stated, however, that nearly every expression has been either derived or checked, sometimes through much effort. A typical analytical and numerical check might involve taking a limit on one or more variables in an expression and then comparing it to a reliable result. The number of approximate expressions contained in this book prohibited an indication of their relative error. Sets of approximations are provided, in part, to help show the relationship of the individual expressions to each other.

As in the preface of *Electromagnetic Compatibility Handbook*, I again thank my friend, mentor, and colleague Professor James C. McLaughlin (Kettering University) for his careful review of these chapters, his many suggestions, and his steadfast support.

Your comments, suggestions, and corrections are most welcomed and encouraged. They are invaluable to the current edition and any possible future editions of this book. At the site http://www.klkaiser.com, student problems, errata, and a few extras are provided for the *Electromagnetic Compatibility Handbook*. I sincerely hope this book is helpful and inspirational to you.Unless specified otherwise, Système International (SI) units are used throughout this book. Enjoy.

Kenneth L. Kaiser
Electrical and Computer Engineering Department
Kettering University

The Author

Kenneth L. Kaiser's interest in electrical engineering began in high school with his involvement in amateur radio. While obtaining a solid theoretical background in a number of fields in electrical engineering, he obtained additional inspiration and practical experience working in several nonacademic positions. Since obtaining a Ph.D. in electrical engineering from Purdue University, he has focused his attention on the researching of topics of personal and industrial interest, on effective teaching methods, and on the writing of this book. He is currently at Kettering University (formerly GMI Engineering & Management Institute), continuing with his research and stimulating his students to excel.

Contents

1
Electrical Length

In many electrical engineering problems, frequently the electrical length, not the physical length, is the most important parameter of interest. The concept of electrical length is used throughout this book.

1.1 Electrical Length vs. Physical Length

What is the relationship between the electrical and physical length of a conductor? How does the permittivity and permeability affect this relationship?

The electrical length of a conductor is a function of its actual physical length and the electrical wavelength of the signal on the conductor. The relationship between the wavelength, λ, of a signal traveling in the positive z direction, its frequency, f, and its velocity, v, is shown in Figure 1.1. As time progresses, the wave travels in the positive z direction. The wavelength is the *spatial* distance between two successive maximums or minimums. For this reason, the wavelength is often used to describe the "size" of a signal. The period, T, describes how frequently the signal repeats itself in *time*. That is, if the amplitude of the signal is examined at one particular location in space (e.g., $z = 0$), the strength of the signal will appear with the same amplitude after T seconds. As seen in Figure 1.1, after $t = T/2$ seconds, or one-half of a period, the signal is 180° out of phase (i.e., same strength but opposite in sign) with its value at $z = 0$.

Although the frequency and period of a signal are not a function of the medium surrounding the signal (or the conductor carrying the signal), the velocity of the signal and its wavelength are a function of the electrical properties of the medium. If the medium is the same everywhere (i.e., homogeneous) and is lossless (i.e., zero conductivity), then the wavelength, λ, is

$$\lambda = \frac{v}{f} = \frac{\frac{1}{\sqrt{\mu\varepsilon}}}{f} \tag{1.1}$$

where μ is the electrical permeability and ε is the electrical permittivity of the surrounding lossless medium. The electrical length or electrical dimension is sometimes given as the ratio of the physical length to the wavelength:

$$\frac{l_{th}}{\lambda} = \frac{l_{th}}{\frac{v}{f}} = \frac{l_{th}}{\frac{1}{f\sqrt{\mu\varepsilon}}} = l_{th} f\sqrt{\mu\varepsilon} \tag{1.2}$$

where l_{th} is the length of the object (or physical dimension of interest). The frequency, f, is usually the highest frequency of interest of the signal in the medium.

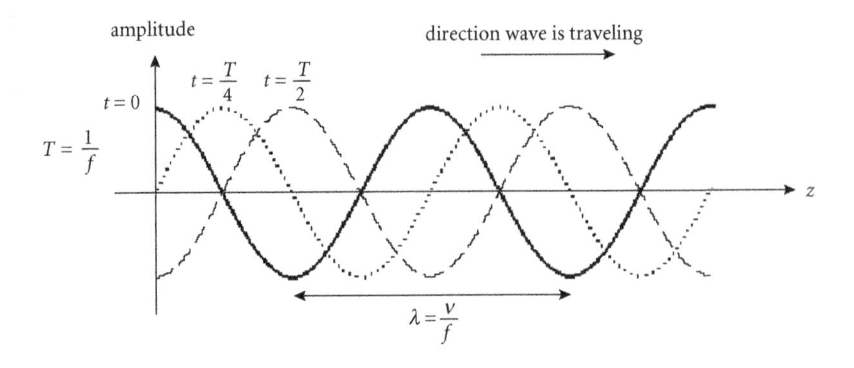

FIGURE 1.1 Wave traveling in the +z direction at three different times.

TABLE 1.1 Free-Space Wavelength and One-Tenth of This Wavelength at Various Frequencies

Frequency (f)	Free-Space Wavelength (λ_o)	$\lambda_o/10$
3 kHz	100 km	10 km
30 kHz	10 km	1 km
300 kHz	1 km	100 m
3 MHz	100 m	10 m
30 MHz	10 m	1 m
300 MHz	1 m	10 cm
3 GHz	10 cm	1 cm
30 GHz	1 cm	1 mm

For a fixed physical length, as the frequency increases, the electrical length increases. As the permeability or permittivity increases, the electrical length also increases since both the velocity of propagation and wavelength decreases. A handy reference relationship for free-space conditions is $\lambda = 1$ m for $f = 300$ MHz. Table 1.1 can be quickly generated from this one relationship.

It is important to remember that a conductor is considered electrically "long" or "large" when its length (or any dimension of interest) is a large fraction of a wavelength, and that a conductor is electrically "short" or "small" when its length (or any dimension of interest) is a small fraction of a wavelength. *In this book, a small fraction of a wavelength will be considered about one-tenth of a wavelength or less.*

1.2 Standing Waves

A ribbon cable is one meter in length. At what frequency could "full" standing waves begin to appear on the cable? [Matisoff]

When the characteristic impedance of a transmission line (e.g., 75 Ω for a common coaxial cable) is not equal to the impedance of the load connected to the line, the line is not matched to its load. Under this condition, and if the line is electrically long, one or more "full" standing waves will appear on the line. More specifically, if the line is at least one-half wavelength long, at least one peak and one valley of the rms values of the voltage and current will appear along an unmatched line (in steady-state conditions for a sinusoidal input signal). One "full" standing wave will be defined as one complete cycle of the rms value of the voltage or current along a line.

In Figure 1.2, the voltage and current waveforms are shown along a transmission line that is electrically long. The 0 Ω load presents a mismatch to the transmission line. The voltage is zero at the load, and it varies along the transmission line in a periodic manner. The current is a maximum at the short, and it

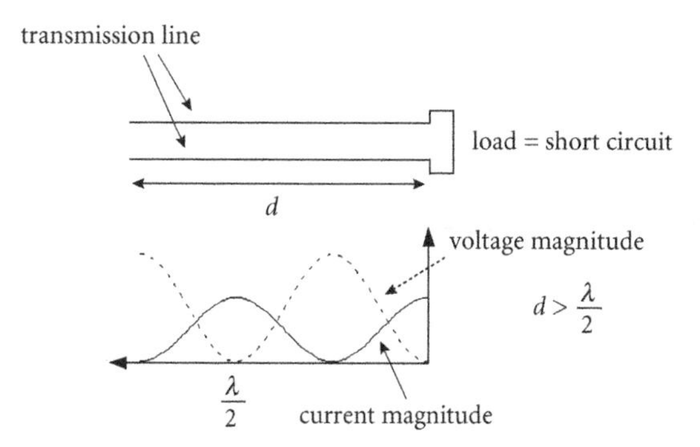

FIGURE 1.2 Voltage and current standing waves along an electrically-long transmission line with a 0 Ω load.

varies in a periodic manner along the transmission line. The peaks and valleys of the voltage and current are clearly present along the line.

Setting one meter equal to $\lambda/2$, λ is equal to 2 m for a full standing wave on the one meter line. The frequency of a signal corresponding to a wavelength of 2 m is

$$2 = \frac{v}{f} = \frac{\frac{1}{\sqrt{\mu\varepsilon}}}{f} \Rightarrow f = \frac{1}{2\sqrt{\mu\varepsilon}} \tag{1.3}$$

Both the surrounding air and ribbon cable are nonmagnetic so the permeability, μ, corresponds to the free-space value, $\mu_o = 4\pi \times 10^{-7}$ H/m. However, the permittivity, ε, of the ribbon cable is *not* equal to air's free-space permittivity, $\varepsilon_o = 8.854 \times 10^{-12}$ F/m. An effective value for the permittivity in (1.3) is often required or at least used. However, for wires in ribbon cable that are surrounded by thick insulative material, the effective value for the permittivity for two adjacent wires is nearly that of the permittivity of the insulation. Polyester, PVC, and mylar are insulators used for ribbon cables. If the insulative material is PVC, the permittivity is approximately $2.7\varepsilon_o$. The frequency corresponding to a wavelength of 2 m is then about 91 MHz:

$$f = \frac{1}{2\sqrt{\mu\varepsilon}} = \frac{1}{2\sqrt{(4\pi \times 10^{-7})(2.7)(8.854 \times 10^{-12})}} \approx 91\,\text{MHz}$$

At least one full standing wave will be present at this frequency and any frequency above 91 MHz for an unmatched line. A variation in the voltage and current along an unmatched line will still be present at frequencies less than 91 MHz but not a full standing wave. (At 18 MHz, where the line is about one-tenth of a wavelength, the line is considered electrically short.)

1.3 Antenna Effects and Effective Permittivity

A trace on an instrument cluster board is 20" long. At what frequency would this trace begin to act like an antenna? [ASM, '89; Paul, '92(b); Coombs; Harper, '00]

Assuming that the trace is part of a microstrip transmission line, the electric field is partially in the dielectric substrate between the trace and ground plane and partially in the air above the trace. The effective relative permittivity, ε_{reff}, or effective dielectric constant is approximately the average of the substrate's and air's relative permittivities:

$$\varepsilon_{reff} \approx \frac{1}{2}(\varepsilon_r + 1) \tag{1.4}$$

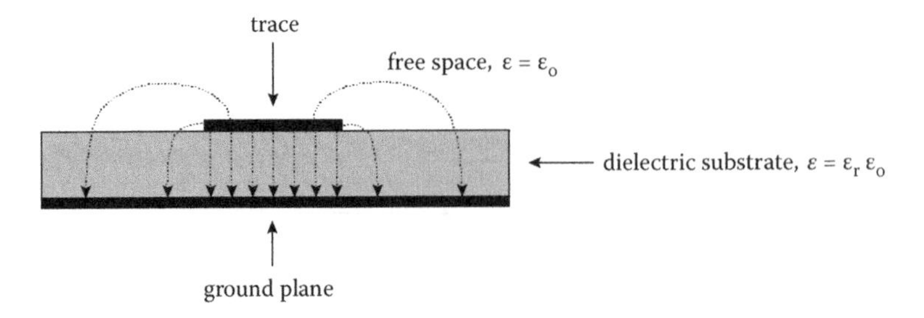

FIGURE 1.3 Rough sketch of the electric field distribution for a microstrip transmission line.

where ε_r is the relative permittivity, or dielectric constant, of the substrate. The effective permittivity, ε_{eff}, is equal to $\varepsilon_{eff} = \varepsilon_{reff}\varepsilon_o$. A rough *sketch* of the electric field distribution for a microstrip line is shown in Figure 1.3, assuming that the trace is at a positive voltage relative to the ground plane. The effective permittivity approaches the permittivity of the substrate material when the total "number"[1] of fields in the substrate is large relative to the total "number" of fields in the air above the substrate.

Currently, for many *printed circuit boards* (PCBs), ε_r ranges from about 2 to 5 (e.g., for the commonly used substrate material FR-4, the low-frequency ε_r is 4.8). The *effective* relative permittivity is about 2.9 for many microstrip lines. The frequency corresponding to a wavelength of 20" (≈ 0.51 m) is

$$f = \frac{\frac{1}{\sqrt{\mu\varepsilon}}}{\lambda} = \frac{1}{0.51\sqrt{(4\pi\times10^{-7})(2.9)(8.854\times10^{-12})}} \approx 350 \text{ MHz}$$

This trace length is considered electrically small for signal wavelengths greater than 5.1 m or frequencies less than 35 MHz. For frequencies much greater than 35 MHz, this trace will begin to act like an (effective) antenna.

The permittivity and loss of materials vary with frequency. For example, FR-4 has a relative permittivity of 4.8 with a dissipation factor (a measure of the loss of the material) defined as the ratio

$$\frac{\sigma}{\omega\varepsilon} = \frac{\sigma}{\omega\varepsilon_r\varepsilon_o} \quad \text{where } \sigma = \text{effective conductivity of the material} \tag{1.5}$$

equal to 0.009 at 100 Hz. At 10 GHz, the dielectric constant of FR-4 is 4.4 with a dissipation factor of 0.025. Notice that the loss of this material increased with frequency. (Ideally, insulators have zero conductivity.) For high-frequency applications, materials such as GT, GX, polystyrene, and cross-linked polystyrene are used with dielectric constants of 2.8, 2.8, 2.5, and 2.6, respectively. They all have dissipation factors less than 0.005.[2]

As the relative or effective permittivity increases, the velocity[3] of propagation decreases:

$$v = \frac{1}{\sqrt{\mu\varepsilon}} = \frac{1}{\sqrt{\mu\varepsilon_{reff}\varepsilon_o}} \tag{1.6}$$

[1]This total "number" is a function of the electric flux density and electric flux.

[2]Rogers Corporation (RO4003), Allied Signal (FR-408), GE Electromaterials (GETEK), and Polycad (PCL-LD-621) offer dielectric materials with low losses at these microwave frequencies.

[3]Actually, this velocity is an "average" result since a signal traveling in a nonhomogeneous medium will have multiple different velocities based on the electrical properties of the medium. For example, in a microstrip line, the velocity of that portion of the signal in the air above the trace is greater than that portion of the signal in the substrate between the trace and ground plane.

For high-speed applications, such as in the GHz range, the velocity of propagation is important. The degree of capacitive crosstalk or coupling between conductors, which can be a source of interference, is also a function of this effective permittivity.

1.4 Unshielded Conductor Radiation

One *electrostatic discharge* (ESD) guideline is that an unshielded portion of a cable conductor should be less than 4 cm long. Determine the frequency at which this unshielded portion begins to act like an antenna.

The frequency corresponding to a wavelength of 0.04 m in free space is

$$f = \frac{\frac{1}{\sqrt{\mu\varepsilon}}}{\lambda} = \frac{1}{0.04\sqrt{(4\pi\times10^{-7})(8.854\times10^{-12})}} \approx 7.5\,\text{GHz} = 7{,}500\,\text{MHz}$$

The unshielded portion of the cable is considered electrically small for signal wavelengths greater than 0.4 m (0.4/10 = 0.04) or frequencies less than 750 MHz. For frequencies greater than 750 MHz, the unshielded portion begins to act like an (effective) antenna. The one-tenth wavelength guideline used in this book to determine whether a conductor is electrically small is only an approximation. Other books use guidelines such as one-twentieth of the wavelength.

1.5 PCB Trace Radiation

A properly designed PCB begins to radiate (efficiently) at a frequency near 100 MHz. What is the length of the trace?

Assuming an effective dielectric constant of 2.9, the wavelength corresponding to 100 MHz is

$$\lambda = \frac{\frac{1}{\sqrt{\mu\varepsilon}}}{f} = \frac{1}{100\times10^{6}\sqrt{(4\pi\times10^{-7})(2.9)(8.854\times10^{-12})}} \approx 1.8\,\text{m}$$

A trace with a length less than 1.8/10 = 0.18 m = 18 cm should not radiate very effectively. If the trace is effectively radiating, it is probably greater than 18 cm in length.

1.6 Electrically-Large Car

If a mid-size automobile is exposed to a radar signal (for speed detection), will it appear electrically small? [Liao]

There are several radar bands or frequency ranges. The X Band is from 8–12 GHz, and the K Band is from 18–27 GHz (these relationships are the IEEE designations, which are different from the U.S. military designations). Selecting f = 25 GHz, the free-space wavelength is

$$\lambda = \frac{\frac{1}{\sqrt{\mu\varepsilon}}}{f} = \frac{1}{25\times10^{9}\sqrt{(4\pi\times10^{-7})(8.854\times10^{-12})}} \approx 0.012\,\text{m}$$

Automobiles are much larger than 0.012/10 = 0.12 cm; thus, an automobile will appear electrically large to the radar signal (i.e., its radar cross section will be large). Furthermore, if the radius of curvature of the automobile at the location of radar contact is much greater than 1.2 cm, the surface will appear flat to the incident wave at that location.

1.7 Properties of Electrically-Small Metallic Objects

If a metallic object, such as a segment of wire or chassis, is electrically small, how will the voltage and current vary along its dimensions?

For an electrically-small metallic object, the variation of the voltage and current on the exterior of the object is often small. Voltage and current variations due to standing waves should not be significant if the object is electrically small at the highest frequency of interest. However, the variation of the electric and magnetic fields (and, hence, voltage and current) into the metallic object can be substantial due to the skin effect and the conductive nature of the object.

2 Cable Modeling

When cables, also referred to as transmission lines, are electrically short, they can be modeled using discrete resistors, capacitors, and inductors. Such a lumped model is typically much simpler to analyze than a distributed model. Also, for individuals without a background in transmission lines, the lumped model can be readily understood.

2.1 Purpose of a Cable

What is the basic purpose of a cable, and what field components are dominant between the conductors of the cable? [Ramo; Haus]

Before delving into the details of cable modeling, it is important to state that transmission lines or cables are usually designed to transmit or carry electromagnetic energy from their input to their output (or from a driver to a receiver) without picking up energy from sources outside of the cable. Also, they generally are not designed to radiate energy like an antenna.[1] The conductors in a cable guide or channel the energy. In many cases, it is desirable to know the actual electric and magnetic field distributions between the conductors of a cable. To determine the reasonableness of these distributions, the expected directions of the electric and magnetic fields should be fully understood. (The following discussion assumes the cable consists of two parallel conductors.)

The electric field is mainly perpendicular to the surfaces of the conductors in a cable. These perpendicular fields are denoted as $E_{\perp}$. For lines with reasonably small loss, the largest component of the electric field starts on one conductor and terminates on the other conductor. The voltage difference between the conductors is the source of this electric field between the conductors. There is also a component of the electric field parallel to the conductors, denoted as $E_{\|}$, but it is small for low-loss lines:

$$|E_{\perp}| \gg |E_{\|}| \tag{2.1}$$

When the conductors are lossless (i.e., infinite conductivity), then the electric field tangential to the conductors must be zero: the electric field along the surface of the conductors is essentially the change in voltage along the conductors. When the lines are not lossless, the tangential field is not zero. For lossy lines, there are components of the electric field in the direction of the conductors both inside and outside the conductors. The attenuation of the fields inside the conductors is a function of the skin depth of the conductors. As the skin depth decreases, the fields attenuate more rapidly when passing from the surface to the inside of the conductor.

The major component of the magnetic field is around each conductor. This component of the magnetic field is produced from the conduction current through each conductor, and the magnetic field's direction is determined by the right-hand rule. The conduction current, of course, is in the direction of the

[1] An exception is a leaky cable that is used for wireless communication systems.

conductor's axis. It is in one direction for the signal conductor and the opposite direction for the return conductor. There is also a displacement current between the conductors defined as

$$\frac{d\vec{D}}{dt} = \varepsilon\frac{d\vec{E}}{dt} \tag{2.2}$$

generated by the electric field between the two conductors (where ε is the permittivity of the material between the conductors). This displacement current does *not* represent the flow of free charges or electrons in the insulator or space between the conductors. (If the insulator is lossy, which *can* imply actual charge flow, the lossy aspect of the insulation can be modeled as a resistance or a conductance, G.) The flow of induced charge (from attraction or repulsion) on one conductor due to charge buildup on the other conductor is the source of this displacement current. The displacement current produces a magnetic field, which encircles an imaginary line directed from one conductor to the other. The magnetic field due to this displacement current is generally small compared to the magnetic field generated from the conduction current. Displacement current helps explain the loss due to radiation of some of the energy carried by a cable. For most cables, the displacement current, and resultant radiation, is small. For lossy lines, an electric field is present both inside and outside the conductors parallel to the axes of the conductors. A magnetic field is also present inside the conductors but encircling the axes of the conductors. As with the electric field inside the conductor, the attenuation of the magnetic field when passing from the surface to the core of the conductor increases with decreasing skin depth. If the conductors were perfect or lossless, then the skin depth would be zero and no electric or magnetic fields would be present inside the conductors.

The direction of the major components of the electric and magnetic fields in a cable can be visualized by recalling the field distribution of two static situations: the electric field of a charged, parallel plate capacitor; and the magnetic field of a current-carrying wire. For a parallel plate capacitor, the electric field is directed from the positively charged plate to the negatively charged plate. For a current-carrying wire, the magnetic field circles the wire, and its direction is obtained from the right-hand rule: when the thumb of the right hand is in the direction of the current, the curling of the remaining digits of the hand represents the direction of the magnetic field. Nontime-varying electric and magnetic fields are not coupled and do not interact. Although both of these situations assume static fields, these field distributions can also be used for time-varying fields in some cases. The time-varying electric and magnetic fields in a given plane along the line can be viewed as quasistatic when the losses are small, spacing between the conductors is electrically small, and medium between the conductors is electrically homogeneous or the same everywhere. In this case, the electric and magnetic fields are perpendicular to each other and transverse (perpendicular) to the direction of the resultant power flow, which is in the direction of the conductors. The electric and magnetic fields for a situation such as this are referred to as being transverse electromagnetic (TEM).

In Figure 2.1, the external electric fields are roughly sketched for a parallel plate transmission line for one instant of time. The length of the line is l_{th}, and its axis is in the z direction. This length is assumed

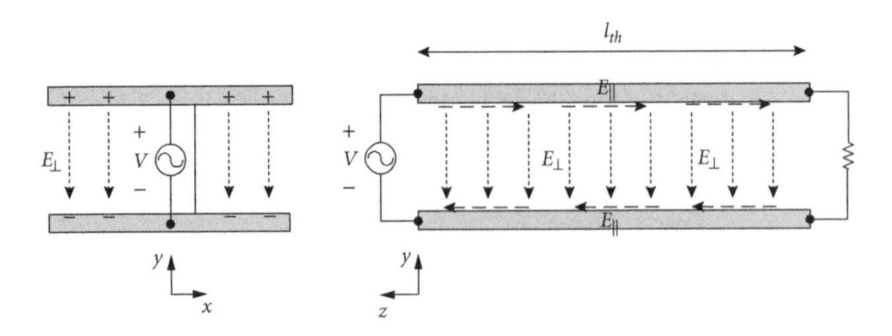

FIGURE 2.1 Electric fields between two closely spaced, parallel plate conductors.

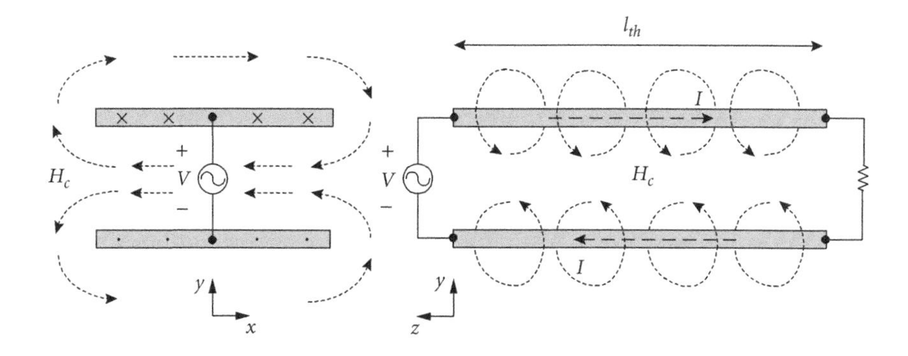

FIGURE 2.2 Magnetic fields from two closely spaced, parallel plate conductors.

short compared to a wavelength (otherwise, the electric field would change in magnitude and direction along the length of the line). The "fringing" fields, or variation of the electric field at the edges of the conductors, are not shown. The perpendicular, electric field component, $E_{\perp}$, is normally much greater than the parallel, electric field component, $E_{\parallel}$. The parallel field component is in the direction of the current along the conductors. The electric field components inside the conductors are not shown in Figure 2.1. Their attenuation into the conductors increases as the skin depth decreases.

In Figure 2.2, the external magnetic fields are roughly sketched for this same parallel plate transmission line for one instant of time. Although not shown in this figure, above the top conductor and below the bottom conductor, the magnetic fields from each of the conductors tends to cancel. The magnetic field components inside the conductors are not shown in Figure 2.2. As with the electric fields, their attenuation into the conductors increases as the skin depth decreases.

2.2 High-Fidelity Speaker Wire Candidates

A few possible high-fidelity speaker wire candidates and their resistance, capacitance, and inductance per foot are

Cable A: Jumper Wire (two parallel 0.375" diameter conductors)
2 mΩ/ft, 10 pF/ft, 0.35 μH/ft

Cable B: #18 Zip/Lamp Cord (two parallel #18 AWG conductors)
13 mΩ/ft, 10 pF/ft, 0.35 μH/ft

Cable C: Flat Ribbon Cable (36 parallel wires each #28 AWG, each conductor consists of 18 alternate wires)
7 mΩ/ft, 300 pF/ft, 0.05 μH/ft

Cable D: #12 Zip-Speaker Wire (two parallel #12 AWG conductors)
3 mΩ/ft, 20 pF/ft, 0.24 μH/ft.

Compare the inductive reactance, capacitive reactance, and ac resistance at 20 kHz for each of the given cables. The resistance value stated for each cable is the *total* dc resistance per foot determined by short circuiting the cable at the load, measuring the input resistance of the cable, and dividing by the total length of the *cable* (not the round-trip length). [Davis; Ballou]

In the world of high fidelity, there is a constant search for the "best" speaker cable. Several possible candidates are provided in this example. It is a common mistake for budding engineers to ignore the skin effect and just assume that the resistance of the conductors is equal to the dc resistance. In some cases, this approximation is valid. To eliminate any uncertainty, the skin depth is calculated for each of the given cables at 20 kHz. It is also a common mistake to assume that the cables are constructed of copper. This assumption must also be checked.

For cable A, assuming a circular cross section, the dc resistance per unit length is equal to

$$r_{DC} = \frac{1}{\sigma \pi r_w^2} \tag{2.3}$$

Solving for the conductivity,

$$\frac{2\times 10^{-3}\,\Omega/\text{ft}}{2}\,\frac{1\,\text{ft}}{0.305\,\text{m}} = \frac{1}{\sigma\pi(4.8\times 10^{-3})^2} \Rightarrow \sigma \approx 4.3\times 10^{6}\ 1/\Omega\text{-m}$$

This value is less than copper's conductivity. The skin depth at 20 kHz for this good conductor is

$$\delta = \frac{1}{\sqrt{\pi f \mu \sigma}} = \frac{1}{\sqrt{\pi(20\times 10^{3})(4\pi\times 10^{-7})(4.3\times 10^{6})}} \approx 1.7\ \text{mm}$$

Since the wire radius is about twice the skin depth, the following ac resistance equation may be used with less than 6% error:

$$r_{AC} = r_{DC}\left[1+\frac{1}{48}\left(\frac{r_w}{\delta}\right)^4\right] = \frac{2\times 10^{-3}}{2}\,\frac{1}{0.305}\left[1+\frac{1}{48}\left(\frac{4.8}{1.7}\right)^4\right] \approx 7.6\ \text{m}\Omega/\text{m}$$

This value is the resistance per meter for one of the conductors.

For cable B, the conductivity is

$$\frac{13\times 10^{-3}\,\Omega/\text{ft}}{2}\,\frac{1\,\text{ft}}{0.305\,\text{m}} = \frac{1}{\sigma\pi(5.1\times 10^{-4})^2} \Rightarrow \sigma \approx 5.7\times 10^{7}\ 1/\Omega\text{-m}$$

It appears that this wire is constructed of copper. The skin depth corresponding to this very good conductor is

$$\delta = \frac{1}{\sqrt{\pi f \mu \sigma}} = \frac{1}{\sqrt{\pi(20\times 10^{3})(4\pi\times 10^{-7})(5.7\times 10^{7})}} \approx 0.47\ \text{mm}$$

Since the wire radius is about the same as the skin depth, the dc resistance equation may be used to approximate the ac resistance with less than 6% error:

$$r_{DC} = \frac{1}{\sigma\pi r_w^2} = \frac{1}{(5.7\times 10^{7})\pi(5.1\times 10^{-4})^2} \approx 21\ \text{m}\Omega/\text{m}$$

This value is the resistance per meter for one of the conductors.[2]

For cable C, the conductivity is

$$18\times\frac{7\times 10^{-3}\,\Omega/\text{ft}}{2}\,\frac{1\,\text{ft}}{0.305\,\text{m}} = \frac{1}{\sigma\pi(1.6\times 10^{-4})^2} \Rightarrow \sigma \approx 6.0\times 10^{7}\ 1/\Omega\text{-m}$$

It appears that this wire is constructed of silver. (More likely, the conductors are constructed of copper, and this greater conductivity is a result of measurement error.) The factor of 18 in this expression is a result

[2]It is (mildly) interesting that when zip wire is partially pulled apart into two separate conductors it has the appearance of an open zipper.

of the 18 identical conductors in parallel. The resistance of 18 identical resistors, R, in parallel is $R/18$ (neglecting the proximity effect). The skin depth corresponding to this very good conductor is

$$\delta = \frac{1}{\sqrt{\pi f \mu \sigma}} = \frac{1}{\sqrt{\pi (20 \times 10^3)(4\pi \times 10^{-7})(6.0 \times 10^7)}} \approx 0.46 \text{ mm}$$

Since the wire radius is about one-third the skin depth, the dc resistance equation may be used to approximate the ac resistance with less than 1% error:

$$r_{DC} = \frac{1}{\sigma \pi r_w^2} = \frac{1}{(6.0 \times 10^7)\pi (1.6 \times 10^{-4})^2} \approx 210 \text{ m}\Omega/\text{m}$$

This value is the resistance per meter for one of the 36 conductors.

For cable D, the conductivity is

$$\frac{3 \times 10^{-3}\ \Omega/\text{ft}}{2} \frac{1\,\text{ft}}{0.305\,\text{m}} = \frac{1}{\sigma \pi (1.0 \times 10^{-3})^2} \quad \Rightarrow \sigma \approx 6.5 \times 10^7 \ 1/\Omega\text{-m}$$

Unless this conductor corresponds to super-cooled wire, this conductor is most likely copper. The skin depth corresponding to this very good conductor is

$$\delta = \frac{1}{\sqrt{\pi f \mu \sigma}} = \frac{1}{\sqrt{\pi (20 \times 10^3)(4\pi \times 10^{-7})(6.5 \times 10^7)}} \approx 0.44 \text{ mm}$$

Since the wire radius is about twice the skin depth, the given ac resistance equation may be used with less than 6% error:

$$r_{AC} = r_{DC}\left[1 + \frac{1}{48}\left(\frac{r_w}{\delta}\right)^4\right] = \frac{3 \times 10^{-3}}{2} \frac{1}{0.305}\left[1 + \frac{1}{48}\left(\frac{1.0}{0.44}\right)^4\right] \approx 7.7 \text{ m}\Omega/\text{m}$$

This value is the resistance per meter for one of the conductors.

The capacitive reactance for any of the cables is

$$X_C = \frac{-1}{\omega C}$$

and the inductive reactance for any of the cables is

$$X_L = \omega L$$

The inductive reactance, capacitive reactance, dc resistance, and ac resistance per meter at 20 kHz for each of the cables are provided in Table 2.1. Notice that the skin effect has noticeably increased the resistance for cables A and D. For these cables, the inductive reactance is greater than the ac resistance. The ideal cable would have zero ac resistance, zero inductive reactance, and infinite capacitive reactance. It appears that cables A and D are most ideal-like. Considering issues such as cost and size, cable D is probably the better choice.

TABLE 2.1 Impedance Properties of Four Cables at dc and 20 kHz

Cable	Inductive Reactance	Capacitive Reactance	dc Resistance	ac Resistance
A: Jumper	140 mΩ/m	−240 kΩ/m	6.6 mΩ/m	15 mΩ/m
B: #18 Zip	140 mΩ/m	−240 kΩ/m	43 mΩ/m	43 mΩ/m
C: Ribbon	20 mΩ/m	−8.1 kΩ/m	23 mΩ/m	23 mΩ/m
D: #12 Zip	99 mΩ/m	−120 kΩ/m	9.8 mΩ/m	15 mΩ/m

2.3 Selecting the Cable Model

Cables are sometimes designed to have a low impedance compared to the speaker's impedance to permit a "flatter" transmission to the speaker. Explain why. The impedance of a real speaker is frequency dependent, and for one particular speaker its magnitude varies from 5 to 25 Ω. If the speaker wire is 10 ft long, which of the speaker cables referred to in Table 2.1 are electrically short? For any one of the cables that is electrically short, use both the low-impedance and high-impedance lumped-circuit models to represent 10 ft of the cable. Using the extreme values of the speaker's impedance (assuming the speaker is resistive, although this is not true), determine the effect of the cable on low-frequency and high-frequency audio signals using both of these models. Which of these cable models is most accurate for this speaker? Determine whether the capacitive reactance, inductive reactance, or ac resistance has the greatest effect. [Paul, '92(b)]

Most speaker cables or transmission lines are designed to transfer information from the receiver to the speakers with negligible distortion and attenuation. By modeling the cable, it can be determined whether the cable is properly delivering its input signal to the load (i.e., speaker). When the cable is electrically short, an *RLC* model can be used to represent the cable. When discrete components are used to model the transmission line (or any electrical device), it is referred to as a lumped-circuit model. One possible lumped-circuit model for the cable is shown in Figure 2.3.

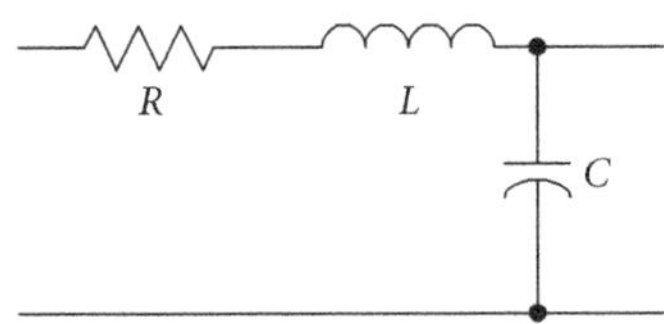

FIGURE 2.3 Simple *RLC* model for an electrically-short lossy cable. (Dielectric losses are not included in this model.)

Not surprisingly, the input impedance of an electrically-short cable is a function of the R, L, and C of the cable. The input impedance is also a function of load impedance connected across the cable, as well as the position of the elements in the model. The impedance of the cable, which should not be confused with the characteristic impedance of the cable to be discussed later, is sometimes defined assuming a zero impedance, or short-circuited load. To state that the cable impedance is low, probably implies the cable characteristics are mainly determined by R and L, and the reactance of the capacitor is large. In this case, the impedance of the cable, $R + j\omega L$, should be small compared to the impedance of the speaker, Z_L, if the distortion and loss due to the cable are to be small. The voltage across the speaker, V_L, given the input signal, V_i, is (ignoring the capacitance, C, of the cable)

$$V_L = V_i \frac{Z_L}{Z_L + R + j\omega L} \tag{2.4}$$

For an ideal speaker cable, $V_L = V_i$ for all frequencies. This equality is only true when the impedance of the cable is zero. However, when the cable impedance is small compared to the speaker impedance,

$$R + j\omega L \ll Z_L \tag{2.5}$$

the loss and distortion are small.[3]

Most audiophiles are not overly concerned with attenuation as long as the attenuation is constant, or flat, at all frequencies of interest. The impedance of an ideal resistor is flat with frequency, but the impedance of an ideal inductance is not flat with frequency. Speaker cable manufacturers often strive to minimize inductance for this and other reasons. Another approach to minimizing the effect of the

[3] It could be argued that a non-neglible cable impedance is desirable. For example, the cable impedance could be "properly" selected so that the cable functions as a predistortion filter to compensate for the distortion introduced by the speaker.

frequency-dependent inductive reactance is to cancel it with capacitive reactance (referred to as power factor correction in power systems).

Ten feet of any of the cables given in Table 2.1 is electrically short for audio frequencies (i.e., from 20 Hz to 20 kHz). The free-space wavelength at the highest frequency of interest, 20 kHz, is

$$\lambda_o = \frac{c}{f} = \frac{3\times 10^8}{20\times 10^3} = 15\text{ km}$$

For any reasonable dielectric constant, this length of cable is much less than one-tenth of a wavelength.

The resistance, inductance, and capacitance were provided for the speaker cables. There are many possible ways of arranging these elements to generate an *RLC* cable model, and the impedances of the source and load are a factor in the selection of the "best" model. The models given in Figures 2.4–2.7 are all appropriate for electrically-short lines, but some are better than others at higher frequencies and

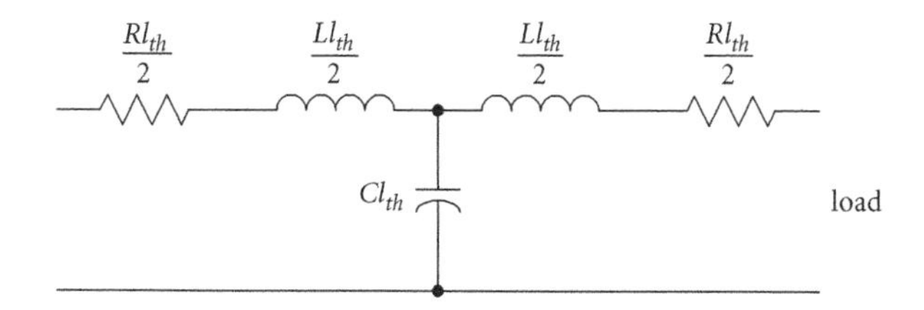

FIGURE 2.4 Lumped-T model for an electrically-short cable.

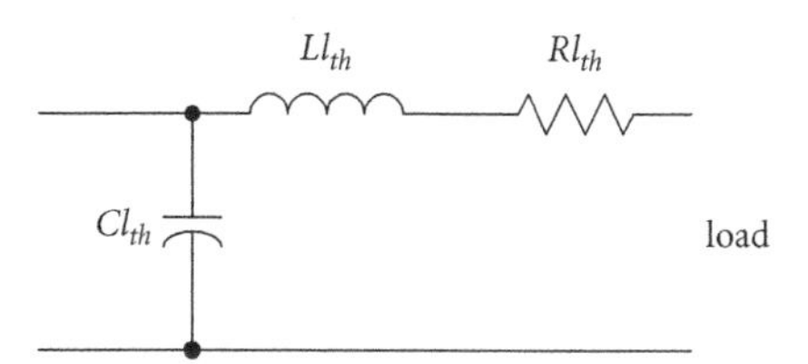

FIGURE 2.5 Lumped-Γ model for an electrically-short cable.

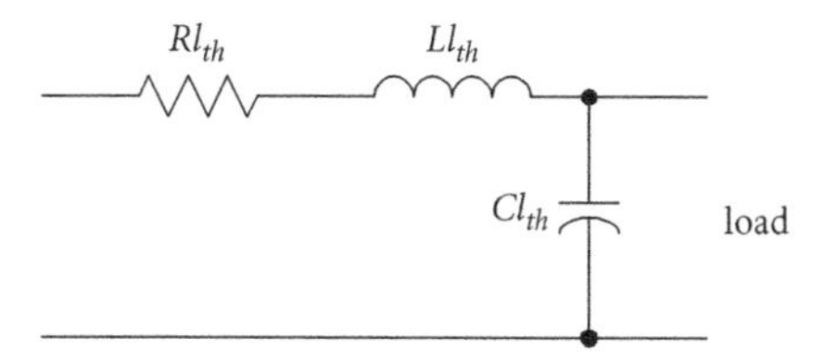

FIGURE 2.6 Lumped-backward Γ model for an electrically-short cable.

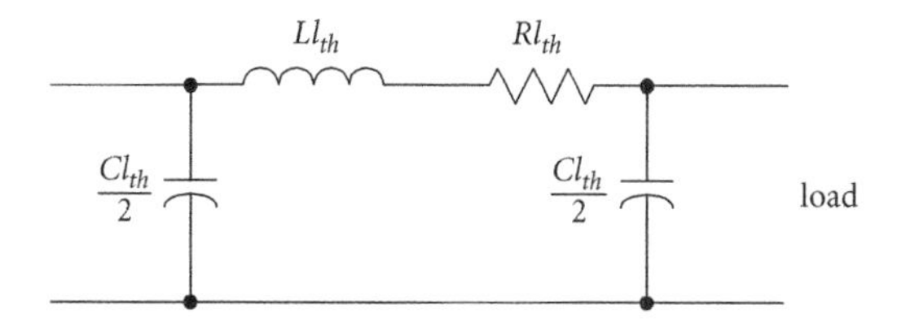

FIGURE 2.7 Lumped-π model for an electrically-short cable.

for low-impedance and high-impedance loads. The lumped-T and lumped-Γ models (l_{th} = line length and R, L, and C are per unit length) are best for low-impedance loads. A low-impedance load is less likely than a high-impedance load to mask the effect of the series inductance of the cable. The lumped-backward Γ and lumped-π models are best for high-impedance loads. Briefly, a high-impedance load is less likely than a low-impedance load to mask the effect of the shunt capacitance of the cable.

The response of the lumped-Γ and lumped-backward Γ model for cable D is given in Mathcad 2.1. The internal inductance is ignored, the ac resistance is set to its maximum value, and the resistance of the receiver

$$l_{th} := 10 \quad L := 0.24\cdot 10^{-6}\cdot l_{th} \quad C := 20\cdot 10^{-12}\cdot l_{th} \quad R := 15\cdot 10^{-3}\cdot 0.305\cdot l_{th} \quad R_s := 0.01$$

$$x := 5, 5.1 .. 60 \qquad j := \sqrt{-1}$$

$$\omega(x) := \left(x + 1 - 10\cdot \text{floor}\left(\frac{x}{10}\right)\right)\cdot 10^{\text{floor}\left(\frac{x}{10}\right)} \qquad X_C(\omega) := \frac{1}{j\cdot\omega\cdot C}$$

$$\text{Gain}_G(\omega, R_L) := \frac{\dfrac{(R_L + R + j\cdot\omega\cdot L)\cdot X_C(\omega)}{R_L + R + j\cdot\omega\cdot L + X_C(\omega)}}{\dfrac{(R_L + R + j\cdot\omega\cdot L)\cdot X_C(\omega)}{R_L + R + j\cdot\omega\cdot L + X_C(\omega)} + R_s}\cdot\frac{R_L}{R_L + R + j\cdot\omega\cdot L}$$

$$\text{Gain}_{IG}(\omega, R_L) := \frac{\dfrac{R_L\cdot X_C(\omega)}{R_L + X_C(\omega)}}{\dfrac{R_L\cdot X_C(\omega)}{R_L + X_C(\omega)} + R + j\cdot\omega\cdot L + R_s}$$

G = Gamma Model
IG = Inverse Gamma Model
l_{th} is in feet

MATHCAD 2.1 Gain vs. frequency for cable D using the lumped-Γ and backward lumped-Γ model.

$l_{th} := 10 \quad L := 0.24 \cdot 10^{-6} \cdot l_{th} \quad C := 20 \cdot 10^{-12} \cdot l_{th} \quad R := 15 \cdot 10^{-3} \cdot 0.305 \cdot l_{th} \quad R_s := 0.01 \quad R_L := 5$

$x := 5, 5.1 .. 60 \qquad j := \sqrt{-1}$

$$\omega(x) := \left(x + 1 - 10 \cdot \text{floor}\left(\frac{x}{10}\right)\right) \cdot 10^{\text{floor}\left(\frac{x}{10}\right)}$$

$$\text{Gain}(\omega, R, L, C) := \frac{\dfrac{(R_L + R + j \cdot \omega \cdot L) \cdot \dfrac{1}{j \cdot \omega \cdot C}}{R_L + R + j \cdot \omega \cdot L + \dfrac{1}{j \cdot \omega \cdot C}}}{\dfrac{(R_L + R + j \cdot \omega \cdot L) \cdot \dfrac{1}{j \cdot \omega \cdot C}}{R_L + R + j \cdot \omega \cdot L + \dfrac{1}{j \cdot \omega \cdot C}} + R_s} \cdot \frac{R_L}{R_L + R + j \cdot \omega \cdot L}$$

The Gamma model is used in this program and l_{th} is in feet.

MATHCAD 2.2 Gain vs. frequency for cable D when the inductance, resistance, or capacitance is individually set to zero.

is set to 0.01 Ω. The response is clearly flat up to 10 kHz for a 25 Ω speaker load. For heavier loading, when the speaker's impedance is 5 Ω, the response begins to deteriorate slightly before 10 kHz. As expected, the gain of the cable changes with the load. What may be somewhat surprising is that the low-impedance and high-impedance models have identical gain responses over the given frequency ranges: the line is electrically short over the plotted range.

To determine whether the inductance, capacitance, or ac resistance was most responsible for the higher frequency deterioration, each of these parameters was separately set to zero, and the 5 Ω low-impedance model responses examined. The results in Mathcad 2.2 clearly indicate that the inductance is most responsible for the frequency response roll off. The capacitance has little effect on the response, and the resistance is responsible for the frequency-independent distortionless loss. The cable capacitance could have more effect on the response if the speaker was modeled more realistically as an *RLC* circuit instead of a pure resistance.

2.4 Failure of the Lumped-Circuit Model

Why are the lumped-circuit representations not valid in the upper microwave range?

Most engineering students are familiar with resistors, inductors, and capacitors. However, many students are unaware that these simple elements are often inadequate for modeling circuits that are not electrically short. For elements that are not electrically short, including wires, traces, and leads, transmission line concepts (and sometimes waveguide concepts) must be evoked.

In most courses in basic circuit theory, when a voltage or current is modified in a system at some time t_x, the voltages and currents throughout the entire circuit are assumed to be immediately affected at t_x. The time delay associated with the wires is assumed zero. (Transient delay, however, can be introduced through capacitance and inductance.) If the distance between two elements in a circuit is electrically short, this is a reasonable assumption. If the distance is not electrically short, this is often not a good assumption. Electromagnetic fields travel at a finite speed.

In the upper microwave range, for example 100 GHz, the free-space wavelength is

$$\lambda_o = \frac{c}{f} = \frac{3\times 10^8}{100\times 10^9} = 3\text{ mm}$$

If the distance between circuit elements is greater than approximately 3/10 = 0.3 mm, then the lumped-circuit model should not be used at this frequency. This distance is quite small and easily exceeded by many lower frequency circuits.

2.5 Characteristic Impedance

Determine the characteristic impedance for each of the transmission lines given in Table 2.1 at 20 Hz and 20 kHz. For which of these cables would a 10 Ω load be considered low impedance? When should the characteristic impedance be used in analysis? When is the characteristic impedance complex for a real transmission line?

For lines that are electrically short, the previously given lumped-circuit models are usually adequate when working with transmission lines. However, if the line is not electrically short, then a transmission line is frequently described by its characteristic impedance, Z_o. It is defined as

$$Z_o = \sqrt{\frac{R + j\omega L}{G + j\omega C}} \tag{2.6}$$

and it has units of ohms. The characteristic impedance is in many cases the most important parameter of a transmission line. The variables R, L, G, and C represent the resistance, inductance, conductance, and capacitance per unit length of the transmission line. Sometimes, the characteristic impedance of a transmission line is referred to as the surge impedance, line impedance, or just impedance of the line.[4] Note that the characteristic impedance of a transmission line is a frequency-dependent complex quantity. It is actually the ratio of the voltage across the line (at a particular position) to the current through one of the conductors (at the same position) for a single traveling wave on the line. A single traveling wave is present on a transmission line when the load at the output of the line is matched or equal to the characteristic impedance of the line. It is also interesting to note that the initial impedance seen by a transient signal source (e.g., a digital signal rising from a low to high voltage) connected to a transmission

[4] The term line impedance or just impedance is confusing since it does *not* refer to $R + j\omega L$, $G + j\omega C$, or $R + j\omega L + G + j\omega C$.

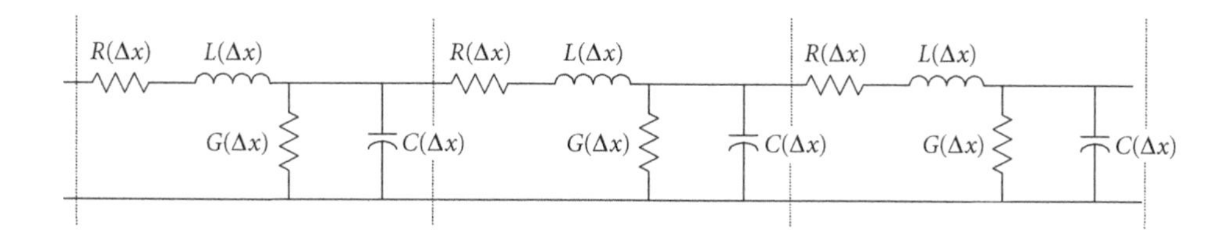

FIGURE 2.8 Transmission line model using three identical, electrically short lumped circuits each of length Δx.

line is equal to Z_o, not $R + j\omega L$ or other simple combination of impedances present in the lumped-circuit models.

When a line is electrically short, it can be represented by a lumped-circuit model. When it is not electrically short, then a simple lumped model is usually inadequate. A transmission line can be modeled, however, as a large number of lumped-circuits connected in cascade where the length of each of the lumped segments is very small. Figure 2.8 illustrates how a transmission line of total length $3(\Delta x)$ is modeled using three identical lumped circuits connected in cascade. As the size of each segment, Δx, becomes smaller, the number of segments must increase for a given overall line length. In addition, as the number of segments increases, the model improves. For this three-segment circuit to be a reasonable model for the transmission line, the length Δx should be electrically small.

The exact equation for the characteristic impedance given in (2.6), which can have both a real and an imaginary term, is not the one that is frequently seen. For many situations, especially at radio frequencies, the inductive reactance, ωL, is much greater than the series resistance, R, and the magnitude of the capacitive reactance, $1/(\omega C)$, is much less than $1/G$:

$$\omega L \gg R \quad \text{and} \quad \frac{1}{\omega C} \ll \frac{1}{G} \tag{2.7}$$

For these conditions, the expression for the characteristic impedance reduces to

$$Z_o \approx \sqrt{\frac{j\omega L}{j\omega C}} = \sqrt{\frac{L}{C}} \tag{2.8}$$

Equation (2.8) is the commonly seen and used high-frequency expression for the characteristic impedance of a transmission line. For many cables, this simplified equation can be used for frequencies above 10 kHz. Obviously, for low "voice" frequencies this simplified equation should not be used. (When the characteristic impedance is entirely real, it can be referred to as the characteristic resistance of the line.) Equation (2.8) is only a function of the inductance and capacitance per unit length of the cable. For example, for one RG/8U type cable, $L = 0.077$ μH/ft and $C = 30.8$ pF/ft. The characteristic impedance of this cable is 50 Ω. Many students incorrectly believe that a 50 Ω cable represents or is modeled by a 50 Ω resistor. (However, the input impedance of a transmission line with a characteristic impedance of Z_o is equal to Z_o when the impedance of the load at its output is also Z_o. Also, for transient signals, the input impedance of the cable is Z_o.) The characteristic impedance typically represents the square root of the ratio of the inductance per unit length to the capacitance per unit length of the cable.

When a cable has low losses or is lossless, the resistance and conductance of the cable are negligible. The expression for the characteristic impedance is again

$$Z_o \approx \sqrt{\frac{L}{C}} \quad \text{if } R \ll \omega L,\ G \ll \omega C \tag{2.9}$$

Since cable inductance increases and cable capacitance decreases as the spacing between the conductors increases, the characteristic impedance increases with conductor spacing; in other words, widely spaced conductors have a greater characteristic impedance than closely spaced conductors. As the conductor size increases, however, the inductance decreases, capacitance increases, and characteristic impedance decreases. In some situations, such with telephone and digital lines, the losses are not negligible, and the fundamental definition for Z_o should be used. When G and R are small but not negligible, the expression for the characteristic impedance can be reduced to one that does not involve the square root of a complex number:

$$
\begin{aligned}
Z_o &= \sqrt{\frac{R+j\omega L}{G+j\omega C}} = \sqrt{\frac{L}{C}}\sqrt{\frac{\dfrac{R}{j\omega L}+1}{\dfrac{G}{j\omega C}+1}} = \sqrt{\frac{L}{C}}\left(\frac{R}{j\omega L}+1\right)^{\frac{1}{2}}\left(\frac{G}{j\omega C}+1\right)^{-\frac{1}{2}} \\
&\approx \sqrt{\frac{L}{C}}\left(1+\frac{1}{2}\frac{R}{j\omega L}\right)\left(1-\frac{1}{2}\frac{G}{j\omega C}\right) \quad \text{if } R \ll \omega L,\ G \ll \omega C \\
&= \sqrt{\frac{L}{C}}\left(1+j\frac{1}{2}\frac{G}{\omega C}-j\frac{1}{2}\frac{R}{\omega L}+\frac{RG}{4\omega^2 LC}\right) \quad \text{if } R \ll \omega L,\ G \ll \omega C \\
&\approx \sqrt{\frac{L}{C}}\left[1+j\left(\frac{G}{2\omega C}-\frac{R}{2\omega L}\right)\right] \quad \text{if } R \ll \omega L,\ G \ll \omega C
\end{aligned}
\qquad (2.10)
$$

The first two terms of the binomial expansion

$$
(1+x)^n = 1+nx+\frac{n(n-1)}{2}x^2+\frac{n(n-1)(n-2)}{6}x^3+\cdots
$$

were used to approximate the two square roots, where $x = R/(j\omega L)$ and $G/(j\omega C)$ and n = 1/2 and −1/2, respectively. Note that the characteristic impedance contains an imaginary term that is a function of the conductor and dielectric resistances of the line.

For the speaker cables given in Table 2.1, the conductance, G, was not provided. It will be assumed negligible, which is reasonable for the 20 Hz–20 kHz frequency range. The low-loss approximation for the characteristic impedance could be used for the jumper, #18 zip, ribbon, and #12 zip cables, but the unapproximated version will be used for all the cables since the numerical package can easily handle complex numbers. As shown in Mathcad 2.3, of the four cables, the characteristic impedance of the flat ribbon cable is the lowest at both 20 Hz (305 – j305 Ω) and 20 kHz (14 – j6 Ω). For all of the cables, the characteristic impedance at 20 kHz is similar to their respective high-frequency low-loss value, $\sqrt{L/C}$. In Mathcad 2.3, the resistance, capacitance, and inductance are per foot, but the characteristic impedances are still in ohms. (Converting R, C, and L to per meter will not affect the final answer since the conversions multipliers will cancel in the characteristic impedance expression.)

At the highest frequency of interest, 20 kHz, the speaker cables should be electrically short. Therefore, providing the characteristic impedance of the cable is of limited value (at least for sinusoidal steady-state analysis). Providing values for R, L, G, and C is of greater importance. If noise transients or other high-frequency signals are also present on the speaker cable, then knowledge of the characteristic impedance may be helpful if the line is not electrically short at these higher frequencies. Whether a load impedance is low or high is sometimes relative to the characteristic impedance of the line. If the load impedance is much less than the characteristic impedance of the line, it is considered a low-impedance load. If the load impedance is much greater than the characteristic impedance of the line, it is considered a high-impedance load. Therefore, at 20 kHz, 10 Ω is a low-impedance load for all of the cables except cable C.

$$G := 0 \quad j := \sqrt{-1}$$

$$Z_O(f, R, L, C) := \sqrt{\frac{R + j \cdot 2 \cdot \pi \cdot f \cdot L}{G + j \cdot 2 \cdot \pi \cdot f \cdot C}} \qquad \sqrt{\frac{L}{C}}$$

Cable A: Jumper

$$Z_O\left(20, 2 \cdot 10^{-3}, 0.35 \cdot 10^{-6}, 10 \cdot 10^{-12}\right) = 902 - 882j$$

$$Z_O\left(20 \cdot 10^{-3}, 4.5 \cdot 10^{-3}, 0.35 \cdot 10^{-6}, 10 \cdot 10^{-12}\right) = 187 - 10j \qquad \sqrt{\frac{0.35 \cdot 10^{-6}}{10 \cdot 10^{-12}}} = 187$$

Cable B: #18 Zip

$$Z_O\left(20, 13 \cdot 10^{-3}, 0.35 \cdot 10^{-6}, 10 \cdot 10^{-12}\right) = 2 \times 10^3 - 2j \times 10^3$$

$$Z_O\left(20 \cdot 10^{3}, 13 \cdot 10^{-3}, 0.35 \cdot 10^{-6}, 10 \cdot 10^{-12}\right) = 189 - 27j \qquad \sqrt{\frac{0.35 \cdot 10^{-6}}{10 \cdot 10^{-12}}} = 187$$

Cable C: Ribbon

$$Z_O\left(20, 7 \cdot 10^{-3}, 0.05 \cdot 10^{-6}, 300 \cdot 10^{-12}\right) = 305 - 305j$$

$$Z_O\left(20 \cdot 10^{3}, 7 \cdot 10^{-3}, 0.05 \cdot 10^{-6}, 300 \cdot 10^{-12}\right) = 14 - 6j \qquad \sqrt{\frac{0.05 \cdot 10^{-6}}{300 \cdot 10^{-12}}} = 13$$

Cable D: #12 Zip

$$Z_O\left(20, 3 \cdot 10^{-3}, 0.24 \cdot 10^{-6}, 20 \cdot 10^{-12}\right) = 776 - 769j$$

$$Z_O\left(20 \cdot 10^{3}, 4.7 \cdot 10^{-3}, 0.24 \cdot 10^{-6}, 20 \cdot 10^{-12}\right) = 110 - 9j \qquad \sqrt{\frac{0.24 \cdot 10^{-6}}{20 \cdot 10^{-12}}} = 110$$

MATHCAD 2.3 Characteristic impedances of four different cables. Also given is the characteristic impedance using the high-frequency approximation.

It is occasionally helpful to know the characteristic impedance of a line even when the line is electrically short. When a lossless line is electrically short, with a load resistance R_L small compared to the line's characteristic impedance, the input impedance of the line is approximately

$$Z_{in} = R_L + j\omega L_T \left(1 - \frac{R_L^2 C_T}{L_T}\right) \tag{2.11}$$

The variables L_T and C_T represent the *total* inductance and capacitance of the transmission line, respectively. For most of the given speaker wire candidates connected to an 8 Ω speaker, all of the assumptions are satisfied. According to (2.11), the input reactance of the cable is then inductive. Notice that when the capacitance of the cable increases, the total reactance decreases. This decreases the distortion associated with the cable. Perhaps, this is the reason why some cable manufacturers state that cable capacitance can actually be beneficial! (However, for the cables listed in Table 2.1 connected to a low-impedance load, the capacitance must be quite large before it can have an impact on the inductive reactance.)

Four per-unit-length parameters are used to describe a transmission line: R, G, L, and C. As previously stated, usually the L and C are used to determine a cable's characteristic impedance. However, at low frequencies where $R \approx \omega L$ or $R > \omega L$, the resistance of the cable is important. The resistance of the conductors used in the cable will increase with frequency because of the skin effect, but beyond a certain frequency, the inductive reactance will dominate. (The inductance of the cable will also vary with frequency, but the capacitance will not vary with frequency.) When R is significant and G is negligible, the characteristic impedance will be complex:

$$Z_o \approx \sqrt{\frac{R + j\omega L}{j\omega C}} = \sqrt{-j\frac{R}{\omega C} + \frac{L}{C}} \quad \text{if } G \approx 0 \tag{2.12}$$

The characteristic impedance contains a negative imaginary term in this case. At higher frequencies, inductive reactance dominates over the ac resistance, and the characteristic impedance is real with no imaginary component:

$$Z_o \approx \sqrt{\frac{j\omega L}{j\omega C}} = \sqrt{\frac{L}{C}} \quad \text{if } \omega L \gg R,\ G \approx 0 \tag{2.13}$$

At even higher frequencies, assuming that the cross-sectional dimensions of the cable are still electrically small so that other waveguide-like modes do not begin to appear between the conductors, the characteristic impedance is still real and given by the previous expression. However, sometimes the G can be significant. Recall that G is the conductance per unit length (the conductance is the inverse of the resistance) *between* the conductors. The loss due to the ac resistance of the conductors is not a function of G. The losses in the dielectric or insulation used between the conductors, sometimes due to contamination, are a function of G. Often, the losses of a dielectric will increase with frequency. When G is significant at higher frequencies, the characteristic impedance is again complex:

$$Z_o \approx \sqrt{\frac{j\omega L}{G + j\omega C}} = \sqrt{\frac{j\omega L}{G + j\omega C}\frac{G - j\omega C}{G - j\omega C}} = \sqrt{\frac{\omega^2 LC + j\omega GL}{G^2 + (\omega C)^2}} \quad \text{if } R \ll \omega L \tag{2.14}$$

This characteristic impedance contains a positive imaginary term. Sometimes, for lower quality lines using stranded wire for one or more of the conductors, the R can also be significant at higher frequencies. This greater resistance is not due to the skin effect: a larger-than-expected high-frequency R can be due to the increase in the contact resistance due to corrosion.

2.6 Characteristic Impedance of a dc Power Bus

For dc power distribution, a characteristic impedance near 0 Ω is desirable. Often, an impedance less than 10 Ω is considered good, near 100 Ω is poor, and greater than 1,000 Ω is not acceptable. Why is the impedance of the dc *power* distribution of interest? Why should the characteristic impedance be kept low? What are the expressions for the characteristic impedance of typical dc busses? [Gupta, '79; Walker]

The dc power distribution or bus is designed to carry dc power to various systems, circuits, and components. Actually, since the characteristic impedance at dc, $\omega = 0$, is

$$Z_o = \sqrt{\frac{R}{G}}$$

and all lines are electrically small at dc, the dc characteristic impedance of the dc power distribution is not of interest (the resistive losses are of importance, however, especially at higher current levels). The characteristic impedance of the power bus is of interest at higher frequencies. These higher frequency signals are present on the bus because the current demands of the various devices change with time, and the ground is also a return for some signals.

To a beginner, it may not be obvious why a small characteristic impedance is desirable for dc buses. Obviously, a small R is desirable since the ohmic power loss decreases as R decreases. The purpose of the power bus is to deliver power to the various devices connected to the bus. In addition, the same voltage level is usually desired everywhere along the bus. If the resistance is high, then the voltage along the bus will strongly vary as a function of the current through the bus. For example, the dc voltage across the bus nearest to the supply source will be greater than the dc voltage across the bus farthest from the supply source. (The maximum dc voltage drop across the conductors is a function of the length and *thickness* of the bus conductors.) At high frequencies, this voltage drop is directly related to the characteristic impedance of the bus. Recall that the characteristic impedance is defined as the ratio of the voltage to the current of a single wave traveling along a line. Therefore,

$$V = Z_o I \tag{2.15}$$

for a single wave or transient signal traveling along the bus. Again, V is the voltage for a single wave traveling along the bus. The characteristic impedance should be small, if the ac voltage across the bus is to be small. At high frequencies,

$$Z_o \approx \sqrt{\frac{L}{C}}$$

A large bus capacitance and small bus inductance are thus desirable. In other words, the power conductor and its return should be large and close, which is essentially a physically large capacitor with low "lead" inductance. The high-frequency characteristic impedance is not a function of the resistance of the conductors. Therefore, even if the conductors were perfect, the voltage would still vary along the conductors. One disadvantage of a small characteristic impedance for high-voltage dc buses is the greater probability of breakdown between the conductors.

The high-frequency characteristic impedance of several common dc distribution lines will now be provided. The characteristic impedance for a microstrip line, shown in Figure 2.9, is given by the expression (see Appendix A for a more accurate and complex expression)

$$Z_o = \begin{cases} \dfrac{60}{\sqrt{\varepsilon_{reff}}} \ln\left(\dfrac{8h}{w} + 0.25\dfrac{w}{h}\right) & \text{if } \dfrac{w}{h} \leq 1 \\ \dfrac{377}{\sqrt{\varepsilon_{reff}}\left[\dfrac{w}{h} + 1.393 + 0.667\ln\left(\dfrac{w}{h} + 1.444\right)\right]} & \text{if } \dfrac{w}{h} \geq 1 \end{cases} \tag{2.16}$$

where ε_{reff} is the effective relative permittivity or effective dielectric constant of the board defined as

$$\varepsilon_{reff} \approx 0.475\varepsilon_r + 0.67 \approx \frac{1}{2}(\varepsilon_r + 1) \tag{2.17}$$

w is the land width, h is the land height above the ground plane, and ε_r is the dielectric constant of the substrate or board. These equations assume that the strip has zero thickness. If the ratio of the strip thickness, t, to the board thickness, h, is small, then the effect of the strip thickness is negligible. (For 1 ounce copper, a common thickness for the lands is t = 1.38 mils.) However, the strip thickness is an important factor when determining ohmic losses. For a double-sided (or two-layer board) where one side is the ground plane, the strip of width w represents the power strip, power bus, or V_{cc} land.

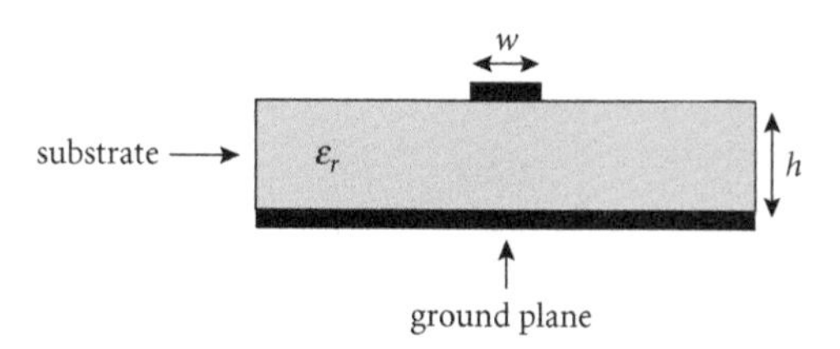

FIGURE 2.9 Cross-sectional view of a microstrip transmission line.

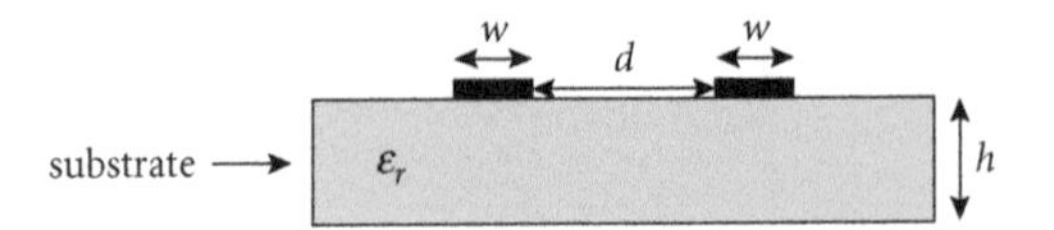

FIGURE 2.10 Cross-sectional view of a coplanar, strip transmission line.

For a single-side (one layer) board where both the ground and power strips are located on the same side, as shown in Figure 2.10, the following characteristic impedance expressions for coplanar strips can be used:

$$Z_o = \begin{cases} \dfrac{377}{\sqrt{\varepsilon_{reff}}} \dfrac{1}{\pi} \ln\left(2\dfrac{1+\sqrt{k}}{1-\sqrt{k}}\right) & \text{if } \dfrac{1}{\sqrt{2}} \leq k \leq 1 \\ \dfrac{377}{\sqrt{\varepsilon_{reff}}} \dfrac{\pi}{\ln\left[2\dfrac{1+(1-k^2)^{\frac{1}{4}}}{1-(1-k^2)^{\frac{1}{4}}}\right]} & \text{if } 0 \leq k \leq \dfrac{1}{\sqrt{2}} \end{cases} \quad \text{where } k = \frac{d}{d+2w} \tag{2.18}$$

and d is the distance between the strips and

$$\varepsilon_{reff} = \frac{\varepsilon_r + 1}{2} \left\{ \begin{array}{l} \tanh\left[0.775 \ln\left(\dfrac{h}{w}\right) + 1.75\right] \\ + \dfrac{kw}{h}\left[0.04 - 0.7k + 0.01(1 - 0.1\varepsilon_r)(0.25 + k)\right] \end{array} \right\} \tag{2.19}$$

The error associated with this expression is less than 1.5% for $h/w \geq 1$. These equations are assuming that the strips have zero thickness. For a nonzero strip thickness, replace k with k_e everywhere *except* the effective relative permittivity equation:

$$k_e \approx k - (1-k^2)\left[\frac{1.25t}{2\pi w}\left(1 + \ln\left(\frac{4\pi w}{t}\right)\right)\right] \tag{2.20}$$

The effective relative permittivity ε_{reff} should be replaced with ε_{ereff}:

$$\varepsilon_{ereff} = \begin{cases} \varepsilon_{reff} - \dfrac{1.4(\varepsilon_{reff}-1)\dfrac{t}{d}}{\dfrac{\pi}{\ln\left(2\dfrac{1+\sqrt{k}}{1-\sqrt{k}}\right)} + \dfrac{1.4t}{d}} & \text{if } \dfrac{1}{\sqrt{2}} \leq k \leq 1 \\ \varepsilon_{reff} - \dfrac{1.4(\varepsilon_{reff}-1)\dfrac{t}{d}}{\dfrac{\ln\left[2\dfrac{1+(1-k^2)^{\frac{1}{4}}}{1-(1-k^2)^{\frac{1}{4}}}\right]}{\pi} + \dfrac{1.4t}{d}} & \text{if } 0 \leq k \leq \dfrac{1}{\sqrt{2}} \end{cases} \tag{2.21}$$

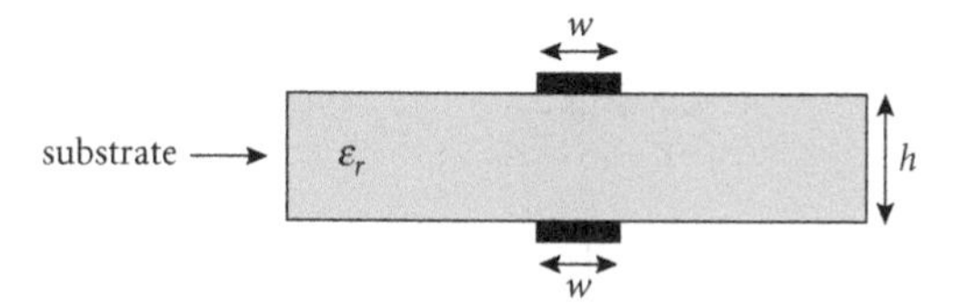

FIGURE 2.11 Cross-sectional view of a parallel, strip transmission line.

Finally, for double-sided boards where traces exist on both sides, or the back side is not just the ground plane, as shown in Figure 2.11, the following characteristic impedance equation for strips separated by a substrate can be used:

$$Z_o = \begin{cases} \dfrac{377}{\sqrt{\varepsilon_r}\left\{\dfrac{w}{h} + 0.441 + \dfrac{\varepsilon_r + 1}{2\pi\varepsilon_r}\left[\ln\left(\dfrac{w}{h} + 0.94\right) + 1.451\right] + 0.082\left(\dfrac{\varepsilon_r - 1}{\varepsilon_r^2}\right)\right\}} & \text{if } \dfrac{w}{h} > 1 \\ \dfrac{377\sqrt{2}}{\pi\sqrt{\varepsilon_r + 1}}\left[\ln\left(\dfrac{4h}{w}\right) + \dfrac{1}{8}\left(\dfrac{w}{h}\right)^2 - \dfrac{1}{2}\left(\dfrac{\varepsilon_r - 1}{\varepsilon_r + 1}\right)\left(0.452 + \dfrac{0.242}{\varepsilon_r}\right)\right] & \text{if } \dfrac{w}{h} < 1 \end{cases} \tag{2.22}$$

When the width of the traces, w, is much greater than the height of the board, h, the common approximation for the characteristic impedance is

$$Z_o \approx \frac{377}{\sqrt{\varepsilon_r}\left\{\dfrac{w}{h} + \dfrac{\varepsilon_r + 1}{2\pi\varepsilon_r}\left[\ln\left(\dfrac{w}{h} + 0.94\right) + 1.451\right]\right\}} \approx \frac{377}{\sqrt{\varepsilon_r}}\frac{h}{w} \quad \text{if } \frac{w}{h} \gg 1 \tag{2.23}$$

This approximate expression is just the square root of the ratio of the inductance to the capacitance of two closely spaced, wide parallel plates, shown in Figure 2.12:

$$Z_o = \sqrt{\frac{L}{C}} = \sqrt{\frac{\dfrac{\mu_o hl}{w}}{\dfrac{\varepsilon_r\varepsilon_o wl}{h}}} = \sqrt{\frac{\mu_o}{\varepsilon_o}}\sqrt{\frac{h^2}{\varepsilon_r w^2}} = \frac{377}{\sqrt{\varepsilon_r}}\frac{h}{w} \tag{2.24}$$

For four or more layer boards, several layers are often dedicated to the power distribution. Equation (2.24) can often be used to model the characteristic impedance of the dc bus for these boards. As expected, to produce a small characteristic impedance, the dielectric constant should be large, the width of the conductors should be large, and the height (or conductor spacing) should be small.

Because of the complexity of most of these equations, it may not be obvious which structure provides the lowest characteristic impedance. In Mathcads 2.4 and 2.5, the characteristic impedance is plotted for these transmission lines as a function of the ratio of the strip width to board thickness. As seen, a very wide, parallel plate structure (i.e., wide parallel strips) provides the lowest characteristic impedance, followed by the microstrip, parallel strips, and coplanar strips. Notice that parallel strips provide a lower

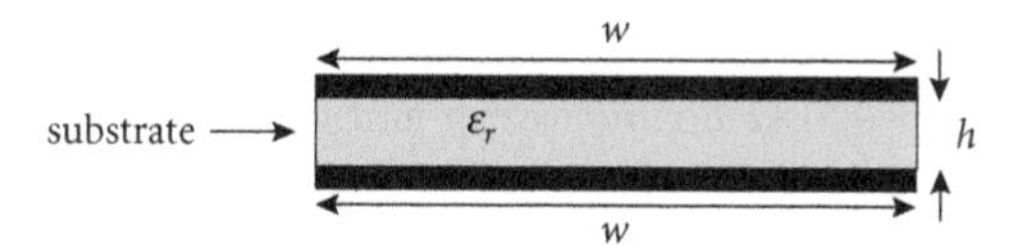

FIGURE 2.12 Cross-sectional view of a parallel plate transmission line.

$$\varepsilon_r := 4.7 \qquad \varepsilon_{reff1}(w,h) := \frac{\varepsilon_r + 1}{2} + \frac{\varepsilon_r - 1}{2\cdot\sqrt{1 + 10\cdot\frac{h}{w}}}$$

$$Z_{o1a}(w,h) := \frac{60}{\sqrt{\varepsilon_{reff1}(w,h)}}\cdot\ln\left(\frac{8\cdot h}{w} + 0.25\cdot\frac{w}{h}\right)$$

$$Z_{o1b}(w,h) := \frac{377}{\sqrt{\varepsilon_{reff1}(w,h)}\cdot\left(\frac{w}{h} + 1.393 + 0.667\cdot\ln\left(\frac{w}{h} + 1.444\right)\right)}$$

$$Z_{o1}(w,h) := \text{if}\left(\frac{w}{h} \le 1, Z_{o1a}(w,h), Z_{o1b}(w,h)\right)$$

$$Z_{o3a}(w,h) := \frac{377\cdot\sqrt{2}}{\pi\cdot\sqrt{\varepsilon_r + 1}}\cdot\left[\ln\left(\frac{4\cdot h}{w}\right) + \frac{1}{8}\cdot\left(\frac{w}{h}\right)^2 - \frac{1}{2}\cdot\left(\frac{\varepsilon_r - 1}{\varepsilon_r + 1}\right)\cdot\left(0.452 + \frac{0.242}{\varepsilon_r}\right)\right]$$

$$Z_{o3b}(w,h) := \frac{377}{\sqrt{\varepsilon_r}\cdot\left[\frac{w}{h} + 0.441 + \frac{\varepsilon_r + 1}{2\cdot\pi\cdot\varepsilon_r}\cdot\left(\ln\left(\frac{w}{h} + 0.94\right) + 1.451\right) + 0.082\cdot\frac{\varepsilon_r - 1}{\varepsilon_r^{\,2}}\right]}$$

$$Z_{o3}(w,h) := \text{if}\left(\frac{w}{h} \le 1, Z_{o3a}(w,h), Z_{o3b}(w,h)\right)$$

Z_{o1} = strip line
Z_{o3} = equal width strips on opposite sides of board
h = board height or thickness
w = strip width

$w := 20, 21..\ 500$

$h := 59$

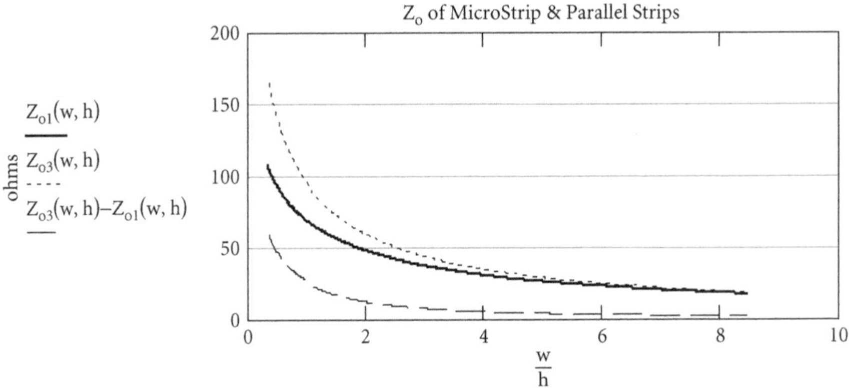

MATHCAD 2.4 Characteristic impedance of the microstrip and opposite side, parallel strip line as a function of the ratio of the strip width to board thickness.

characteristic impedance than coplanar strips when the spacing between the strips is equal to the board thickness. As expected, as the quantity of metal (e.g., copper) on the board increases, the characteristic impedance decreases. This decrease in the impedance is one reason why it is recommended that the copper not be removed on unused portions of a printed circuit board. For printed circuit boards, characteristic impedances in the 50–70 Ω range are common. It is (currently) more difficult to obtain impedances lower than about 30–40 Ω for microstrip and other transmission lines that are a strong function of the substrate thickness (vs. for example, the trace width). To reduce Z_o, a substrate material with a larger dielectric constant (e.g., alumina) or smaller thickness (e.g., a few mils) can be used. If the board thickness and dielectric constant are fixed, then the impedance can be reduced by increasing the width of the traces.

$$\varepsilon_r := 4.7 \qquad k(d, w) := \frac{d}{d + 2\cdot w}$$

$$\varepsilon_{reff2}(w, h, d) := \frac{\varepsilon_r + 1}{2}\cdot\left[\tanh\left(0.775\cdot\ln\left(\frac{h}{w}\right) + 1.75\right)\ldots \atop + \frac{k(d, w)\cdot w}{h}\cdot\left[0.04 - 0.7\cdot k(d, w) + 0.01\cdot(1 - 0.1\cdot\varepsilon_r)\cdot(0.25 + k(d, w))\right]\right]$$

$$Z_{o2a}(w, h, d) := \frac{377}{\sqrt{\varepsilon_{reff2}(w, h, d)}}\cdot\frac{\pi}{\ln\left(2\cdot\frac{1 + \sqrt[4]{1 - k(d, w)^2}}{1 - \sqrt[4]{1 - k(d, w)^2}}\right)}$$

$$Z_{o2b}(w, h, d) := \frac{377}{\sqrt{\varepsilon_{reff2}(w, h, d)}}\cdot\frac{1}{\pi}\cdot\ln\left(2\cdot\frac{1 + \sqrt{k(d, w)}}{1 - \sqrt{k(d, w)}}\right)$$

$$Z_{o2}(w, h, d) := \mathrm{if}\left(k(d, w) \le \frac{1}{\sqrt{2}}, Z_{o2a}(w, h, d), Z_{o2b}(w, h, d)\right)$$

$$Z_{o3a}(w, h) := \frac{377\cdot\sqrt{2}}{\pi\cdot\sqrt{\varepsilon_r + 1}}\cdot\left[\ln\cdot\left(\frac{4\cdot h}{w}\right) + \frac{1}{8}\cdot\left(\frac{w}{h}\right)^2 - \frac{1}{2}\cdot\left(\frac{\varepsilon_r - 1}{\varepsilon_r + 1}\right)\cdot\left(0.452 + \frac{0.242}{\varepsilon_r}\right)\right]$$

$$Z_{o3b}(w, h) := \frac{377}{\sqrt{\varepsilon_r}\cdot\left[\frac{w}{h} + 0.441 + \frac{\varepsilon_r + 1}{2\cdot\pi\cdot\varepsilon_r}\cdot\left(\ln\left(\frac{w}{h} + 0.94\right) + 1.451\right) + 0.082\cdot\frac{\varepsilon_r - 1}{\varepsilon_r^{\,2}}\right]}$$

$$Z_{o3}(w, h) := \mathrm{if}\left(\frac{w}{h} \le 1, Z_{o3a}(w, h), Z_{o3b}(w, h)\right)$$

$w := 10, 11..\ 100$

$h := 59 \qquad d := 59$

Z_{o2} = coplanar strips

Z_{o3} = opposite strips

d = distance between coplanar strips

w = strip width

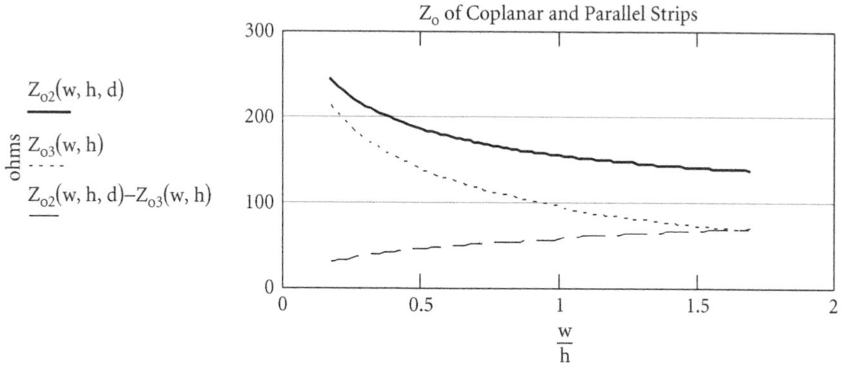

Z_o of Coplanar and Parallel Strips

ohms

$Z_{o2}(w, h, d)$

$Z_{o3}(w, h)$

$Z_{o2}(w, h, d) - Z_{o3}(w, h)$

300, 200, 100, 0

0, 0.5, 1, 1.5, 2

$\frac{w}{h}$

MATHCAD 2.5 Characteristic impedance of the coplanar and opposite side, parallel strip line as a function of the ratio of the strip width to board thickness.

The guidelines given for good, poor, and unacceptable characteristic impedances are based mostly on trial and error (i.e., experience). The actual, acceptable voltage change on a power bus is a function of the noise margin or sensitivity of the various components. For example, digital devices have dc noise margins, and for some logic families, if the supply voltage falls too low, the device will not function. A High voltage could be interpreted as a Low voltage and vice versa if the supply voltage changes too much. For analog devices, as the supply voltage changes, the dc biasing can change, the dynamic range of the output can change, and positive feedback (e.g., oscillation) can occur. To determine analytically the acceptable characteristic impedance of a power bus is not necessarily a trivial task. The following example, though, will provide some insight into the relationship between the characteristic impedance of a power bus and current demands of a device connected across this bus. If one TTL gate is driving several gates and the transient drive current is about 20 mA, then the voltage drop across the dc bus is

$$\Delta V \approx Z_o(20 \times 10^{-3}) = \begin{cases} 0.1\text{ V} & \text{if } Z_o = 5\,\Omega \\ 0.2\text{ V} & \text{if } Z_o = 10\,\Omega \\ 2\text{ V} & \text{if } Z_o = 100\,\Omega \end{cases}$$

It is immediately obvious that the voltage drop is a function of Z_o and the current demand. The Thévenin equivalent seen by the various active devices on a board is a strong function of Z_o. However, the impedance seen by a device is also affected by nearby components, including bypass capacitors.

2.7 Reducing the Characteristic Impedance

For dc power distributions, characteristic impedances less than 1 Ω are difficult to obtain. Sometimes bypass capacitors are added in shunt with the power bus. Why is this done, and what problems can occur? What other methods are available for reducing the characteristic impedance? [Laport]

A small characteristic impedance is desirable to minimize the voltage change on the dc power bus due to time-varying currents. As seen in Mathcads 2.4 and 2.5, Z_o's below 1 Ω are not easily obtained unless large, closely spaced parallel plates are used. Currently, a board thickness of 62 mil is very common for single-sided and double-sided boards, but the thickness can range from about 4 to 125 mils. By reducing the board thickness, Z_o is reduced. In addition, Z_o is reduced by increasing the dielectric constant (and cost) of the substrate. Although there is no substitute for a board with a low Z_o bus, bypass or decoupling capacitors can be added across the power bus to increase the net capacitance of the board. These additional capacitors are in parallel with the dc bus: capacitors in parallel add like resistors in series. The useable frequency range of the capacitors (e.g., their resonant frequency) must be considered since the transient currents on the bus may have very high-frequency components.

In addition to adding lumped capacitors across the power bus, the characteristic impedance of the bus (or other transmission lines) can also be reduced by

1. increasing the conductor size (e.g., width) while keeping the center-to-center distance between the conductors the same
2. decreasing the center-to-center distance between the conductors
3. increasing the number of conductors for each feed side, including placing cables in parallel
4. increasing the effective relative permittivity between the conductors.

2.8 Influence of Dielectric Constant

Using real materials, locate an insulating material for a cable that has a dielectric constant close to one. When would this material be helpful?

Before answering this question, the difference between a good insulator and good dielectric will be briefly explained. A good electrical insulator will allow very little conduction current to pass through it. If a voltage source is applied across a good insulator, the current through the insulator is small. The dielectric constant of a good insulator, however, is not necessarily high. A good dielectric, on the other hand, has a high dielectric constant or relative permittivity, but it is not necessarily a good insulator. Water, for example, has a high dielectric constant (around 80), but it is a poor insulator at lower frequencies. For capacitors, a *high* dielectric constant, good-insulating material between the plates is nearly always desirable. As will be seen, for many signal cables, a good-insulating, *low* dielectric constant medium between the conductors is often desirable (a high dielectric constant is desirable for some types of delay lines).

The dielectric constant of a vacuum, an excellent insulator, is one. (A good vacuum has very little free charge in it to allow for conduction.) A very common insulating material is PVC. It has a dielectric constant that ranges from 4 to 8. Solid Teflon has a dielectric constant of 2, while foamed Teflon has a dielectric constant of around 1.6 (the air in the Teflon lowers the dielectric constant).[5] Of course, air or free space has a dielectric constant of one. However, the conductors of a cable require support or spacers. These spacers have dielectric constants greater than one. The reason a low dielectric constant is usually desirable is that the velocity of a signal on the line increases as the dielectric constant decreases. The equation for the velocity of a signal on a lossless transmission line with an inductance per unit length of L and capacitance per unit length of C is[6]

$$v = \frac{1}{\sqrt{LC}} = \frac{1}{\sqrt{\mu_{eff}\varepsilon_{eff}}} \tag{2.25}$$

where μ_{eff} and ε_{eff} are the *effective* (like a weighted average) permeability and permittivity, respectively, of the materials between the conductors. Since in most applications the material between and near the conductors of a transmission line is nonmagnetic, the effective permeability is equal to the free-space permeability, $4\pi \times 10^{-7}$ H/m. The effective permittivity, however, is affected by the dielectric constant of the material between and near the conductors of a transmission line: the capacitance is affected by the permittivity, but the inductance is *not* affected by the permittivity. For high-speed transmission lines, a dielectric constant near one can be desirable since it allows for a greater signal velocity along the line and smaller line delay. Unlike dc buses where a large capacitance is desirable, a large capacitance on a high-speed line is not always sought-after.

2.9 Coax and Twin-Lead

What are the expressions for the characteristic impedance of a coaxial cable (or coax) and twin-lead cable? When is the velocity of propagation for a plastic-coated twin-lead cable greater than the velocity of a coaxial cable? Assume the dielectrics used for both cables are identical. [Walker; Belden; ARRL, '96; Andrew; Johnson, '50; Skilling; Harper, '72; Wood]

As will be discussed, the velocity of a signal that travels down a twin-lead line is not necessarily equal to the signal velocity for coax even when the dielectric material is the same. The velocity of a signal on a lossless line is a function of the capacitance and inductance per unit length, which are a function of the *effective* permittivity or *effective* dielectric constant. The classical relationships for the characteristic impedance of and signal velocity on coax will first be presented.

The high-frequency characteristic impedance of the coaxial line shown in Figure 2.13 is

$$Z_o = \sqrt{\frac{L}{C}} = \sqrt{\frac{\dfrac{\mu_r\mu_o}{2\pi}\ln\left(\dfrac{a}{b}\right)}{\dfrac{2\pi\varepsilon_r\varepsilon_o}{\ln\left(\dfrac{a}{b}\right)}}} = \frac{1}{2\pi}\sqrt{\frac{\mu_o}{\varepsilon_o}}\sqrt{\frac{\mu_r}{\varepsilon_r}}\ln\left(\frac{a}{b}\right) \approx 60\sqrt{\frac{\mu_r}{\varepsilon_r}}\ln\left(\frac{a}{b}\right)\ \Omega$$

[5]For coaxial cable, foam dielectric also has a lower loss compared to solid dielectric. The general wisdom is that the lower loss is not due to the low-loss air in the foam but is mostly due to the larger inner conductor. For a given characteristic impedance and fixed outer diameter, as the dielectric constant is decreased, the inner diameter must increase. Often, most of the resistive loss in the coax is from the inner conductor, which has a smaller surface area than the outer conductor.

[6]Actually, for nonhomogeneous mediums, the velocity of propagation is not a single value. A signal will travel at a different velocity in each of the mediums.

where a is the inner radius of the *outer* conductor and b is the outer radius of the *inner* conductor. In most cases, the medium between the inner and outer conductor is nonmagnetic so that $\mu_r = 1$:

$$Z_o = 60\frac{1}{\sqrt{\varepsilon_r}}\ln\left(\frac{a}{b}\right)$$

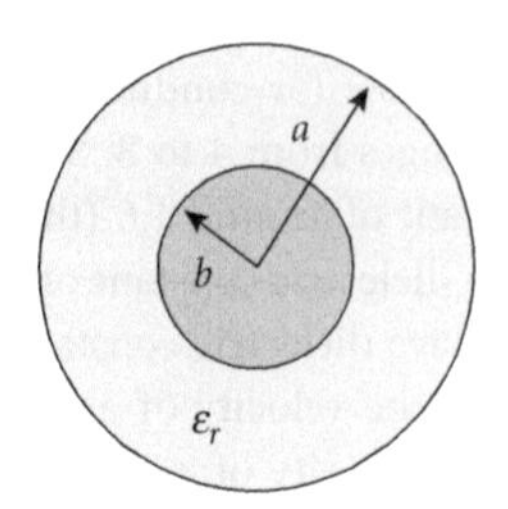

FIGURE 2.13 Cross-sectional view of a coaxial transmission line.

The voltage source is applied between the inner and outer conductor, and the field lines are contained within the outer conductor. The electric field passes entirely through the one dielectric medium between the two conductors, thus this medium is considered homogeneous. Since the medium is homogeneous and nonmagnetic, the equation for the velocity of propagation at higher frequencies is

$$v = \frac{1}{\sqrt{LC}} = \frac{1}{\sqrt{\mu_r\mu_o\varepsilon_r\varepsilon_o}} = \frac{1}{\sqrt{\mu_r\varepsilon_r}}\frac{1}{\sqrt{\mu_o\varepsilon_o}} = \frac{3\times10^8}{\sqrt{\varepsilon_r}} \text{ m/s} \tag{2.26}$$

In this case, L is "replaced" by μ and C is "replaced" by ε. If the dielectric constant or relative permittivity of the medium between the conductors is provided, the velocity can be determined. To repeat, it is not necessary to know the capacitance or inductance per unit length to determine the characteristic impedance for a coaxial cable if the dielectric constant is known. However, it is necessary to know the dimensions of the cable and dielectric constant if Z_o is required. Also, note that

$$Z_o = \sqrt{\frac{L}{C}} = 60\frac{1}{\sqrt{\varepsilon_r}}\ln\left(\frac{a}{b}\right) \neq \sqrt{\frac{\mu_r\mu_o}{\varepsilon_r\varepsilon_o}} \approx \frac{377}{\sqrt{\varepsilon_r}} \tag{2.27}$$

In addition to the dielectric constant, the characteristic impedance of a cable is also a function of the shape, size, and position of the conductors.

It might initially seem that smaller, miniature, or micro coaxial cables would have smaller characteristic impedances since the distance between the inner and outer conductors is smaller. However, the expression for Z_o is only a function of the ratio of the outer to inner radii of the two conductors. Therefore, if this ratio is kept constant for a coaxial cable, the characteristic impedance does not change. The current handling capability, though, is less for smaller conductor sizes. Occasionally, inquisitive students question the constancy of Z_o as the cable size is reduced because they believe that the capacitance should increase as the spacing between the inner and outer conductors decreases. However, the capacitance is a function of both the distance between two conductors and surface area of the conductors. As the distance between the conductors decreases and the cable cross section decreases, the surface area of the inner and outer conductors also decreases.

The high-frequency characteristic impedance of the twin-lead (or parallel conductor) line shown in Figure 2.14 (taking into account the proximity effect) is

$$Z_o = \frac{120}{\sqrt{\varepsilon_{reff}}}\ln\left\{\frac{d}{2r_w}\left[1+\sqrt{1-\left(\frac{2r_w}{d}\right)^2}\right]\right\}$$

$$\approx \frac{120}{\sqrt{\varepsilon_{reff}}}\ln\left[\frac{d}{2r_w}\left\{1+\left[1-\frac{1}{2}\left(\frac{2r_w}{d}\right)^2\right]\right\}\right] \approx \frac{120}{\sqrt{\varepsilon_{reff}}}\ln\left(\frac{d}{r_w}\right) \quad \text{if } \frac{2r_w}{d} \ll 1$$

where r_w is the wire radius and d is the center-to-center distance between the wires (the binomial expansion was used to obtain the given approximation). If the medium surrounding and between the

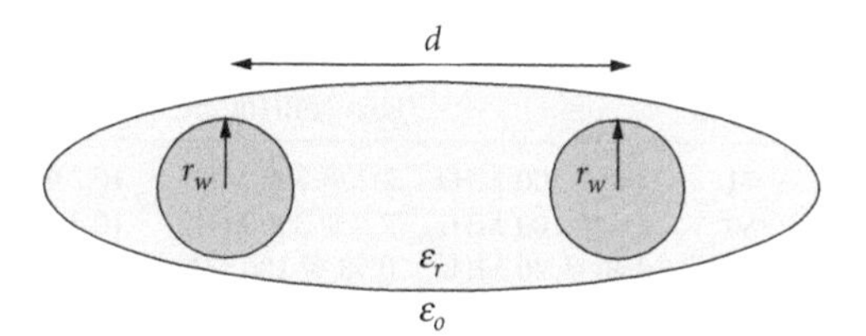

FIGURE 2.14 Cross-sectional view of a twin-lead transmission line with surrounding insulation. The insulation is of finite thickness around the conductors.

two conductors is free space, then the effective relative permittivity, ε_{reff}, is one and the line is referred to as open-wire or "window" line. (Of course, spacers are required along an air-insulated twin-lead line to support the conductors and control their spacing. The local capacitance of the line is increased at these spacers.) If the insulation surrounding the two conductors is thick so that most of the electric field lines are contained within the insulation, then ε_{reff} is about equal to the relative permittivity of the insulation. When the insulation is not thick, ε_{reff} is greater than one but less than the relative permittivity of the insulation. In this case, the fringing electric fields outside the insulation are important.

For thick insulation, the equation for the velocity of propagation for a twin-lead line at higher frequencies (so that the R term can be neglected) is given in (2.26). Again, the insulating material is assumed nonmagnetic. The velocity of propagation is actually a function of the shape, size, and position of the wires if the insulation surrounding the conductors is not thick. This was not the case for the coaxial line. For twin-lead with thin insulation, the air surrounding the line tends to increase the "average" velocity of propagation and decrease the capacitance. The velocity of propagation on a twin-lead line with finite insulation thickness is always greater than or equal to the velocity on a coaxial line if the dielectric constants of the insulations are the same. Furthermore, if the characteristic impedances of the twin-lead and coax are the same, as well as the dielectric constant of the insulation used, then the capacitance of the twin-lead must be smaller if its velocity is greater:

$$v = \frac{1}{\sqrt{LC}} = \frac{1}{Z_o C} \tag{2.28}$$

To avoid complicating the previous discussions, the effect of a nonhomogeneous medium around the conductors was simply modeled as an effective permittivity. Actually, the field patterns do change from the standard *transverse electromagnetic* (TEM) patterns for nonhomogeneous medium. For homogeneous medium between two lossless conductors that are electrically close, the electric and magnetic fields are perpendicular, or transverse, to the direction of propagation, along the axes of the conductors. Most of the expressions for the electric and magnetic fields for transmission lines are derived based on TEM conditions. The expression for the velocity of a signal on a line, which is a function of the inductance and capacitance per unit length,

$$v = \frac{1}{\sqrt{LC}}$$

is also based on TEM conditions. However, when the medium is nonhomogeneous (e.g., plastic-coated twin-lead line) additional "modes," or field components, can exist on the line. For example, an electric field component *in* the direction of propagation can exist in addition to the standard transverse electric field component directed between the conductors. This complicates the analysis and the expression for the velocity. Frequently, if the distance between the conductors is electrically small (and the width of the strip for microstrip lines is electrically small), the nontransverse components are negligible. The error introduced by neglecting these other field components is usually small. However, when the dimensions are not electrically small, the transmission line can have waveguide-like modes, and many of the previous expressions must be modified.

TABLE 2.2 Modest Listing of Transmission Lines and a Few of Their Properties

Line	Z_o (Ω)	C (pF/m)	Vel %	Loss (dB/100 m)			Comments
Twin-lead	75	20	71	3.6 @ 100 MHz	7.2 @ 300 MHz	10.2 @ 500 MHz	—
Twin-lead	300	11.8	80	3.6 @ 100 MHz	7.2 @ 300 MHz	10.2 @ 500 MHz	—
Open-wire	300		95	0.46 @ 20 MHz	0.98 @ 100 MHz	1.4 @ 200 MHz	Fast, environment sensitive, low loss, low frequencies
RG-8/U	50	85.3	78	1.6 @ 10 MHz	5.2 @ 100 MHz	11.5 @ 400 MHz	Fast, common, foam type
RG-58/U	53.5	93.5	66	3.9 @ 10 MHz	13.8 @ 100 MHz	29.9 @ 400 MHz	—
RG-59/U	75	56.8	78	3.0 @ 10 MHz	9.8 @ 100 MHz	21.7 @ 400 MHz	Flexible version
RG-213U	50	101	66	2.0 @ 10 MHz	6.9 @ 100 MHz	15.7 @ 400 MHz	Modern 8U
RG-214U	50	101	66	1.8 @ 10 MHz	6.2 @ 100 MHz	13.4 @ 400 MHz	Silver-coated copper outer braid
RG-174U	50	101.8	66	10.8 @ 10 MHz	27.6 @ 100 MHz	62.3 @ 400 MHz	Small diameter
1/2" hardline	50	23.1	88	0.69 @ 10 MHz	2.2 @ 100 MHz	4.7 @ 400 MHz	Solid outer conductor, low loss, stiff, rugged

Properties of several coaxial and twin-lead lines are given in Table 2.2. The "Vel %" is relative to the speed of light, 3×10^8 m/s.[7] The losses shown are those for a line terminated in Z_o. Although the labels given for two cables may be the same, the properties of the cables might be different. For example, there are several types of RG-8/U cable, some with high velocities and low losses. Interestingly, the characteristic impedance and velocity of propagation of a coaxial cable can even vary with the type of the outer conductor. For a particular type of coaxial cable, for example, the measured velocity for a spiral-wound outer-conductor version of the cable is 62% (of the speed of light), while for a solid or braided outer-conductor version the velocity is 69%. A reasonable explanation for this difference in velocity is the increase in the inductance resulting from the spiraling. (The "length" of the current path through the spiral is greater than through a solid outer conductor.) It is best to consult a cable catalog for specific information such as characteristic impedance, dielectric and ohmic losses, maximum operating voltage, shielding effectiveness, military specifications, flexibility, jacket material, temperature range, and size.

The characteristic impedance is probably the most important single parameter of a cable (followed by signal velocity and attenuation). The characteristic impedance usually ranges from 30 to 600 Ω. For twin-lead line, 300 Ω is common probably because it is about the impedance of a folded dipole antenna used for FM and TV reception. Typically, there is a trade-off between attenuation and power handling capacity for cables. The attenuation due to conductor resistance in coax (neglecting the proximity effect) is related to the frequency, the relative permittivity (or dielectric constant), and the inner and outer radii according to the expression

$$\alpha \propto \frac{\sqrt{f\varepsilon_r}}{a}\frac{\left(1+\frac{a}{b}\right)}{\log\left(\frac{a}{b}\right)} \tag{2.29}$$

The optimum value for *a*/*b* is about 3.6 for a fixed outer radius from an *attenuation* standpoint. If the radius of the inner conductor is less than this optimum value, its resistance is greater. However, if the radius of the inner conductor is greater than this optimum value, the capacitance is greater, the characteristic impedance

[7]Sometimes the velocity factor is given for a transmission line. The velocity factor is the Vel % divided by 100 or the velocity relative to the speed of light as a fraction.

is smaller, and, for a given voltage, the current and I^2R losses are greater. The characteristic impedance at this optimum value for free space is about 77 Ω:

$$Z_o = 60\frac{1}{\sqrt{1}}\ln(3.6) \approx 77\ \Omega \tag{2.30}$$

(Some individuals argue that this free-space impedance value is the reason 75 Ω line is commonly used in CATV systems, even though an air dielectric is not used.) For a dielectric constant of 2.3, the impedance is about 51 Ω. Interestingly, the geometric mean of the input impedance of a resonant half-wave dipole (73 Ω) and resonant quarter-wave monopole (37 Ω) is about $\sqrt{73 \times 37} \approx 52\ \Omega$. When using 52 Ω coax, the VSWR for either antenna is about 1.4.

The maximum field strength is an important factor for power cables. If the outer radius of a coaxial line is fixed, over a range of inner radii and for a fixed voltage, as the inner radius decreases, the field strength decreases. This decrease in the field strength is to be expected since the distance between the inner and outer conductors is increasing. However, if the radius of the inner conductor is too small, the electric field along this high-curvature sharp conductor will intensify, and the surrounding medium can break down. The power carried by the cable is related to V^2/Z_o. For air dielectric, the optimum value from a *power transmission* standpoint for a/b is 1.65. This corresponds to a characteristic impedance of 30 Ω. If a cable is designed for maximum *voltage*, the optimum value for a/b is about 2.72 or $Z_o = 60\ \Omega$ for air dielectric. (Frequently before the maximum voltage is reached, the maximum operating temperature of the cable is reached and is thus the limiting factor.)

As was seen, the "best" value for Z_o is dependent on many factors. For standardization purposes, 50 Ω and 75 Ω coaxial cables are commonly used. The 50 Ω coaxial cable is usually used for precision work and signal operation above 30 MHz. The 75 Ω coax is usually used for frequencies below 30 MHz, long lines, and low-loss applications such as with CATV.

2.10 Thinly Coated Twin-Lead

Determine an approximation for the capacitance between two parallel wires of equal diameter and equal insulation thickness. Assume that the distance between the wires is large compared to the insulation thickness. [Smythe; Das, '95; Walker]

Before this problem is solved, it is helpful to determine the capacitance between two closely spaced parallel plates when the dielectric between the plates varies as shown in Figure 2.15. Since the interfaces between the insulators and air occur along equipotentials, the overall capacitance can be modeled as three capacitors in series. (The equipotentials for a parallel plate capacitor are parallel to the plates. For example, if the lower plate is at 0 V and the upper plate is at 10 V, the 5 V equipotential plane is midway between the plates and parallel to them. Obviously, the bottom plate, being a conductor, is the 0 V equipotential plane, and the upper plate, also being a conductor, is the 10 V equipotential plane.) It may

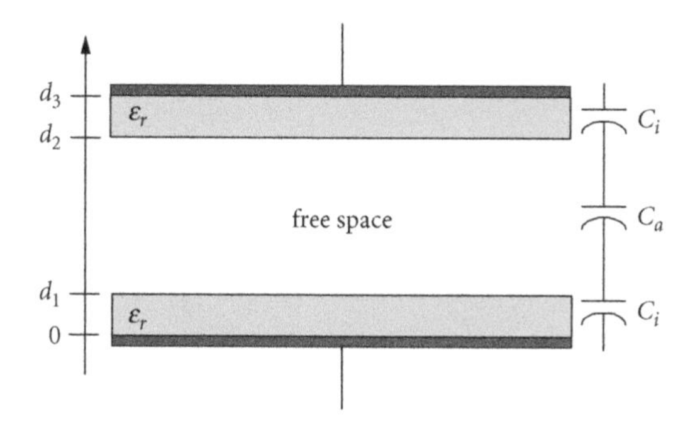

FIGURE 2.15 Flat dielectric coating on both conductors of a parallel plate capacitor.

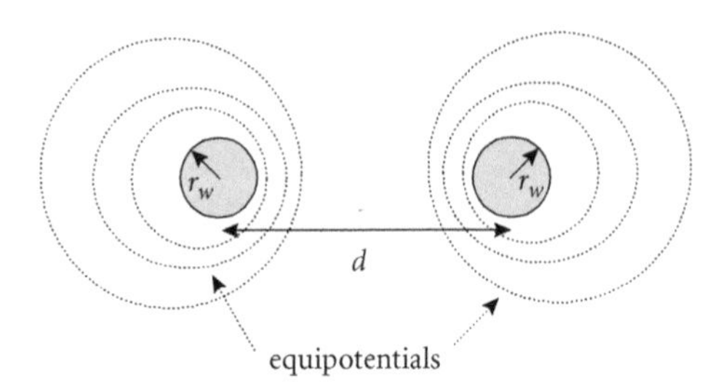

FIGURE 2.16 Sketch of the equipotentials between two long, parallel circular conductors with an applied voltage across them.

initially seem strange to think about the capacitance between either of the conducting plates and insulating interfaces or the capacitance between the two insulating interfaces. However, imagine that an infinitely thin conducting plate is placed along any of the insulating interfaces. Because the plate is infinitely thin and along an equipotential, it will not affect the field or potential distributions. This "floating" plate will assume the single potential of the interface. By placing an infinitely thin plate along each of the insulating interfaces, the capacitances between the various plates can be visualized and modeled. The ubiquitous expression for the capacitance between two closely spaced parallel plates of area A and separation distance d

$$C = \frac{\varepsilon A}{d}$$

is used to determine the net capacitance for the given structure consisting of three capacitors in series:

$$C_T = \frac{1}{\frac{1}{C_i} + \frac{1}{C_a} + \frac{1}{C_i}} = \frac{1}{\frac{d_1}{\varepsilon_r \varepsilon_o A} + \frac{d_2 - d_1}{\varepsilon_o A} + \frac{d_3 - d_2}{\varepsilon_r \varepsilon_o A}}$$

For two parallel circular conductors, the equipotentials consist of circles. Unfortunately, the centers of these equipotential circles do not coincide with the center of the conductors as shown in Figure 2.16. Therefore, if the insulation and corresponding conductor are concentric and the insulation is of uniform thickness, the outer surfaces of the insulation are *not* along equipotentials. (In the previous parallel plate example, the insulation interfaces were along the equipotentials.) However, if the insulation thickness, Δ, is small compared to the distance between the conductors, d, and the conductors are not too close, then the outer surfaces of the insulation are approximately along equipotentials. If the insulation thickness is permitted to vary, then the air-insulation interface can exist along an equipotential surface as is shown in Figure 2.17.

To estimate the net capacitance between two coated conductors that are far apart, the equation for the capacitance per unit length between two cylinders is required:

$$C = \frac{2\pi\varepsilon}{\cosh^{-1}\left(\pm\frac{d^2 - r_1^2 - r_2^2}{2r_1 r_2}\right)}$$

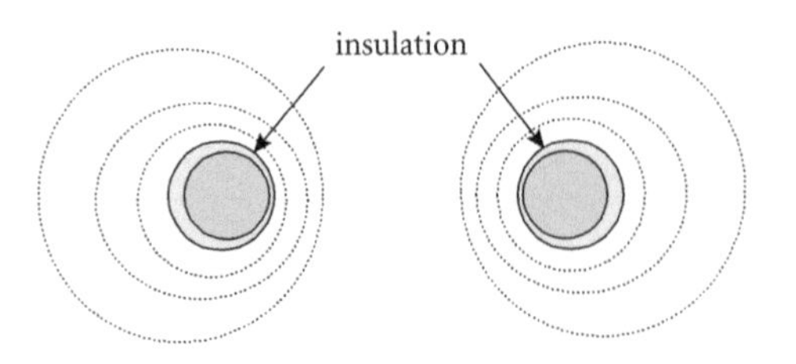

FIGURE 2.17 Variable insulation thickness.

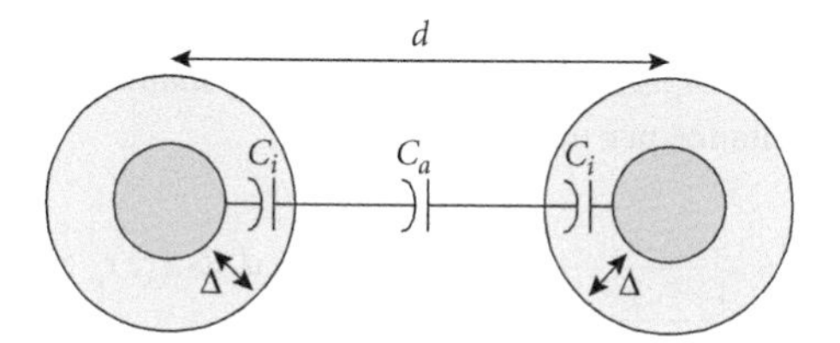

FIGURE 2.18 Model of the capacitances between the two conductors assuming that the thickness, Δ, is small.

where d is the center-to-center distance between the centers of the two cylinders of radii r_1 and r_2. The negative sign multiplier (the –) in the argument of the inverse hyperbolic cosine term is used when one cylinder is inside the other. Using this equation and referring to Figure 2.18, the capacitance between a conductor and outer surface of a concentric uniform-thickness insulator is

$$C_i = \frac{2\pi\varepsilon_r\varepsilon_o}{\cosh^{-1}\left[-\dfrac{0-r_w^2-(r_w+\Delta)^2}{2r_w(r_w+\Delta)}\right]} = \frac{2\pi\varepsilon_r\varepsilon_o}{\cosh^{-1}\left[\dfrac{2r_w^2+2r_w\Delta+\Delta^2}{2r_w(r_w+\Delta)}\right]}$$
$$= \frac{2\pi\varepsilon_r\varepsilon_o}{\cosh^{-1}\left[1+\dfrac{\Delta^2}{2r_w(r_w+\Delta)}\right]} \tag{2.31}$$

$$C_i \approx \frac{2\pi\varepsilon_r\varepsilon_o}{\cosh^{-1}\left(1+\dfrac{\Delta^2}{2r_w^2}\right)} = \frac{2\pi\varepsilon_r\varepsilon_o}{\ln\left[1+\dfrac{\Delta^2}{2r_w^2}+\sqrt{\left(1+\dfrac{\Delta^2}{2r_w^2}\right)^2-1}\right]}$$
$$\approx \frac{2\pi\varepsilon_r\varepsilon_o}{\ln\left(1+\dfrac{\Delta^2}{2r_w^2}+\dfrac{\Delta}{r_w}\right)} \approx \frac{2\pi\varepsilon_r\varepsilon_o}{\ln\left(1+\dfrac{\Delta}{r_w}\right)} \approx \frac{2\pi\varepsilon_r\varepsilon_o}{\dfrac{\Delta}{r_w}} \quad \text{if } r_w \gg \Delta \tag{2.32}$$

where ε_r is the relative permittivity or dielectric constant of the insulation.[8] The capacitance per unit length between the two insulator surfaces is

$$C_a = \frac{2\pi\varepsilon_o}{\cosh^{-1}\left[\dfrac{d^2-(r_w+\Delta)^2-(r_w+\Delta)^2}{2(r_w+\Delta)(r_w+\Delta)}\right]} = \frac{2\pi\varepsilon_o}{\cosh^{-1}\left[\dfrac{d^2-2(r_w+\Delta)^2}{2(r_w+\Delta)^2}\right]} \tag{2.33}$$

$$C_a \approx \frac{2\pi\varepsilon_o}{\cosh^{-1}\left[\dfrac{d^2}{2(r_w+\Delta)^2}\right]} = \frac{2\pi\varepsilon_o}{\ln\left[\dfrac{d^2}{2(r_w+\Delta)^2}+\sqrt{\left(\dfrac{d^2}{2(r_w+\Delta)^2}\right)^2-1}\right]}$$
$$\approx \frac{2\pi\varepsilon_o}{\ln\left[\dfrac{d^2}{(r_w+\Delta)^2}\right]} = \frac{\pi\varepsilon_o}{\ln\left(\dfrac{d}{r_w+\Delta}\right)} \quad \text{if } d \gg \max(r_w,\Delta) \tag{2.34}$$

[8]This approximation can be obtained more readily by using the common expression for the capacitance of a coaxial cable, $2\pi\varepsilon_r\varepsilon_o/\ln[(r_w+\Delta)/r_w] \approx 2\pi\varepsilon_r\varepsilon_o/(\Delta/r_w)$.

Since the insulation thickness is small compared to the distance between the wires, C_a is much less than C_i. It follows that since the three capacitances are in series, C_a mainly determines the net capacitance. The equation for the net capacitance per unit length is

$$C_T \approx \frac{1}{\dfrac{1}{C_i}+\dfrac{1}{C_a}+\dfrac{1}{C_i}} = \frac{C_i C_a}{C_i + 2C_a} \quad \text{if } d \gg r_w,\ r_w \gg \Delta \tag{2.35}$$

Substituting the given approximations,

$$C_T = \frac{\dfrac{2\pi\varepsilon_r\varepsilon_o}{\dfrac{\Delta}{r_w}} \dfrac{\pi\varepsilon_o}{\ln\left(\dfrac{d}{r_w+\Delta}\right)}}{\dfrac{2\pi\varepsilon_r\varepsilon_o}{\dfrac{\Delta}{r_w}} + 2\dfrac{\pi\varepsilon_o}{\ln\left(\dfrac{d}{r_w+\Delta}\right)}} = \frac{\pi\varepsilon_o}{\ln\left(\dfrac{d}{r_w+\Delta}\right)+\dfrac{\Delta}{\varepsilon_r r_w}} \quad \text{if } d \gg r_w,\ r_w \gg \Delta \tag{2.36}$$

For comparison, the standard equation for the capacitance per unit length between two parallel circular conductors surrounded by free space is

$$C_{wo} = \frac{\pi\varepsilon_o}{\ln\left\{\dfrac{d}{2r_w}\left[1+\sqrt{1-\left(\dfrac{2r_w}{d}\right)^2}\right]\right\}} \approx \frac{\pi\varepsilon_o}{\ln\left(\dfrac{d}{r_w}\right)} \quad \text{if } d \gg r_w$$

Notice that the free-space capacitance remains constant as long as the ratio of the center-to-center distance to the conductor radius remains constant. Comparing this expression with the C_T approximation, it is seen that the thin insulation does increase the capacitance:

$$\frac{C_T}{C_{wo}} = \frac{\dfrac{\pi\varepsilon_o}{\ln\left(\dfrac{d}{r_w+\Delta}\right)+\dfrac{\Delta}{\varepsilon_r r_w}}}{\dfrac{\pi\varepsilon_o}{\ln\left(\dfrac{d}{r_w}\right)}} = \frac{\varepsilon_r r_w \ln\left(\dfrac{d}{r_w}\right)}{\varepsilon_r r_w \ln\left(\dfrac{d}{r_w+\Delta}\right)+\Delta} \quad \text{if } d \gg r_w,\ r_w \gg \Delta$$
$$\approx \frac{\ln\left(\dfrac{d}{r_w}\right)}{\ln\left(\dfrac{d}{r_w+\Delta}\right)} > 1 \quad \text{if } d \gg r_w,\ r_w \gg \Delta \tag{2.37}$$

If the insulation were present everywhere around and between the two conductors, the insulation would increase the capacitance by ε_r.

As seen in Mathcad 2.6, the rough approximation of assuming the insulation interfaces are along equipotentials is reasonable over a limited range. The first graph compares the total capacitance expression based on the three-series capacitors with the standard equation for the capacitance per unit length between two parallel conductors in free space. The second graph shows the total capacitance, capacitance of the air region, and capacitance of one of the insulation regions. As expected, the total capacitance is only slightly affected by the thin insulation since the air capacitance is small. (Recall that capacitors in series add like resistors in parallel; therefore, the smallest capacitance determines the largest possible value of the net capacitance.)

$\varepsilon_o := 8.854 \cdot 10^{-12}$ $\quad r_w := 5 \cdot 10^{-3}$ $\quad d := 5 \cdot 10^{-3}$

$$C_i(r_w, \Delta, \varepsilon_r) := \frac{2 \cdot \pi \cdot \varepsilon_r \cdot \varepsilon_o}{\operatorname{acosh}\left[\dfrac{2 \cdot r_w^2 + 2 \cdot r_w \cdot \Delta + \Delta^2}{2 \cdot r_w \cdot (r_w + \Delta)}\right]} \qquad C_a(r_w, \Delta, d) := \frac{2 \cdot \pi \cdot \varepsilon_o}{\operatorname{acosh}\left[\dfrac{d^2 - 2 \cdot (r_w + \Delta)^2}{2 \cdot (r_w + \Delta)^2}\right]}$$

$$C_T(r_w, \Delta, d, \varepsilon_r) := \frac{C_i(r_w, \Delta, \varepsilon_r) \cdot C_a(r_w, \Delta, d)}{C_i(r_w, \Delta, \varepsilon_r) + 2 \cdot C_a(r_w, \Delta, d)} \qquad C_{wo}(r_w, d) := \frac{\pi \cdot \varepsilon_o}{\ln\left[\dfrac{d}{2 \cdot r_w} \cdot \left[1 + \sqrt{1 - \left(\dfrac{2 \cdot r_w}{d}\right)^2}\right]\right]}$$

$$C_{app}(r_w, \Delta, d, \varepsilon_r) := \frac{\pi \cdot \varepsilon_o}{\ln\left(\dfrac{d}{r_w + \Delta}\right) + \dfrac{\Delta}{\varepsilon_r \cdot r_w}}$$

$\Delta := 1 \cdot 10^{-5}, 1.5 \cdot 10^{-5} .. 0.6 \cdot 10^{-3}$

MATHCAD 2.6 Capacitance of two coated parallel conductors and two parallel uncoated conductors. Also shown is the capacitance of the air and insulation components of the coated conductors vs. coating thickness.

2.11 Beads in Coax

Instead of completely filling in the region between the inner and outer conductors of a coaxial cable with a dielectric foam, a dielectric bead of relative permittivity ε_r and width w meters is placed around the inner conductor every s meters (center-to-center distance) to support the inner conductor. What is the major electrical advantage in this arrangement over the complete filling? Determine the effective dielectric constant of this arrangement. [King; Skilling]

A cross-sectional view of the coax is shown in Figure 2.19. If the beads were not present, then the relative permittivity between the conductors would be one, which is that of free space. The velocity of propagation would be its maximum possible value, and the dielectric resistance (modeled through the conductance, G, in the cable model) would be high. As material is placed between the two conductors, to support the inner conductor and provide a controlled spacing between the conductors, the effective relative permittivity would increase, the velocity of propagation would decrease, and the dielectric losses would increase. The supporting material can be in the shape of a disc, bead, or helix. Of course, as with most flexible coaxial cables, the supporting material can be in the form of a solid dielectric material. Beads are used to reduce dielectric losses and increase the signal velocity.

If the spacing between the beads is electrically short, the insertion of the beads merely increases the relative effective permittivity of the cable above that of free space:

$$\varepsilon_{reff} \approx 1 + \frac{w}{s}(\varepsilon_r - 1) \tag{2.38}$$

The relative effective permittivity is a measure of the fraction of the field lines in the dielectric beads vs. the air. When the spacing between the beads is small, the addition of the beads increases the capacitance of the cable at those specific locations. The various equations for a transmission line that is not electrically short are obtained from the summation of an infinite number of small segments of the lumped-circuit model. It follows that the increased capacitance at the specific locations from the beads will add directly to the capacitance of the air-filled version of the cable. Also, note that the characteristic impedance will decrease with the insertion of these beads:

$$Z_o = 60 \frac{1}{\sqrt{1 + \frac{w}{s}(\varepsilon_r - 1)}} \ln\left(\frac{a}{b}\right) \tag{2.39}$$

where a is the inner radius of the outer conductor and b is the outer radius of the inner conductor.

When the spacing between the beads is not electrically short, then each bead acts like a mismatched load along the line. A fraction of the signal will transmit through the bead and a fraction will be reflected. If a two-bead line is matched, the beads are electrically thin, and the line and beads are low loss, then the beads will have a negligible effect on the line when the center-to-center distance between the two beads is

$$s = \frac{\lambda_o}{4} - \frac{w}{2}(\varepsilon_r - 1) \tag{2.40}$$

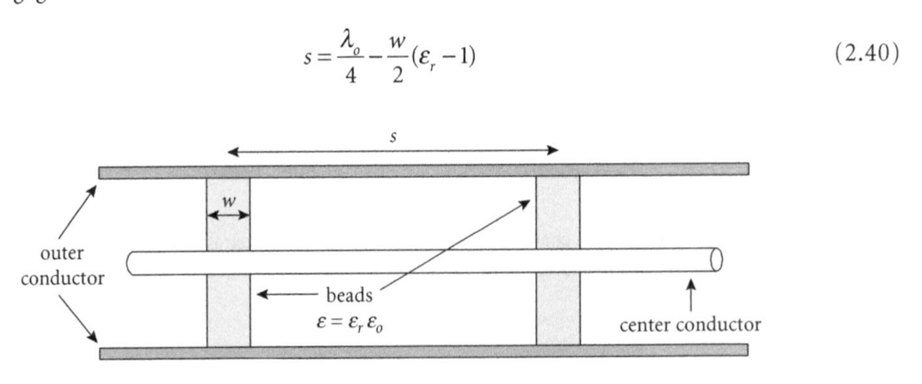

FIGURE 2.19 Dielectric beads spaced a distance of s along a coaxial cable.

When the spacing between the beads is slightly less than $\lambda_o/4$, the various reflections and transmissions will combine such that the effect of the beads should not be significant. However, this technique is frequency specific and not broadbanded: a spacing corresponding to a quarter-wave length at one frequency does not correspond to a quarter-wave length at other nearby frequencies. When the spacing is $\lambda_o/2$, the reflections will add and the beads will have a strong effect on the signal traveling on the line.

2.12 Dielectric Resistance and Insulators

In an attempt to measure the dc voltage across two conductors separated by a slab of an excellent insulator (Teflon), an electrometer is used. A coaxial cable connects the electrometer to the source. Will the leakage (or dielectric) resistance between the center conductor and shield (the dielectric resistance) of the cable affect the performance of the electrometer? Explain. [Keithley; Belden; Rochester; Hippel]

For most decent transmission lines, the dielectric conductance (modeled as G in the lumped-circuit model), or the dielectric resistance ($1/G$), and the charge flow between the conductors of a transmission line are usually negligible. There are situations, however, when the dielectric resistance should not be neglected. For example, at very high frequencies or when the dielectric is wet or dirty, the losses may become substantial. In addition, as will be seen, when a source or load resistance connected to a cable is comparable to the dielectric resistance, the dielectric resistance cannot be neglected. (When measuring the voltage across a very high resistance with a standard multimeter, the input resistance of the multimeter can load down the measurement.)

A practical engineer should have a "feeling" for insulating materials and their respective resistances. There are several properties of interest when working with insulating materials: volume resistivity, surface resistivity, water absorption, piezoelectric effects, triboelectric effects, dielectric constant and absorption, and breakdown strength. The volume resistivity is a measure of the current through the bulk of the material when a voltage is applied across the insulator. For a perfect insulator, the current would be zero and the volume resistivity would be infinite. The surface resistivity is a measure of the current along the surface of the material with an applied voltage. This surface resistivity is mainly a function of surface contaminants (e.g., finger oils). The water absorption is a measure of the water absorbed by the insulator. The piezoelectric effect is the ability of some materials to deform mechanically when a voltage is placed across it. Conversely, when a piezoelectric material is mechanically deformed, a voltage is generated across the material. Triboelectric charging is the tendency of a material to pick up or give up charge when in contact with other objects. Tribocharging generally increases with contact area and speed of rubbing. Dielectric absorption is the tendency of an insulator to store and release charge over extended time periods. Finally, the breakdown strength is a measure of the maximum voltage that can be applied across the insulator before it breaks down or its properties drastically change.

Although obvious to an experienced engineer, the manner in which these properties vary with frequency is important. For example, if the resistance of a dielectric used in a cable varies dramatically with frequency, the cable can introduce significant distortion to a signal that has multiple frequency components.

An uncharacteristically short list (at least for this book) of several very good insulators and their important properties follows:

Sapphire: $\rho = 10^{14}$–10^{16} Ω-m, one of the best insulators, costly, difficult to machine and form

Teflon PTFE: $\rho > 10^{16}$ Ω-m, chemically inert, flameproof, easy to clean, charges internally generated when deformed producing spurious voltages and currents, not suitable in nuclear radiation or in high electric fields

Polystyrene: $\rho > 10^{14}$ Ω-m, less costly than Teflon, easily machined but internal crazing can develop, surface water film at high humidity

Polyethylene: $\rho > 10^{14}$ Ω-m, similar to polystyrene, flexible, used in coaxial cables, low melting point, flammable, large temperature coefficient, good at high frequencies and high voltages, susceptible to corona

Polypropylene: $\rho > 10^{14}$ Ω-m, similar to polyethylene but harder, flexible, used in coaxial cables, low melting point
Ceramic: $\rho = 10^{12}$–10^{13} Ω-m, poor surface properties at high humidity, difficult to machine
PVC (polyvinyl chloride): $\rho = 5\times10^{11}\ \Omega\text{-m}$, less expensive than Teflon, wide temperature range not available, high losses at high frequencies ($f > 500$ MHz), little elasticity
Glass Epoxy and Phenolic: $\rho = 10^{11}$ Ω-m, poor surface properties at high humidity, difficult to machine
Mica: $\rho = 10^{11}$ Ω-m, thin sheets of uniform thickness possible, excellent high-frequency response

An electrometer is used to measure the voltage across the two conductors because of its high sensitivity and special input characteristics. An electrometer can measure voltage, current, resistance, and charge. With its very high input resistance (greater than 100 TΩ or 10^{14} Ω), in many situations an electrometer has an insignificant loading effect on the circuit under test. Also, the input offset current is very small (less than 50 fA or 5×10^{-14} A). Furthermore, an electrometer can even measure extremely small charge levels around 800 aC or 8×10^{-16} C. Because the voltage across the two conductors is separated by an excellent insulator, Teflon, the test instrument and connecting cable should have a very large shunt resistance. The electrometer and dielectric resistance of the cable should be large compared to the resistance of the device under test. The dielectric resistance of the cable is shown in shunt with both the device under test and electrometer in Figure 2.20

Normally, the dielectric resistance, $1/G$, is neglected because the source resistance is much less than $1/G$. The resistance of the slab material is a function of the resistivity of the material between the two conductors and the shape and size of the material. The common expression for the dc resistance between two closely spaced parallel, flat conductors is

$$R = \frac{\rho d}{A}$$

where d is the distance between the conductors and A is the area (or effective area) of the conductors in contact with the insulation. Assuming that the dc volume resistivity, ρ, is 10^{16} Ω-m,

$$R = 10^{16}\frac{d}{A}\ \Omega$$

The dimensions of the slab were not provided in the initial question statement. However, if $d/A < 100$, then $R < 100{,}000$ GΩ. The dielectric resistance of the cable varies with the radii of the inner and outer conductors, composition of the insulation separating the two conductors, and frequency of operation. Typically, dc values for $1/G$ are 100 GΩ-m or more. If the dielectric material of the coaxial cable is assumed to be polyethylene, which has a lower resistivity than Teflon, the resistance of the dielectric between the conductors (where a and b are the radii of the outer and inner conductors, respectively, and l_{th} is the length of the cable)

$$R_{diel} = \frac{\rho \ln\left(\frac{a}{b}\right)}{2\pi l_{th}}\ \Omega \tag{2.41}$$

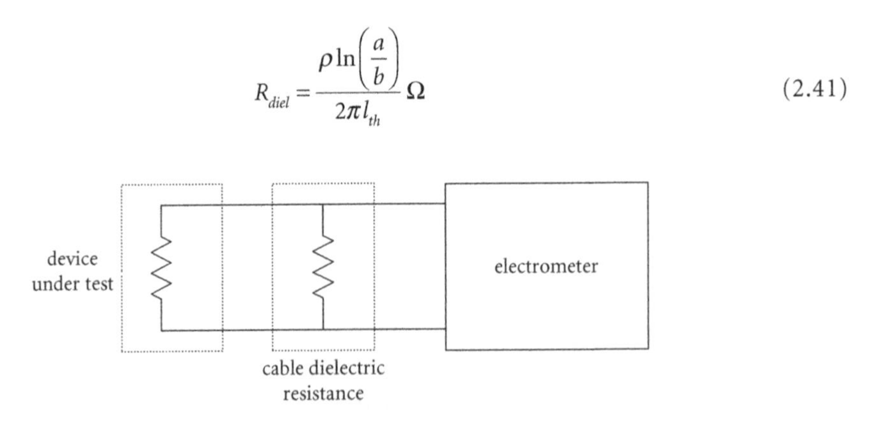

FIGURE 2.20 Model of the cable used to connect an electrometer to a device under test.

may be less than the resistance of the source. (Notice that (2.41) is different from the resistance between two closely spaced plates. For a coaxial structure, the current between the inner and outer conductors is not uniform.) In conclusion, the dielectric resistance of the cable can significantly influence a measurement.

2.13 Cable Capacitance and Audio Cables

Why is the cable capacitance typically only important for audio cables when the source impedance is high?

For most cables used in the audio frequency range (i.e., less than 20 kHz), the cable itself is electrically short, and the lumped model for the cable may be used. The total capacitance of the cable can be determined by multiplying the length of the cable, l_{th}, by the capacitance per unit length, C. Furthermore, when the source or input impedance is high, such as with certain microphones, the resistance and inductance of the cable can often be neglected. The cable can then be simply modeled as a shunt capacitor as shown in Figure 2.21. The ratio of the output voltage to the input voltage in the frequency domain is

$$\frac{V_o}{V_s} = \frac{\dfrac{1}{j\omega Cl_{th}}}{\dfrac{1}{j\omega Cl_{th}} + Z_s} = \frac{1}{1 + j\omega Cl_{th}Z_s}$$

The loading effect of the amplifier is assumed small. When the source impedance is entirely real and equal to R_s, then

$$\frac{V_o}{V_s} = \frac{1}{1 + \dfrac{j\omega}{\dfrac{1}{R_s Cl_{th}}}} \tag{2.42}$$

This transfer function corresponds to a low-pass filter with a cutoff frequency of

$$\omega_c = \frac{1}{R_s Cl_{th}} \tag{2.43}$$

(In other words, the time constant of the system is $R_s Cl_{th}$.) Clearly, for a large source resistance, the cutoff frequency of the low-pass filter is small. The inverse relationship between the cutoff frequency and capacitance is one reason why the length of the cable between a high-impedance microphone and its amplifier is kept short. For example, if the source resistance is 50 kΩ and the capacitance per unit length is 100 pF/m, then the cable length, l_{th}, should be limited to about 1.6 m for high-fidelity audio:

$$\omega_c = 2\pi(20\times 10^3) = \frac{1}{R_s Cl_{th}} = \frac{1}{50\times 10^3(100\times 10^{-12})l_{th}} \quad \Rightarrow l_{th} \approx 1.6\text{ m}$$

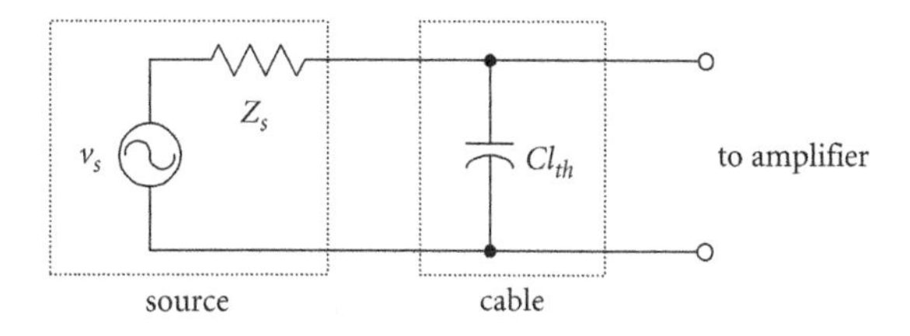

FIGURE 2.21 Single-capacitor model of an electrically-short cable connected between a source and an amplifier.

A maximum length guideline of 15 ft or 4.6 m is seen in some audio handbooks. For some communication systems, the length of the cable can be greater since the highest frequency of interest is about 3 kHz. For professional systems, either the microphone is low impedance, referred to as Low-Z (e.g., 200 Ω), or an amplifier with a low output impedance is used between the microphone and transmission line. When the source impedance is low, the cutoff frequency is high and the filtering effect of the cable will be negligible for reasonable cable lengths.

The impedance of high-impedance, or High-Z, microphones is commonly stated as 60–100 kΩ for crystal microphones and in the MΩ range for capacitor microphones. However, these values correspond to the magnitude of the impedance. For capacitor microphones, the microphone is obviously modeled as a capacitor, and the impedance is reactive not resistive. The ratio of the output to the input voltage corresponds to a capacitor divider:

$$\frac{V_o}{V_s} = \frac{\frac{1}{j\omega Cl_{th}}}{\frac{1}{j\omega Cl_{th}} + \frac{1}{j\omega C_s}} = \frac{C_s}{C_s + Cl_{th}} \tag{2.44}$$

A typical value for C_s is about 50 pF. Since a value of 50 pF/m for the cable is not uncommon, the length of the cable should be short for low loss.

Piezoelectric crystals, when used well below their fundamental resonant frequency, are essentially capacitive. They can be modeled as a very good capacitor of high impedance. Crystals are used as transducers for converting a mechanical force to an electrical voltage. The cable capacitance has a similar effect on the overall response as with other high-source impedance devices. A high-input impedance field-effect transistor can be used directly after the crystal, before connecting to a cable, to reduce some of the cable's loading effects.

2.14 Grounding Strap Impedance

A chassis is connected to ground via a conducting strap. The strap is modeled as a *RLC* circuit. Physically, what does the *R*, *L*, and *C* represent? At low frequencies, the strap is mainly resistive. As the frequency is increased, the *first* resonance occurs. Will the value of *R* remain the same, increase, or decrease with frequency? Will the *first* resonance correspond to a parallel *LC* circuit or series *LC* circuit? Assuming that it is a parallel *LC* circuit, before resonance occurs, is the circuit mainly resistive, inductive, or capacitive? [Mardiguian, '88(b); Johnson, '50]

It is common for safety or electrical noise reasons to ground a chassis. The impedance of the grounding strap should be small if the voltage difference between the ground and chassis is to be small. To model the strap as an *RLC* circuit, the dimensions of the strap, specifically the length of the strap, should be electrically small. The ohmic losses of the strap, and possibly connector and corrosion-related losses, could be modeled through *R*. As the frequency increases, the ohmic losses will increase due to the skin effect: the effective cross-sectional area of the strap will decrease since the current will tend to crowd toward the outer surface. The inductance, *L*, represents both the internal and external inductance of the strap. Since inductance is a closed-loop concept, this inductance is probably the self *partial* inductance of the strap. The capacitance, *C*, represents the capacitance of the strap. Since capacitance must involve two or more objects, *C* is the capacitance between the strap and nearby objects such as the ground plane and chassis. In many situations, it is difficult to determine analytically this capacitance.

A first-order model for the grounding strap could consist of a single *R*, *L*, and *C*. There are several possible methods of modeling this short transmission line. Since the load is the ground or a short circuit, the model in Figure 2.22 is selected (with other models, one of the elements could be shorted out). The return path to the chassis may be through the ground of the power cord. With this model

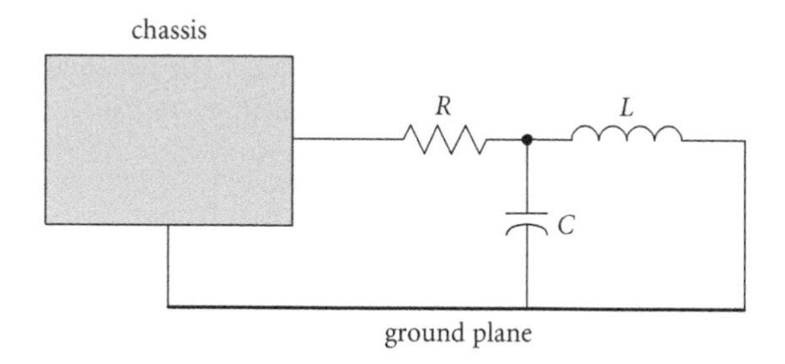

FIGURE 2.22 *RLC* model of a grounding strap connected to a chassis.

for the grounding strap, the first resonance would correspond to a parallel *LC* circuit:

$$Z_{in} = R + \frac{j\omega L \frac{1}{j\omega C}}{j\omega L + \frac{1}{j\omega C}} = R + \frac{j\omega L}{1 - \omega^2 LC} \tag{2.45}$$

At the frequency

$$\omega = \frac{1}{\sqrt{LC}} \tag{2.46}$$

the input impedance is infinite. This frequency is referred to as an antiresonant frequency since the impedance is infinite. For frequencies slightly less than this frequency, the imaginary portion of the impedance is positive, which corresponds to inductive reactance. However, for frequencies slightly greater than this antiresonant frequency, the imaginary portion of the impedance is negative, which corresponds to capacitive reactance. With this model for the strap, at low frequencies the strap is resistive in nature. As the frequency increases, the magnitude of the impedance increases because of the inductance. After resonance, the magnitude of the impedance decreases because the capacitive reactance cancels (some of) the inductive reactance. Again, this lumped model for the grounding strap should only be used when the grounding strap is electrically small. Near or above this resonant frequency, the lumped-circuit model should be used with caution.

If the strap were modeled as a series *RLC* circuit, then the first resonance would correspond to a series *RLC* resonance. The impedance of a series *RLC* circuit at resonance is equal to *R*. In most situations, the magnitude of the impedance of a grounding strap increases from its dc value, similar to a parallel *RLC* circuit not a series *RLC* circuit. The capacitance of the strap is mostly shunt capacitance to ground and other objects and not a series capacitance (across the strap?). However, as will be seen later, multiple series-like and parallel-like resonances can occur after the first antiresonance of the strap. (There may be situations where series capacitance is present such as with a spiraling strap. In this case, there is capacitance between the "turns" of the spiral similar to the capacitance between the turns of an inductor.)

In reference to the previous grounding strap discussion, the broader the "passband" of the strap (i.e., the low-impedance frequency range of the strap) beginning at dc, the lower the impedance and the better the ground. Why is increasing the cross-sectional area of the grounding strap desirable from the "passband" viewpoint? Engineers sometimes place several different ground straps in parallel and connect them to one point at the ground (i.e., single-point ground). How is the "passband" affected by this arrangement? What if all the ground straps were connected in series? [Corp; Paul, '92(b); Grover; Amin; Mardiguian, '88(b)]

The resistance and inductance of the strap usually dominate before the first resonant frequency. The resistance of the strap at dc is a function of the cross-sectional area and length of the strap, as well as the conductivity of the strap. As the cross-sectional area of the strap increases and length of the strap decreases, the dc resistance of the strap decreases. For higher frequencies where the skin effect is important, the larger area will imply a larger strap circumference and lower ac resistance.

The self partial inductance of a strap will also increase with the length of the strap but not linearly. The inductance will also decrease with the cross-sectional area. For example, the external self partial inductance of a circular wire of length l_{th} and radius r_w where $l_{th}/r_w \gg 1$ is

$$L_p \approx 2\times 10^{-7} l_{th}\left[\ln\left(\frac{2l_{th}}{r_w}\right)-1\right]$$

As the wire radius increases, the self partial inductance decreases. For a rectangular grounding strap of width w, thickness t, and length l_{th}, where $l_{th}/(t+w) \gg 1$, the self partial inductance is approximately (assuming a uniform current distribution over the strap)

$$L_p \approx 2\times 10^{-7} l_{th}\left[\ln\left(\frac{2l_{th}}{w+t}\right)+0.5\right]$$

Again, as the cross-sectional area, wt, of the rectangular strap increases, the inductance decreases. The impedance of the strap well before the first antiresonance is

$$Z_{in} = R + \frac{j\omega L}{1-\omega^2 LC} = R + \frac{j\omega L}{1-\left(\dfrac{\omega}{\dfrac{1}{\sqrt{LC}}}\right)^2} \approx R + j\omega L \tag{2.47}$$

By decreasing both the resistance and inductance of the strap, the impedance of the strap will decrease, and the frequency range of the low-impedance "passband" will increase.

By placing several similar-length grounding straps in parallel, the overall impedance will decrease. It is better, however, to place several grounding straps of different lengths in parallel from the chassis to the ground plane. The resonant frequencies of the straps will (normally) be different when their lengths are different. Therefore, when one strap is at an antiresonant frequency, which implies a very high impedance to ground, the other lines are not at antiresonance. (As will be seen from transmission line theory, when the length of a short-circuited lossless line is one-quarter wavelength, the input impedance is infinite. Therefore, the length of the straps should ideally not correspond to one-quarter of a wavelength to avoid this antiresonance condition.) By placing the straps at different locations along the chassis, the inductive and capacitive coupling between the straps is reduced. Furthermore, various locations along the chassis are more likely to be at one single ground reference when the strap connections to the chassis are not electrically close. The single-point grounding ensures that the reference for the straps is the same. If single-point grounding is not used and significant current exists along the ground reference, the reference for the straps may be different because of the voltage drop along the ground.

Bonding straps are available as a solid conductor or collection of wires woven together. The solid straps are usually preferable but are more rigid. The braided versions are sometimes not recommended because the oxides that form between the conductors can be troublesome. One problem is the noise that can be generated by the nonlinear junctions formed by the conductors and oxides. Another problem is the noise that can be received or transmitted from broken wires in the braid.

If a rectangular grounding strap (not close to large metal objects) is l_{th} meters long with an impedance of $Z = R + j\omega L$, it was stated in a book that "standing waves increase the effective impedance by $\tan(2\pi l_{th}/\lambda)$," so the actual impedance seen is

$$|Z|\left[1+\tan\left(\frac{2\pi l_{th}}{\lambda}\right)\right] \tag{2.48}$$

Is this statement reasonable?

This equation *appears* similar to the input impedance expression for a lossless transmission line. Of course, the characteristic impedance of a transmission line is not equal to *Z*. The capacitance is required. The capacitance is difficult to determine analytically for many grounding straps. Although the equation for the "actual impedance," (2.48), does not contain capacitance, a capacitance is required if the system is to resonate resulting in the large values of "actual impedance" at odd multiples of $\pi/2$:

$$\frac{2\pi l_{th}}{\lambda} = \frac{\pi}{2} \Rightarrow \frac{l_{th}}{\lambda} = \frac{1}{4}$$

Without this capacitance, the *RL* circuit cannot resonate. The impedance of the strap merely increases with frequency.

An equipment chassis is grounded by a wire that runs parallel to a grounding surface for a distance of one meter. The far end of the wire is connected to this ground. Determine the equation for the input impedance seen by the chassis. The per-unit-length values for the resistance, inductance, and capacitance are *R*, *L*, and *C*, respectively. Using a typical set of values for *R*, *L*, *C*, and an appropriate lumped-circuit model, use a numerical package to plot the input impedance of this grounding wire vs. frequency. Then, plot the input impedance vs. frequency of two cascaded *RLC* circuits (in this case, the *R*, *L*, and *C* values must be appropriately reduced by a factor of two). Repeat this for three cascaded *RLC* circuits (in this case, the *R*, *L*, and *C* values must be appropriately reduced by a factor of three). Plot the input impedance vs. frequency using the standard, sinusoidal steady-state, transmission line equation for input impedance. Compare all of these plots. What is the input impedance when the grounding wire resonates?

For the lumped-circuit model shown in Figure 2.23, the length, l_{th}, is one meter. A lumped model should only be used for frequencies where the line is electrically short. The frequency corresponding to a free-space wavelength of one meter is 300 MHz. Therefore, this strap is electrically short for frequencies up to 30 MHz. The input impedance is

$$Z_{in} = R + \frac{j\omega L \frac{1}{j\omega C}}{j\omega L + \frac{1}{j\omega C}} = R + \frac{j\omega L}{1-\omega^2 LC} \tag{2.49}$$

If the lumped-circuit model is broken into two segments, each of 0.5 m length, as shown in Figure 2.24, then the upper frequency range of the two cascaded *LC* circuits is extended to about 60 MHz. For three cascaded sections, each corresponding to a length of about 0.33 m, the upper frequency range extends to about 90 MHz.

In the limit as the number of sections approaches infinity and the length of each section approaches zero, the resultant equations for the voltage and current along the line become differential. Since the grounding strap runs parallel with the ground plane, the capacitance between the grounding strap and ground plane should remain relatively constant over the length of the strap. This grounding strap and

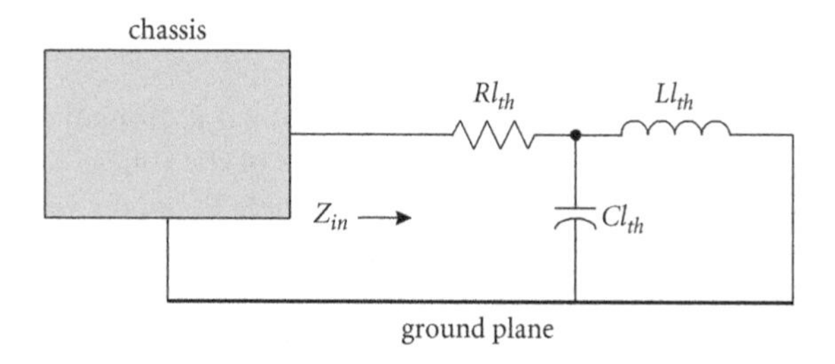

FIGURE 2.23 Grounding strap modeled using one lumped-circuit section.

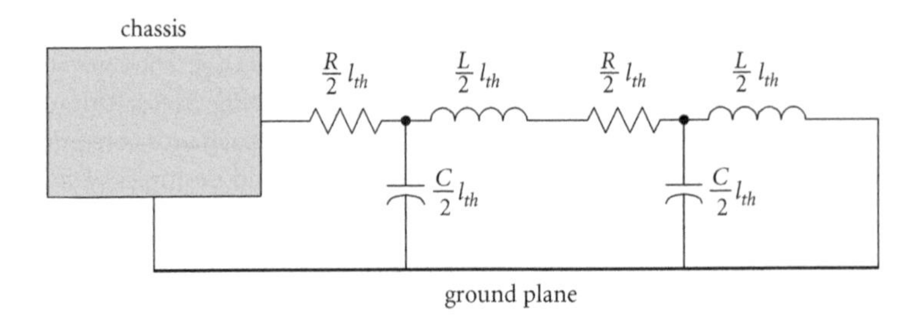

FIGURE 2.24 Grounding strap modeled using two lumped-circuit sections.

ground plane can be modeled as a transmission line. The sinusoidal input impedance for a transmission line is given by the expression

$$Z_{in} = Z_o \frac{Z_L + jZ_o \tanh(\gamma l_{th})}{Z_o + jZ_L \tanh(\gamma l_{th})} \tag{2.50}$$

where l_{th} is the distance to the load, Z_o is the characteristic impedance of the line, Z_L is the load impedance, and γ is the propagation constant:

$$\gamma = \sqrt{(R + j\omega L)(G + j\omega C)} = \sqrt{(R + j\omega L)j\omega C} \quad \text{if } G = 0$$

The per-unit-length values for R, L, and C are for the grounding wire above a large ground plane. For a short-circuited load, the input impedance is

$$Z_{in} = jZ_o \tanh(\gamma l_{th})$$

This sinusoidal expression can be used for any frequency in which the strap and ground plane can be viewed as a transmission line. If the distance *between* the strap and plane is electrically short, and the ohmic losses are low, then the fields between these these two conductors are essentially TEM. TEM fields are the standard field distribution for transmission lines. At the first resonance, the impedance of the strap is infinite, corresponding to antiresonance.

As seen in Mathcad 2.7, for frequencies corresponding to an electrically short strap, the input impedance of the strap is similar for any of the models, including the transmission line equation. Although these plots show the magnitude of the impedance for frequencies up to 100 MHz, each model should only be used for the frequencies previously stated. Notice that as the complexity of the model increases, the number of resonant points (related to the number of maximums and minimums) increases. Also note that the location of these resonances varies with the model used. To simplify the numerical analysis, the resistance per unit length, R, of the transmission line is assumed constant. Actually, for the frequencies plotted, which are greater than 1 MHz, the inductive reactance is much greater than the ac resistance.

To improve the ground of an FM transmitter, the diameter of a 3 ft grounding strap is changed from #10 AWG to #0000 (or #4/0) AWG. Determine the magnitude of the voltage drop across each of these wires at 100 MHz if the current magnitude is 1 mA. Use a simple *RL* model without the capacitance to ground. Is the extra cost associated with the much larger grounding strap worthwhile? Is modeling the strap as an *RL* (or even *RLC*) circuit legitimate? Explain.

Before determining the voltage drop across the grounding strap, the electrical length of this strap will be determined. For free-space conditions, the wavelength corresponding to 100 MHz is 3 m. Distances less than about 0.3 m are electrically short at this frequency. Therefore, a 3 ft strap is probably

$l_{th} := 1 \quad L := 0.6 \cdot 10^{-6} \quad C := 400 \cdot 10^{-12} \quad R := 0.5 \cdot 10^{-3} \quad L_T := L \cdot l_{th} \quad C_T := C \cdot l_{th} \quad R_T := R \cdot l_{th} \quad j := \sqrt{-1}$

$x := 50, 50.01 .. 90$

$$\omega(x) := \left(x + 1 - 10 \cdot \mathrm{floor}\left(\frac{x}{10}\right)\right) \cdot 10^{\mathrm{floor}\left(\frac{x}{10}\right)}$$

$$Z_{in}(\omega, N) := \frac{\left(0 + j \cdot \omega \cdot \frac{L_T}{N}\right) \cdot \frac{1}{j \cdot \omega \cdot \frac{C_T}{N}}}{\left(0 + j \cdot \omega \cdot \frac{L_T}{N}\right) + \frac{1}{j \cdot \omega \cdot \frac{C_T}{N}}} + \frac{R_T}{N}$$

$$Z_{in1}(\omega) := Z_{in}(\omega, 1) \qquad Z_{in2a}(\omega) := Z_{in}(\omega, 2)$$

$$Z_{in3a}(\omega) := Z_{in}(\omega, 3)$$

$$Z_{in2}(\omega) := \frac{\left(Z_{in2a}(\omega) + j \cdot \omega \cdot \frac{L_T}{2}\right) \cdot \frac{1}{j \cdot \omega \cdot \frac{C_T}{2}}}{\left(Z_{in2a}(\omega) + j \cdot \omega \cdot \frac{L_T}{2}\right) + \frac{1}{j \cdot \omega \cdot \frac{C_T}{2}}} + \frac{R_T}{2}$$

$$Z_{in3b}(\omega) := \frac{\left(Z_{in3a}(\omega) + j \cdot \omega \cdot \frac{L_T}{3}\right) \cdot \frac{1}{j \cdot \omega \cdot \frac{C_T}{3}}}{\left(Z_{in3a}(\omega) + j \cdot \omega \cdot \frac{L_T}{3}\right) + \frac{1}{j \cdot \omega \cdot \frac{C_T}{3}}} + \frac{R_T}{3}$$

$$Z_{in3}(\omega) := \frac{\left(Z_{in3b}(\omega) + j \cdot \omega \cdot \frac{L_T}{3}\right) \cdot \frac{1}{j \cdot \omega \cdot \frac{C_T}{3}}}{\left(Z_{in3b}(\omega) + j \cdot \omega \cdot \frac{L_T}{3}\right) + \frac{1}{j \cdot \omega \cdot \frac{C_T}{3}}} + \frac{R_T}{3}$$

$$Z_{in}(\omega) := j \cdot \sqrt{\frac{R + j \cdot \omega \cdot L}{j \cdot \omega \cdot C}} \cdot \tanh\left[\sqrt{(R + j \cdot \omega \cdot L) \cdot j \cdot \omega \cdot C} \cdot l_{th}\right]$$

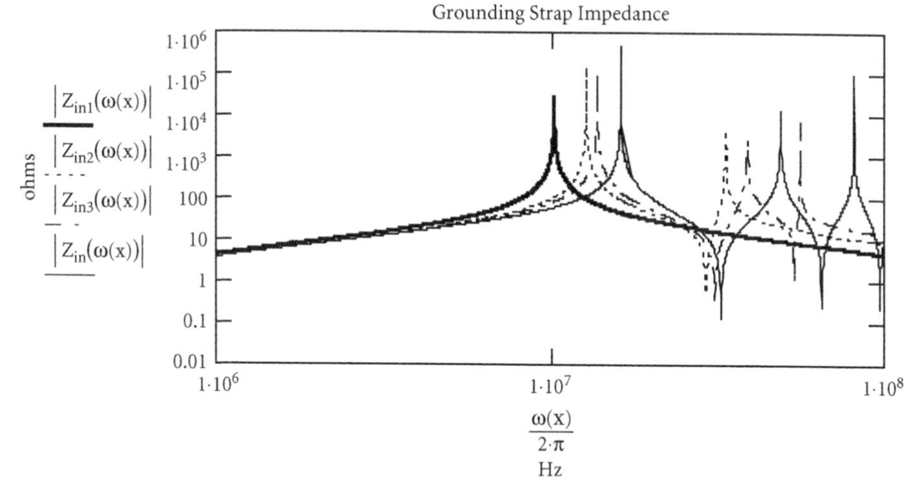

MATHCAD 2.7 Magnitude of the grounding strap impedance vs. frequency using one-section, two-section, and three-section lumped-circuit models of the strap. The impedance using a transmission line model is also shown.

not electrically short. However, without additional information, including the capacitance, the lumped *RL* model will have to suffice.

The wire radii corresponding to #10 and #4/0 AWG wire are 1.29 and 5.84 mm, respectively. Because the skin depth for copper at 100 MHz

$$\delta = \frac{1}{\sqrt{\pi f \mu \sigma}} \approx 6.6\ \mu\mathrm{m}$$

is much less than either wire's radius, the very high-frequency approximation for the ac resistance may be used:

$$r_{AC} = r_{DC}\frac{r_w}{2\delta} = \frac{l_{th}}{\sigma\pi r_w^2}\frac{r_w}{2\delta} \approx \begin{cases} 0.29\,\Omega & \#10\text{ AWG} \\ 0.065\,\Omega & \#4/0\text{ AWG} \end{cases}$$

where $l_{th} \approx 0.91$ m. These resistances are insignificant compared to the inductive impedances at 100 MHz. The self partial inductances of the circular wires are

$$L_p \approx 2\times10^{-7} l_{th}\left[\ln\left(\frac{2l_{th}}{r_w}\right)-1\right] \approx \begin{cases} 1.1\,\mu\text{H} & \#10\text{ AWG} \\ 0.87\,\mu\text{H} & \#4/0\text{ AWG} \end{cases}$$

Notice that the inductances of the wires are about the same even though one wire has a much larger cross section: the wire radius is in the argument of the natural logarithm and logarithms "compress." To reduce significantly the inductance of a wire, it is usually best to decrease its length, l_{th}. The magnitude of the impedance at 100 MHz follows:

$$|Z| = |r_{AC} + j\omega L_p| = \sqrt{r_{AC}^2 + (2\pi f L_p)^2} \approx \begin{cases} 690\,\Omega & \#10\text{ AWG} \\ 550\,\Omega & \#4/0\text{ AWG} \end{cases}$$

(When the wire is not electrically short, the actual magnitude of the impedance may be much greater or smaller than these values.) The impedances are quite similar since the inductive reactances, which dominate at these frequencies, are similar. The magnitude of the voltage drop at one particular frequency, 100 MHz, is the magnitude of the impedance multiplied by the magnitude of the current:

$$|V| = |I||Z| \approx \begin{cases} 0.69\text{ V} & \#10\text{ AWG} \\ 0.55\text{ V} & \#4/0\text{ AWG} \end{cases}$$

It appears that at this frequency the extra cost associated with the larger wire is probably not worthwhile. Again, this model assumes the capacitance is not significant and 100 MHz is well below the first resonance of the strap. Otherwise, the impedance, and hence voltage drop, could be significantly different from the stated values.

2.15 ESD Signal Wire Guideline

One ESD design guideline is that signal lines less than 30 cm long must be within 13 mm of a ground conductor. Determine a typical range for the characteristic impedance of this line. Is this characteristic impedance helpful? [Boxleitner; Walker]

As the distance between two conductors increase, the inductance increases and the capacitance decreases. The net result is an increase in the characteristic impedance of the line. Referring to Figure 2.25, the high-frequency characteristic impedance for a thin strip of width w a distance h above a large ground plane is

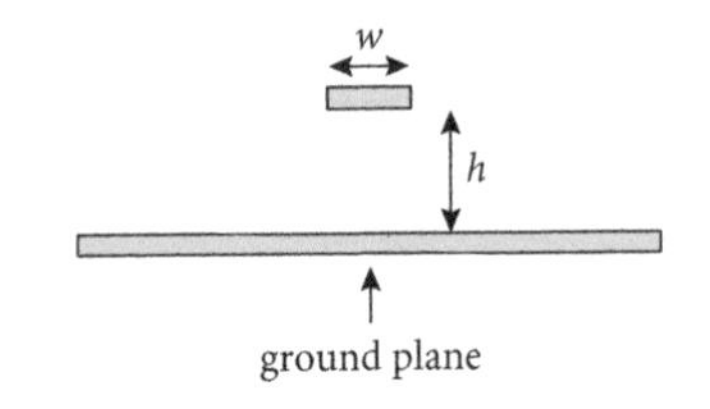

FIGURE 2.25 Thin strap above a large ground plane.

$$Z_o = \begin{cases} 60\ln\left(\frac{8h}{w}+0.25\frac{w}{h}\right) & \text{for } \frac{w}{h}\le 1 \\ \dfrac{377}{\left[\frac{w}{h}+1.393+0.667\ln\left(\frac{w}{h}+1.444\right)\right]} & \text{for } \frac{w}{h}\ge 1 \end{cases} \tag{2.51}$$

While, referring to Figure 2.26, the characteristic impedance for a round wire of radius r_w a distance h above a large ground plane is

$$Z_o = 60 \ln\left\{\frac{h}{r_w}\left[1+\sqrt{1-\left(\frac{r_w}{h}\right)^2}\right]\right\} \approx 60\ln\left(\frac{2h}{r_w}\right) \quad \text{for } \frac{r_w}{h} \ll 1 \tag{2.52}$$

In Mathcad 2.8, the characteristic impedance for both lines 13 mm above a large ground plane is plotted. Unless the wire diameter or strip width is very wide (e.g., greater than about 1 cm), the characteristic impedance will be about 100 Ω or greater. If the line is not electrically short, then this information may be helpful in determining the voltage drop across the line. The smaller the characteristic impedance, the smaller the transient voltage drop across the line.

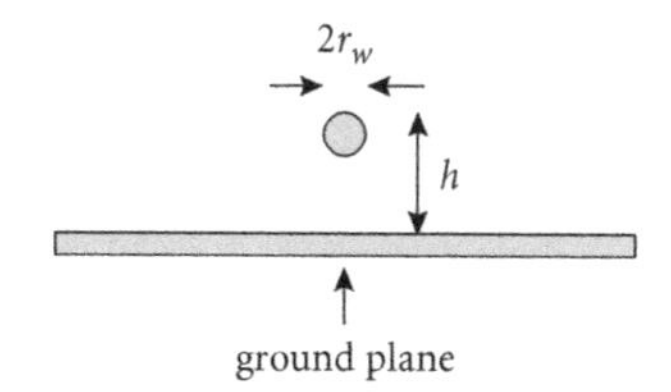

FIGURE 2.26 Round wire above a large ground plane.

If the line is electrically small, which corresponds to frequencies less than 100 MHz, the impedance of the line will be mainly resistive and inductive. The external inductance of a long circular wire ($l_{th} \gg r_w$) above a large ground plane is

$$L_p \approx 2\times10^{-7} l_{th} \ln\left(\frac{2h}{r_w}\right)$$

If the frequency is high so that $R \ll \omega L$, then the impedance is dominated by the inductive reactance. This reactance is plotted from 1 to 100 MHz in Mathcad 2.9 for round wire with a gauge of about #30,

$h := 0.013$

$w := 0.005, 0.01 .. 0.025$

$$Z_{o1}(w) := 60\cdot\ln\left(\frac{8\cdot h}{w} + 0.25\cdot\frac{w}{h}\right) \quad Z_{o2}(w) := 377\cdot\left(\frac{w}{h} + 1.393 + 0.677\cdot\ln\left(\frac{w}{h} + 1.444\right)\right)^{-1}$$

$$Z_{oms}(w) := \text{if}\left(\frac{w}{h} < 1, Z_{o1}(w), Z_{o2}(w)\right) \quad r_w(w) := \frac{w}{2}$$

$$Z_{ow}(w) := 60\cdot\ln\left[\frac{h}{r_w(w)}\cdot\left[1+\sqrt{1-\left(\frac{r_w(w)}{h}\right)^2}\right]\right]$$

Strip and Wire Characteristic Z

ohms — $Z_{oms}(w)$ (solid), $Z_{ow}(w)$ (dashed)

$1\cdot10^3$, 100, 10

0.5, 1, 1.5, 2, 2.5

$w\cdot100$

cm

MATHCAD 2.8 Variation of the characteristic impedance of a thin strip and round wire above a large ground plane as the width (or wire radius) increases. For the round wire, the wire radius is assumed equal to $w/2$.

$h := 0.013$

$x := 60, 60.1 .. 80$

$$\omega(x) := \left(x + 1 - 10 \cdot \mathrm{floor}\left(\frac{x}{10}\right)\right) \cdot 10^{\mathrm{floor}\left(\frac{x}{10}\right)}$$

$$L(r_w, l_{th}) := 2 \cdot 10^{-7} \cdot l_{th} \cdot \ln\left(\frac{2 \cdot h}{r_w}\right)$$

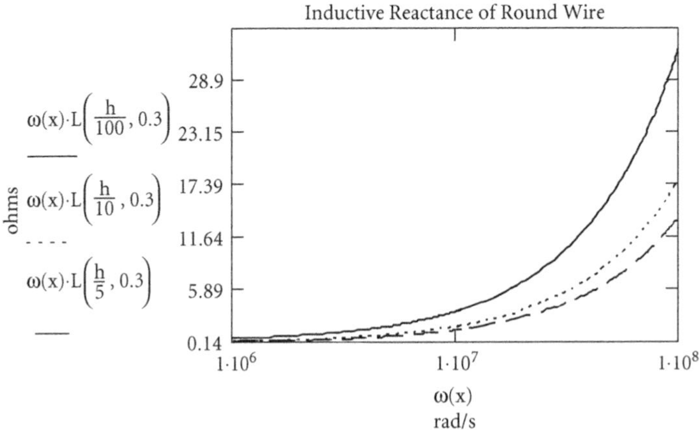

h/100 corresponds to the #30 AWG wire, h/10 corresponds to the #10 AWG wire, and h/5 corresponds to the #4 AWG wire

MATHCAD 2.9 Inductive reactance of three different diameter, 30 cm long, round wires as a function of frequency.

#10, and #4. For frequencies less than about 10 MHz, the reactance is low. Therefore, the ESD guideline seems reasonable if the line is electrically short.

2.16 Twisted Pair

Coated wire of #16 gauge is used to construct three types of transmission lines: twisted at five turns per inch, tightly twisted, and not twisted (parallel lines). The characteristic impedances measured are 35, 40, and 50 Ω (not necessarily respectively). Match the impedances with the lines and explain your reasoning. [Freeman, '95; Lefferson; Krauss, '92]

Twisted pair consists of two insulated wires that are twisted together with a regular spiral pattern. Twisted-pair transmission lines are commonly used. Many years ago, they replaced open-air parallel lines for telephone wires for several reasons including their reduced environmental sensitivity (e.g., to icing across the lines) and increased line density. Unlike the twin-lead line and coax, however, the capacitance and inductance along the twisted pair are not constant.

The twisted-pair cable is considered a nonuniform line since the cross-sectional dimensions vary along the line. Fortunately, the line can be modeled using identical cascaded sections if the nonuniform characteristics are periodic along the line (i.e., the twist rate and spacing do not vary along the line) and the size of the sections is electrically small (i.e., the spacing between twists is electrically small). The element labels $C(x)$ and $L(x)$ in Figure 2.27 emphasize the variation of the capacitance and inductance along each of the twist segments. To simplify the analysis, an average value over each half-twist is often used for these parameters.

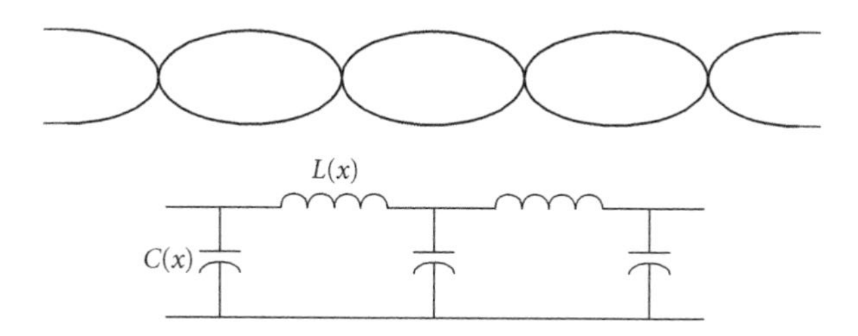

FIGURE 2.27 Simple model of twisted-pair transmission line.

To determine the qualitative relationship between the twist rate and characteristic impedance, a simple understanding of capacitance and inductance is required. Recall that the capacitance between two objects increases as the distance between the objects decreases. Also, the inductance of a closed loop decreases as the loop area decreases. Since the high-frequency characteristic impedance is defined as

$$Z_o = \sqrt{\frac{L}{C}}$$

the parallel lines probably correspond to the impedance of 50 Ω. The distance between the conductors is the largest for the parallel wires, therefore the inductance is the largest, the capacitance is the smallest, and the impedance is the largest. As the wires approach each other and the twisting rate increases, the capacitance will increase and the inductance will decrease. The characteristic impedance should generally decrease with an increase in the number of twists per centimeter (or tighter twisting). Therefore, the five-twist-per-inch line probably correspond to 40 Ω, and the tightly twisted line probably corresponds to 35 Ω. The approximate twist rate vs. the measured characteristic impedance for several coated wires is provided in Table 2.3. One twist is defined as two loops while half of a twist is defined as one loop.

The twisting, in addition to reducing the loop area, has the effect of reducing the overall inductance since the phases of neighboring loops are opposite. Thus, the voltages induced in neighboring loops from an external field source are of opposite sign. This can be modeled with mutual inductance. The positions of the polarity or dot markers for the mutual inductance are opposing for neighboring loops.

Since the wires are coated with an insulative material with a dielectric constant greater than one, it is difficult to determine analytically an expression for the capacitance. There has been little work in this area. Restrictive, empirically fitted, expressions for the effective dielectric constant as a function of the wire pitch are available.

There are two major types of twisted pair: *unshielded twisted pair* (UTP) and *shielded twisted pair* (STP). A common characteristic impedance for UTP is 100 Ω and for STP is 150 Ω. However, as with many transmission lines, these values are the high-frequency impedances. For example, for some telephone UTP, the characteristic impedance can vary from 600 Ω at 1 kHz to 100 Ω at 1 MHz. Common gauges used for the conductors are #19 through #26.

Before the various categories of twisted pair are introduced, bandwidth and capacity will be discussed. The bandwidth provided can be relative to a center frequency such as for a TV transmission centered at

TABLE 2.3 Characteristic Impedance of Twisted-Pair Line as a Function of the Twist Rate

Wire Type	Z_o @ Twists/cm	Z_o @ Twists/cm	Z_o @ Twists/cm
#30 AWG vinyl coated	127 Ω @ 1	125 Ω @ 1.2	115 Ω @ 2.5
#30 AWG enamel insulation	72 Ω @ 0.9	64 Ω @ 1.3	60 Ω @ 1.8
#22 AWG enamel insulation	53 Ω @ 0.65	51 Ω @ 1	45 Ω @ 1.8
#20 AWG enamel insulation	42 Ω @ 0.8	40 Ω @ 1.1	30 Ω @ 2

channel 12 or it can correspond to the highest frequency that can be acceptably received at the output of a cable of a given length. In many cases involving transmission lines, the second definition was probably used. Furthermore, it is likely that the bandwidth was measured with the load matched to the cable (so that reflections do not complicate the measurement). If the bandwidth was determined by measuring the output level of a signal and comparing it to the input signal level as the frequency is increased, then losses on the cable probably limit the upper frequency or bandwidth. This bandwidth is not necessarily the 3 dB bandwidth. In some telephone lines, 1 dB/km is the acceptable loss level. There are two sources of cable loss: dielectric loss in the insulation and ohmic loss in the conductors. The dielectric losses, which can be frequency dependent, are usually very small except at very high frequencies. The ohmic losses are a function of the wire size (and spacing) and the skin depth. The bandwidth of twisted pair can also be a function of the crosstalk between different twisted pairs in the same cable. For example, for full duplex operation (transmitting and receiving at the same time), two pairs of twisted pair are usually required. If the two pairs are relatively close, a fraction of the signal on the transmitting twisted pair will be picked up by the receiving twisted pair. This crosstalk can be sufficiently strong to limit the operating frequency and bandwidth of the cable. Shielding each pair can reduce crosstalk.

In some books it is stated that twisted pairs (UTP or STP) have low bandwidths and are typically less desirable than coax. These statements are not necessarily true. For a given dielectric material, the dielectric losses are similar for twisted pair and coax. If the wire used for the twisted pair is not too small, then the skin effect losses for the twin-lead and coax can be comparable. (However, since the twisted-pair wires are usually in closer proximity than the center and outer conductors of a coaxial cable, the proximity effect is likely to have a greater effect on the ac resistance and loss of the twisted pair.) If the twisting of the wires is tight or finely controlled, then the impedance of the twisted pair should be uniform along the cable. The problems associated with reflections from a change in the impedance along the line should be small. At very high frequencies, both twin-lead and coax suffer from impedance variation along the line: the spacing between the conductors is not exactly constant over distances that are electrically not small. However, the attenuation rather than reflections typically limit the speed or bandwidth of a cable.

The bandwidth of many telephone lines is limited to around 3.5 kHz. However, this lower bandwidth is not necessarily because of the limitations of the twisted pair or other cable used. Instead, it is due to the low-pass filtering and sampling that is performed.

Frequently, the capacity of a cable is given in addition to the bandwidth of a cable. Shannon's formula relates the capacity, *bandwidth* (*BW*), and *signal-to-noise ratio* (*SNR*):

$$\text{capacity} = BW \log_2(1+SNR) = BW \frac{\log_{10}(1+SNR)}{\log_{10}(2)} \tag{2.53}$$

The *SNR* ratio is a function of factors such as the signal power level, modulation scheme, crosstalk, line length, and temperature. The capacity is typically given in *bits per second* (bps) or Mbps. This equation states that for BW = 1 MHz and SNR = 100, the capacity is about 6.7 Mbps. Note that the capacity and *BW* are not necessarily equal. As expected, the capacity of a cable will decrease as the length of the line increases. There are several categories for UTP. Category 1 and 2 are for voice and low data-rate signals such as alarm systems. Category 3 is for data rates of up to 10 Mbps. Category 4 is for rates up to 16 Mbps. The popular category 5, used for computer networks, is for rates up to 100 Mbps. The capacitance per meter for category 4 and 5 UTP is 56 pF/m. Even higher data rates are possible over short distances.

The method of feeding signals into the twisted pair will affect the emissions level of the cable and its capacity. For example, when the currents in the two conductors are equal in magnitude but oppositely directed, referred to as balanced, the line emissions are generally less than when the currents are not equal in magnitude, referred to as unbalanced. When one conductor of the twisted pair is connected to ground, the line is considered unbalanced.

2.17 When the Line Can Be Ignored

Using the Z_{in} equation for a transmission line, determine the *approximate* input impedance of a line of length $\lambda/100$ with a load Z_L. State all assumptions. When is the characteristic impedance of the line important?

When a line is electrically short, the R, L, and C of the line are often times ignored, and the input impedance is assumed equal to the load impedance. This discussion will examine the validity of this common assumption. First, a line one-hundredth of a wavelength is considered electrically short in this book. Second, the sinusoidal input impedance for a transmission line is given by the expression

$$Z_{in} = Z_o \frac{Z_L + jZ_o \tanh(\gamma l_{th})}{Z_o + jZ_L \tanh(\gamma l_{th})}$$

where l_{th} is the distance to the load, Z_o is the characteristic impedance of the line, Z_L is the load impedance, and γ is the propagation constant:

$$\gamma = \sqrt{(R + j\omega L)(G + j\omega C)} = \sqrt{(R + j\omega L) j\omega C}$$

The conductance, G, of the line is assumed negligible. Since the line is electrically short, γl_{th} is small, and assuming R is small,

$$\tanh(\gamma l_{th}) \approx \gamma l_{th} \quad \text{if } |\gamma l_{th}| \ll 1$$

since the series representation of a hyperbolic tangent is

$$\tanh(x) = x - \frac{x^3}{3} + \frac{2x^5}{15} - \cdots$$

When x is small, the first term of the series dominates. Therefore, the sinusoidal input impedance of the electrically short cable is

$$Z_{in} \approx Z_o \frac{Z_L + jZ_o \gamma l_{th}}{Z_o + jZ_L \gamma l_{th}} = Z_L \frac{1 + j\dfrac{Z_o}{Z_L}\gamma l_{th}}{1 + j\dfrac{Z_L}{Z_o}\gamma l_{th}} \tag{2.54}$$

Notice that

$$Z_{in} \approx Z_L \quad \text{if } \left|\frac{Z_o}{Z_L}\gamma l_{th}\right| \ll 1 \quad \text{and} \quad \left|\frac{Z_L}{Z_o}\gamma l_{th}\right| \ll 1 \tag{2.55}$$

That is, the properties of a short line should be taken into account when the characteristic impedance of the line and the load impedance are much different. For example, if the load is an ammeter with a very small impedance, then the characteristics of a typical line connected to it should be taken into account.

These inequalities can be more easily understood by assuming that R is negligible so that

$$\gamma \approx \sqrt{(j\omega L) j\omega C} = j\omega\sqrt{LC} = j2\pi f \frac{1}{v} = j\frac{2\pi}{\lambda} \quad \text{if } R \ll \omega L \tag{2.56}$$

Then,

$$\left|\frac{Z_o}{Z_L}\gamma l_{th}\right| = 2\pi\left|\frac{Z_o}{Z_L}\right|\frac{l_{th}}{\lambda} \ll 1 \quad \text{and} \quad \left|\frac{Z_L}{Z_o}\gamma l_{th}\right| = 2\pi\left|\frac{Z_L}{Z_o}\right|\frac{l_{th}}{\lambda} \ll 1 \tag{2.57}$$

or

$$\frac{2\pi l_{th}}{\lambda} \ll \left|\frac{Z_o}{Z_L}\right| \ll \frac{\lambda}{2\pi l_{th}} \quad (2.58)$$

For a line length of $\lambda/100$,

$$\frac{1}{16} \ll \left|\frac{Z_o}{Z_L}\right| \ll 16$$

For example, for $Z_o = 200\ \Omega$, the magnitude of Z_L should be much less than 3.2 kΩ but much greater than 13 Ω if the connecting line is to be ignored. This may be surprising to some students. Realize, however, that a characteristic impedance of 200 Ω is approximately the impedance of two wires loosely twisted. In many student laboratory settings, the wires are neither close together nor twisted. The characteristic impedance for these wires is much larger than 200 Ω (e.g., 600–1,000 Ω). Also, in many undergraduate laboratories, the operating frequency is usually a few MHz or less. The free-space wavelength at 3 MHz is 100 m. One-hundredth of a wavelength is still 1 m, which is usually much greater than the wires lengths between components on a breadboard.

Obviously, for many breadboard situations, the capacitance and inductance of the connecting wires are not constant over their length. However, when the wire is electrically short, one of the previously given lumped models may be used. The line inductance and capacitance can typically be neglected if the load impedance is not an extreme value such as that corresponding to a short circuit or an open circuit.

2.18 Line Resonance

The power cord for the "docking station" of a cordless telephone is about 7 ft in length. Show that the "resonant" frequency of this line is about 30 MHz. Why is this important? [ARRL, '96]

It is extremely common to discuss the "resonant" frequency of a line or system. At this "resonant" frequency, typically the amplitude of some signal is unexpectedly large or small, similar to a signal near or at the resonant frequency of a circuit. For transmission lines, the meaning of "resonant" length is not always clear. However, a line that is one-half wavelength in size (or integer multiples of one-half wavelength) has particular properties that are often considered "resonant-like." If 7 ft (≈ 2.1 m) corresponds to one-half of a wavelength, then the frequency corresponding to 7 ft of #18 Zip/Lamp Cord (C = 10 pF/ft ≈ 33 pF/m, L = 0.35 μH/ft ≈ 1.1 μH/m) is

$$f = \frac{v}{\lambda} = \frac{\frac{1}{\sqrt{LC}}}{2\times 2.1} = \frac{1.65\times 10^8\ \text{m/s}}{4.2\ \text{m}} \approx 40\ \text{MHz}$$

It is important to remember that L and C are per length (length is in meters in the SI system); otherwise, the velocity of a signal on a cable would be a function of the length of the cable! Since the electrical properties of the power cord are not provided, 40 MHz is a reasonable first estimate for one of the resonant frequencies.

It is well known that the impedance repeats itself on a line every half wavelength. If the load is an open circuit, then the input impedance of the half-wavelength lossless line is infinite. Similarly, if the load is a short circuit, then the input impedance is zero. This infinite and zero input impedance is similar to the impedance of a parallel and series resonant circuit, respectively. The label "resonant" line might be based on these impedance relationships. A possible consequence of line resonance is that the current levels and radiation of this power cord/antenna may be significant and different at this frequency (and integer multiples of this frequency). For example, the source *may* see a very low impedance and the current levels *may* consequently increase at these frequencies. (It should also be noted that the source and load impedances would slightly affect the resonant frequency of the power cord similar to the inductive loading of an antenna.)

Actually, the manner in which the line is connected to the source and load (e.g., grounding location) and the impedances of the source and load determine the resonant length and the degree to which the line radiates.

The properties of a half-wavelength structure are also very well known. It is both an excellent receiver and transmitter of energy at this frequency. Often in reverberation chamber tests, harnesses are used to connect sources and loads (e.g., a battery to an automobile cluster). The resonant frequencies of these harnesses are frequently determined and recorded. The results of both emission and susceptibility tests in these chambers must be cautiously interpreted at these resonant frequencies.

Although wires of lengths less than and greater than a half of a wavelength also radiate, their properties (e.g., input impedance and radiation pattern) are usually not as "convenient." For example, the load impedance of many transmission lines is in the 75 Ω range, the radiation resistance of a half-wavelength antenna. Many transmitters are designed to expect a load of 75 Ω. If the length of the antenna deviates significantly from a half of a wavelength, its impedance will be much different from 75 Ω. Hence, the transmitter is less likely to transfer energy efficiently into these other antenna sizes.

2.19 Multiple Receiver Loading

Transmission lines are frequently tapped at several locations by similar receivers. Often these receivers are capacitive in nature. If the receivers are uniformly distributed along the transmission line and their spacing is electrically small, then a very good approximation for the impedance of the line is

$$Z_o = \sqrt{\frac{L}{C + \frac{NC_R}{l_{th}}}}$$

where L and C are the original parameters for the line, C_R is the capacitance of each receiver, N is the number of receivers, and l_{th} is the total length of the line. Using a numerical program, partially verify the validity of this result by first assuming that the load is matched to this new characteristic impedance and then plotting the input impedance of the line as a function of the distance from the load along the line. Let

$$\begin{gathered} L = 300\ \text{nH/m},\ C = 120\ \text{pF/m},\ C_R = 20\ \text{pF} \\ N = 10,\ \omega = 100\ \text{Mrad/s},\ l_{th} = 10\ \text{cm} \end{gathered} \tag{2.59}$$

[Johnson, '93; King; Rosenstark; Montrose, '00]

The common high-frequency expression for the characteristic impedance of a line is

$$Z = \sqrt{\frac{L}{C}}$$

where L is the inductance per unit length and C is the capacitance per unit length. When N receivers are placed across the line of length l_{th} at uniform intervals, each with a capacitance of C_R, and the spacing between the receivers is electrically small, then the capacitance per unit length increases to

$$C + \frac{NC_R}{l_{th}} \tag{2.60}$$

and the characteristic impedance reduces to

$$Z_o = \sqrt{\frac{L}{C + \frac{NC_R}{l_{th}}}} \tag{2.61}$$

This impedance is also referred to as the modified or effective characteristic impedance. Although a receiver is an obvious source of loading along a transmission line, a less obvious source is a component pad on a printed circuit board.

Transmission lines with large characteristic impedances are more affected by capacitive loading than lines with small values, assuming the large values for Z_o are mainly due to small values of C (instead of large values of L). Lines with small characteristic impedances are less affected by capacitive loading than lines with large values, assuming the small values of Z_o are mainly due to large values of C (instead of small values of L). For this reason, it is sometimes stated that low-impedance systems (transmission lines?) are less sensitive to the effects of capacitive loading. For example, a 1 m long 50 Ω line with $C = 120$ pF/m and $L = 300$ nH/m will decrease to only about 46 Ω for $N = 10$, 2 pF loads (a 7% drop). However, a 1 m long 71 Ω line with $C = 60$ pF/m and $L = 300$ nH/m will decrease to about 61 Ω for $N = 10$, 2 pF loads (a 13% drop).[9]

If the receivers are not uniformly spaced along the line, then this characteristic impedance equation cannot be used since it was derived based on uniform L and C along the line. Furthermore, if the spacing between the receivers is not electrically short, then the per-unit-length capacitance, C, cannot merely be increased as shown in (2.60). When the spacing between receivers is not electrically small, the receivers are not "part" of the transmission line but are shunt loads across the line. Each of these shunt loads, if they generate an impedance mismatch, will produce reflections. If the receivers cannot be placed uniformly along the line, sometimes referred to as unbalanced loading, then Z_o can sometimes be adjusted (e.g., via the trace width for printed circuit boards) between the loads to reduce the impedance mismatch.

For the parameters given in (2.59), the transmission line impedance without the receivers is

$$Z = \sqrt{\frac{300 \times 10^{-9}}{120 \times 10^{-12}}} = 50\ \Omega$$

The corresponding velocity of a signal along the line (assuming the line is lossless) is

$$v = \frac{1}{\sqrt{LC}} \approx 1.7 \times 10^{8}\ \text{m/s}$$

The wavelength of a 100 Mrad/s signal on the line is then

$$\lambda = \frac{v}{f} = \frac{v}{\frac{\omega}{2\pi}} \approx 11\ \text{m}$$

The spacing between the receivers should be less than 1.1 m for the distance to be electrically short. Because the length of the line is 10 cm and there are 10 receivers, the spacing is 1 cm, which is obviously electrically small. The modified characteristic impedance is therefore

$$Z_o = \sqrt{\frac{L}{C + \frac{NC_R}{l_{th}}}} \approx 12\ \Omega$$

The characteristic impedance has dropped rather dramatically. This will affect the impedance matching and increase the transient current demand. Also, note that the velocity of a signal on the line will be much less with this additional capacitance:

$$v = \frac{1}{\sqrt{L\left(C + \frac{NC_R}{l_{th}}\right)}} \approx 0.40 \times 10^{8}\ \text{m/s}$$

[9] If the 1 m long 71 Ω line has a $C = 120$ pF/m and an $L = 600$ nH/m, then the effective loaded impedance would be about 65 Ω, which corresponds to a change of only ≈ 7%.

Thus, the additional capacitance increases the delay of a signal on the line. The wavelength of a 100 Mrad/s signal on this loaded line is 2.5 m. However, the spacing between the receivers is still electrically small. This difference in velocity can be significant for high-speed systems. For example, if the distance between two receivers is about 1 cm, the delay between receivers for the unloaded system is

$$\Delta t = \frac{d}{v} = \frac{0.01}{1.7\times 10^8} \approx 59 \text{ ps}$$

and for the loaded system

$$\Delta t = \frac{d}{v} = \frac{0.01}{0.40\times 10^8} \approx 250 \text{ ps}$$

The last receiver will receive the signal about 0.53 ns after the first receiver for the unloaded line while about 2.3 ns for the loaded line. This may affect the timing for high-speed systems.

In Mathcad 2.10, the periodic nature of the impedance for a line with 150 receivers and a length of 1.5 m is illustrated. The expression for the input impedance of a lossless transmission line is

$$Z_{in} = Z_o \frac{Z_L + jZ_o \tan(\beta d)}{Z_o + jZ_L \tan(\beta d)}$$

$$l_{th} := 1.5 \quad L := 300\cdot 10^{-9} \quad C := 120\cdot 10^{-12} \quad C_R := 20\cdot 10^{-12} \quad N := 150 \quad \omega := 100\cdot 10^{6} \quad j := \sqrt{-1}$$

$$Zm := \sqrt{\frac{L}{C + \frac{C_R\cdot N}{l_{th}}}} \quad Zm = 11.896 \quad d := \frac{l_{th}}{N} \quad d = 0.01 \quad Z_o := \sqrt{\frac{L}{C}} \quad Z_o = 50 \quad \beta := \omega\cdot\sqrt{L\cdot C} \quad \beta = 0.6$$

$$Zin_0 := Zm$$

$$i := 0, 1 .. 149$$

$$Zin_{i+1} := Z_o\cdot\frac{\left(\text{if}\left(i<1, Zin_0, \frac{Zin_i\cdot\frac{1}{j\cdot\omega\cdot C_R}}{Zin_i + \frac{1}{j\cdot\omega\cdot C_R}}\right) + j\cdot Z_o\cdot\tan(\beta\cdot d)\right)}{Z_o + j\cdot\text{if}\left(i<1, Zin_0, \frac{Zin_i\cdot\frac{1}{j\cdot\omega\cdot C_R}}{Zin_i + \frac{1}{j\cdot\omega\cdot C_R}}\right)\cdot\tan(\beta\cdot d)}$$

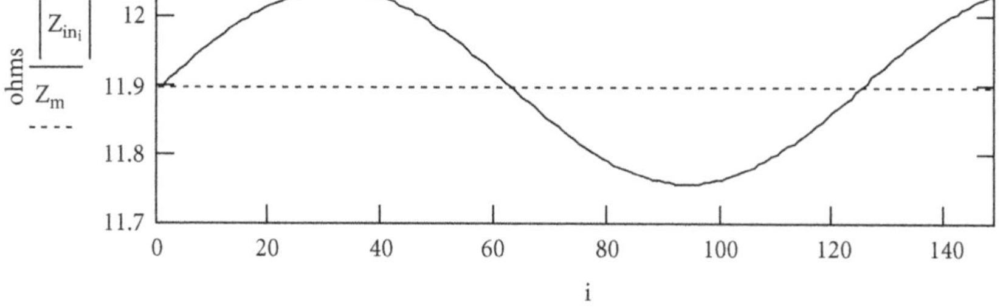

MATHCAD 2.10 Variation in the input impedance along a line with a matched load and uniformly distributed capacitance along its length.

where d is the distance to the load, Z_o is the characteristic impedance of the line, Z_L is the load impedance, and $\beta = \omega\sqrt{LC}$ is the phase constant of the line. In the solution shown, the input impedance is determined for each segment of line between the capacitive loads or receivers. The load, Z_L, however, changes for each segment of length d since the load for one line is a function of the input impedance of the prior line segment(s). If the capacitance of the receivers were uniformly and *continuously* distributed over the entire length of the line instead of just at discrete locations, then the input impedance over the line should be equal to (2.61). (The impedance of the line's last load was set equal to the impedance of the modified characteristic impedance. Therefore, since the line is matched, the impedance looking toward the load should be the same everywhere and equal to the modified characteristic impedance.) Notice that the input impedance does not vary much from 12 Ω. Also, the period of the impedance variation is about 125(0.01) = 1.25 m. This corresponds to one-half wavelength of a 100 Mrad/s signal on the loaded line. For a transmission line, the input impedance repeats every $\lambda/2$ while the current and voltage along the line repeat every λ. (The *amplitude* of the current and voltage repeat every $\lambda/2$.)

In conclusion, multiple receivers closely and uniformly spaced along a transmission line do increase the line loading and decrease the line's effective characteristic impedance. Furthermore, as fanout or increased loading increases, the propagation delay increases since the velocity decreases with increased capacitance. Thus, the maximum recommended line length decreases with fanout. In many cases, it is better to spread out the loading over the length of the line rather than lumping it at the end of the line. (The line then appears more uniform.)

2.20 Proximity Effect

Discuss the proximity effect for twin-lead transmission line. Discuss the proximity effect for a coaxial transmission line. [Chipman; Arnold; Welsby; Magnusson; Smith, '72; Watt; Sim; Perry, '85]

When two current-carrying conductors are in close proximity, the electromagnetic fields from one wire affect the current distribution in the other wire. This effect is in addition to the skin effect. Unlike the skin effect that crowds the current uniformly to the outside of an isolated wire, the proximity effect for two parallel conductors causes a nonuniform distribution of the current in the wires. For two parallel conductors with currents in the same direction, the currents are most concentrated where the two conductor surfaces are the farthest. When the currents are in opposite directions, the currents are most concentrated where the two conductor surfaces are the closest. The distribution in this latter case is roughly illustrated in Figure 2.28.

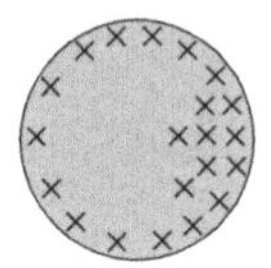
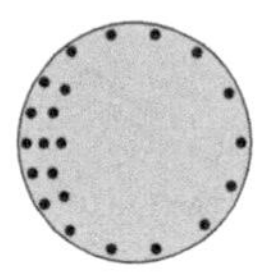

FIGURE 2.28 Current distribution for two closely spaced wires carrying equal magnitude but oppositely directed currents.

For high frequencies where the skin depth is small compared to the wire radius, the increase in resistance due to the proximity effect for currents in opposite directions is 1.09 (9%) when the ratio of the center-to-center spacing to the wire radius is five. If this ratio is ten, the increase in resistance is 1.02 (2%). An easy way to remember a ratio of four is if a wire of the same diameter can be inserted between the two wires, then the ratio is at least four. A ratio of four or greater is satisfied by many transmission lines.

The proximity effect also influences the ac resistance of coils. It should not be too surprising that the ac resistance of a straight piece of wire is nearly the same as the ac resistance of a single loop formed from this wire (assuming the diameter of the loop is much greater than the wire's radius). However, when a tightly wound coil is formed out of the wire, the ac resistance increases. Because of the proximity effect, the electromagnetic fields generated from each loop in the coil affect the current distribution in the other loops. In the design of high-Q coils, the loops are separated to minimize the increase in the resistance due to the proximity effect. Depending on factors such as the number of turns, spacing between turns, and wire radius, the proximity effect can increase the resistance by a factor of three or more.

In this case, the currents in neighboring wires are in the same direction and the skin depth is assumed small compared to the wire radius. The increase in resistance due to the proximity effect is 1.08 (8%) when the ratio of the center-to-center spacing to the wire radius is five. If this ratio is ten, the increase in resistance is 1.02 (2%).

An approximation for the total, high-frequency ac resistance, r_{AC}, for two parallel wires of radius r_w with a center-to-center separation of d and oppositely directed currents is

$$r_{AC} = r_{DC}\left[\frac{K}{2}\frac{r_w}{\delta} + \frac{K(2-K^2)}{4} + \frac{K(9-10K^2+4K^4)}{32}\frac{\delta}{r_w}\right]$$

$$\text{where } K = \frac{1}{\sqrt{1-4\left(\frac{r_w}{d}\right)^2}}, \quad \delta = \frac{1}{\sqrt{\pi f \mu \sigma}} \tag{2.62}$$

The dc resistance of the wires is r_{DC}. This equation for the total ac resistance includes both the skin and proximity effects and is valid when $r_{AC}/r_{DC} > 2$. The error associated with this equation is less than 10% when $d/r_w \geq 2.3$ and $r_w \geq 3.5\delta$. Equation (2.62) reduces to the high-frequency approximation for the ac resistance due to the skin effect alone when $d \to \infty$ (i.e., $K \to 1$):

$$r_{AC} \approx r_{DC}\left(\frac{1}{2}\frac{r_w}{\delta} + \frac{1}{4} + \frac{3}{32}\frac{\delta}{r_w}\right) \quad \text{if } k \approx 1 \tag{2.63}$$

The error associated with this high-frequency equation using only the first term is less than 10% if $r_w \geq 5\delta$. This error decreases as the second and third terms are added. Notice also that K is an approximate indication of the effect of the wire spacing on the ac resistance due to the proximity effect if the skin depth is small compared to the wire radius and the wire spacing is not too small:

$$r_{AC} \approx r_{DC}\frac{K}{2}\frac{r_w}{\delta}$$

When $d = 5r_w$, $K \approx 1.09$, and when $d = 10r_w$, $K \approx 1.02$, which are in agreement with the previous statements. Mathcad 2.11 illustrates the importance of the proximity effect for close wire spacing. The ac resistance increases with the proximity effect since the current in the wires is passing through a smaller effective area.

Although the proximity effect is present in coaxial lines when the inner and outer conductors are closely spaced, the current distribution remains essentially only a function of the radial distance from the axis: the inner and outer conductors are concentric.

A factor that is not normally mentioned is the effect of the magnitude of the current in the wires. When the current in the wires is very large, the ac resistance of the wires can be a function of the magnitude of the current in the wires in addition to the wire radius and spacing. This relationship between the current amplitude and resistance is a nonlinear effect since this phenomenon is a function of the amplitude of the current and superposition fails. Although the resistance can also change because of an increase in temperature, this change in resistance with current is not due to a temperature change. It is due to the strong magnetic fields generated by the high currents in the wire and their interaction with the electrons in the "other" wire. For coils and transformers with high current levels, this current dependency of the resistance can be important. At dc, it is unclear what influence the proximity effect has, especially at high current levels and with close conductor spacings.

$s := 2.25, 2.30 .. 20$

$$K(s) := \frac{1}{\sqrt{1 - \frac{4}{s^2}}} \qquad K(2.25) := 2.183 \qquad K(10) := 1.021$$

$$R(K, x) := \frac{K}{2} \cdot x + \frac{K \cdot \left(2 - K^2\right)}{4} + \frac{K \cdot \left(9 - 10 \cdot K^2 + 4 \cdot K^4\right)}{32} \cdot \frac{1}{x}$$

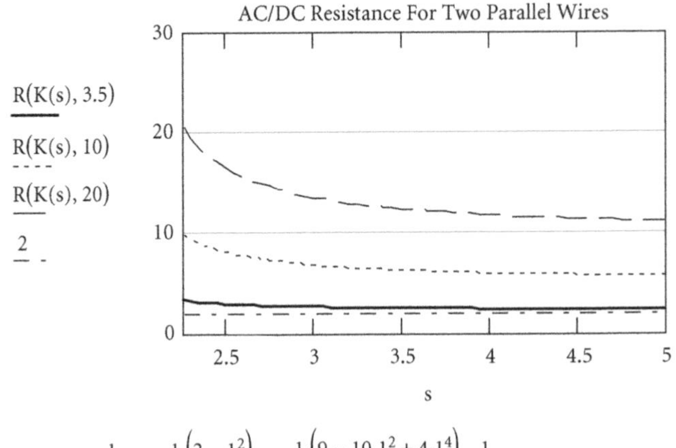

$$RSkin(x) := \frac{1}{2} \cdot x + \frac{1 \cdot \left(2 - 1^2\right)}{4} + \frac{1 \cdot \left(9 - 10 \cdot 1^2 + 4 \cdot 1^4\right)}{32} \cdot \frac{1}{x}$$

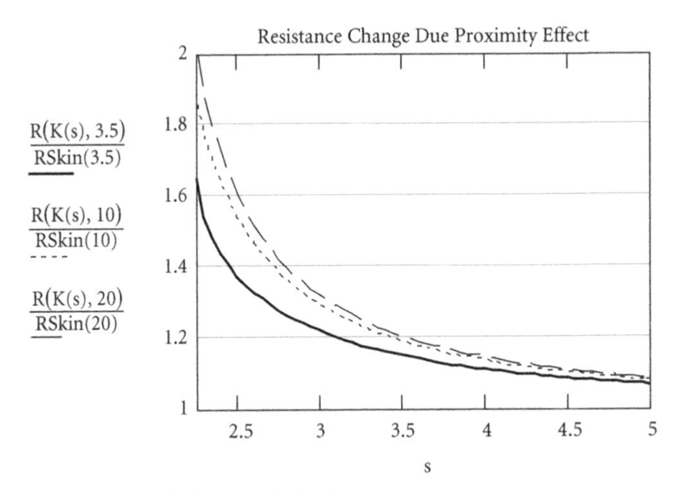

x = wire radius divided by the skin depth
s = center-to-center spacing divided by the wire radius

MATHCAD 2.11 Ratio of the ac resistance (including the proximity effect) to the dc resistance vs. the ratio of the wire spacing to the wire radius. Also, the ratio of the total ac resistance to the ac resistance due to the skin effect alone is plotted.

2.21 Characteristic Impedance Formula

Tabulate all of the common expressions for the characteristic impedance of transmission lines. How is the characteristic impedance defined for transmission lines containing three conductors? [Magnusson]

The characteristic impedance, Z_o, of a transmission line or cable is defined as

$$Z_o = \sqrt{\frac{R + j\omega L}{G + j\omega C}}$$

The characteristic impedance has units of ohms. The characteristic impedance is probably the most important parameter for a transmission line. The variables R, L, G, and C represent the resistance, inductance, conductance, and capacitance per unit length of the transmission line, respectively. The characteristic impedance of a transmission line is also referred to as the surge impedance, line impedance, or just impedance of the line. The expressions given in Appendix A are for low-loss and mainly high-frequency situations. Under these conditions, the characteristic impedance expression reduces to

$$Z_o \approx \sqrt{\frac{L}{C}}$$

which is only a function of the inductance and capacitance of the transmission line. Although this expression is not the general definition for characteristic impedance, it is usually the one used by most individuals when referring to the Z_o of a cable.

In many of the expressions given for Z_o in the table, the medium surrounding the conductors is assumed equal to free space. Often, however, the medium surrounding the conductors is not free space. The expressions can be easily modified when the medium surrounding the conductors is homogeneous and the relative permittivity, ε_r, and relative permeability, μ_r, of the surrounding medium are known. Homogeneous implies that the medium everywhere surrounding the conductors has the same electrical properties. To be more precise, the medium must be homogeneous where the electric and magnetic fields are generated by the conductors. (For an ideal coaxial cable, for example, the fields are completely between the inner and outer conductors and, therefore, the medium surrounding the outer conductor does not matter.) Both a homogeneous and nonhomogeneous situation are shown in Figure 2.29. In the nonhomogeneous situation, the substrate material is only between the narrow strip conductor and the larger return plane conductor, but some of the electric and magnetic field lines are outside this substrate.

In practice, only the medium near the conductors where most of the electric and magnetic energy exists needs to be homogeneous (to use the medium's permittivity and permeability). It is simple to account for a material between the conductors for homogeneous or nearly homogeneous mediums. Because the capacitance is proportional to the relative permittivity and the inductance is proportional to the relative permeability, if the surrounding medium is homogeneous, the high-frequency, free-space characteristic impedance can be multiplied by the following expression:

$$\sqrt{\frac{\mu_r}{\varepsilon_r}} \tag{2.64}$$

For free space, this quantity is equal to one. The relative permeability, μ_r, of nonmagnetic materials is one. The electrical parameters given in (2.64) are not those of the conductors themselves of the transmission

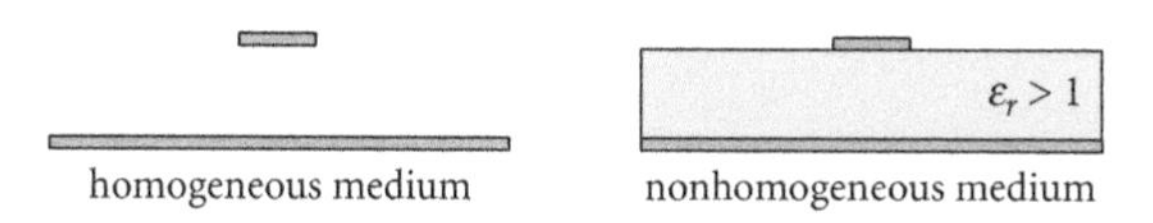

FIGURE 2.29 Homogeneous and nonhomogeneous medium between two identical conductors.

line but of the medium between the conductors.[10] If the medium near the conductors is not homogeneous, then an estimate for the characteristic impedance can be obtained by using a weighted-like average of the surrounding relative permittivities and permeabilities. These weighted-like averages are sometimes referred to as an effective relative permittivity, ε_{reff}, and effective relative permeability, μ_{reff}. The free-space characteristic impedance can then be multiplied by

$$\sqrt{\frac{\mu_{reff}}{\varepsilon_{reff}}} \tag{2.65}$$

The characteristic impedances of some nonhomogeneous structures are given in Appendix A. It is probably obvious that the effective value for the relative permittivity is between the smallest and largest values of the relative permittivities between the conductors. In some cases, the characteristic impedance for a line with a nonhomogeneous dielectric can be obtained by using a capacitance expression. For example, the capacitance for the two-layer dielectric coaxial cable shown in Figure 2.30 is given as

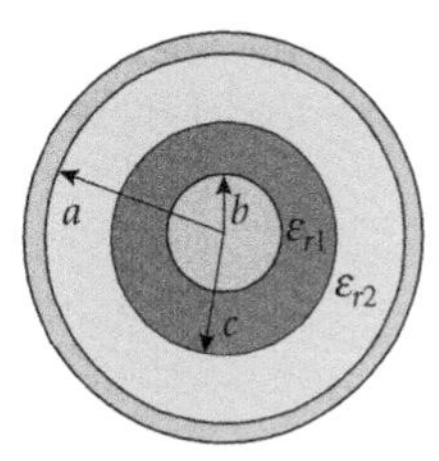

FIGURE 2.30 Coaxial cable with two different dielectrics.

$$C = \frac{2\pi\varepsilon_{r1}\varepsilon_{r2}\varepsilon_o l_{th}}{\varepsilon_{r2}\ln\left(\frac{c}{b}\right) - \varepsilon_{r1}\ln\left(\frac{c}{a}\right)}$$

where l_{th} is the length of the cable, which is much greater than its cross-sectional dimensions. Since the inductance of this cable is not affected by these nonmagnetic dielectrics, its inductance is given by

$$L = 2\times10^{-7} l_{th} \ln\left(\frac{a}{b}\right)$$

Therefore, the high-frequency characteristic impedance is

$$Z_o = \sqrt{\frac{L}{C}} \approx 60\sqrt{\frac{\left[\varepsilon_{r2}\ln\left(\frac{c}{b}\right) - \varepsilon_{r1}\ln\left(\frac{c}{a}\right)\right]\ln\left(\frac{a}{b}\right)}{\varepsilon_{r1}\varepsilon_{r2}}} \tag{2.66}$$

If $\varepsilon_{r2} = 1$ and $\varepsilon_{r2} = \varepsilon_r$, (2.66) reduces to

$$Z_o = 60\ln\left(\frac{a}{b}\right)\sqrt{\frac{\ln\left(\frac{c}{b}\right) + \varepsilon_r \ln\left(\frac{a}{c}\right)}{\varepsilon_r \ln\left(\frac{a}{b}\right)}} \tag{2.67}$$

When $\varepsilon_{r2} = 1$, (2.67) further reduces to the standard result for an air-filled coax.

[10]The relative permittivity of most conductors is nearly one and overshadowed by the high conductivity of the conductor. The relative permeability of magnetic conductors can affect the internal inductance of the conductors, which is usually negligible at high frequencies.

If the medium surrounding the conductors is homogeneous, then the capacitance or inductance per unit length can be determined from the expression for the characteristic impedance. The inductance and capacitance per unit length of a long transmission line are related to the relative permeability and permittivity of the homogeneous medium:

$$v=\frac{1}{\sqrt{LC}}=\frac{1}{\sqrt{\mu\varepsilon}}=\frac{1}{\sqrt{\mu_r\mu_o\varepsilon_r\varepsilon_o}}=\frac{c}{\sqrt{\mu_r\varepsilon_r}}=\frac{3\times10^8}{\sqrt{\mu_r\varepsilon_r}}\quad\Rightarrow L=\frac{\mu_r\varepsilon_r}{Cc^2},\ C=\frac{\mu_r\varepsilon_r}{Lc^2} \tag{2.68}$$

where v is the velocity of propagation of a signal on the line, c is the speed of light, and L (high-frequency inductance) and C are per unit length. Using the definition for the high-frequency characteristic impedance,

$$\begin{aligned} L&=CZ_o^2=\frac{\mu_r\varepsilon_r}{Lc^2}Z_o^2 \quad\Rightarrow L=\frac{Z_o\sqrt{\mu_r\varepsilon_r}}{c}=\frac{Z_o\sqrt{\mu_r\varepsilon_r}}{3\times10^8}\\ C&=\frac{L}{Z_o^2}=\frac{\mu_r\varepsilon_r}{Cc^2}\frac{1}{Z_o^2} \quad\Rightarrow C=\frac{\sqrt{\mu_r\varepsilon_r}}{cZ_o}=\frac{\sqrt{\mu_r\varepsilon_r}}{(3\times10^8)Z_o}\end{aligned} \tag{2.69}$$

Therefore, the formulas for the high-frequency inductance and capacitance *per unit length* can be easily obtained from the high-frequency characteristic impedance expressions given in Appendix A.

The given expressions for the characteristic impedance assume (in most cases) that

1. the distance between the conductors is electrically small (i.e., much less than the wavelength of the highest-frequency signal expected on the line),
2. the cross section of the conductors (including the spacing and position of the conductors) is uniform along the axis of the line,
3. the length of the conductors is large compared to any cross-sectional dimensions so that end effects can be ignored,
4. the resistance per unit length, R, of the conductors is small compared to ωL, and the conductance per unit length, G, of the insulation between the conductors is small compared to ωC.

Furthermore, in many cases, the individual conductors of the transmission line are far enough away from each other so that the proximity effect can be neglected.

For those transmission lines containing three conductors, and only one conductor is considered the ground, return, or shield conductor, it is common to define the even-mode (also referred to as the common or push-push mode) and odd-mode (also referred to as the differential, balanced, transmission-line, or push-pull mode) characteristic impedances for the line. These concepts are best understood through an example. For the two wires above a ground plane (a three-conductor system) shown in Figure 2.31, there are several ways of connecting signal sources to the conductors.

In the even or common-mode connection, the same sign voltage is applied to conductors *a* and *b* relative to *c*. This implies that the current through conductors *a* and *b* are in the same direction, and the current returns through conductor *c*. If the same voltage source is actually applied to conductors *a* and *b*, then the wires are essentially connected in parallel. Used in this manner, this configuration is essentially a two-conductor transmission line: *a* and *b* are "one" conductor and *c* is the other conductor. In Figure 2.32, the current at one instant of time is into the page for conductors *a* and *b* and out of the page for conductor *c*.

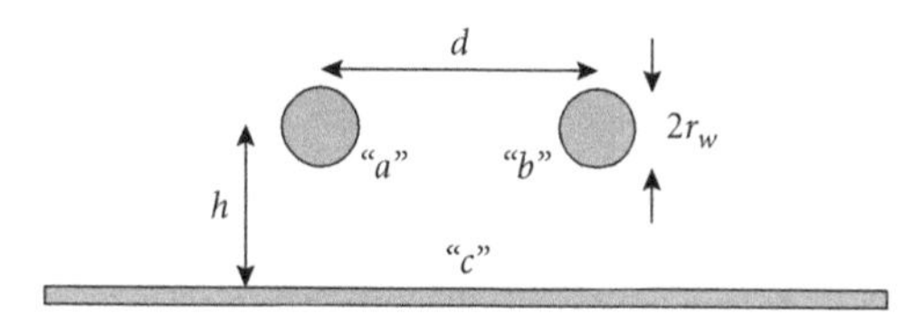

FIGURE 2.31 Three-conductor system.

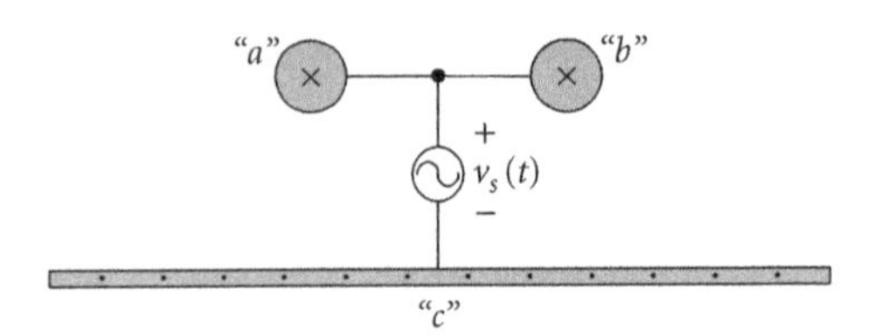

FIGURE 2.32 Even-mode connection. The same voltage source is applied to conductors a and b.

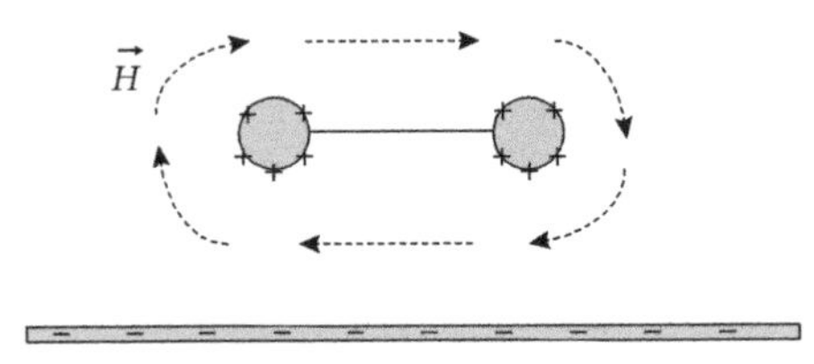

FIGURE 2.33 Sketch of the magnetic field and charge distributions for the even-mode of operation.

The characteristic impedance is a function of the inductance and capacitance. In this even-mode, the inductance is determined by obtaining the total magnetic flux generated by these currents between either conductors a and c or conductors b and c, since a and b are connected. (At high frequencies, the currents on a and b will affect the current distribution on the other conductors.) The capacitance is determined by obtaining the total charge on both conductors a and b or on just conductor c. The magnetic field and charge distributions are crudely shown in Figure 2.33 for one instant in time. The even-mode characteristic impedance for this structure is given by

$$Z_{oe} = 30\ln\left[\frac{2h}{r_w}\sqrt{1+\left(\frac{2h}{d}\right)^2}\right] \quad \text{if } \min(d,h) \gg r_w$$

This equation corresponds to the characteristic impedance that would be seen by a voltage source connected between conductor a (that is also connected to b) and conductor c. Notice that if the distance between the conductors is large compared to their height, then

$$Z_{oe} \approx 30\ln\left(\frac{2h}{r_w}\right) \quad \text{if } \min(d,h) \gg r_w,\ d \gg h$$

This expression is one-half of the characteristic impedance of a single wire above a large flat plane. (The impedance of two identical parallel wires is one-half of a single wire.)

In the odd or differential-mode connection, the voltage applied to conductor a, relative to conductor c, is opposite in sign to the voltage applied to conductor b, relative to conductor c. This implies that the current through conductors a and b are in opposite directions, and both currents return through conductor c. In Figure 2.34, the current at one instant of time is roughly shown.

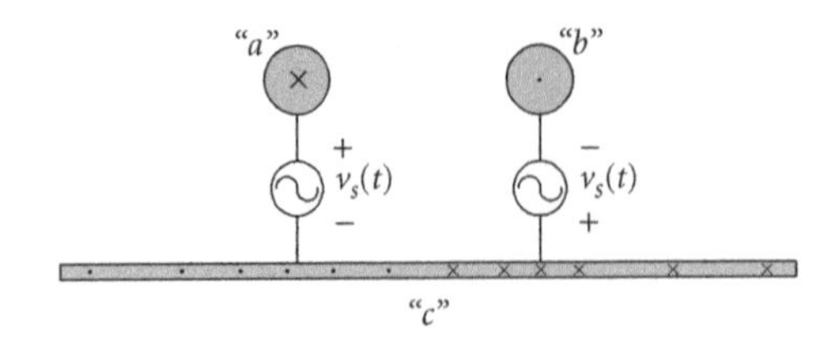

FIGURE 2.34 Odd-mode connection. Equal magnitude but opposite sign voltage sources are applied to conductors a and b.

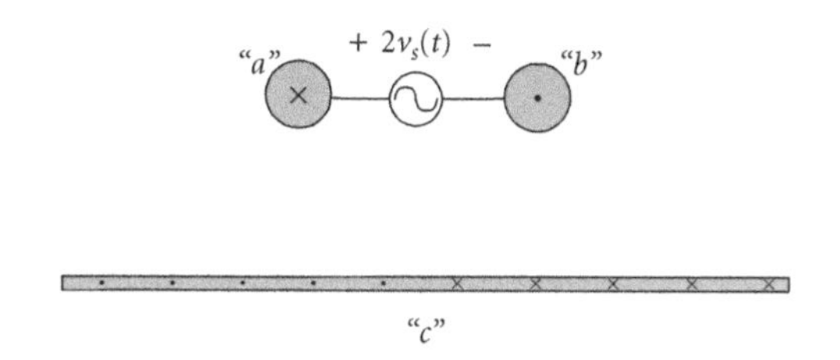

FIGURE 2.35 Odd-mode connection using a single floating voltage supply.

In Figure 2.34, the potential difference between conductors a and b is $2v_s(t)$. These connections correspond to the commonly used mode for a two-wire line near a large ground plane. The currents are equal in magnitude but of opposite sign: the radiation or emissions from this balanced mode are much less than for the even mode for the same current magnitude. If a floating signal source was available (a source with neither side connected to ground or a reference), then it could be connected directly between the conductors a and b as shown in Figure 2.35. If a floating signal source was not available, a balun or transformer could be used to generate a balanced signal source. It is instructive to examine the magnetic field and charge distributions for this odd-mode shown in Figure 2.36. The magnetic field and charge distributions are obviously different from the corresponding even-mode distributions. The inductance would be determined by obtaining the total flux between conductors a and b. The capacitance would be determined by obtaining the total charge on conductor a or conductor b. The odd-mode characteristic impedance for this structure is given by

$$Z_{oo} = 120 \ln\left[\frac{d}{r_w}\frac{1}{\sqrt{1+\left(\frac{d}{2h}\right)^2}}\right] \quad \text{if } d \gg r_w,\ h \gg r_w$$

This characteristic impedance is a function of the wire radius. If a signal source were connected between conductors a and b, it would see Z_{oo}. If the wires are far from the ground or flat plane, then

$$Z_{oo} \approx 120 \ln\left(\frac{d}{r_w}\right) \quad \text{if } d \gg r_w,\ h \gg r_w,\ h \gg d$$

This expression is also the characteristic impedance of two wires with no other nearby conductors. If the wires are far from each other so little coupling occurs between them, then this expression reduces to

$$Z_{oo} \approx 120 \ln\left(\frac{2h}{r_w}\right) \quad \text{if } \min(d, h_a, h_b) \gg r_w,\ h = h_a = h_b,\ d \gg h$$

This corresponds to twice the impedance of a single wire above a flat conducting plane.

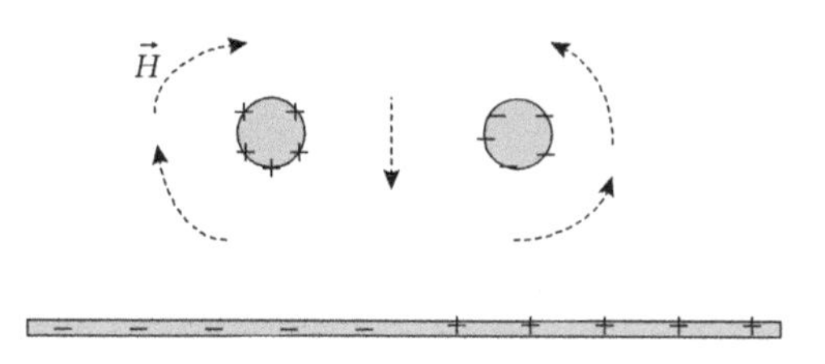

FIGURE 2.36 Sketch of the magnetic field and charge distributions for the odd-mode of operation.

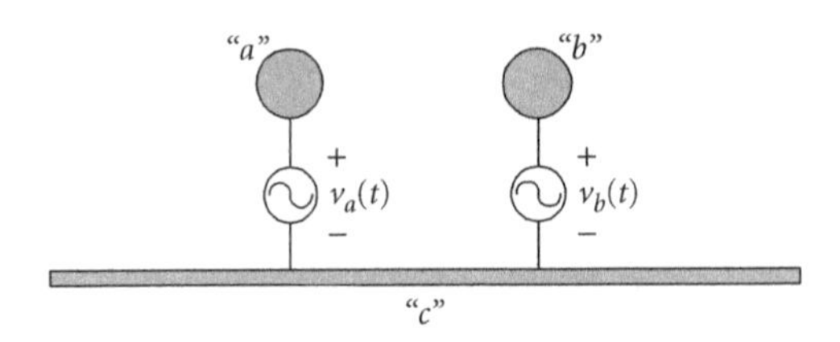

FIGURE 2.37 Different voltages applied to conductors a and b, relative to conductor c.

For systems involving three or more conductors, other types of characteristic impedances can also be defined. Imagine that different voltages are applied to conductors a and b relative to conductor c as shown in Figure 2.37. If $i_a^+(z,t)$ and $i_b^+(z,t)$ are the corresponding currents on the two conductors, traveling in the $+z$ direction, then the voltage anywhere along either conductor, traveling in the $+z$ direction, relative to conductor c can be written in terms of these currents and self and mutual characteristic impedances:

$$\begin{aligned} v_a^+(z,t) &= Z_{oa} i_a^+(z,t) + Z_{oab} i_b^+(z,t) \\ v_b^+(z,t) &= Z_{oab} i_a^+(z,t) + Z_{ob} i_b^+(z,t) \end{aligned} \tag{2.70}$$

where Z_{oa} is the self characteristic impedance of conductor a, Z_{ob} is the self characteristic impedance of conductor b, and Z_{oab} is the mutual characteristic impedance between conductors a and b. The impedance Z_{oa} is the ratio of the voltage on conductor a to the current through conductor a (at the same location) when the current through conductor b is zero:

$$v_a^+(z,t) = Z_{oa} i_a^+(z,t) + Z_{oab}(0) \quad \Rightarrow Z_{oa} = \left.\frac{v_a^+(z,t)}{i_a^+(z,t)}\right|_{i_b^+=0} \tag{2.71}$$

Similarly,

$$Z_{ob} = \left.\frac{v_b^+(z,t)}{i_b^+(z,t)}\right|_{i_a^+=0}, \quad Z_{oab} = \left.\frac{v_a^+(z,t)}{i_b^+(z,t)}\right|_{i_a^+=0} = \left.\frac{v_b^+(z,t)}{i_a^+(z,t)}\right|_{i_b^+=0} \tag{2.72}$$

Writing these voltages in terms of these self and mutual characteristic impedances is reasonable since there is magnetic coupling (via mutual inductance) and electric coupling (via mutual capacitance) between these conductors. If there are n conductors in a system (with one being the return), then there will be $n - 1$ equations. They can be written in matrix form as

$$[v^+(z,t)] = [Z_o][i^+(z,t)] \tag{2.73}$$

These self and mutual characteristic impedances can be written in terms of the even-mode and odd-mode characteristic impedances. The two-conductor voltages and currents can be rewritten as a function of the differential-mode and common-mode voltages and currents:[11]

$$\begin{aligned} v_a^+(z,t) &= \frac{v_D^+(z,t)}{2} + v_C^+(z,t), \quad v_b^+(z,t) = -\frac{v_D^+(z,t)}{2} + v_C^+(z,t) \\ i_a^+(z,t) &= i_D^+(z,t) + \frac{i_C^+(z,t)}{2}, \quad i_b^+(z,t) = -i_D^+(z,t) + \frac{i_C^+(z,t)}{2} \end{aligned} \tag{2.74}$$

[11]The variable v_D was divided by two instead of v_C so that the even-mode connection generates a common voltage of v_s for both conductors while the odd mode generates a differential voltage of $2v_s$ across the conductors.

The common-mode and differential-mode voltages and currents as a function of the positive-traveling waveforms on the line are then

$$v_C^+(z,t) = \frac{v_a^+(z,t) + v_b^+(z,t)}{2}, \quad v_D^+(z,t) = v_a^+(z,t) - v_b^+(z,t)$$
$$i_C^+(z,t) = i_a^+(z,t) + i_b^+(z,t), \quad i_D^+(z,t) = \frac{i_a^+(z,t) - i_b^+(z,t)}{2} \tag{2.75}$$

The common-mode voltages and currents correspond to the even mode previously discussed (the differential-mode voltage and current are zero) while the differential-mode voltages and currents correspond to the odd mode previously discussed (the common-mode voltage and current are zero). In terms of the self and mutual characteristic impedances and the common and differential currents, the common-mode and differential-mode voltages are

$$\begin{aligned} v_C^+(z,t) &= \frac{(Z_{oa} + Z_{oab})i_a^+(z,t) + (Z_{oa} + Z_{oab})i_b^+(z,t)}{2} \\ &= \frac{(Z_{oa} + Z_{oab})\left[i_D^+(z,t) + \frac{i_C^+(z,t)}{2}\right] + (Z_{oa} + Z_{oab})\left[-i_D^+(z,t) + \frac{i_C^+(z,t)}{2}\right]}{2} \\ &= \frac{(Z_{oa} - Z_{ob})}{2} i_D^+(z,t) + \frac{(Z_{oa} + Z_{ob} + 2Z_{oab})}{4} i_C^+(z,t) \end{aligned} \tag{2.76}$$

$$\begin{aligned} v_D^+(z,t) &= (Z_{oa} - Z_{oab})i_a^+(z,t) + (Z_{oab} - Z_{ob})i_b^+(z,t) \\ &= (Z_{oa} - Z_{oab})\left[i_D^+(z,t) + \frac{i_C^+(z,t)}{2}\right] + (Z_{oab} - Z_{ob})\left[-i_D^+(z,t) + \frac{i_C^+(z,t)}{2}\right] \\ &= (Z_{oa} + Z_{ob} - 2Z_{oab})i_D^+(z,t) + \frac{(Z_{oa} - Z_{ob})}{2} i_C^+(z,t) \end{aligned} \tag{2.77}$$

These common-mode and differential-mode voltages can also be written in terms of the even-mode, odd-mode, and even-odd mode characteristic impedances:

$$\begin{aligned} v_C^+(z,t) &= Z_{oeo} i_D^+(z,t) + Z_{oe} i_C^+(z,t) \\ v_D^+(z,t) &= Z_{oo} i_D^+(z,t) + Z_{oeo} i_C^+(z,t) \end{aligned} \tag{2.78}$$

Setting these two set of expressions equal, and solving for the odd, even, and even-odd impedances results in

$$\begin{aligned} Z_{oo} &= Z_{oa} + Z_{ob} - 2Z_{oab} \\ Z_{oe} &= \frac{1}{4}(Z_{oa} + Z_{ob}) + \frac{1}{2} Z_{oab} \\ Z_{oeo} &= \frac{1}{2}(Z_{oa} - Z_{ob}) \end{aligned} \tag{2.79}$$

Assuming these various self and mutual characteristic impedances are known, the odd, even, and even-odd characteristic impedances can be determined. If conductors a and b are symmetrical (or balanced) relative to conductor c, then $Z_{oa} = Z_{ob}$, and

$$\begin{aligned} Z_{oo} &= 2(Z_{oa} - Z_{oab}) \\ Z_{oe} &= \frac{1}{2}(Z_{oa} + Z_{oab}) \\ Z_{oeo} &= 0 \end{aligned} \tag{2.80}$$

The relationship between the voltages and currents are

$$v_C^+(z,t) = \frac{(Z_{oa} + Z_{oab})}{2} i_C^+(z,t) = Z_{oe} i_C^+(z,t) \quad \Rightarrow Z_{oe} = \frac{v_C^+(z,t)}{i_C^+(z,t)} \tag{2.81}$$

$$v_D^+(z,t) = 2(Z_{oa} - Z_{oab}) i_D^+(z,t) = Z_{oo} i_D^+(z,t) \quad \Rightarrow Z_{oo} = \frac{v_D^+(z,t)}{i_D^+(z,t)} \tag{2.82}$$

The even (or common) characteristic impedance, Z_{oe}, is the impedance seen by a common voltage source across conductor a (which is connected to conductor b) and conductor c. The current $i_C^+(z,t) = i_a^+(z,t) + i_b^+(z,t)$ is the current from this supply passing to both conductors a and b. The odd (or differential) characteristic impedance, Z_{oo}, is the impedance seen by a voltage source connected between conductors a and b, where $v_D^+(z,t) = v_a^+(z,t) - v_b^+(z,t)$. The current from the supply is the differential current $i_D^+(z,t) = [i_a^+(z,t) - i_b^+(z,t)]/2$.

Appendix A Characteristic Impedance Formula

Two Concentric Cylindrical Conductors [Howard W. Sams]

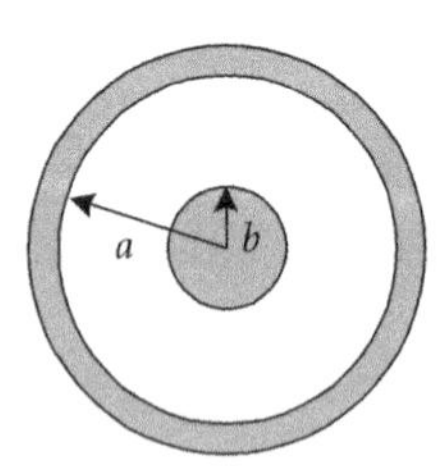

For this transmission line, commonly known as a coaxial cable,

$$Z_o = 60 \ln\left(\frac{a}{b}\right)$$

If insulating beads are used at frequent intervals (i.e., at electrically-short intervals) between the inner conductor and outer conductor, then the new characteristic impedance is

$$Z_o' = \frac{Z_o}{\sqrt{1 + (\varepsilon_r - 1)\frac{w}{s}}}$$

where ε_r is the relative permittivity of the beads, w is the width of the beads, and s is the center-to-center distance between the beads. The currents in the two conductors are equal in magnitude but in opposite directions.

Confocal Elliptical Conductors [Hilberg; Weber; Rudge]

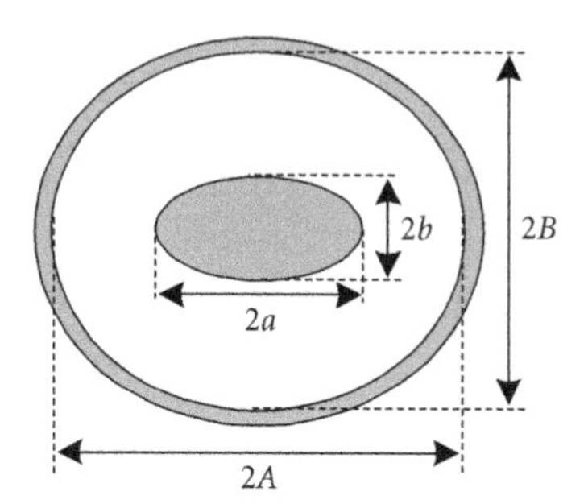

If the focal lengths of the inner and outer elliptical-shape conductors are the same

$$\sqrt{A^2 - B^2} = \sqrt{a^2 - b^2}$$

and $A > B$ and $a > b$, then the characteristic impedance is

$$Z_o = 60\left[\cosh^{-1}\left(\frac{A}{\sqrt{A^2 - B^2}}\right) - \cosh^{-1}\left(\frac{a}{\sqrt{a^2 - b^2}}\right)\right] = 60\left[\tanh^{-1}\left(\frac{B}{A}\right) - \tanh^{-1}\left(\frac{b}{a}\right)\right] = 60\ln\left(\frac{A+B}{a+b}\right)$$

$$Z_o \approx 60\ln\left(\frac{A}{a}\right) + \frac{15(A^2 - B^2)}{A^2}\left(\frac{A^2 - a^2}{a^2}\right) \approx 60\ln\left(\frac{A}{a}\right) \quad \text{if } A \approx B,\ a \approx b$$

The currents in the two conductors are equal in magnitude but in opposite directions.

Flat Conductor Surrounded by Cylindrical Conductor [Hilberg]

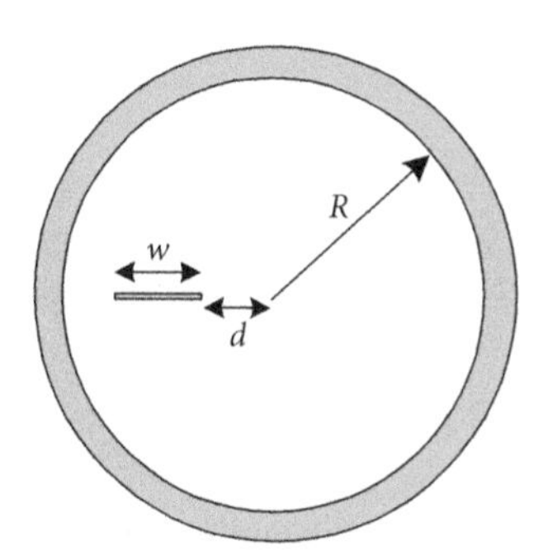

When the flat inner conductor is thin (compared to w and R), then

$$Z_o = \begin{cases} 60\ln\left[2\dfrac{\sqrt{\dfrac{(R-d)(R+d+w)}{(R+d)(R-d-w)}}+1}{\sqrt{\dfrac{(R-d)(R+d+w)}{(R+d)(R-d-w)}}-1}\right] & \text{if } Z_o > 94\,\Omega \\ \dfrac{148}{\ln\left[2\sqrt{\dfrac{(R-d)(R+d+w)}{(R+d)(R-d-w)}}\right]} & \text{if } Z_o < 94\,\Omega \end{cases}$$

$$Z_o = \begin{cases} 60\ln\left(\dfrac{4R}{w}\right) & \text{if } Z_o > 94\,\Omega, d = -\dfrac{w}{2} \\ \dfrac{148}{\ln\left[2\left(\dfrac{\frac{2R}{w}+1}{\frac{2R}{w}-1}\right)\right]} & \text{if } Z_o < 94\,\Omega, d = -\dfrac{w}{2} \end{cases}$$

The currents in the two conductors are equal in magnitude but in opposite directions.

Flat Conductor Surrounded by Elliptical Conductor [Hilberg]

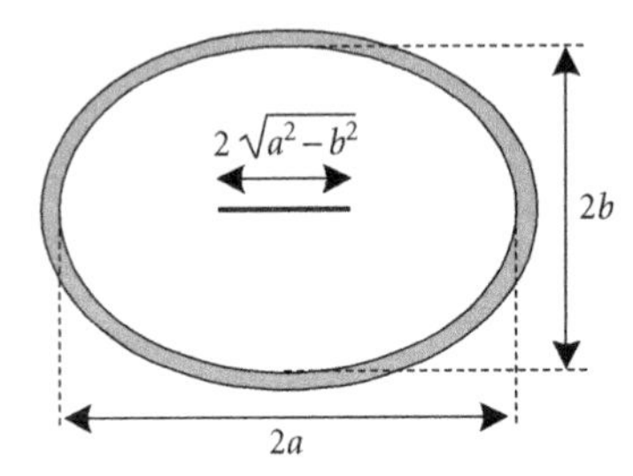

When the flat inner conductor is thin (compared to a and b) and $a > b$, then

$$Z_o = 60\cosh^{-1}\left(\frac{a}{\sqrt{a^2-b^2}}\right) = 60\ln\left[\frac{a}{\sqrt{a^2-b^2}} + \sqrt{\left(\frac{a}{\sqrt{a^2-b^2}}\right)^2 - 1}\right]$$

$$Z_o \approx 60\ln\left[\frac{2}{\sqrt{1-\left(\frac{b}{a}\right)^2}}\right] \quad \text{if } a \approx b$$

The currents in the two conductors are equal in magnitude but in opposite directions.

Two Concentric Conductors with Dielectric Wedge [Howard W. Sams]

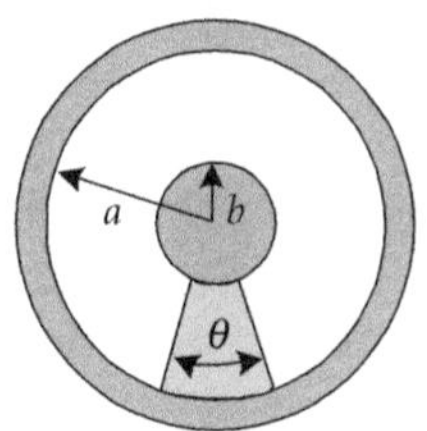

$$Z_o = \frac{60 \ln\left(\frac{a}{b}\right)}{\sqrt{1+(\varepsilon_r - 1)\frac{\theta}{360}}}$$

where ε_r is the relative permittivity of the wedge of angle θ in radians. The currents in the two conductors are equal in magnitude but in opposite directions.

Two Concentric Conductors with Slotted Outer Conductor [Howard W. Sams; Hatsuda]

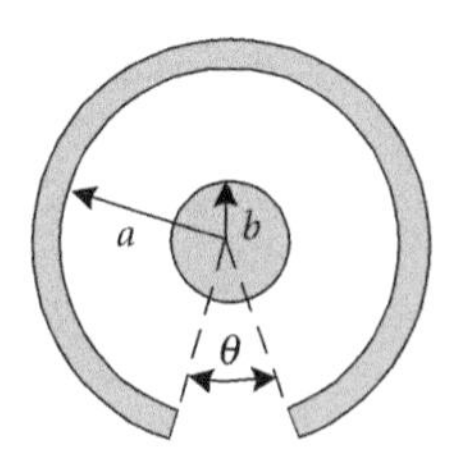

$$Z_o \le 60 \ln\left(\frac{a}{b}\right) + 0.03\theta^2$$

where θ, in radians, is the angle of the slot opening. The currents in the two conductors are equal in magnitude but in opposite directions.

Two Conductors with Displaced Inner Conductor [Howard W. Sams]

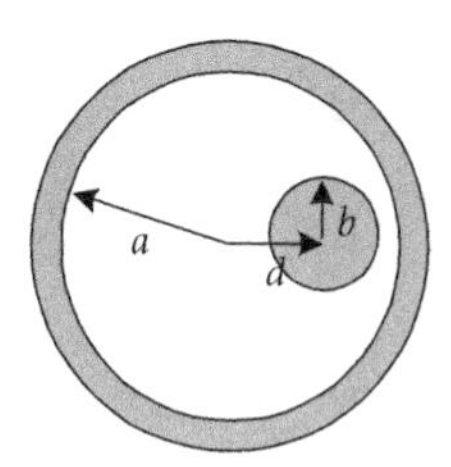

$$Z_o = 60 \cosh^{-1}\left[\frac{1}{2}\left(\frac{a}{b}+\frac{b}{a}-\frac{d^2}{ab}\right)\right] = 60 \ln\left[\frac{1}{2}\left(\frac{a}{b}+\frac{b}{a}-\frac{d^2}{ab}\right)+\sqrt{\left[\frac{1}{2}\left(\frac{a}{b}+\frac{b}{a}-\frac{d^2}{ab}\right)\right]^2 - 1}\right] \quad \text{if } (a-b) > d$$

$$Z_o \approx 60 \ln\left[\left(\frac{a^2+b^2-d^2}{ab}\right) - \left(\frac{ab}{a^2+b^2-d^2}\right)\right] \approx 60 \ln\left(\frac{a^2+b^2-d^2}{ab}\right) \quad \text{if } (a-b) \gg d$$

$$Z_o \approx 60 \ln\left\{\frac{a}{b}\left[1-\left(\frac{d}{a}\right)^2\right]\right\} \quad \text{if } (a-b) \gg d,\ a \gg b$$

The currents in the two conductors are equal in magnitude but in opposite directions.

Split Thin-Walled Cylinder [Hilberg; Howard W. Sams]

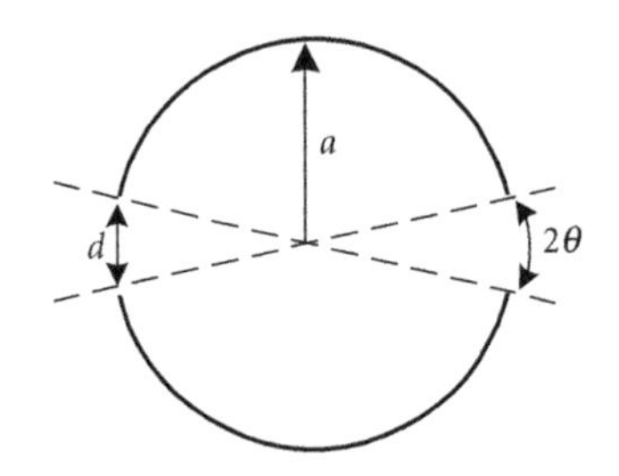

$$Z_o = \begin{cases} 120 \ln\left[2\cot\left(\frac{\pi}{4}-\frac{\theta}{2}\right)\right] & \text{if } Z_o > 189\,\Omega, \frac{\pi}{2} \geq \theta \geq \frac{\pi}{4} \\ \dfrac{296}{\ln\left[2\cot\left(\frac{\theta}{2}\right)\right]} & \text{if } Z_o < 189\,\Omega, \frac{\pi}{4} \geq \theta \geq 0 \end{cases}$$

$$Z_o \approx \frac{297}{\ln\left(\frac{4}{\theta}\right)} = \frac{297}{\ln\left(\frac{8a}{d}\right)} \quad \text{if } \theta \ll 1, d = 2\theta a$$

where θ is in radians. The currents in the two conductors are equal in magnitude but in opposite directions.

Two Conical Conductors [Weeks, '68; Hilberg]

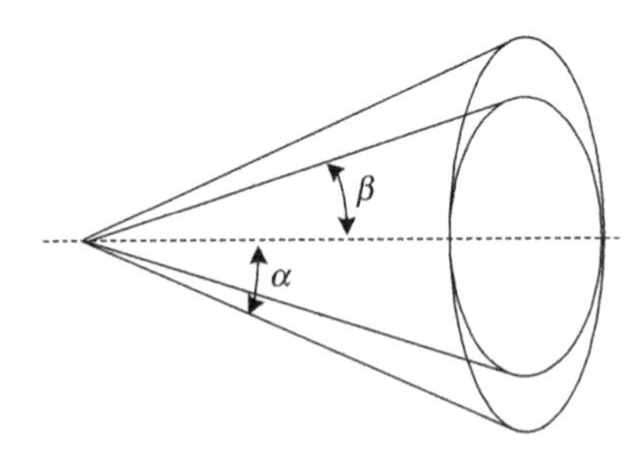

For this coaxial circular-cone transmission line,

$$Z_o = 60 \ln\left[\frac{\tan\left(\frac{\alpha}{2}\right)}{\tan\left(\frac{\beta}{2}\right)}\right]$$

$$Z_o = 60 \ln\left[\cot\left(\frac{\beta}{2}\right)\right] \quad \text{if } \alpha = \frac{\pi}{2}\ (=90^\circ)$$

$$Z_o = 60\ln\left[\cot^2\left(\frac{\beta}{2}\right)\right] \quad \text{if } \alpha = \pi - \beta$$

The currents in the two conductors are equal in magnitude but in opposite directions.

Cylindrical Conductor in Square Enclosure [Lo; Howard W. Sams; Rudge; Wadell; Wheeler, '79]

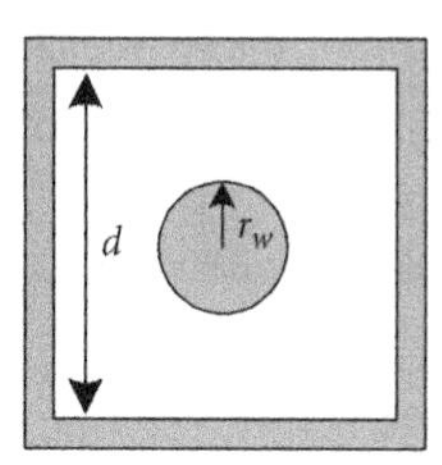

For this square coaxial cable,

$$Z_o = 60\,\ln\left(\frac{d}{2r_w}\right) + 6.48 - 2.34\left[\frac{1+0.405\left(\frac{d}{2r_w}\right)^{-4}}{1-0.405\left(\frac{d}{2r_w}\right)^{-4}}\right] - 0.48\left[\frac{1+0.163\left(\frac{d}{2r_w}\right)^{-8}}{1-0.163\left(\frac{d}{2r_w}\right)^{-8}}\right] - 0.12\left[\frac{1+0.067\left(\frac{d}{2r_w}\right)^{-12}}{1-0.067\left(\frac{d}{2r_w}\right)^{-12}}\right]$$

$$Z_o \approx 60\ln\left(0.54\frac{d}{r_w}\right) \quad \text{if } d \gg r_w$$

$$Z_o \approx 21.2\sqrt{\frac{d}{2r_w} - 1} \quad \text{if } Z_o \le 2\,\Omega$$

The currents in the two conductors are equal in magnitude but in opposite directions.

Cylindrical Conductor in Semicircular Tube [Rudge]

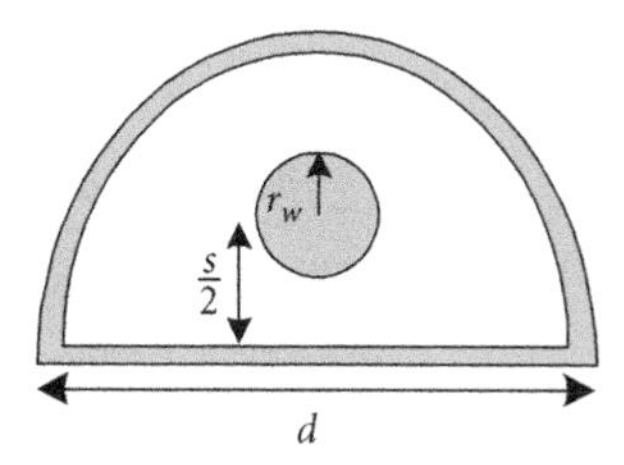

$$Z_o = 60\ln\left\{\frac{1-\left(\frac{s}{d}\right)^2}{\frac{r_w}{s}\left[1+\left(\frac{s}{d}\right)^2\right]}\right\} \quad \text{if } \min(d,s) \gg r_w$$

$$Z_o \approx 60\ln\left(\frac{s}{r_w}\right) \quad \text{if } d \gg s,\ \min(d,s) \gg r_w$$

where $d > s$. The currents in the two conductors are equal in magnitude but in opposite directions.

Square Conductor in Square Enclosure [Lo]

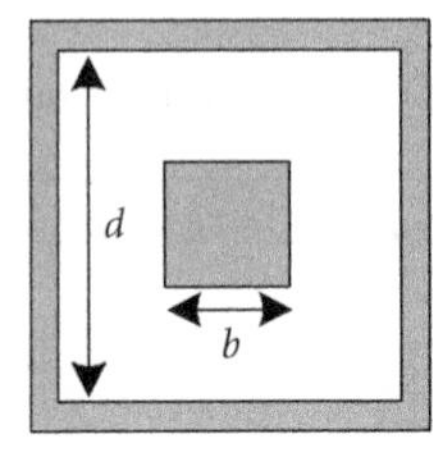

For this square coaxial line with a square center conductor,

$$Z_o = \begin{cases} \dfrac{47.1\left(1-\dfrac{b}{d}\right)}{0.279+0.721\dfrac{b}{d}} & \text{if } \dfrac{d}{4} \le b \le \dfrac{d}{2} \\ 59.4\ln\left(0.956\dfrac{d}{b}\right) & \text{if } b \le \dfrac{d}{2} \end{cases}$$

The currents in the two conductors are equal in magnitude but in opposite directions.

Rectangular Conductor in Rectangular Enclosure [Lo; Tippet]

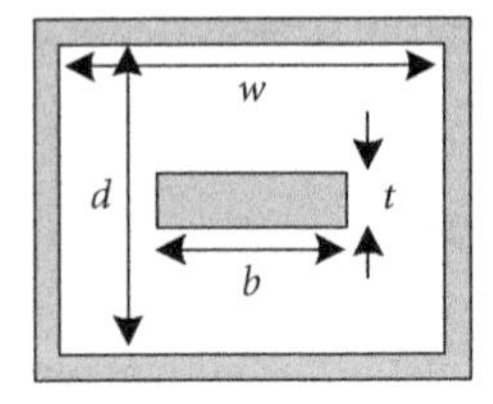

For this rectangular coaxial cable,

$$Z_o = 60\ln\left(\frac{1+\dfrac{w}{d}}{\dfrac{b}{d}+\dfrac{t}{d}}\right) \quad \text{if } d > 3t,\ w > 1.25b$$

$$Z_o = 60\ln\left(\frac{d}{b}\right) \quad \text{if } d > 3t,\ w > 1.25b,\ w = d,\ b = t$$

$$Z_o \approx 60\ln\left(\frac{d+w}{b}\right) \quad \text{if } d \gg t,\ w > 1.25b$$

$$Z_o = \frac{94}{\dfrac{b}{d}+\dfrac{2}{\pi}\ln\left[1+\coth\left(\dfrac{\pi w}{2d}\right)\right]} \quad \text{if } d \gg t,\ w > 2.86d$$ [12]

The currents in the two conductors are equal in magnitude but in opposite directions.

[12] A more accurate expression, used for TEM cells, is available when the distance $w - b$ is small.

Cylindrical Conductor in Trough [Lo; Howard W. Sams; Wheeler, '79]

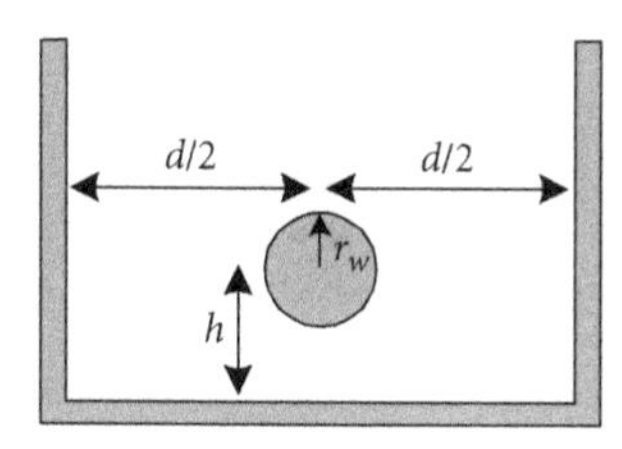

For this trough or channel transmission line,

$$Z_o = 60\ln\left[\tanh\left(\frac{\pi h}{d}\right)\coth\left(\frac{\pi r_w}{2d}\right)\right] \quad \text{if } d > 8r_w,\ h > 3r_w$$

$$Z_o \approx 60\ln\left[\frac{2d}{\pi r_w}\tanh\left(\frac{\pi h}{d}\right)\right] \quad \text{if } d \gg r_w,\ h \gg r_w$$

The currents in the two conductors are equal in magnitude but in opposite directions.

Cylindrical Conductor in Corner Trough [Hilberg; Johnson, '61; Wheeler, '79]

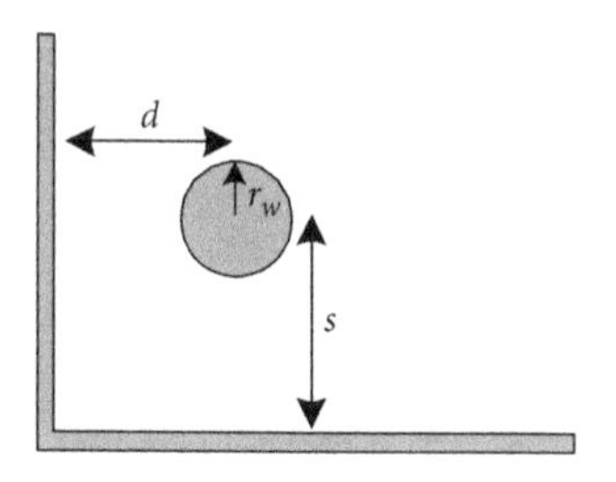

$$Z_o = 60\cosh^{-1}\left(\frac{ds}{r_w\sqrt{s^2+d^2}}\right) = 60\ln\left[\frac{ds}{r_w\sqrt{s^2+d^2}} + \sqrt{\left(\frac{ds}{r_w\sqrt{s^2+d^2}}\right)^2 - 1}\right] \quad \text{if } \min(d,s) \gg r_w$$

$$Z_o = 60\cosh^{-1}\left(\frac{d}{r_w\sqrt{2}}\right) \approx 60\ln\left(\frac{d\sqrt{2}}{r_w}\right) \quad \text{if } \min(d,s) \gg r_w,\ s = d$$

$$Z_o \approx 60\ln\left[\frac{d}{r_w} + \sqrt{\left(\frac{d}{r_w}\right)^2 - 1}\right] \quad \text{if } \min(d,s) \gg r_w,\ s \gg d$$

The current in the cylindrical conductor is equal in magnitude but in opposite direction to the current in the right-angled conductor. The width of each side of the flat walls of the trough is large compared to d and s.

Cylindrical Conductor in Corner Trough above Large Flat Conductor [Laport]

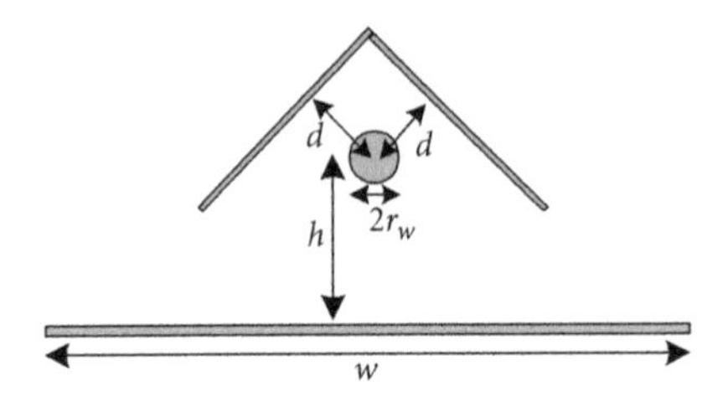

For the mode where the current in the center conductor of radius r_w returns through the corner trough and large flat plane, which are connected together,

$$Z_o = 30\ln\left[\frac{2hd(h+2d\sqrt{2})}{r_w\sqrt{2}\,(h+d\sqrt{2})^2}\right] \quad \text{if } d \gg r_w$$

$$Z_o \approx 30\ln\left(\frac{d\sqrt{2}}{r_w}\right) \quad \text{if } d \gg r_w,\ h \gg d$$

The width of each side of the flat walls of the trough is large compared to d. The width, w, of the large flat conductor is large compared to h.

Cylindrical Conductor near Concave Cylindrical Conductor [Hilberg]

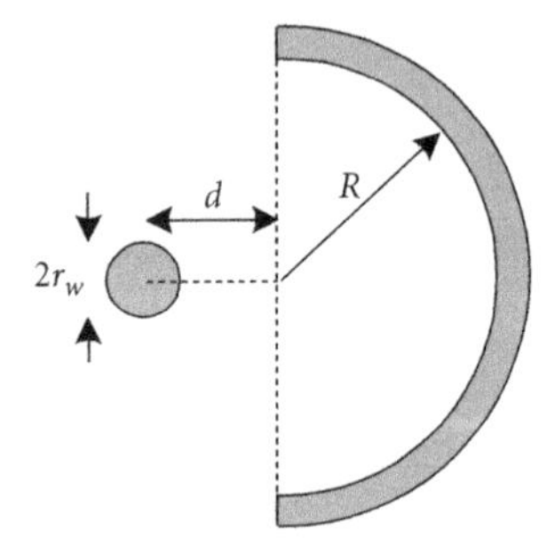

For these two conductors,

$$Z_o = 60\cosh^{-1}\left\{\frac{\sin\left[\frac{\pi}{4}+\frac{1}{2}\tan^{-1}\left(\frac{2Rd}{R^2-d^2+r_w^2}\right)\right]}{\sin\left[\frac{1}{2}\tan^{-1}\left(\frac{2Rr_w}{R^2+d^2-r_w^2}\right)\right]}\right\} \quad \text{if } R \gg r_w$$

$$Z_o \approx 60\cosh^{-1}\left(\frac{R}{r_w\sqrt{2}}\right) \approx 60\ln\left(\frac{1.4R}{r_w}\right) \quad \text{if } R \gg r_w,\ R \gg d$$

where $d \geq 0$. The currents in the two conductors are equal in magnitude but in opposite directions.

Open Two Conductors [Howard W. Sams; Lo]

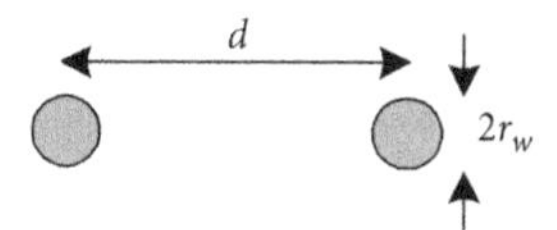

For these two conductors, referred to as (open-air) twin-lead,

$$Z_o = 120\cosh^{-1}\left(\frac{d}{2r_w}\right) = 120\ln\left[\frac{d}{2r_w} + \sqrt{\left(\frac{d}{2r_w}\right)^2 - 1}\right]$$

$$Z_o \approx 120\ln\left(\frac{d}{r_w} - \frac{r_w}{d}\right) \approx 120\ln\left(\frac{d}{r_w}\right) \quad \text{if } d \gg r_w$$

where d is the center-to-center distance between the conductors. The currents in the two conductors are equal in magnitude but in opposite directions.

Open Two Conductors with Different Radii [Zahn; Howard W. Sams]

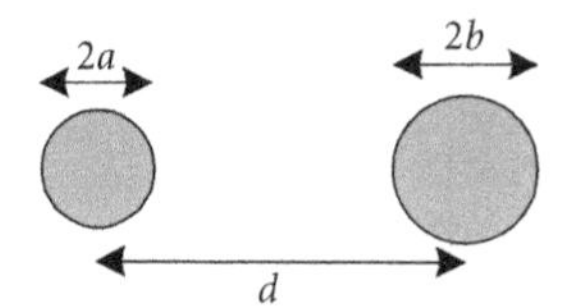

$$Z_o = 60\cosh^{-1}\left[\frac{1}{2}\left(\frac{d^2}{ab} - \frac{a}{b} - \frac{b}{a}\right)\right] = 60\ln\left[\frac{1}{2}\left(\frac{d^2}{ab} - \frac{a}{b} - \frac{b}{a}\right) + \sqrt{\left[\frac{1}{2}\left(\frac{d^2}{ab} - \frac{a}{b} - \frac{b}{a}\right)\right]^2 - 1}\right]$$

$$Z_o \approx 60\ln\left(\frac{d^2}{ab}\right) = 120\ln\left(\frac{d}{\sqrt{ab}}\right) \quad \text{if } d \gg \max(a,b)$$

where d is the center-to-center distance between the conductors. The currents in the two conductors are equal in magnitude but in opposite directions.

Two Rectangular Conductors [Paul, '92(b); Lo]

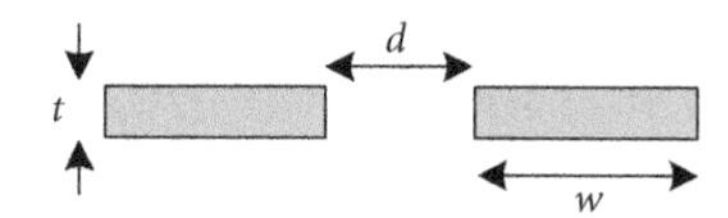

$$Z_o = \begin{cases} 120\ln\left[2\dfrac{(1+\sqrt{k})}{(1-\sqrt{k})}\right] & \text{if } \dfrac{1}{\sqrt{2}} \le k \le 1,\ w \gg t \\[2ex] \dfrac{1184}{\ln\left[2\dfrac{1+(1-k^2)^{\frac{1}{4}}}{1-(1-k^2)^{\frac{1}{4}}}\right]} & \text{if } 0 \le k \le \dfrac{1}{\sqrt{2}},\ w \gg t \end{cases}$$

$$Z_o \approx 120\ln\left(\frac{4d}{w}-2\right) \approx 120\ln\left(\frac{4d}{w}\right) \quad \text{if } d \gg w$$

$$Z_o \approx \frac{296}{\ln\left(2\sqrt{\frac{2w+d}{d}}\right)} \approx \frac{296}{\ln\left(2\sqrt{\frac{2w}{d}}\right)} \quad \text{if } w \gg \max(d,t)$$

where $k = d/(d+2w)$. The currents in the two conductors are equal in magnitude but in opposite directions.

Two Coplanar Flat Conductors on Dielectric [Paul, '92(b); Gupta, '79; Walker]

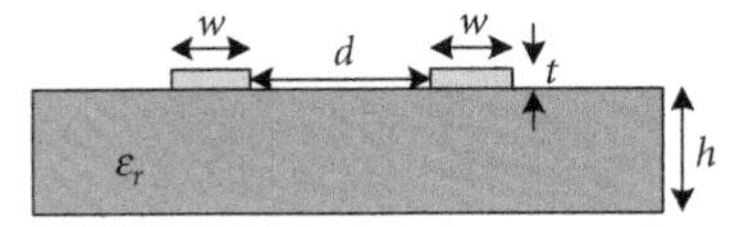

For this transmission line, also known as a slotline, or coplanar line,

$$Z_o = \begin{cases} \dfrac{120}{\sqrt{\varepsilon_{reff}}}\ln\left(2\dfrac{1+\sqrt{k}}{1-\sqrt{k}}\right) & \text{if } \dfrac{1}{\sqrt{2}} \le k \le 1 \\ \dfrac{1{,}184}{\sqrt{\varepsilon_{reff}}}\dfrac{1}{\ln\left[2\dfrac{1+(1-k^2)^{\frac{1}{4}}}{1-(1-k^2)^{\frac{1}{4}}}\right]} & \text{if } 0 \le k \le \dfrac{1}{\sqrt{2}} \end{cases}$$

where $w \gg t$ (i.e., thin conductors)

$$k = \frac{d}{d+2w}$$

$$\varepsilon_{reff} = \frac{\varepsilon_r+1}{2}\left\{\tanh\left[0.775\ln\left(\frac{h}{w}\right)+1.75\right]+\frac{kw}{h}[0.04-0.7k+0.01(1-0.1\varepsilon_r)(0.25+k)]\right\}$$

If the thickness, t, is not small, then the k should be replaced by k_e only in the Z_o expression where

$$k_e \approx k-(1-k^2)\left\{\frac{1.25t}{2\pi w}\left[1+\ln\left(\frac{4\pi w}{t}\right)\right]\right\}$$

Furthermore, the effective relative permittivity ε_{reff} should be replaced with ε_{ereff}.

$$\varepsilon_{ereff} = \begin{cases} \varepsilon_{reff} - \dfrac{1.4(\varepsilon_{reff}-1)\dfrac{t}{d}}{\dfrac{\pi}{\ln\left(2\dfrac{1+\sqrt{k}}{1-\sqrt{k}}\right)}+\dfrac{1.4t}{d}} & \text{if } \dfrac{1}{\sqrt{2}} \le k \le 1 \\ \varepsilon_{reff} - \dfrac{1.4(\varepsilon_{reff}-1)\dfrac{t}{d}}{\dfrac{\ln\left[2\dfrac{1+(1-k^2)^{\frac{1}{4}}}{1-(1-k^2)^{\frac{1}{4}}}\right]}{\pi}+\dfrac{1.4t}{d}} & \text{if } 0 \le k \le \dfrac{1}{\sqrt{2}} \end{cases}$$

A much simpler, very approximate expression is

$$Z_o \approx \frac{120}{\sqrt{\varepsilon_{ereff}}}\ln\left[\frac{\pi(d-w)}{w+t}+1\right]$$

$$\approx \frac{120}{\sqrt{\varepsilon_{ereff}}}\ln\left(\frac{\pi d}{w+t}\right) \quad \text{if } w \geq t,\ d \gg w$$

The currents in the two conductors are equal in magnitude but in opposite directions.

Two Parallel Conductors [Lo; Montrose, '99]

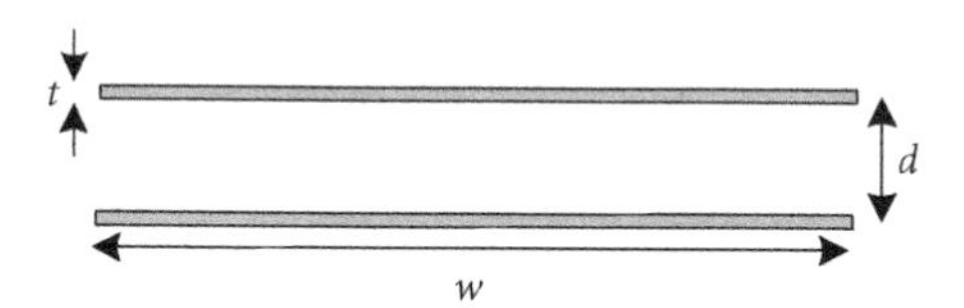

$$Z_o = 120\ln\left[\frac{6}{u}+\frac{2\pi-6}{u}e^{-\left(\frac{30.7}{u}\right)^{0.7528}}+\sqrt{1+\left(\frac{2}{u}\right)^2}\right] \quad \text{if } w > 3d$$

$$Z_o \approx 377\frac{d}{w} \quad \text{if } w \gg d,\ d > 3t$$

where

$$u = \frac{2w}{d}+\frac{2t}{\pi d}\ln\left(1+\frac{d}{2t}\frac{4e^1}{\coth^2\sqrt{13\frac{w}{d}}}\right)$$

The currents in the two conductors are equal in magnitude but in opposite directions.

Dielectric Sandwiched between Two Parallel Strips [Paul, '92(b)]

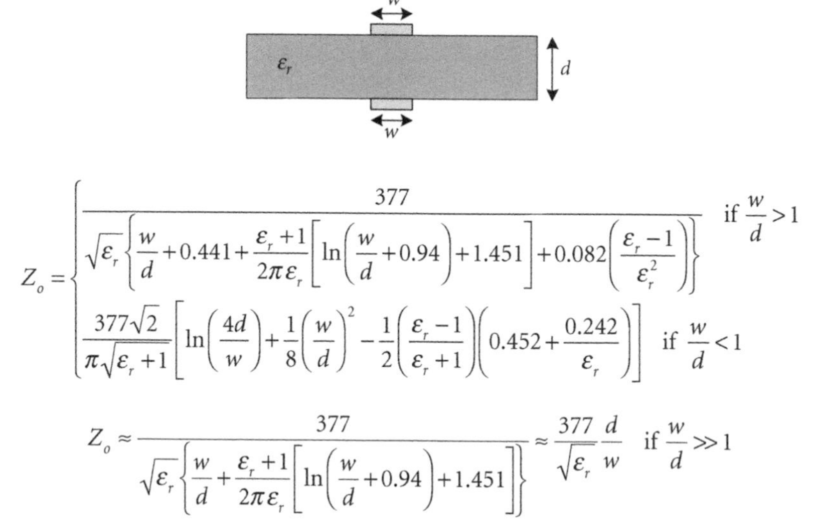

$$Z_o = \begin{cases} \dfrac{377}{\sqrt{\varepsilon_r}\left\{\dfrac{w}{d}+0.441+\dfrac{\varepsilon_r+1}{2\pi\varepsilon_r}\left[\ln\left(\dfrac{w}{d}+0.94\right)+1.451\right]+0.082\left(\dfrac{\varepsilon_r-1}{\varepsilon_r^2}\right)\right\}} & \text{if } \dfrac{w}{d} > 1 \\ \dfrac{377\sqrt{2}}{\pi\sqrt{\varepsilon_r+1}}\left[\ln\left(\dfrac{4d}{w}\right)+\dfrac{1}{8}\left(\dfrac{w}{d}\right)^2-\dfrac{1}{2}\left(\dfrac{\varepsilon_r-1}{\varepsilon_r+1}\right)\left(0.452+\dfrac{0.242}{\varepsilon_r}\right)\right] & \text{if } \dfrac{w}{d} < 1 \end{cases}$$

$$Z_o \approx \frac{377}{\sqrt{\varepsilon_r}\left\{\frac{w}{d}+\frac{\varepsilon_r+1}{2\pi\varepsilon_r}\left[\ln\left(\frac{w}{d}+0.94\right)+1.451\right]\right\}} \approx \frac{377}{\sqrt{\varepsilon_r}}\frac{d}{w} \quad \text{if } \frac{w}{d} \gg 1$$

The currents in the two conductors are equal in magnitude but in opposite directions.

Dielectric Sandwiched between Slotted, Parallel Flat Conductors [Hilberg]

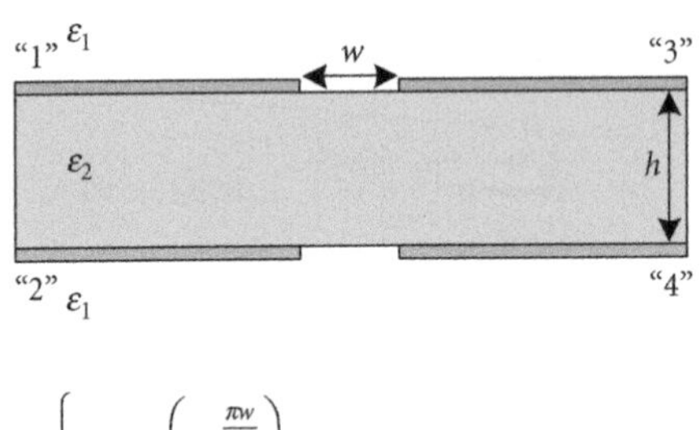

$$Z_o = \begin{cases} 240\ln\left(2e^{\frac{\pi w}{2h}}\right) & \text{if } Z_o > 377\,\Omega,\ \varepsilon_2 \gg \varepsilon_1 \\ \dfrac{592}{\ln\left[2\coth\left(\dfrac{\pi w}{4h}\right)\right]} & \text{if } Z_o < 377\,\Omega,\ \varepsilon_2 \gg \varepsilon_1 \end{cases}$$

$$Z_o \approx \frac{592}{\ln\left(2\dfrac{4h}{\pi w}\right)} \quad \text{if } \varepsilon_2 \gg \varepsilon_1,\ h \gg w$$

The currents in conductors “1” and “2” return in conductors “3” and “4.” The width of the four conductors is large compared to w and h.

Cylindrical Conductor above Flat Conductor [Hilberg; Haus; Howard W. Sams]

$$Z_o = 60\cosh^{-1}\left\{\frac{\sin\left[\dfrac{1}{2}\tan^{-1}\left(\dfrac{4wd}{w^2 - 4d^2 + 4r_w^2}\right)\right]}{\sin\left[\dfrac{1}{2}\tan^{-1}\left(\dfrac{4wr_w}{w^2 + 4d^2 - 4r_w^2}\right)\right]}\right\} \quad \text{if } d \gg r_w$$

$$Z_o \approx 60\cosh^{-1}\left(\frac{2d^2}{wr_w}\right) \approx 120\ln\left(\frac{2d}{\sqrt{wr_w}}\right) \quad \text{if } d \gg r_w,\ d \gg w$$

$$Z_o \approx 60\cosh^{-1}\left(\frac{d}{r_w}\right) = 60\ln\left[\frac{d}{r_w} + \sqrt{\left(\frac{d}{r_w}\right)^2 - 1}\right] \quad \text{if } d \gg r_w,\ w \gg d$$

$$Z_o \approx 60\ln\left(\frac{2d}{r_w} - \frac{r_w}{2d}\right) \approx 60\ln\left(\frac{2d}{r_w}\right) \quad \text{if } d \gg r_w,\ w \gg d$$

The current in the cylindrical conductor is equal in magnitude but in opposite direction to the current in the flat plane.

Cylindrical Conductor above Slotted Flat Conductor [Hilberg]

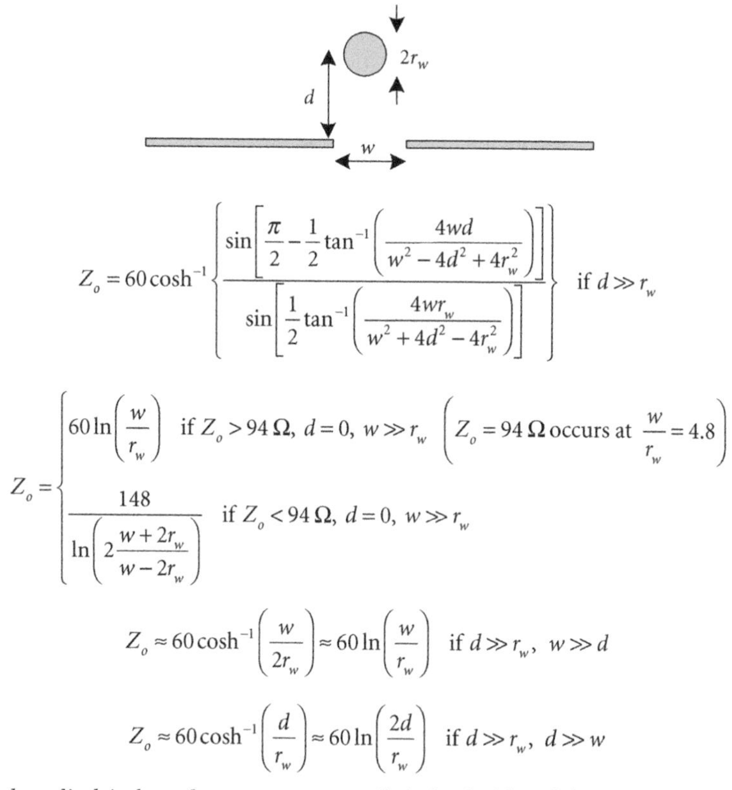

$$Z_o = 60\cosh^{-1}\left\{\frac{\sin\left[\frac{\pi}{2}-\frac{1}{2}\tan^{-1}\left(\frac{4wd}{w^2-4d^2+4r_w^2}\right)\right]}{\sin\left[\frac{1}{2}\tan^{-1}\left(\frac{4wr_w}{w^2+4d^2-4r_w^2}\right)\right]}\right\} \quad \text{if } d \gg r_w$$

$$Z_o = \begin{cases} 60\ln\left(\frac{w}{r_w}\right) & \text{if } Z_o > 94\,\Omega,\ d=0,\ w \gg r_w \quad \left(Z_o = 94\,\Omega \text{ occurs at } \frac{w}{r_w}=4.8\right) \\ \dfrac{148}{\ln\left(2\dfrac{w+2r_w}{w-2r_w}\right)} & \text{if } Z_o < 94\,\Omega,\ d=0,\ w \gg r_w \end{cases}$$

$$Z_o \approx 60\cosh^{-1}\left(\frac{w}{2r_w}\right) \approx 60\ln\left(\frac{w}{r_w}\right) \quad \text{if } d \gg r_w,\ w \gg d$$

$$Z_o \approx 60\cosh^{-1}\left(\frac{d}{r_w}\right) \approx 60\ln\left(\frac{2d}{r_w}\right) \quad \text{if } d \gg r_w,\ d \gg w$$

The current in the cylindrical conductor returns equally in both sides of the slotted plane. The cylindrical conductor is centered over a slot in a very wide flat plane conductor (compared to d and w).

Cylindrical Conductor near Conductor Edge/Corner [Hilberg]

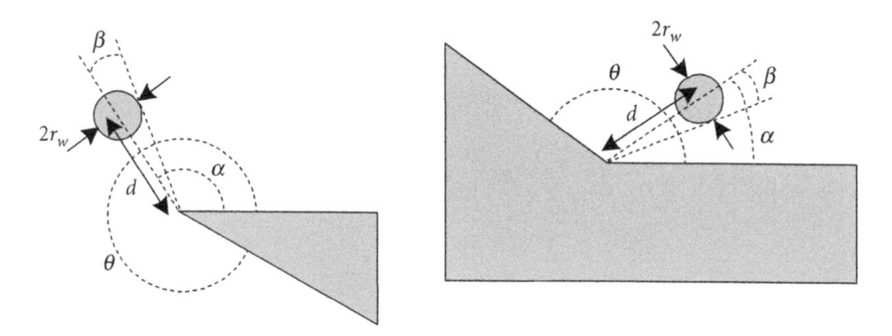

In the left figure, θ is roughly 330° and α is roughly 130°, while in the right figure θ is roughly 145° and α is roughly 45°.

$$Z_o = 60\cosh^{-1}\left[\frac{\sin\left(\pi\frac{\alpha}{\theta}\right)}{\sin\left(\pi\frac{\beta}{\theta}\right)}\right] \quad \text{if } (\theta-\beta) > \alpha > \beta$$

$$Z_o = 60\cosh^{-1}\left[\frac{\sin\left(\pi\frac{\alpha}{\theta}\right)}{\sin\left(\pi\frac{r_w}{d\theta}\right)}\right] \quad \text{if } \left(\theta - \frac{r_w}{d}\right) > \alpha > \frac{r_w}{d},\ r_w \approx \beta d$$

$$Z_o = 60\cosh^{-1}\left[\frac{\sin\left(\frac{\alpha}{2}\right)}{\sin\left(\frac{\beta}{2}\right)}\right] \quad \text{if } \theta = 2\pi\ (=360°),\ (2\pi - \beta) > \alpha > \beta$$

$$= 60\cosh^{-1}\left[\frac{1}{\sin\left(\frac{\beta}{2}\right)}\right] \quad \text{if } \theta = 2\pi\ (=360°),\ \alpha = \pi\ (=180°),\ (2\pi - \beta) > \pi > \beta$$

$$\approx 60\cosh^{-1}\left(\frac{2}{\beta}\right) \approx 60\ln\left(\frac{4}{\beta}\right) \approx 60\ln\left(\frac{4d}{r_w}\right)$$

$$\text{if } \theta = 2\pi\ (=360°),\ \alpha = \pi\ (=180°),\ (2\pi - \beta) > \pi > \beta,\ 1 >> \beta,\ r_w \approx \beta d$$

$$Z_o = 60\cosh^{-1}\left[\frac{\sqrt{3}}{2\sin\left(\frac{2\beta}{3}\right)}\right] \quad \text{if } \theta = \frac{3\pi}{2}\ (=270°),\ \alpha = \frac{\pi}{2}\ (=90°),\ \left(\frac{3\pi}{2} - \beta\right) > \frac{\pi}{2} > \beta$$

$$\approx 60\cosh^{-1}\left(\frac{3\sqrt{3}}{4\beta}\right) \approx 60\ln\left(\frac{3\sqrt{3}}{2\beta}\right) \quad \text{if } \theta = \frac{3\pi}{2},\ \alpha = \frac{\pi}{2},\ 1 >> \beta$$

$$\approx 60\ln\left(\frac{3\sqrt{3}d}{2r_w}\right) \approx 60\ln\left(\frac{d}{r_w}\right) + 57 \quad \text{if } \theta = \frac{3\pi}{2},\ \alpha = \frac{\pi}{2},\ 1 \gg \beta,\ r_w \approx \beta d$$

$$Z_o = 60\cosh^{-1}\left[\frac{1}{\sin(\beta)}\right] \quad \text{if } \theta = \pi\ (=180°),\ \alpha = \frac{\pi}{2}\ (=90°),\ (\pi - \beta) > \frac{\pi}{2} > \beta$$

$$\approx 60\cosh^{-1}\left(\frac{1}{\beta}\right) \approx 60\ln\left(\frac{2}{\beta}\right) \quad \text{if } \theta = \frac{3\pi}{2},\ \alpha = \frac{\pi}{2}, 1 \gg \beta$$

$$\approx 60\ln\left(\frac{2d}{r_w}\right) \quad \text{if } \theta = \frac{3\pi}{2},\ \alpha = \frac{\pi}{2},\ 1 \gg \beta,\ r_w \approx \beta d$$

$$Z_o = 60\cosh^{-1}\left[\frac{1}{\sin(2\beta)}\right] \quad \text{if } \theta = \frac{\pi}{2}\ (=90°),\ \alpha = \frac{\pi}{4}\ (=45°),\ \left(\frac{\pi}{2} - \beta\right) > \frac{\pi}{4} > \beta$$

$$\approx 60\cosh^{-1}\left(\frac{1}{2\beta}\right) \approx 60\ln\left(\frac{1}{\beta}\right) \quad \text{if } \theta = \frac{3\pi}{2},\ \alpha = \frac{\pi}{2},\ 1 \gg \beta$$

$$\approx 60\ln\left(\frac{d}{r_w}\right) \quad \text{if } \theta = \frac{3\pi}{2},\ \alpha = \frac{\pi}{2},\ 1 \gg \beta,\ r_w \approx \beta d$$

The currents in the two conductors are equal in magnitude but in opposite directions. All angles are in radians. The dimensions of the edge/corner conductor are large compared to d and r_w. Also, the closest distance between the edge/corner conductor and cylindrical conductor is large compared to r_w.

Flat Conductor Perpendicular to Flat Conductor[Hilberg]

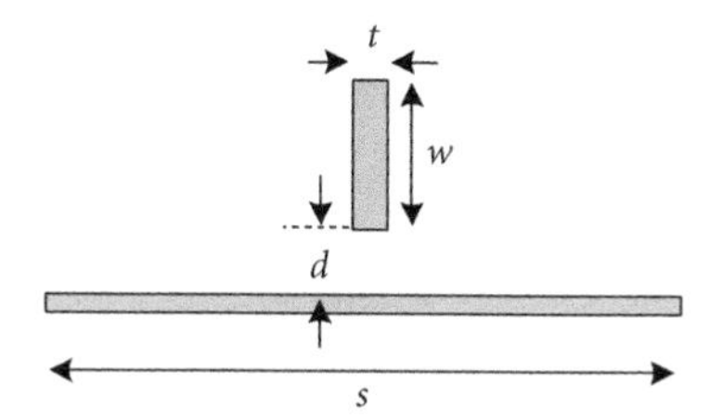

For these two conductors,

$$Z_o = \begin{cases} 60\ln\left\{2\dfrac{\sqrt{\left(1+\dfrac{w}{d}\right)\left[\dfrac{1+\sqrt{1+\left(\dfrac{2d}{s}\right)^2}}{1+\sqrt{1+\left(\dfrac{2d+2w}{s}\right)^2}}\right]}+1}{\sqrt{\left(1+\dfrac{w}{d}\right)\left[\dfrac{1+\sqrt{1+\left(\dfrac{2d}{s}\right)^2}}{1+\sqrt{1+\left(\dfrac{2d+2w}{s}\right)^2}}\right]}-1}\right\} & \text{if } Z_o > 94\,\Omega \\ \dfrac{148}{\ln\left\{2\sqrt{\left(1+\dfrac{w}{d}\right)\left[\dfrac{1+\sqrt{1+\left(\dfrac{2d}{s}\right)^2}}{1+\sqrt{1+\left(\dfrac{2d+2w}{s}\right)^2}}\right]}\right\}} & \text{if } Z_o < 94\,\Omega \end{cases}$$

$$Z_o \approx \begin{cases} 60\ln\left\{2\dfrac{\sqrt{\dfrac{s}{2d}\left[1+\sqrt{1+\left(\dfrac{2d}{s}\right)^2}\right]}+1}{\sqrt{\dfrac{s}{2d}\left[1+\sqrt{1+\left(\dfrac{2d}{s}\right)^2}\right]}-1}\right\} & \text{if } Z_o > 94\,\Omega,\ w \gg \max(s,d) \\ \dfrac{148}{\ln\left\{2\sqrt{\dfrac{s}{2d}\left[1+\sqrt{1+\left(\dfrac{2d}{s}\right)^2}\right]}\right\}} & \text{if } Z_o < 94\,\Omega,\ w \gg \max(s,d) \end{cases}$$

$$Z_o \approx \begin{cases} 60\ln\left(2\dfrac{\sqrt{1+\dfrac{w}{d}}+1}{\sqrt{1+\dfrac{w}{d}}-1}\right) & \text{if } Z_o > 94\,\Omega,\ s \gg \max(w,d)\ \left(Z_o = 94\,\Omega \text{ occurs at } \dfrac{w}{d} = 4.8\right) \\ \dfrac{148}{\ln\left(2\sqrt{1+\dfrac{w}{d}}\right)} & \text{if } Z_o < 94\,\Omega,\ s \gg \max(w,d) \end{cases}$$

The currents in the two conductors are equal in magnitude but in opposite directions. The conductor of width w is centered about the plane of width s. The thickness, t, is small compared to w and s.

Flat Conductor Perpendicular to Slotted Flat Conductor [Hilberg]

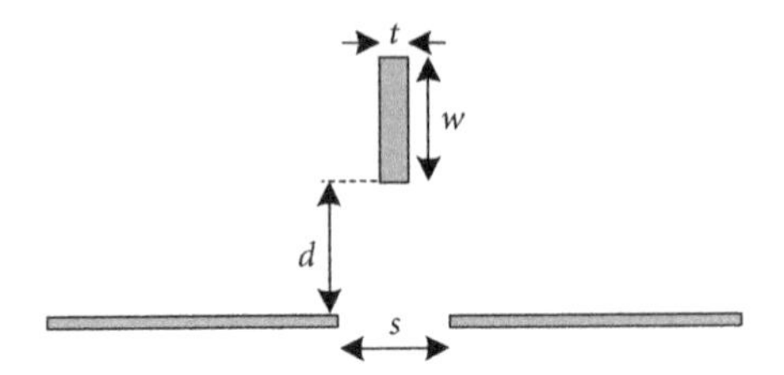

For these two conductors,

$$Z_o = \begin{cases} 60\ln\left[2\dfrac{\sqrt{\dfrac{d+w+\sqrt{\left(\frac{s}{2}\right)^2+(d+w)^2}}{d+\sqrt{\left(\frac{s}{2}\right)^2+d^2}}}+1}{\sqrt{\dfrac{d+w+\sqrt{\left(\frac{s}{2}\right)^2+(d+w)^2}}{d+\sqrt{\left(\frac{s}{2}\right)^2+d^2}}}-1}\right] & \text{if } Z_o > 94\,\Omega \\[2ex] \dfrac{148}{\ln\left[2\sqrt{\dfrac{d+w+\sqrt{\left(\frac{s}{2}\right)^2+(d+w)^2}}{d+\sqrt{\left(\frac{s}{2}\right)^2+d^2}}}\right]} & \text{if } Z_o < 94\,\Omega \end{cases}$$

$$Z_o \approx \begin{cases} 60\ln\left(2\dfrac{\sqrt{1+\frac{w}{d}}+1}{\sqrt{1+\frac{w}{d}}-1}\right) & \text{if } Z_o > 94\,\Omega,\ w \gg s,\ d \gg s \\[2ex] \dfrac{148}{\ln\left(2\sqrt{1+\frac{w}{d}}\right)} & \text{if } Z_o < 94\,\Omega,\ w \gg s,\ d \gg s \end{cases}$$

$$Z_o \approx 60\ln\left(2\dfrac{\sqrt{1+\frac{2w}{s}}+1}{\sqrt{1+\frac{2w}{s}}-1}\right) \quad \text{if } s \gg \max(w,d)$$

The current in the conductor of width w returns equally in both sides of the slotted plane. The conductor of width w is centered about the slot of the plane with a width that is large compared to d, s, and w. The thickness, t, is small compared to w and s.

Flat Conductor near Conductor Edge/Corner [Hilberg]

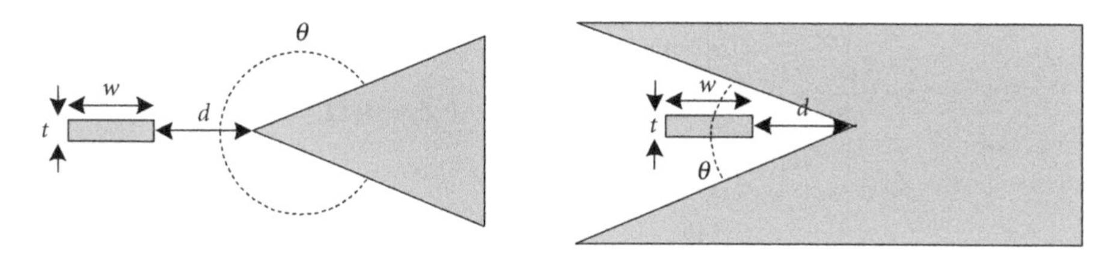

In the left figure, θ is roughly 320° while in the right figure θ is roughly 40°. For these conductors, where θ is in radians,

$$Z_o = \begin{cases} 60\ln\left[2\dfrac{\left(1+\dfrac{w}{d}\right)^{\frac{1}{2\left(\frac{\theta}{\pi}\right)}}+1}{\left(1+\dfrac{w}{d}\right)^{\frac{1}{2\left(\frac{\theta}{\pi}\right)}}-1}\right] & \text{if } Z_o > 94\,\Omega \\ \dfrac{148}{\ln\left[2\left(1+\dfrac{w}{d}\right)^{\frac{1}{2\left(\frac{\theta}{\pi}\right)}}\right]} & \text{if } Z_o < 94\,\Omega \end{cases}$$

$$Z_o = \begin{cases} 60\ln\left(2\dfrac{\sqrt{1+\dfrac{w}{d}}+1}{\sqrt{1+\dfrac{w}{d}}-1}\right) & \text{if } Z_o > 94\,\Omega,\ \theta = \pi\,(=180°) \\ \dfrac{148}{\ln\left(2\sqrt{1+\dfrac{w}{d}}\right)} & \text{if } Z_o < 94\,\Omega,\ \theta = \pi\,(=180°) \end{cases}$$

$$Z_o = \begin{cases} 60\ln\left(\dfrac{4d}{w}+2\right) & \text{if } Z_o > 94\,\Omega,\ \theta = \dfrac{\pi}{2}\,(=90°) \\ \dfrac{148}{\ln\left[2\left(1+\dfrac{w}{d}\right)\right]} & \text{if } Z_o < 94\,\Omega,\ \theta = \dfrac{\pi}{2}\,(=90°) \end{cases}$$

The currents in the two conductors are equal in magnitude but in opposite directions. The flat conductor of width w is centered about the "corner" conductor that has dimensions that are large compared to d and w. The thickness, t, is small compared to w.

Cylindrical Conductor along Edge of Flat Conductor [Hilberg]

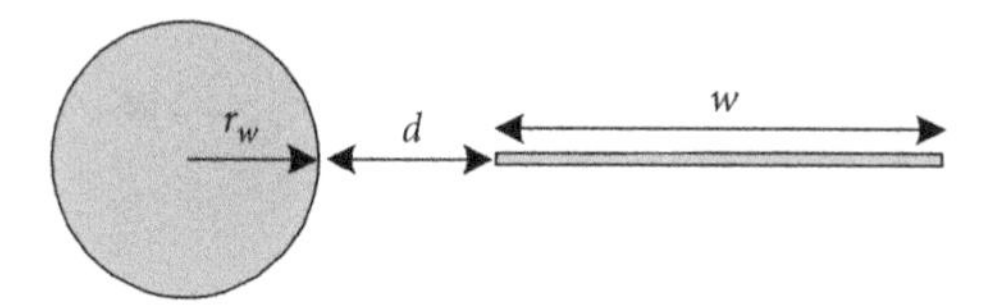

For these two conductors,

$$Z_o = \begin{cases} 60\ln\left[2\dfrac{\sqrt{\dfrac{(d+w)(2r_w+d)}{(2r_w+w+d)d}}+1}{\sqrt{\dfrac{(d+w)(2r_w+d)}{(2r_w+w+d)d}}-1}\right] & \text{if } Z_o > 94\,\Omega \\ \dfrac{148}{\ln\left[2\sqrt{\dfrac{(d+w)(2r_w+d)}{(2r_w+w+d)d}}\right]} & \text{if } Z_o < 94\,\Omega \end{cases}$$

$$Z_o \approx \begin{cases} 60\ln\left(2\dfrac{\sqrt{\dfrac{2r_w}{d}+1}+1}{\sqrt{\dfrac{2r_w}{d}+1}-1}\right) & \text{if } Z_o > 94\,\Omega,\ w \gg \max(r_w, d) \\ \dfrac{148}{\ln\left(2\sqrt{\dfrac{2r_w}{d}+1}\right)} & \text{if } Z_o < 94\,\Omega,\ w \gg \max(r_w, d) \end{cases}$$

$$Z_o \approx 60\ln\left(\frac{4d}{r_w}\right) \quad \text{if } w \gg d,\ d \gg r_w$$

$$Z_o \approx 60\ln\left(\frac{4d^2}{r_w w}\right) \quad \text{if } d \gg \max(r_w, w)$$

The currents in the two conductors are equal in magnitude but in opposite directions. The thickness of the plane conductor is small compared to r_w.

Two Coplanar Flat Conductors [Hilberg]

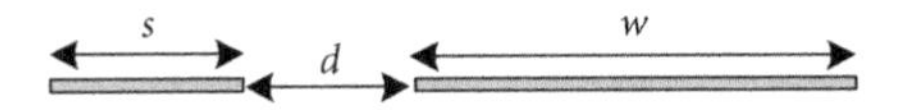

For these two conductors,

$$Z_o = \begin{cases} 60\ln\left[2\dfrac{\left(\dfrac{d+s}{w+d+s}\,\dfrac{w+d}{d}\right)^{\frac{1}{4}}+1}{\left(\dfrac{d+s}{w+d+s}\,\dfrac{w+d}{d}\right)^{\frac{1}{4}}-1}\right] & \text{if } Z_o > 94\,\Omega \\ \dfrac{148}{\ln\left[2\left(\dfrac{d+s}{w+d+s}\,\dfrac{w+d}{d}\right)^{\frac{1}{4}}\right]} & \text{if } Z_o < 94\,\Omega \end{cases}$$

$$Z_o \approx \begin{cases} 60\ln\left[2\dfrac{\left(1+\dfrac{s}{d}\right)^{\frac{1}{4}}+1}{\left(1+\dfrac{s}{d}\right)^{\frac{1}{4}}-1}\right] & \text{if } Z_o > 94\,\Omega,\ w \gg \max(d,s) \\ \dfrac{148}{\ln\left[2\left(1+\dfrac{s}{d}\right)^{\frac{1}{4}}\right]} & \text{if } Z_o < 94\,\Omega,\ w \gg \max(d,s) \end{cases}$$

$$Z_o = \begin{cases} 60\ln\left[2\dfrac{\left(\dfrac{d+w}{d+2w}\dfrac{w+d}{d}\right)^{\frac{1}{4}}+1}{\left(\dfrac{d+w}{d+2w}\dfrac{w+d}{d}\right)^{\frac{1}{4}}-1}\right] \approx 120\ln\left(2\dfrac{1+\sqrt{\dfrac{d}{d+2w}}}{1-\sqrt{\dfrac{d}{d+2w}}}\right) & \text{if } Z_o > 94\,\Omega,\ s = w \\ \dfrac{148}{\ln\left[2\left(\dfrac{d+w}{d+2w}\dfrac{w+d}{d}\right)^{\frac{1}{4}}\right]} \approx \dfrac{296}{\ln\left(2\sqrt{\dfrac{2w+d}{d}}\right)} & \text{if } Z_o < 94\,\Omega,\ s = w \end{cases}$$

The currents in the two conductors are equal in magnitude but in opposite directions. The thickness of both conductors is small compared to d, s, and w.

Dielectric Sandwiched between Strip and Large Flat Conductor [Janssen; Schneider; Cahill; White, '82]

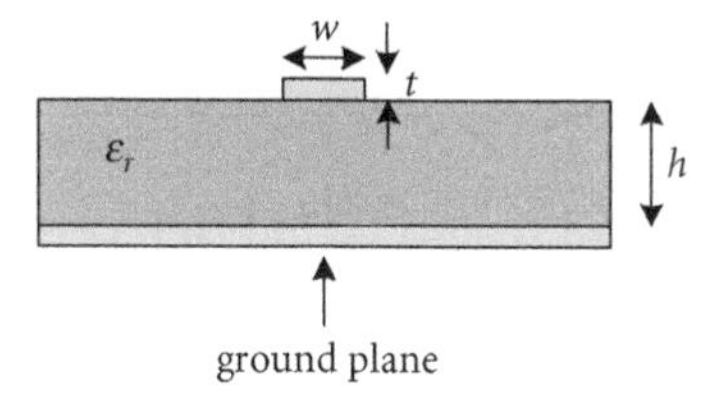

For this transmission line, also known as the standard or surface microstrip,

$$Z_o = \begin{cases} \dfrac{60}{\sqrt{\varepsilon_{reff}}}\ln\left(\dfrac{8h}{w}+0.25\dfrac{w}{h}\right) & \text{if } \dfrac{w}{h} \le 1 \\ \dfrac{377}{\sqrt{\varepsilon_{reff}}\left[\dfrac{w}{h}+1.393+0.667\ln\left(\dfrac{w}{h}+1.444\right)\right]} & \text{if } \dfrac{w}{h} \ge 1 \end{cases}$$

$$Z_o \approx \begin{cases} \dfrac{60}{\sqrt{\varepsilon_{reff}}}\ln\left(\dfrac{8h}{w}\right)+\dfrac{1.88}{\sqrt{\varepsilon_{reff}}}\left(\dfrac{w}{h}\right)^2 \approx \dfrac{60}{\sqrt{\varepsilon_{reff}}}\ln\left(\dfrac{8h}{w}\right) & \text{if } h \gg w \\ \dfrac{377}{\sqrt{\varepsilon_{reff}}\left[\dfrac{w}{h}+2.42-0.44\dfrac{h}{w}+\left(1-\dfrac{h}{w}\right)^6\right]} \approx \dfrac{377}{\sqrt{\varepsilon_{reff}}\left(\dfrac{w}{h}+1.393\right)} & \text{if } w \gg h \end{cases}$$

where

$$\varepsilon_{reff} = \begin{cases} \dfrac{\varepsilon_r + 1}{2} + \dfrac{\varepsilon_r - 1}{2}\left[\dfrac{1}{\sqrt{1+12\dfrac{h}{w}}} + 0.04\left(1-\dfrac{w}{h}\right)^2\right] & \text{if } \dfrac{w}{h} \le 1 \\ \dfrac{\varepsilon_r + 1}{2} + \dfrac{\varepsilon_r - 1}{2\sqrt{1+12\dfrac{h}{w}}} \quad \text{if } \dfrac{w}{h} > 1 \end{cases}$$

$$\varepsilon_{reff} = 1 \quad \text{if } \varepsilon_r = 1$$

$$\varepsilon_{reff} \approx \frac{1}{2}(\varepsilon_r + 1) \quad \text{if } h \gg w$$

The thickness of the strip is considered small compared to its width w. If the thickness is not negligible, it can be incorporated into these expressions by replacing the real width w with an effective width w_{eff}:

$$w_{eff} = \begin{cases} w + \dfrac{t}{\pi}\left[1 + \ln\left(\dfrac{4\pi w}{t}\right)\right] & \text{if } \dfrac{1}{2\pi} > \dfrac{w}{h} > 2\dfrac{t}{h} \\ w + \dfrac{t}{\pi}\left[1 + \ln\left(\dfrac{2h}{t}\right)\right] & \text{if } \dfrac{w}{h} > \dfrac{1}{2\pi} > 2\dfrac{t}{h} \end{cases}$$

Actually, these correction terms for finite thickness straps assume that the $\varepsilon_r = 1$. It has been recommended that one-half of the *correction* term added to w be used when $\varepsilon_r = 10$ and proportional adjustments for $1 < \varepsilon_r < 10$. For larger strip thicknesses,

$$Z_o \approx \frac{87}{\sqrt{\varepsilon_r + 1.41}} \ln\left(\frac{5.98h}{0.8w + t}\right) \quad \text{if } 0.1 \le \frac{t}{w} \le 0.8$$

The currents in the two conductors are equal in magnitude but in opposite directions. The width of the larger flat "ground" plane is large compared to h.

Embedded Strip above Large Flat Conductor[13] [Montrose, '99; Wadell]

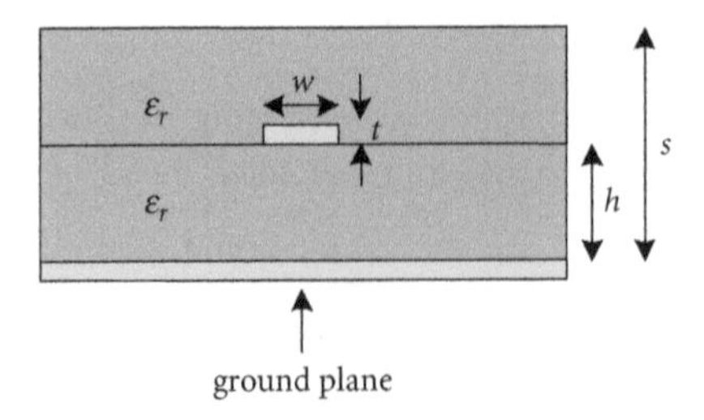

[13]So that this equation (when certain limits were taken) was consistent with other equations in this book, the constant multiplier given in the reference was changed to 56.

For this transmission line, also known as the embedded microstrip, the characteristic impedance is roughly

$$Z_o = \frac{56}{\sqrt{\varepsilon_{reff}}}\ln\left(\frac{5.98h}{0.8w+t}\right)$$

where

$$\varepsilon_{reff} = \varepsilon_r\left(1-e^{-1.55\frac{s}{h}}\right)$$

The thickness of the upper coating, $s-(t+h)$, should be comparable or large compared to h. If $s < 1.2h$, then the microstrip expressions without the upper coating can be used. The currents in the two conductors are equal in magnitude but in opposite directions. The width of the larger "ground" plane is large compared to h, w, and t.

Cylindrical Conductor between Two Large Flat Conductors [Howard W. Sams; Frankel; Johnson, '61]

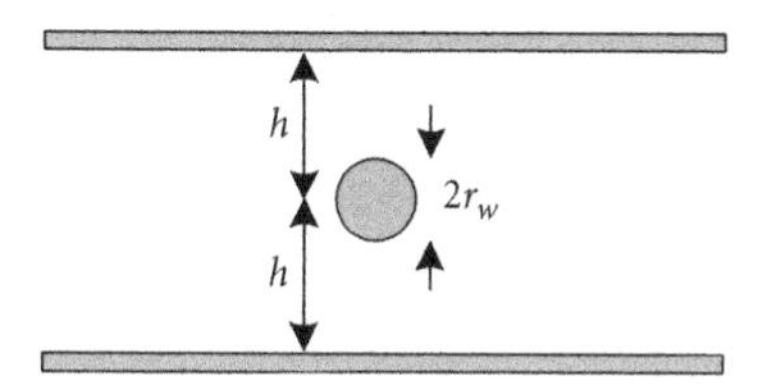

For this transmission line,

$$Z_o = 15\ln\left[1+1.314u+\sqrt{(1.314u)^2+2u}\right] \approx 60\ln\left(\frac{4h}{\pi r_w}\right) - \frac{5.06\left(\frac{r_w}{h}\right)^4}{1-0.355\left(\frac{r_w}{h}\right)^4}$$

$$Z_o \approx 60\ln\left(\frac{4h}{\pi r_w}\right) \quad \text{if } h \gg r_w$$

where

$$u = \left(\frac{h}{r_w}\right)^4 - 1$$

The current in the cylindrical conductor returns equally in the two large flat conductors. The width of the flat conductors is large compared to h.

Flat Conductor between Two Large Flat Conductors [Lo; Montrose, '99; Schneider; Hilberg; White, '82]

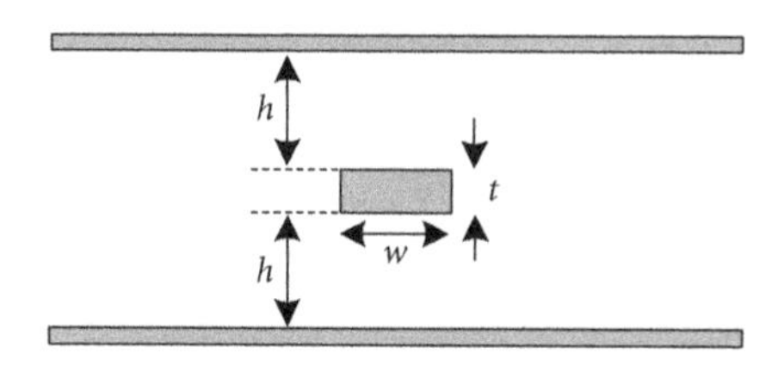

The characteristic impedance for this shielded centered strip line, sandwich line, or Triplate stripline, is

$$Z_o = 30\ln\left\{1+\frac{8h}{\pi u}\left[\frac{16h}{\pi u}+\sqrt{\left(\frac{16h}{\pi u}\right)^2+6.27}\right]\right\}$$

$$Z_o \approx 30\ln\left\{1+\frac{8h}{\pi w}\left[\frac{16h}{\pi w}+\sqrt{\left(\frac{16h}{\pi w}\right)^2+6.27}\right]\right\} \quad \text{if } w \gg t$$

$$Z_o \approx \begin{cases} 60\ln\left[2\coth\left(\dfrac{\pi w}{8h}\right)\right] & \text{if } Z_o > 94\,\Omega,\ h \gg t \\ \dfrac{148}{\ln\left(2e^{\frac{\pi w}{4h}}\right)} & \text{if } Z_o < 94\,\Omega,\ h \gg t \end{cases}$$

$$Z_o \approx 60\ln\left(\frac{3.8h}{0.8w+t}\right) \approx 60\ln\left(\frac{16h}{\pi w}\right) \quad \text{if } h \gg w \gg t$$

$$Z_o \approx \frac{240h}{\pi w}\left(2.5+\frac{16h}{\pi w}\right) \approx \frac{600h}{\pi w} \quad \text{if } w \gg h \gg t$$

where

$$u = w+\frac{t}{\pi}-\frac{t}{2\pi}\ln\left[\left(\frac{1}{\frac{4h}{t}+1}\right)^2+\left(\frac{\frac{\pi}{4}}{\frac{w}{t}+1.1}\right)^{\frac{2}{1+\frac{t}{3h}}}\right]$$

The current in the conductor of thickness t returns equally in the two large flat conductors. The width of the larger "ground" planes is large compared to h, w, and t.

Flat Conductor Perpendicular to Two Large Flat Conductors [Hilberg]

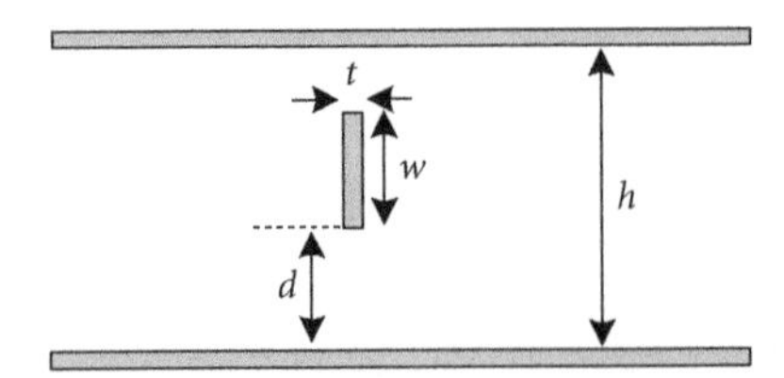

The characteristic impedance for this shielded vertical strip line is

$$Z_o = \begin{cases} 60\ln\left\{2\dfrac{\sqrt{\dfrac{\tan\left[\dfrac{\pi}{2}\left(\dfrac{d+w}{h}\right)\right]}{\tan\left(\dfrac{\pi}{2}\dfrac{d}{h}\right)}+1}}{\sqrt{\dfrac{\tan\left[\dfrac{\pi}{2}\left(\dfrac{d+w}{h}\right)\right]}{\tan\left(\dfrac{\pi}{2}\dfrac{d}{h}\right)}-1}}\right\} & \text{if } Z_o > 94\,\Omega \\ \dfrac{148}{\ln\left\{2\sqrt{\dfrac{\tan\left[\dfrac{\pi}{2}\left(\dfrac{d+w}{h}\right)\right]}{\tan\left(\dfrac{\pi}{2}\dfrac{d}{h}\right)}}\right\}} & \text{if } Z_o < 94\,\Omega \end{cases}$$

$$Z_o = \begin{cases} 60\ln\left[2\cot\left(\dfrac{\pi}{4}\dfrac{w}{h}\right)\right] & \text{if } Z_o > 94\,\Omega,\ d=\dfrac{h-w}{2} \\ \dfrac{148}{\ln\left\{2\cot\left[\dfrac{\pi}{4}\left(1-\dfrac{w}{h}\right)\right]\right\}} & \text{if } Z_o < 94\,\Omega,\ d=\dfrac{h-w}{2} \end{cases}$$

$$Z_o \approx \begin{cases} 60\ln\left(2\dfrac{\sqrt{1+\dfrac{w}{d}}+1}{\sqrt{1+\dfrac{w}{d}}-1}\right) & \text{if } Z_o > 94\,\Omega,\ h \gg \max(w,d) \\ \dfrac{148}{\ln\left(2\sqrt{1+\dfrac{w}{d}}\right)} & \text{if } Z_o < 94\,\Omega,\ h \gg \max(w,d) \end{cases}$$

The current in the conductor of thickness t returns in the two large plane conductors. The thickness of the vertical flat conductor is small compared to w. The width of the two parallel flat conductors is large compared to h.

Three Equal Cylindrical Conductors [Johnson, '61; Hilberg]

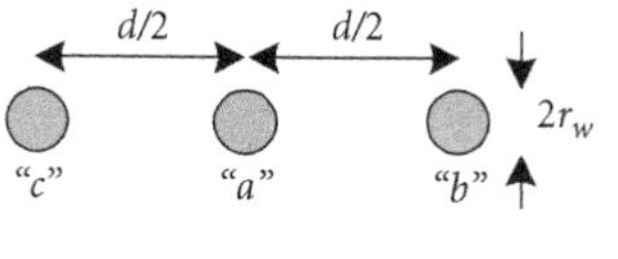

$d/2$ $d/2$ $2r_w$

"c" "a" "b"

$$Z_o = 90\ln\left(0.398\frac{d}{r_w}\right) \quad \text{if } d \gg r_w$$

The current in the center conductor "a" returns equally in the two outer conductors, "b" and "c."

Four Cylindrical Conductors [Johnson, '61; Lo; Hilberg; Laport]

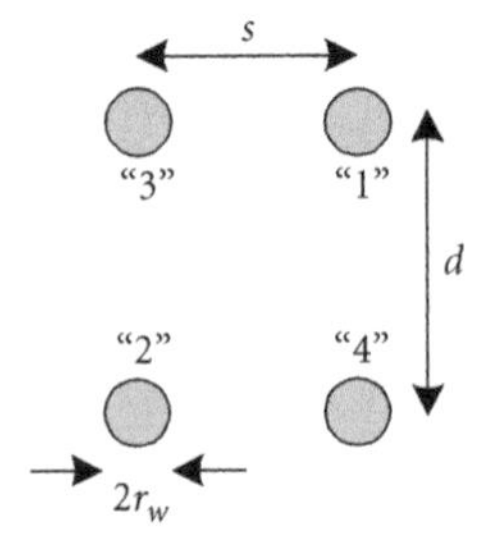

For this transmission line, also referred to as a balanced four-wire,

$$Z_o = 60\ln\left[\frac{d}{r_w\sqrt{1+\left(\frac{d}{s}\right)^2}}\right] \quad \text{if } \min(d,s) \gg r_w$$

$$Z_o \approx 60\ln\left(\frac{d}{r_w}\right) \quad \text{if } \min(d,s) \gg r_w,\ s \gg d$$

The currents in conductors "1" and "2" are equal in magnitude and sign. These currents return equally in conductors "3" and "4." When the currents in conductors "1" and "3" are equal in magnitude and sign, and they return equally through conductors "2" and "4," the impedance is

$$Z_o = 60\ln\left[\frac{d}{r_w}\sqrt{1+\left(\frac{d}{s}\right)^2}\right] \quad \text{if } \min(d,s) \gg r_w$$

$$Z_o = 60\ln\left(\frac{d\sqrt{2}}{r_w}\right) \quad \text{if } d = s,\ \min(d,s) \gg r_w$$

$$Z_o \approx 60\ln\left(\frac{d}{r_w}\right) \quad \text{if } \min(d,s) \gg r_w,\ s \gg d$$

Six Cylindrical Conductors [Laport]

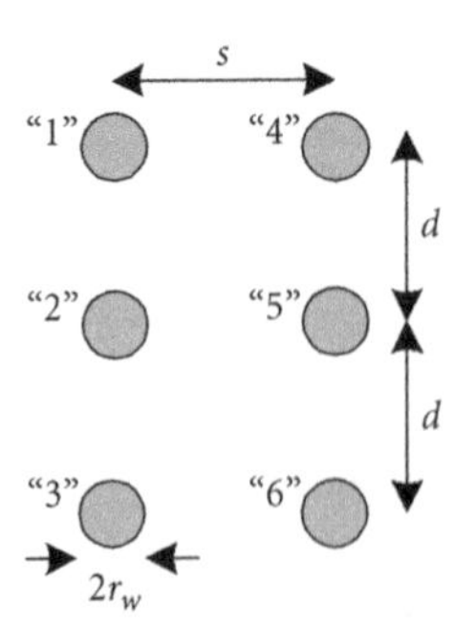

For this transmission line, also referred to as a balanced six-wire,

$$Z_o = \frac{120}{2+A}\left\{\ln\left(\frac{s}{r_w}\right)+\ln\left[\sqrt{\left(\frac{s}{2d}\right)^2+1}\right]+A\ln\left[\sqrt{\left(\frac{s}{d}\right)^2+1}\right]\right\} \quad \text{if } \min(d,s) \gg r_w$$

$$\text{where } A = \frac{\ln\left[\dfrac{sd\sqrt{s^2+4d^2}}{2r_w(s^2+d^2)}\right]}{\ln\left[\dfrac{s}{r_w\sqrt{\left(\dfrac{s}{d}\right)^2+1}}\right]}$$

$$Z_o \approx \frac{120}{2+A}\left[\ln\left(\frac{s}{r_w}\right)+\ln\left(\frac{s}{2d}\right)+A\ln\left(\frac{s}{d}\right)\right] \quad \text{if } \min(d,s) \gg r_w,\ s \gg d$$

$$\text{where } A \approx \frac{\ln\left(\dfrac{d}{2r_w}\right)}{\ln\left(\dfrac{d}{r_w}\right)}$$

$$Z_o \approx 40\left\{\ln\left(\frac{s}{r_w}\right)+\ln\left[1+\frac{1}{2}\left(\frac{s}{2d}\right)^2\right]+\ln\left[1+\frac{1}{2}\left(\frac{s}{d}\right)^2\right]\right\} \quad \text{if } \min(d,s) \gg r_w,\ d \gg s$$

$$\approx 40\ln\left(\frac{s}{r_w}\right)$$

The currents in conductors "1," "2," and "3" are equal in sign. These currents return equally in conductors "4," "5," and "6."

Three Cylindrical Conductors [Hilberg]

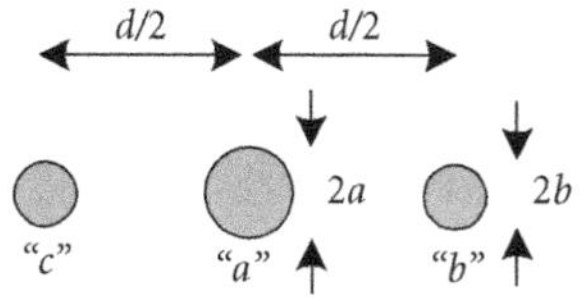

d/2 d/2
2a 2b
"c" "a" "b"

For the odd mode where the currents in conductors "b" and "c" are equal in magnitude but in opposite directions (and conductor "a" is the voltage reference),

$$Z_{oo} = 120\cosh^{-1}\left\{\frac{\sin\left[2\tan^{-1}\left(\dfrac{ad}{a^2-\dfrac{d^2}{4}+b^2}\right)\right]}{\sin\left[2\tan^{-1}\left(\dfrac{-2ab}{a^2+\dfrac{d^2}{4}-b^2}\right)\right]}\right\} \quad \text{if } d \gg \max(a,b)$$

$$Z_{oo} \approx 120 \ln\left(\frac{d}{b}\right) \quad \text{if } d \ggg \max(a,b)$$

The radii of conductors "b" and "c" are the same.

Five Cylindrical Conductors [Johnson, '61; Howard W. Sams; Lo]

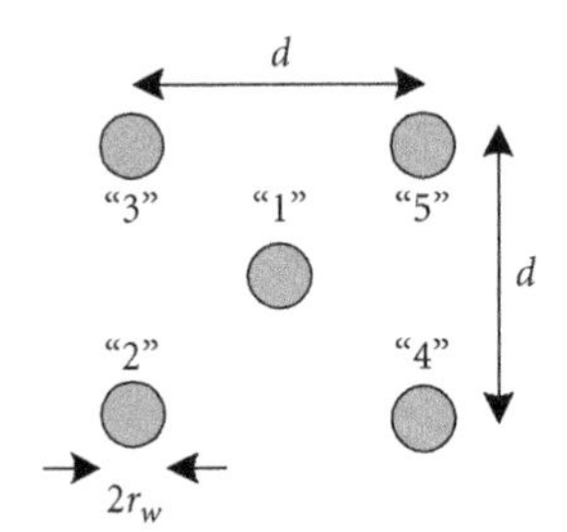

$$Z_o = 75 \ln\left(\frac{d}{1.9 r_w}\right) \quad \text{if } d \gg r_w$$

The current in conductor "1" returns equally in conductors "2" through "5." Conductor "1" is at the center of the surrounding conductors.

Two Cylindrical Conductors Surrounded by Four Cylindrical Conductors [Howard W. Sams]

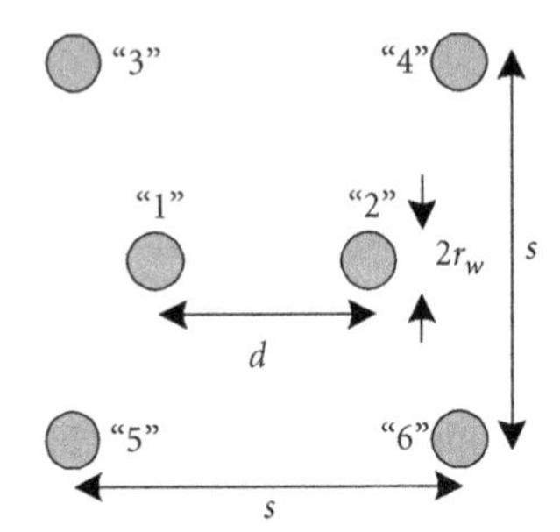

For the odd mode where the currents in conductors "1" and "2" are equal in magnitude but in opposite directions (and conductors "3" through "6" are the reference conductors),

$$Z_{oo} = 120 \left\{ \ln\left(\frac{d}{r_w}\right) - \frac{\left[\ln \dfrac{1+\left(1+\dfrac{d}{s}\right)^2}{1+\left(1-\dfrac{d}{s}\right)^2} \right]^2}{\ln\left(\dfrac{s\sqrt{2}}{r_w}\right)} \right\}$$

$$Z_{oo} \approx 120 \ln\left(\frac{d}{r_w}\right) \quad \text{if } s \gg d$$

2N Cylindrical Conductors [Hilberg; Laport]

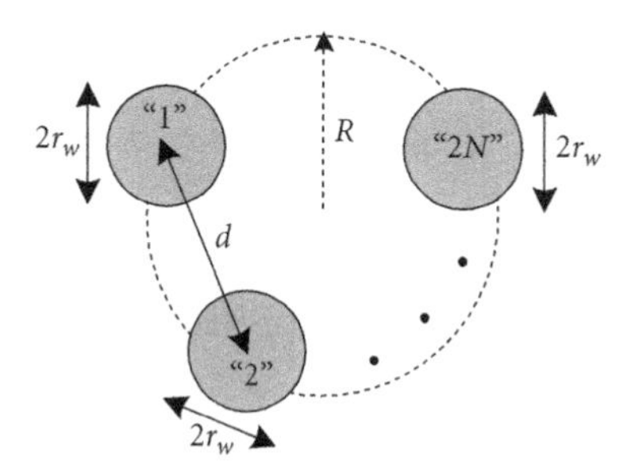

For $2N$ parallel, identical cylindrical conductors uniformly spaced on the perimeter of a circle of radius R,

$$Z_o = \frac{120}{N}\cosh^{-1}\left\{\frac{1}{\sin\left[N\sin^{-1}\left(\frac{r_w}{R}\right)\right]}\right\} \quad \text{if } R \gg r_w$$

$$Z_o \approx \frac{120}{N}\cosh^{-1}\left(\frac{R}{Nr_w}\right) \approx \frac{120}{N}\ln\left(\frac{2R}{Nr_w}\right) \quad \text{if } R \gg Nr_w$$

$$Z_o \approx 120\ln\left(\frac{2R}{r_w}\right) \quad \text{if } R \gg r_w,\ N = 1$$

The $N = 1$ situation corresponds to two conductors. The currents in conductors "1," "3," . . ., "$2N - 1$" are equal in magnitude but in opposite direction to the currents in conductors "2," "4," . . ., "$2N$."

Two Cylindrical Conductors above Large Flat Conductor [Johnson, '61; Howard W. Sams; Laport]

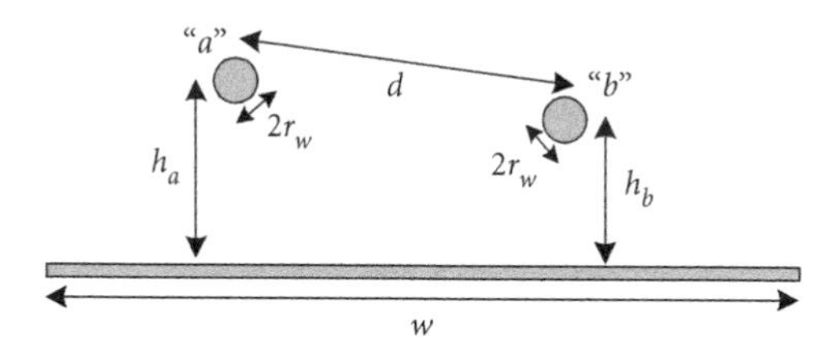

For the odd mode where the currents in conductors "a" and "b" are equal in magnitude but in opposite directions (also referred to as balanced two-wire),

$$Z_{oo} = 120\ln\left(\frac{d}{r_w\sqrt{1+\frac{d^2}{4h_a h_b}}}\right) \quad \text{if } \min(d, h_a, h_b) \gg r_w$$

$$= 120\ln\left[\frac{d}{r_w\sqrt{1+\left(\frac{d}{2h}\right)^2}}\right] = 120\ln\left[\frac{2h}{r_w\sqrt{1+\left(\frac{2h}{d}\right)^2}}\right] \quad \text{if } \min(d, h_a, h_b) \gg r_w,\ h = h_a = h_b$$

$$Z_{oo} \approx 120 \ln\left(\frac{2h}{r_w}\right) \quad \text{if } \min(d, h_a, h_b) \gg r_w,\ h = h_a = h_b,\ d \gg h$$

$$Z_{oo} \approx 120 \ln\left(\frac{d}{r_w}\right) \quad \text{if } \min(d, h_a, h_b) \gg r_w,\ h = h_a = h_b,\ h \gg d$$

For the even mode where the currents in conductors "*a*" and "*b*" are equal in magnitude and in the same direction and their current returns through the large flat plane (also referred to as an unbalanced parallel two-wire),

$$Z_{oe} = 30 \ln\left[\frac{2h}{r_w}\sqrt{1+\left(\frac{2h}{d}\right)^2}\right] \quad \text{if } \min(d, h_a, h_b) \gg r_w,\ h = h_a = h_b$$

$$Z_{oe} \approx 30 \ln\left(\frac{2h}{r_w}\right) \quad \text{if } \min(d, h_a, h_b) \gg r_w,\ h = h_a = h_b,\ d \gg h$$

$$Z_{oe} \approx 60 \ln\left(\frac{2h}{\sqrt{r_w d}}\right) \quad \text{if } \min(d, h_a, h_b) \gg r_w,\ h = h_a = h_b,\ h \gg d$$

When the current through conductor "*a*" returns through both conductor "*b*" and the large flat plane, which are connected together, the impedance for this mode is

$$Z_o = 60 \ln\left(\frac{2h}{r_w}\right) - 60\frac{\left[\ln\left(\frac{2h}{d}\right)\right]^2}{\ln\left(\frac{2h}{r_w}\right)} \quad \text{if } \min(d, h_a, h_b) \gg r_w,\ h \gg d,\ h = h_a = h_b$$

$$\approx 120 \ln\left(\frac{d}{r_w}\right) \quad \text{if } \min(d, h_a, h_b) \gg r_w,\ h \ggg d,\ h = h_a = h_b$$

The width, w, of the large flat conductor is large compared to h.

Four Cylindrical Conductor Square above Large Flat Conductor[14] [Laport]

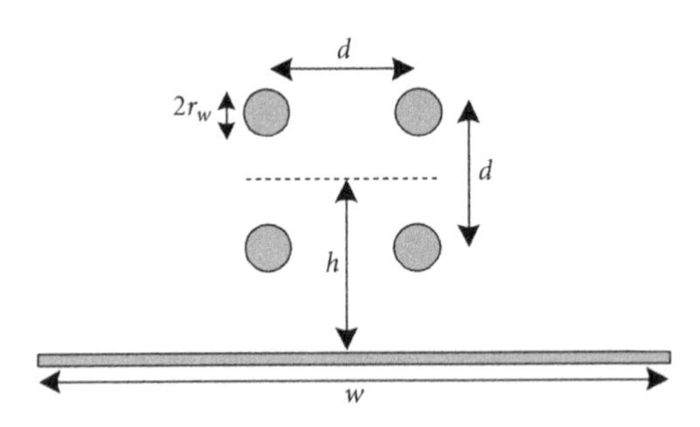

[14]This equation does not appear to agree with an approximate result obtained using the method of images.

For the mode where the currents in all four cylindrical conductors are equal and in the same direction and their current returns through the large flat plane (also referred to as an unbalanced parallel four wire),

$$Z_{oe} = 15\ln\left(\frac{16h^4}{r_w d^3\sqrt{2}}\right) \quad \text{if } d \gg r_w,\ h \gg d$$

The width, w, of the large flat conductor is large compared to h.

Three Cylindrical Conductors above Large Flat Conductor [Laport]

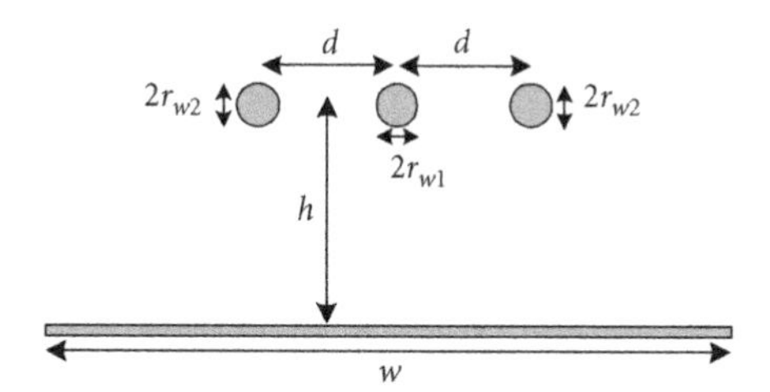

For the mode where the current in the center conductor of radius r_{w1} returns through the two outer conductors of radius r_{w2} and the large flat plane, which are connected together,

$$Z_o = 60\ln\left(\frac{2h}{r_{w1}}\right) - 120\frac{\left[\ln\left(\frac{2h}{d}\right)\right]^2}{\ln\left(\frac{2h^2}{r_{w2}d}\right)} \quad \text{if } d \gg \max(r_{w1}, r_{w2}),\ h \gg d$$

$$Z_o \approx 60\ln\left(\frac{d^{\frac{3}{2}}}{r_{w1}\sqrt{2r_{w2}}}\right) \quad \text{if } d \gg \max(r_{w1}, r_{w2}),\ h \ggg d$$

The width, w, of the large flat conductor is large compared to h.

Cylindrical Conductor near Three Cylindrical Conductors and above Large Flat Conductor [Laport]

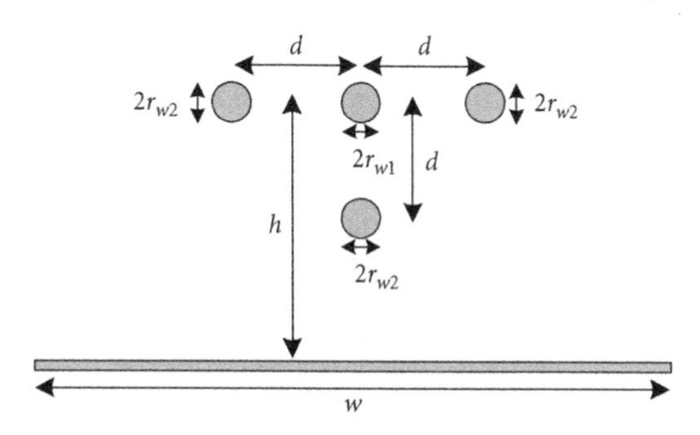

For the mode where the current in the center conductor of radius r_{w1} returns through the three outer conductors of radius r_{w2} and the large flat plane, which are connected together,

$$Z_o = 60\ln\left(\frac{2h}{r_{w1}}\right) - 180\frac{\left[\ln\left(\frac{2h}{d}\right)\right]^2}{2\ln\left(\frac{2h}{d}\right) + \ln\left(\frac{h}{r_{w2}}\right)} \quad \text{if } d \gg \max(r_{w1}, r_{w2}),\ h \gg d$$

$$Z_o \approx 20\ln\left(\frac{d^4}{2r_{w2}r_{w1}^3}\right) \quad \text{if } d \gg \max(r_{w1}, r_{w2}),\ h \ggg d$$

The width, w, of the large flat conductor is large compared to h.

Cylindrical Conductor Surrounded by Four Cylindrical Conductors and above Large Flat Conductor [Laport]

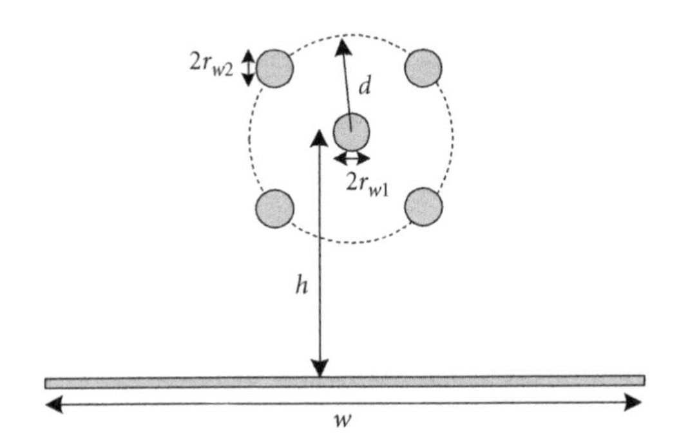

For the mode where the current in the center conductor of radius r_{w1} returns through the four surrounding conductors of radius r_{w2} and the large flat plane, which are connected together,

$$Z_o = 60\ln\left(\frac{2h}{r_{w1}}\right) - 240\frac{\left[\ln\left(\frac{2h}{d}\right)\right]^2}{\ln\left(\frac{2h}{r_{w2}}\right)+\ln\left(\frac{2h^2}{d^2}\right)+\ln\left(\frac{h}{d}\right)} \quad \text{if } d \gg \max(r_{w1}, r_{w2}),\ h \gg d$$

$$Z_o \approx 15\ln\left(\frac{d^5}{4r_{w2}r_{w1}^4}\right) \quad \text{if } d \gg \max(r_{w1}, r_{w2}),\ h \ggg d$$

The width, w, of the large flat conductor is large compared to h.

Two Cylindrical Conductors Surrounded by Four Cylindrical Conductors and above Large Flat Conductor [Laport]

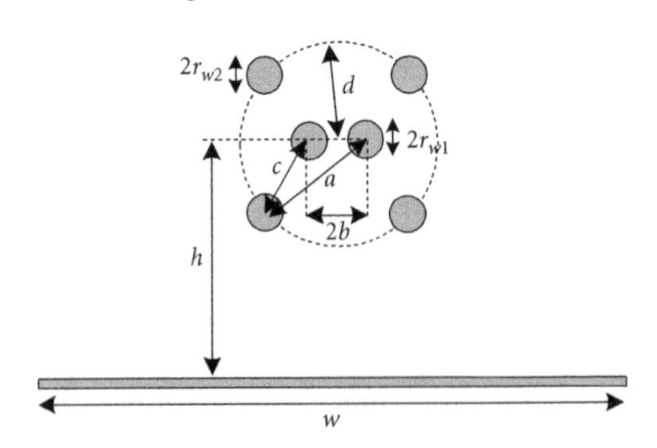

For the mode where the currents in the two inner conductors of radius r_{w1} (evenly spaced about the center) return through the four surrounding conductors of radius r_{w2} and the large flat plane, which are connected together,

$$Z_o = 30\ln\left(\frac{2h^2}{br_{w1}}\right) - 60\frac{\left[\ln\left(\frac{4h^2}{ac}\right)\right]^2}{\ln\left(\frac{2h^2}{dr_{w2}}\right) + \ln\left(\frac{2h^2}{d^2}\right)} \quad \text{if } \min(a,b,c,d) \gg \max(r_{w1}, r_{w2}),\ h \gg d$$

$$Z_o \approx 15\ln\left(\frac{a^4c^4}{16r_{w2}r_{w1}^2b^2d^3}\right) \quad \text{if } \min(a,b,c,d) \gg \max(r_{w1}, r_{w2}),\ h \ggg d$$

The width, *w*, of the large flat conductor is large compared to *h*.

Eight Conductor Square above Large Flat Conductor[15] [Laport]

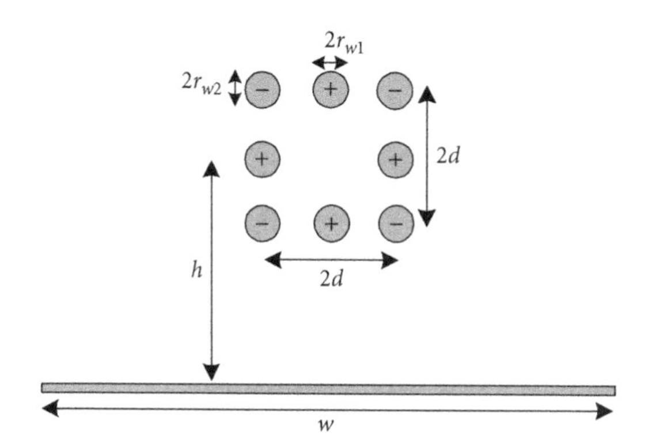

For the mode where the currents in the four conductors of radius r_{w1} (labeled with a "+") return through the four conductors of radius r_{w2} at the corners of the square (labeled with a "–") and the large flat plane, which are connected together,

$$Z_o = 15\ln\left(\frac{2h}{r_{w1}}\right) + 30\ln\left(\frac{2h}{d\sqrt{2}}\right) + 15\ln\left(\frac{h}{d}\right) - 30\left\{\frac{\left[\ln\left(\frac{2h}{d}\right) + \ln\left(\frac{2h}{d\sqrt{5}}\right)\right]^2}{\ln\left(\frac{2h}{r_{w2}}\right) + 2\ln\left(\frac{h}{d}\right) + \ln\left(\frac{h}{d\sqrt{2}}\right)}\right\}$$

$$\text{if } d \gg \max(r_{w1}, r_{w2}),\ h \gg d$$

The width, *w*, of the large flat conductor is large compared to *h*.

[15] This expression from Laport appears to contain one (or more) unknown restrictions since in the limit as $h \to \infty$, the variable h remains in the expression. The expression is more reasonable if –30 is replaced by –60.

Large Cylindrical Conductor Surrounded by Eight Cylindrical Conductors and above Large Flat Conductor [Laport]

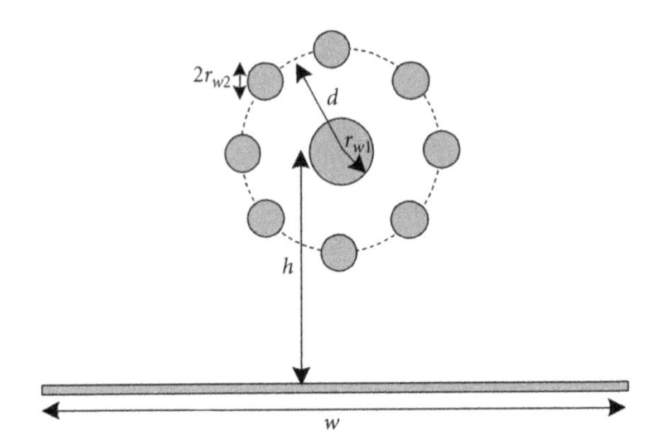

For the mode where the current in the center conductor of radius r_{w1} returns through the eight surrounding uniformly spaced conductors of radius r_{w2} and the large flat plane, which are connected together,

$$Z_o = 60\ln\left(\frac{2h}{r_{w1}}\right) - 480\frac{\left[\ln\left(\frac{2h}{d}\right)\right]^2}{\ln\left(\frac{2h}{r_{w2}}\right) + 2\ln\left(\frac{2h}{0.77d}\right) + 2\ln\left(\frac{2h}{d\sqrt{2}}\right) + 2\ln\left(\frac{2h}{1.86d}\right) + \ln\left(\frac{h}{d}\right)}$$

if $d \gg \max(r_{w1}, r_{w2}),\ h \gg d$

$$Z_o \approx 7.5\ln\left(\frac{d^9}{8.4 r_{w2} r_{w1}^8}\right) \quad \text{if } d \gg \max(r_{w1}, r_{w2}),\ h \ggg d$$

The width, w, of the large flat conductor is large compared to h.

Four Parallel Cylindrical Conductors above Large Flat Conductor [Laport]

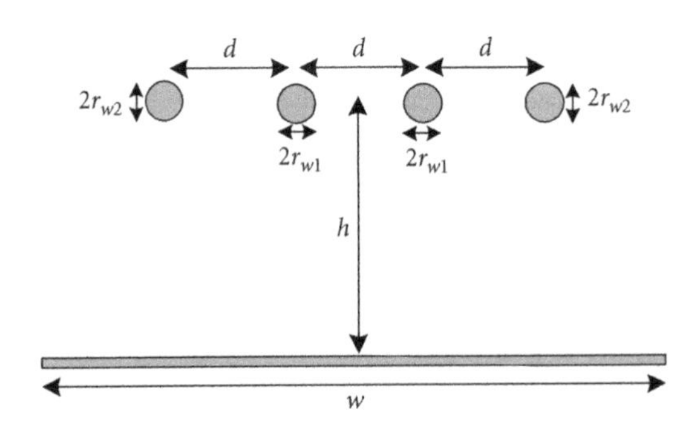

For the mode where the currents in the two inner conductors, which are of equal magnitude and sign, return through the two outer conductors of radius r_{w2} and the large flat plane, which are connected together,

$$Z_o = 30\ln\left(\frac{4h^2}{dr_{w1}}\right) - 30\frac{\left[\ln\left(\frac{2h^2}{d^2}\right)\right]^2}{\ln\left(\frac{4h^2}{3dr_{w2}}\right)} \quad \text{if } d \gg \max(r_{w1}, r_{w2}),\ h \gg d$$

$$Z_o \approx 30\ln\left(\frac{4d^2}{3r_{w1}r_{w2}}\right) \quad \text{if } d \gg \max(r_{w1}, r_{w2}),\ h \ggg d$$

The width, w, of the large flat conductor is large compared to h.

Two Cylindrical Parallel Conductors above Two Cylindrical Conductors and Large Flat Conductor [Laport]

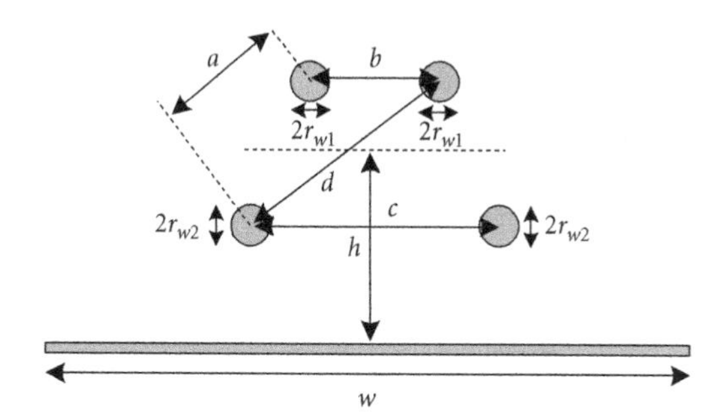

For the mode where the currents in the two upper conductors, which are of equal magnitude and sign, return through the two lower conductors of radius r_{w2} and the large flat plane, which are connected together,

$$Z_o = 30\ln\left(\frac{4h^2}{br_{w1}}\right) - 30\frac{\left[\ln\left(\frac{4h^2}{ad}\right)\right]^2}{\ln\left(\frac{4h^2}{cr_{w2}}\right)} \quad \text{if } \min(a,b,c,d) \gg \max(r_{w1}, r_{w2}),\ h \gg \max(a,b,c,d)$$

$$Z_o \approx 30\ln\left(\frac{a^2d^2}{bcr_{w1}r_{w2}}\right) \quad \text{if } \min(a,b,c,d) \gg \max(r_{w1}, r_{w2}),\ h \ggg \max(a,b,c,d)$$

The width, w, of the large flat conductor is large compared to h.

Two Coplanar Cylindrical Conductors between Two Large Flat Conductors [Lo; Frankel]

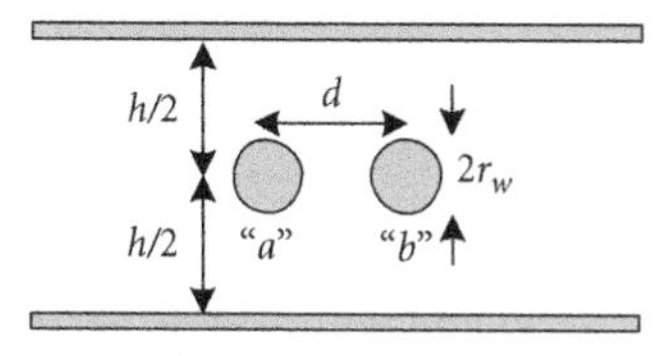

For the odd mode where the currents in conductors "a" and "b" are equal in magnitude but in opposite directions,

$$Z_{oo} = 120\ln\left[\tanh\left(\frac{\pi d}{2h}\right)\coth\left(\frac{\pi r_w}{2h}\right)\right] \quad \text{if } \min(d,h) \gg r_w$$

$$Z_{oo} \approx 120\ln\left[\frac{2h\tanh\left(\frac{\pi d}{2h}\right)}{\pi r_w}\right] \approx 120\ln\left(\frac{d}{r_w}\right) \quad \text{if } \min(d,h) \gg r_w,\ h \gg d$$

For the even mode where the currents in conductors "a" and "b" are equal in magnitude and in the same direction and their current returns equally through the large flat conductors,

$$Z_{oe} = 60\ln\left[\coth\left(\frac{\pi d}{2h}\right)\coth\left(\frac{\pi r_w}{2h}\right)\right] \quad \text{if } \min(d,h) \gg r_w$$

$$Z_{oe} \approx 60\ln\left[\frac{2h\coth\left(\frac{\pi d}{2h}\right)}{\pi r_w}\right] \approx 120\ln\left(\frac{2h}{\pi\sqrt{r_w d}}\right) \quad \text{if } \min(d,h) \gg r_w,\ h \gg d$$

The width of the large flat conductors is large compared to h.[16]

Two Cylindrical Conductors between Two Large Flat Conductors [Frankel]

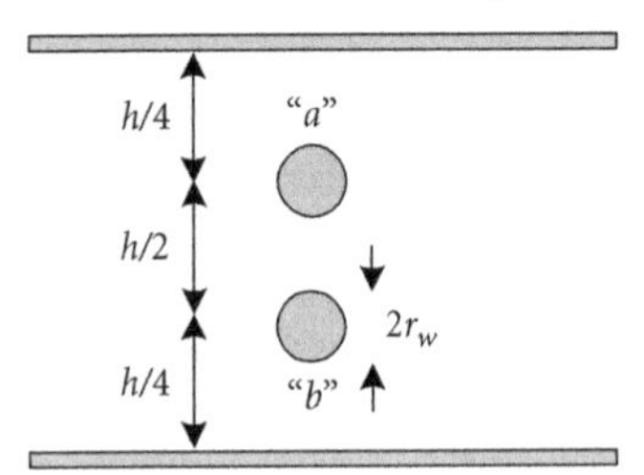

For the odd mode where the currents in conductors "a" and "b" are equal in magnitude but in opposite directions,

$$Z_{oo} = 120\ln\left(\frac{h}{\pi r_w}\right) \quad \text{if } h \gg r_w$$

The width of the large flat conductors is large compared to h.

Three Cylindrical Conductors between Two Large Flat Conductors [Frankel]

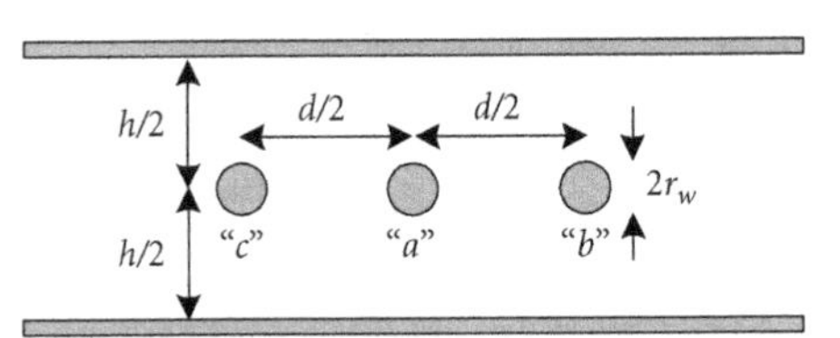

For the balanced mode where the currents in conductors "b" and "c" are equal in magnitude and direction and their current returns through conductor "a,"

$$Z_o = 90\ln\left\{\frac{2h\tanh\left(\frac{\pi d}{2h}\right)}{\pi r_w\left[1+\operatorname{sech}\left(\frac{\pi d}{2h}\right)\right]^{\frac{4}{3}}}\right\} \quad \text{if } \min(d,h) \gg r_w$$

$$Z_o \approx 90\ln\left(0.397\frac{d}{r_w}\right) \quad \text{if } d \gg r_w,\ h \gg d$$

The width of the large flat conductors is large compared to h.

[16]When the two circular conductors are far apart, it would seem that this result *should* be about one-half of the impedance of a single circular conductor between two ground planes.

Two Cylindrical Conductors in Trough [Lo; Howard W. Sams]

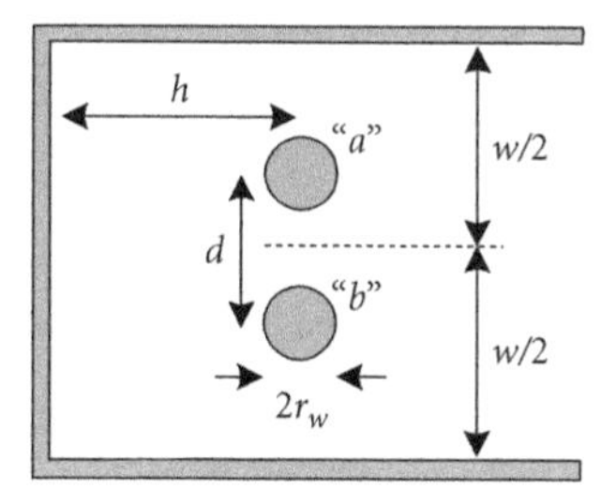

For the odd mode where the currents in conductors "*a*" and "*b*" are equal in magnitude but in opposite directions,

$$Z_{oo} = 120\ln\left[\frac{w}{\pi r_w\sqrt{\csc^2\left(\frac{\pi d}{w}\right)+\operatorname{csch}^2\left(\frac{2\pi h}{w}\right)}}\right] \quad \text{if } \min(d,h,w) \gg r_w$$

$$Z_{oo} \approx 120\ln\left[\frac{d}{r_w\sqrt{1+\left(\frac{d}{2h}\right)^2}}\right] \quad \text{if } \min(d,h,w) \gg r_w,\ w \gg \max(h,d)$$

$$Z_{oo} \approx 120\ln\left[\frac{w}{\pi r_w \csc\left(\frac{\pi d}{w}\right)}\right] \quad \text{if } \min(d,h,w) \gg r_w,\ h \gg w$$

The distance between the two cylindrical conductors and the open end of the trough is large compared to w and h.

Two Cylindrical Conductors in Rectangular Enclosure[17] [Frankel; King; Agarwal]

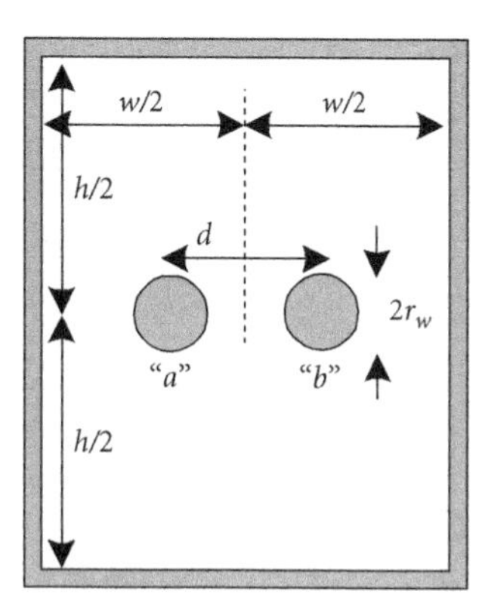

[17]The even-mode impedance expression, from King, was modified by a factor of 1/4 so that it (i) agrees with King's numerical example and (ii) approximately agrees with the even-mode expression for two cylindrical conductors within a circular enclosure.

For the odd mode where the currents in conductors "*a*" and "*b*" are equal in magnitude but in opposite directions,

$$Z_{oo} = 120\left\{\ln\left[\frac{2h\tanh\left(\frac{\pi d}{2h}\right)}{\pi r_w}\right] - \sum_{m=1}^{\infty}\ln\left[\frac{1+\dfrac{\sinh^2\left(\frac{\pi d}{2h}\right)}{\cosh^2\left(\frac{m\pi w}{2h}\right)}}{1-\dfrac{\sinh^2\left(\frac{\pi d}{2h}\right)}{\sinh^2\left(\frac{m\pi w}{2h}\right)}}\right]\right\} \quad \text{if } \min(d,h,w) \gg r_w$$

$$Z_{oo} \approx 120\left\{\ln\left[\frac{2h\tanh\left(\frac{\pi d}{2h}\right)}{\pi r_w}\right] - \ln\left[\frac{1+\dfrac{\sinh^2\left(\frac{\pi d}{2h}\right)}{\cosh^2\left(\frac{\pi w}{2h}\right)}}{1-\dfrac{\sinh^2\left(\frac{\pi d}{2h}\right)}{\sinh^2\left(\frac{\pi w}{2h}\right)}}\right]\right\} \quad \text{if } \min(d,h,w) \gg r_w$$

$$Z_{oo} \approx 120\ln\left[\frac{2h\tanh\left(\frac{\pi d}{2h}\right)}{\pi r_w}\right] \quad \text{if } \min(d,h,w) \gg r_w,\ w \gg \max(h,d)$$

For the even mode where the currents in conductors "*a*" and "*b*" are equal in magnitude and in the same direction and their current returns through the rectangular enclosure,

$$Z_{oe} = 30\left\{\ln\left[\frac{2h\coth\left(\frac{\pi d}{2h}\right)}{\pi r_w}\right] + \sum_{m=1}^{\infty}(-1)^m\ln\left[\frac{1+\dfrac{\cosh^2\left(\frac{\pi d}{2h}\right)}{\sinh^2\left(\frac{m\pi w}{2h}\right)}}{1-\dfrac{\cosh^2\left(\frac{\pi d}{2h}\right)}{\cosh^2\left(\frac{m\pi w}{2h}\right)}}\right]\right\} \quad \text{if } \min(d,h,w) \gg r_w$$

$$Z_{oe} \approx 30\left\{\ln\left[\frac{2h\coth\left(\frac{\pi d}{2h}\right)}{\pi r_w}\right] - \ln\left[\frac{1+\dfrac{\cosh^2\left(\frac{\pi d}{2h}\right)}{\sinh^2\left(\frac{\pi w}{2h}\right)}}{1-\dfrac{\cosh^2\left(\frac{\pi d}{2h}\right)}{\cosh^2\left(\frac{\pi w}{2h}\right)}}\right]\right\} \quad \text{if } \min(d,h,w) \gg r_w$$

Three Cylindrical Conductors in Rectangular Enclosure [Frankel]

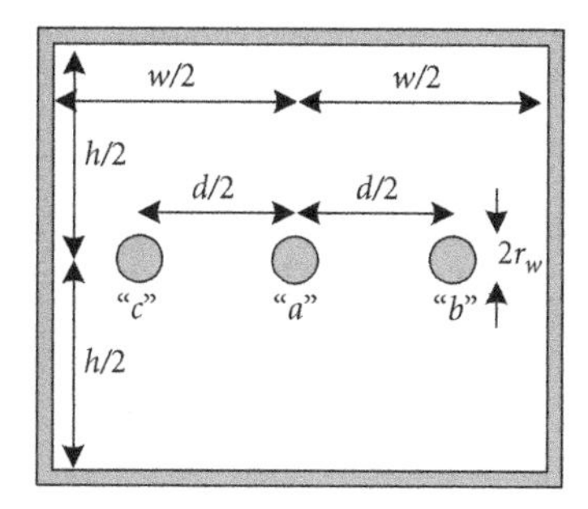

For the balanced mode where the currents in conductors "b" and "c" are equal in magnitude and direction and their current returns through conductor "a,"

$$Z_o = 90\left\{ \ln\left\{ \frac{2h\tanh\left(\frac{\pi d}{2h}\right)}{\pi r_w\left[1+\operatorname{sech}\left(\frac{\pi d}{2h}\right)\right]^{\frac{4}{3}}} \right\} - \frac{2}{3}\sum_{m=1}^{\infty}(-1)^m \ln\left[\frac{1-\dfrac{\sinh^2\left(\frac{\pi d}{2h}\right)}{\sinh^2\left(\frac{m\pi w}{2h}\right)}}{1+\dfrac{\sinh^2\left(\frac{\pi d}{2h}\right)}{\cosh^2\left(\frac{m\pi w}{2h}\right)}} \right]\left[\frac{1+\dfrac{\sinh^2\left(\frac{\pi d}{4h}\right)}{\cosh^2\left(\frac{m\pi w}{2h}\right)}}{1-\dfrac{\sinh^2\left(\frac{\pi d}{4h}\right)}{\sinh^2\left(\frac{m\pi w}{2h}\right)}} \right] \right\} \quad \text{if } \min(d,h,w) \gg r_w$$

$$Z_o \approx 90\ln\left\{ \frac{2h\tanh\left(\frac{\pi d}{2h}\right)}{\pi r_w\left[1+\operatorname{sech}\left(\frac{\pi d}{2h}\right)\right]^{\frac{4}{3}}} \right\} \quad \text{if } \min(d,h,w) \gg r_w,\ w \gg \max(h, d)$$

Two Cylindrical Conductors in Cylindrical Enclosure [Lo; Howard W. Sams]

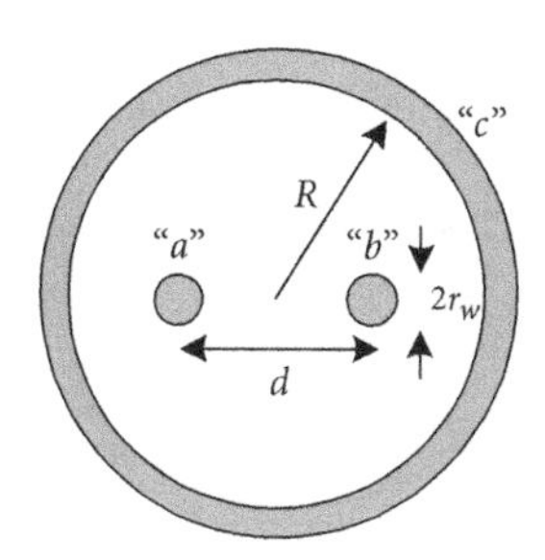

For the odd mode where the currents in conductors "*a*" and "*b*" are equal in magnitude but in opposite directions, the impedance of this shielded pair of conductors is

$$Z_{oo} = 120 \ln \left[\frac{d}{r_w} \frac{1 - \left(\frac{d}{2R}\right)^2}{1 + \left(\frac{d}{2R}\right)^2} \right] \quad \text{if } \min\left(d, R - \frac{d}{2}\right) \gg r_w$$

$$Z_{oo} \approx 120 \ln \left(\frac{d}{r_w} \right) \quad \text{if } \min\left(d, R - \frac{d}{2}\right) \gg r_w,\ R \gg d$$

If insulating beads are used at frequent intervals (i.e., at electrically-short intervals) between the inner conductors and outer conductor, then the new characteristic impedance is

$$Z'_{oo} = \frac{Z_o}{\sqrt{1 + (\varepsilon_r - 1)\frac{w}{s}}}$$

where ε_r is the relative permittivity of the beads, w is the width of the beads, and $s - w$ is the distance between the beads. For the even mode where the currents in conductors "*a*" and "*b*" are equal in magnitude and in the same direction and their current returns through the cylindrical conductor "*c*,"

$$Z_{oe} = 30 \ln \left\{ \frac{R^2}{d r_w} \left[1 - \left(\frac{d}{2R} \right)^4 \right] \right\} \quad \text{if } \min\left(d, R - \frac{d}{2}\right) \gg r_w$$

$$Z_{oe} \approx 30 \ln \left(\frac{R^2}{d r_w} \right) = 60 \ln \left(\frac{R}{\sqrt{d r_w}} \right) \quad \text{if } R \gg d,\ \min\left(d, R - \frac{d}{2}\right) \gg r_w$$

Two Coplanar Flat Conductors in Cylindrical Enclosure [Hilberg]

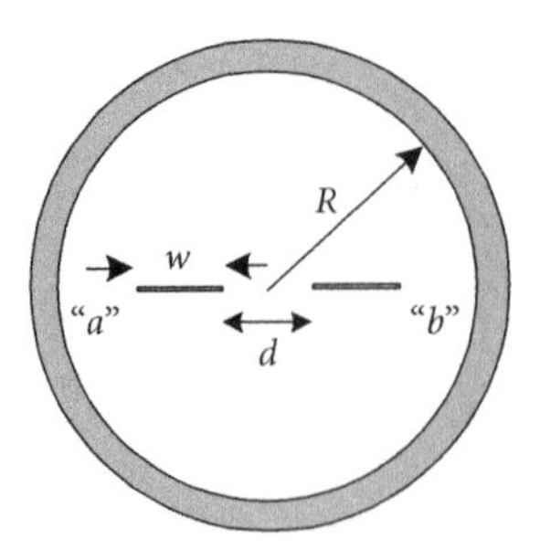

For the odd mode where the currents in conductors "*a*" and "*b*" are equal in magnitude but in opposite directions, the impedance of this shielded pair of conductors is

$$Z_{oo} = \begin{cases} 120\ln\left[2\dfrac{\sqrt{\left(\dfrac{d+2w}{d}\right)\dfrac{R^2-\left(\dfrac{d}{2}\right)^2}{R^2-\left(\dfrac{d}{2}+w\right)^2}}+1}{\sqrt{\left(\dfrac{d+2w}{d}\right)\dfrac{R^2-\left(\dfrac{d}{2}\right)^2}{R^2-\left(\dfrac{d}{2}+w\right)^2}}-1}\right] & \text{if } Z_{oo} > 189\,\Omega \\ \dfrac{296}{\ln\left[2\sqrt{\left(\dfrac{d+2w}{d}\right)\dfrac{R^2-\left(\dfrac{d}{2}\right)^2}{R^2-\left(\dfrac{d}{2}+w\right)^2}}\right]} & \text{if } Z_{oo} < 189\,\Omega \end{cases}$$

$$Z_{oo} \approx 120\ln\left(\frac{4d}{w}-2\right) \approx 120\ln\left(\frac{4d}{w}\right) \quad \text{if } d \gg w,\ R \gg \max(w,d)$$

$$Z_{oo} \approx \begin{cases} 120\ln\left(2\dfrac{1+\sqrt{\dfrac{d}{d+2w}}}{1-\sqrt{\dfrac{d}{d+2w}}}\right) & \text{if } Z_{oo} > 189\,\Omega,\ R \gg \max(w,d) \\ \dfrac{296}{\ln\left(2\sqrt{\dfrac{d+2w}{d}}\right)} & \text{if } Z_{oo} < 189\,\Omega,\ R \gg \max(w,d) \end{cases}$$

The coplanar flat strips are thin compared to w and R.

Two Coplanar Flat Conductors between Two Large Flat Conductors [Hilberg]

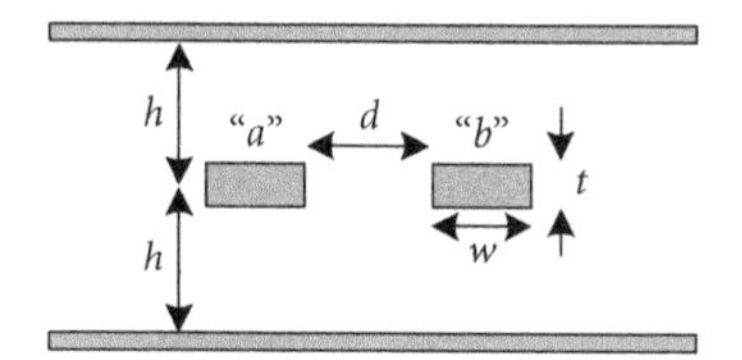

For the odd mode where the currents in conductors "a" and "b" are equal in magnitude but in opposite directions (and the two large flat conductors are the reference conductors) and the strips are thin (i.e., $h \gg t$), the impedance of this edge-coupled stripline is

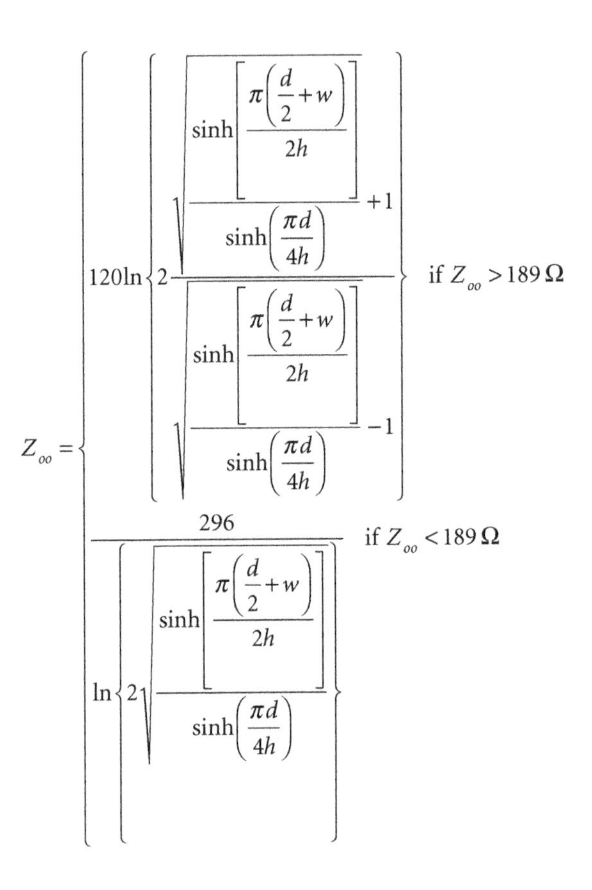

$$Z_{oo} = \begin{cases} 120 \ln\left\{ 2 \dfrac{\sqrt{\dfrac{\sinh\left[\dfrac{\pi\left(\frac{d}{2}+w\right)}{2h}\right]}{\sinh\left(\frac{\pi d}{4h}\right)}} + 1}{\sqrt{\dfrac{\sinh\left[\dfrac{\pi\left(\frac{d}{2}+w\right)}{2h}\right]}{\sinh\left(\frac{\pi d}{4h}\right)}} - 1} \right\} & \text{if } Z_{oo} > 189\,\Omega \\ \dfrac{296}{\ln\left\{ 2\sqrt{\dfrac{\sinh\left[\dfrac{\pi\left(\frac{d}{2}+w\right)}{2h}\right]}{\sinh\left(\frac{\pi d}{4h}\right)}} \right\}} & \text{if } Z_{oo} < 189\,\Omega \end{cases}$$

Two Coplanar Flat Conductors Perpendicular to Two Large Flat Conductors [Hilberg]

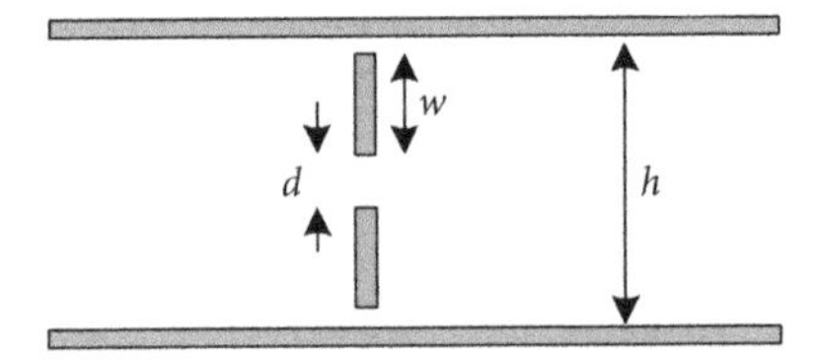

For the odd mode where the currents in the two conductors of width w are equal in magnitude but in opposite directions (and the two large flat conductors are the reference conductors),

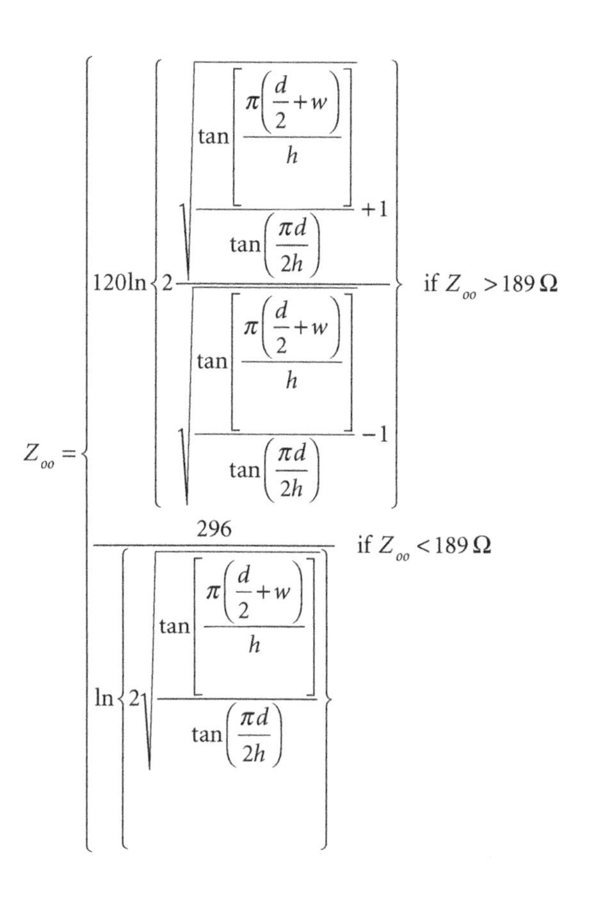

$$Z_{oo} = \begin{cases} 120\ln\left\{2\dfrac{\sqrt{\dfrac{\tan\left[\dfrac{\pi\left(\frac{d}{2}+w\right)}{h}\right]}{\tan\left(\dfrac{\pi d}{2h}\right)}}+1}{\sqrt{\dfrac{\tan\left[\dfrac{\pi\left(\frac{d}{2}+w\right)}{h}\right]}{\tan\left(\dfrac{\pi d}{2h}\right)}}-1}\right\} & \text{if } Z_{oo} > 189\,\Omega \\ \dfrac{296}{\ln\left\{2\sqrt{\dfrac{\tan\left[\dfrac{\pi\left(\frac{d}{2}+w\right)}{h}\right]}{\tan\left(\dfrac{\pi d}{2h}\right)}}\right\}} & \text{if } Z_{oo} < 189\,\Omega \end{cases}$$

$$Z_{oo} \approx \begin{cases} 120\ln\left(2\dfrac{\sqrt{1+\dfrac{2w}{d}}+1}{\sqrt{1+\dfrac{2w}{d}}-1}\right) & \text{if } Z_{oo} > 189\,\Omega,\ h \gg \max(w,d) \\ \dfrac{296}{\ln\left(2\sqrt{1+\dfrac{2w}{d}}\right)} & \text{if } Z_{oo} < 189\,\Omega,\ h \gg \max(w,d) \end{cases}$$

The thickness of the vertical conductors is small compared to w.

Two Broadside Flat Conductors between Two Large Flat Conductors [Cohn, '60; Lo]

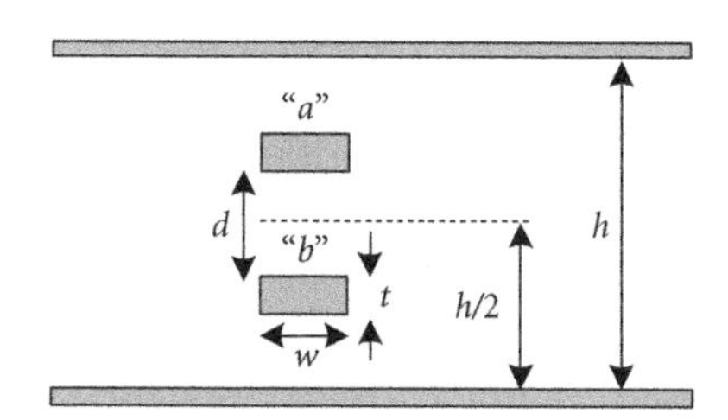

For the odd mode where the currents in conductors "*a*" and "*b*" are equal in magnitude but in opposite directions (and the large flat conductors are the reference conductors) and the dual strips are thin (i.e., $h \gg t$), the impedance of this shielded dual or broadside-coupled stripline is

$$Z_{oo} = 188\left(\frac{1}{\dfrac{w}{h-d}+\dfrac{w}{d}+k}\right) \quad \text{if } w > 0.35d,\ w > 0.35(h-d)$$

where

$$k = \frac{h}{\pi d}\left[\ln\left(\frac{h}{h-d}\right)+\frac{d}{h-d}\ln\left(\frac{h}{d}\right)\right]$$

For the even mode where the currents in conductors "*a*" and "*b*" are equal in magnitude and in the same direction and their current returns through the large flat conductors,

$$Z_{oe} = 188\left(\frac{1}{\dfrac{w}{h-d}+0.4413+k\dfrac{d}{h}}\right) \quad \text{if } w > 0.35d,\ w > 0.35(h-d)$$

where k is as previously defined. The large flat "ground" conductors are large compared to w and h.

Three Coplanar Flat Conductors [Hilberg]

The odd-mode impedance is

$$Z_{oo} = \begin{cases} 120\ln\left\{2\dfrac{\left[\dfrac{\left(\frac{w}{2}+d+n\right)^2-\left(\frac{w}{2}\right)^2}{\left(\frac{w}{2}+d\right)^2-\left(\frac{w}{2}\right)^2}\right]^{\frac{1}{4}}+1}{\left[\dfrac{\left(\frac{w}{2}+d+n\right)^2-\left(\frac{w}{2}\right)^2}{\left(\frac{w}{2}+d\right)^2-\left(\frac{w}{2}\right)^2}\right]^{\frac{1}{4}}-1}\right\} & \text{if } Z_{oo} > 189\,\Omega,\ m=n,\ s=d \\[2em] \dfrac{296}{\ln\left\{2\left[\dfrac{\left(\frac{w}{2}+d+n\right)^2-\left(\frac{w}{2}\right)^2}{\left(\frac{w}{2}+d\right)^2-\left(\frac{w}{2}\right)^2}\right]^{\frac{1}{4}}\right\}} & \text{if } Z_{oo} < 189\,\Omega,\ m=n,\ s=d \end{cases}$$

With the odd mode, the currents in the two outer conductors of width m and n are of equal magnitude but in opposite directions. The center conductor of width w is the reference conductor. The even-mode impedance is

$$Z_{oe}=\begin{cases}30\ln\left\{2\dfrac{\left[\dfrac{\left(\frac{w}{2}+d+n\right)^2-\left(\frac{w}{2}\right)^2}{\left(\frac{w}{2}+d+n\right)^2}\dfrac{\left(\frac{w}{2}+d\right)^2}{\left(\frac{w}{2}+d\right)^2-\left(\frac{w}{2}\right)^2}\right]^{\frac{1}{4}}+1}{\left[\dfrac{\left(\frac{w}{2}+d+n\right)^2-\left(\frac{w}{2}\right)^2}{\left(\frac{w}{2}+d+n\right)^2}\dfrac{\left(\frac{w}{2}+d\right)^2}{\left(\frac{w}{2}+d\right)^2-\left(\frac{w}{2}\right)^2}\right]^{\frac{1}{4}}-1}\right\} & \text{if } Z_{oe}>47\ \Omega,\ m=n,\ s=d\\[2ex] \dfrac{74}{\ln\left\{2\left[\dfrac{\left(\frac{w}{2}+d+n\right)^2-\left(\frac{w}{2}\right)^2}{\left(\frac{w}{2}+d+n\right)^2}\dfrac{\left(\frac{w}{2}+d\right)^2}{\left(\frac{w}{2}+d\right)^2-\left(\frac{w}{2}\right)^2}\right]^{\frac{1}{4}}\right\}} & \text{if } Z_{oe}<47\ \Omega,\ m=n,\ s=d\end{cases}$$

$$Z_{oe}\approx\begin{cases}240\ln\left[2\left(\dfrac{w+s}{w+s+d}\dfrac{w+d}{w}\right)^{\frac{1}{4}}\right] & \text{if } Z_{oe}>377\ \Omega,\ m,n\gg\max(s,w,d)\\[2ex] \dfrac{592}{\ln\left[2\dfrac{\left(\frac{w+s}{w+s+d}\frac{w+d}{w}\right)^{\frac{1}{4}}+1}{\left(\frac{w+s}{w+s+d}\frac{w+d}{w}\right)^{\frac{1}{4}}-1}\right]} & \text{if } Z_{oe}<377\ \Omega,\ m,n\gg\max(s,w,d)\end{cases}$$

$$Z_{oe}\approx\begin{cases}120\ln\left(2\sqrt{1+\dfrac{2d}{w}}\right) & \text{if } Z_{oe}>189\ \Omega,\ m,n\gg\max(s,w,d),\ s=d\\[2ex] \dfrac{296}{\ln\left(2\dfrac{\sqrt{1+\frac{2d}{w}}+1}{\sqrt{1+\frac{2d}{w}}-1}\right)} & \text{if } Z_{oe}<189\ \Omega,\ m,n\gg\max(s,w,d),\ s=d\end{cases}$$

With the even mode, the current in the center conductor of width w returns through the two outer conductors of width m and n. The thickness of the three conductors is small compared to n, w, and m.

Three Coplanar Flat Conductors on a Substrate [Lo; Gupta, '81]

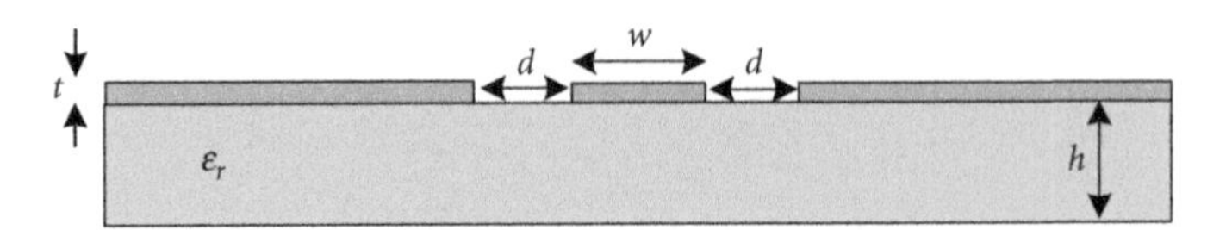

The even-mode impedance for these coplanar strips is

$$Z_{oe} = \begin{cases} \dfrac{296}{\sqrt{\varepsilon_{reff}}} \dfrac{1}{\ln\left(2\dfrac{1+\sqrt{k}}{1-\sqrt{k}}\right)} & \text{if } \dfrac{1}{\sqrt{2}} \le k \le 1 \\ \dfrac{30}{\sqrt{\varepsilon_{reff}}} \ln\left[2\dfrac{1+(1-k^2)^{\frac{1}{4}}}{1-(1-k^2)^{\frac{1}{4}}}\right] & \text{if } 0 \le k \le \dfrac{1}{\sqrt{2}} \end{cases}$$

where

$$k = \frac{w + \dfrac{1.25t}{\pi}\left[1+\ln\left(\dfrac{4\pi w}{t}\right)\right]}{w + 2d - \dfrac{1.25t}{\pi}\left[1+\ln\left(\dfrac{4\pi w}{t}\right)\right]}, \quad \varepsilon_{reff} = \varepsilon_m - \frac{0.7(\varepsilon_m - 1)\dfrac{t}{d}}{\varepsilon_n + 0.7\dfrac{t}{d}}$$

$$\varepsilon_m = \frac{\varepsilon_r + 1}{2}\left\{\tanh\left[0.775\ln\left(\frac{h}{d}\right) + 1.75\right] + \frac{d}{h}\left(\frac{w}{w+2d}\right)\left[0.04 - 0.7\frac{w}{w+2d} + 0.01(1-0.1\varepsilon_r)\left(0.25 + \frac{w}{w+2d}\right)\right]\right\}$$

$$\varepsilon_n = \begin{cases} 0.318\ln\left(2\dfrac{1+\sqrt{\dfrac{w}{w+2d}}}{1-\sqrt{\dfrac{w}{w+2d}}}\right) & \text{if } \dfrac{1}{\sqrt{2}} \le \dfrac{w}{w+2d} \le 1 \\ \dfrac{3.14}{\ln\left\{2\dfrac{1+\left[1-\left(\dfrac{w}{w+2d}\right)^2\right]^{\frac{1}{4}}}{1-\left[1-\left(\dfrac{w}{w+2d}\right)^2\right]^{\frac{1}{4}}}\right\}} & \text{if } 0 \le \dfrac{w}{w+2d} \le \dfrac{1}{\sqrt{2}} \end{cases}$$

With the even mode, the current in the center conductor of width w returns equally through the two outer conductors (wide in comparison to w and d).

Four Coplanar Flat Conductors [Hilberg]

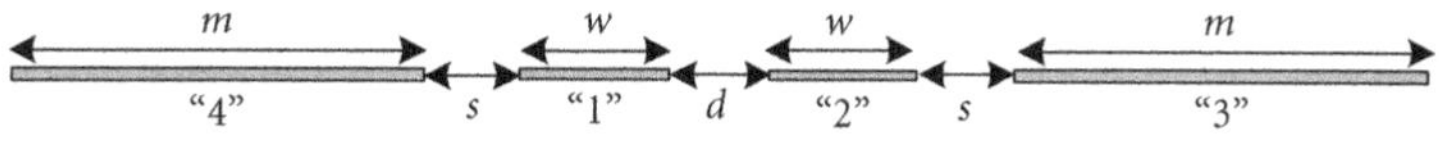

The odd-mode impedance is

$$
Z_{oo}=\begin{cases}
480\ln\left\{2\left[\dfrac{\left(\frac{d}{2}+w+s\right)^2-\left(\frac{d}{2}\right)^2}{\left(\frac{d}{2}+w+s\right)^2}\,\dfrac{\left(\frac{d}{2}+w\right)^2}{\left(\frac{d}{2}+w\right)^2-\left(\frac{d}{2}\right)^2}\right]^{\frac{1}{4}}\right\} & \text{if } Z_{oo}>754\,\Omega,\ m\gg\max(w,s,d)\\[2ex]
\dfrac{1{,}184}{\ln\left\{2\dfrac{\left[\dfrac{\left(\frac{d}{2}+w+s\right)^2-\left(\frac{d}{2}\right)^2}{\left(\frac{d}{2}+w+s\right)^2}\,\dfrac{\left(\frac{d}{2}+w\right)^2}{\left(\frac{d}{2}+w\right)^2-\left(\frac{d}{2}\right)^2}\right]^{\frac{1}{4}}+1}{\left[\dfrac{\left(\frac{d}{2}+w+s\right)^2-\left(\frac{d}{2}\right)^2}{\left(\frac{d}{2}+w+s\right)^2}\,\dfrac{\left(\frac{d}{2}+w\right)^2}{\left(\frac{d}{2}+w\right)^2-\left(\frac{d}{2}\right)^2}\right]^{\frac{1}{4}}-1}\right\}} & \text{if } Z_{oo}<754\,\Omega,\ m\gg\max(w,s,d)
\end{cases}
$$

With the odd mode, the currents in conductors "1" and "2" are of the same magnitude but in opposite directions. Conductors "3" and "4" are the reference conductors. The even-mode impedance is

$$Z_{oe} = \begin{cases} 120\ln\left\{2\left[\dfrac{\left(\dfrac{d}{2}+w+s\right)^2-\left(\dfrac{d}{2}\right)^2}{\left(\dfrac{d}{2}+w\right)^2-\left(\dfrac{d}{2}\right)^2}\right]^{\frac{1}{4}}\right\} & \text{if } Z_{oe} > 189\,\Omega,\ m \gg \max(w,s,d) \\ \dfrac{296}{\ln\left\{2\dfrac{\left[\dfrac{\left(\dfrac{d}{2}+w+s\right)^2-\left(\dfrac{d}{2}\right)^2}{\left(\dfrac{d}{2}+w\right)^2-\left(\dfrac{d}{2}\right)^2}\right]^{\frac{1}{4}}+1}{\left[\dfrac{\left(\dfrac{d}{2}+w+s\right)^2-\left(\dfrac{d}{2}\right)^2}{\left(\dfrac{d}{2}+w\right)^2-\left(\dfrac{d}{2}\right)^2}\right]^{\frac{1}{4}}-1}\right\}} & \text{if } Z_{oe} < 189\,\Omega,\ m \gg \max(w,s,d) \end{cases}$$

$$Z_{oe} \approx \begin{cases} 120\ln\left(2\sqrt{\dfrac{s}{w}}\right) & \text{if } Z_{oe} > 189\,\Omega,\ m \gg \max(w,s,d),\ s \gg \max(w,d),\ w \gg d \\ \dfrac{296}{\ln\left(2\dfrac{\sqrt{\dfrac{s}{w}+1}+1}{\sqrt{\dfrac{s}{w}+1}-1}\right)} & \text{if } Z_{oe} < 189\,\Omega,\ m \gg \max(w,s,d),\ s \gg \max(w,d),\ w \gg d \end{cases}$$

With the even mode, the currents in conductors "1" and "2" are of the same magnitude and direction and they return equally through conductors "3" and "4." The thickness of the four conductors is small compared to w and m.

3

Transmission Lines and Matching

When the length of a pair of conductors connecting a driver to a receiver is not electrically short, the conductors are considered a transmission line. Transmission lines have been and will continue to be extremely valuable "components" in an electrical system. They can generate fascinating, and to those nescient in this subject, unexpected results such as "ringing" and signal distortion.

3.1 Voltage Reflection and Transmission Coefficients

What are the voltage reflection and transmission coefficients for a signal traveling down a transmission line? When is the magnitude of the reflection coefficient approximately one and approximately zero?

Two important concepts when dealing with transmission lines are reflection and transmission. When a signal that is applied to the input of a transmission line appears at the output or load of the line, a fraction of the signal is reflected off the load and a fraction is transmitted into the load. The reflection is analogous to the reflection of sound waves off a wall. The degree of reflection and transmission is a function of the load impedance, Z_L, and the line's characteristic impedance, Z_o. The characteristic impedance is a function of the resistance, inductance, capacitance, and conductance of the line, as well as the operating frequency. The voltage reflection coefficient is defined as

$$\rho = \frac{Z_L - Z_o}{Z_L + Z_o} \tag{3.1}$$

while the voltage transmission coefficient is defined as

$$\mathrm{T} = 1 + \rho = \frac{2Z_L}{Z_L + Z_o} \tag{3.2}$$

As an introduction to the use of these coefficients, assume that both the characteristic impedance of a line and its load are purely resistive (so that ρ and T are real). Referring to Figure 3.1, if a transient signal with a voltage amplitude of A travels down this transmission line, the voltage reflected off the real load is

$$A\rho \tag{3.3}$$

The voltage transmitted into (or beyond) the load is

$$A\mathrm{T} \tag{3.4}$$

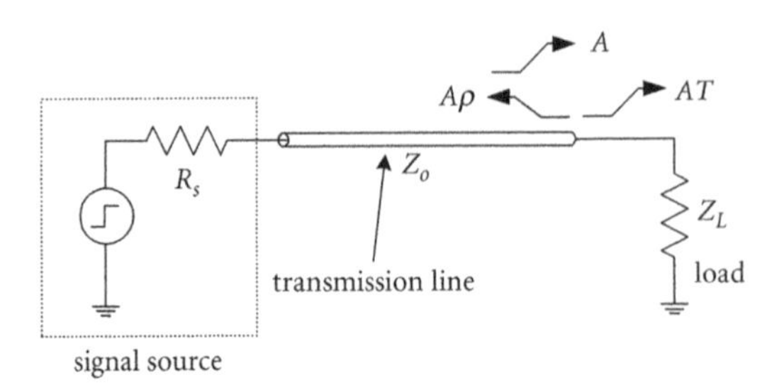

FIGURE 3.1 Incident, transmitted, and reflected voltage signals.

When the load impedance is much different from the characteristic impedance of the line, the magnitude of the reflection coefficient is nearly one: the load is not matched to the line, and a "mismatch" exists. Most of the electrical energy is reflected off the load when the reflection coefficient is nearly one. However, when the load impedance is about equal to the characteristic impedance of the line, the system is closely "matched," and the reflection coefficient is nearly zero.[1] Most of the electrical energy is transmitted into the load when the reflection coefficient is nearly zero.

3.2 Impedance Mismatch

The impedance of a TV antenna is 300 Ω. Instead of using 300 Ω twin-lead line to connect the antenna to the 300 Ω input of the TV, 75 Ω coax is used (fed into the 75 Ω input of the TV). Is there a mismatch on the coaxial line?

In this case, the source of energy is the receiving antenna. The 300 Ω resistor and the voltage source in Figure 3.2 represent the antenna. The load is the 75 Ω input of the TV. The 75 Ω coax is matched to the TV input, so there is no impedance mismatch for the transmission line. The reflection coefficient, ρ, is a function of the *load* impedance and the characteristic impedance of the transmission line.

It is tempting to discuss the "mismatch" at the source between the 300 Ω resistor and the 75 Ω input impedance of the coax or the violation of the maximum power transfer theorem (i.e., a lack of a conjugate match between the antenna and the transmission line).[2] In this particular situation, transferring the largest possible voltage to the TV load is the major goal.

The purpose of a receiving antenna is to collect electromagnetic fields and convert them to power that is delivered to the receiver. The antenna's 300 Ω impedance may be a pure radiation resistance, at a particular frequency near resonance. However, as the frequency deviates from this resonant point, the impedance will vary in both magnitude and phase; that is, the impedance of the antenna will contain both a real and reactive part. Furthermore, the antenna's source strength will also vary with frequency. When selecting an antenna for a particular application, factors such as the radiation pattern, gain, and

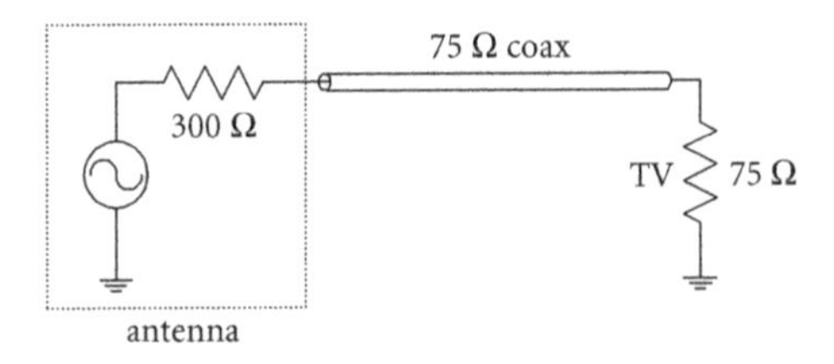

FIGURE 3.2 Antenna connected to a TV through a 75 Ω coaxial transmission line.

[1]Matched, unless stated otherwise, does not indicate a conjugate match in this book.

[2]As with most (all?) real problems, the maximum power transfer theorem has little practical use.

bandwidth are usually of primary importance, while the impedance of the antenna is of secondary importance (if it is not too extreme).

If a transient were induced on the line, such as from a lightning discharge, the mismatch between the antenna and coax would result in a reflection from the antenna when the transient reached the antenna. However, this reflected signal would be absorbed when it reached the properly matched TV/coax interface. (Note that the input impedance of the TV is not necessarily constant at 75 Ω for all frequencies.)

3.3 VSWR and SWR

What is the VSWR? What is the SWR? Are they a function of the source impedance? Explain. [Chipman]

The VSWR is the voltage standing wave ratio. The SWR is the standing wave ratio. The VSWR and SWR are commonly measured quantities that provide an indication of the quality of the match between a transmission line and its load for sinusoidal steady-state signals. (For transient signals, there would not be a *standing* wave on the line.) Surprisingly, the VSWR and SWR are not a function of the source impedance but only the load impedance and line characteristic impedance.[3] The VSWR and SWR are defined in terms of the maximum and minimum amplitudes of the voltage along the line:

$$\text{VSWR} = \frac{|V|_{max}}{|V|_{min}} = \frac{1+|\rho|}{1-|\rho|} = \frac{1+\left|\dfrac{Z_L - Z_o}{Z_L + Z_o}\right|}{1-\left|\dfrac{Z_L - Z_o}{Z_L + Z_o}\right|} \tag{3.5}$$

$$\text{SWR} = 20\log(\text{VSWR})\,\text{dB} \tag{3.6}$$

where $|\rho|$ is the magnitude of the reflection coefficient at the load, Z_L is the load impedance, and Z_o is the characteristic impedance of the line. (In practice, there is sometimes no distinction made between the VSWR and SWR.) The VSWR is equal to the ratio of the maximum magnitude of the voltage along the line to the minimum magnitude of the voltage along the line. As expected, when the line is matched, the magnitude of the voltage along the line is constant, $\rho = 0$, and VSWR = 1. For all mismatch conditions, the VSWR > 1. For a short-circuited or an open-circuited load, the VSWR is very large (theoretically infinite).

The equation for the VSWR involves the magnitudes of the maximum and minimum voltages along a transmission line. In theory, if a thin slot is cut along the length of the outer conductor of a coaxial cable, a probe could be used to measure the rms (root mean square) voltage along the coax. The ratio of the maximum rms voltage to the minimum rms voltage measured along the line would be the VSWR. (In practice, the VSWR for most lines are not measured in this manner.) Time-average measurements instead of rms measurements would not work since the traveling waves along the line are oscillating with time: the time-average value along the line would be the same everywhere. For example, the time-average value of a single traveling wave with zero dc offset is zero everywhere.

The voltage along a specific transmission line, for various times as the reflection coefficient is varied, is shown in Mathcad 3.1. (The line is one meter in length, which is equal to one wavelength.) In addition, the rms value of the signal along the line is shown. Notice that the rms voltage is constant along the line when the reflection coefficient is zero, corresponding to a matched load. When the reflection coefficient is one, corresponding to an open-circuited load, the rms voltage varies along the line as expected and is a maximum at the load. The VSWR, the ratio of the maximum rms voltage to the minimum rms voltage, is infinite since the minimum voltage is zero. For a reflection coefficient of 0.5, the VSWR should be 3:

$$\text{VSWR} = \frac{1+0.5}{1-0.5} = 3$$

[3] For lines where the losses cannot be neglected, the VSWR and SWR are also a function of other parameters such as the position along the line.

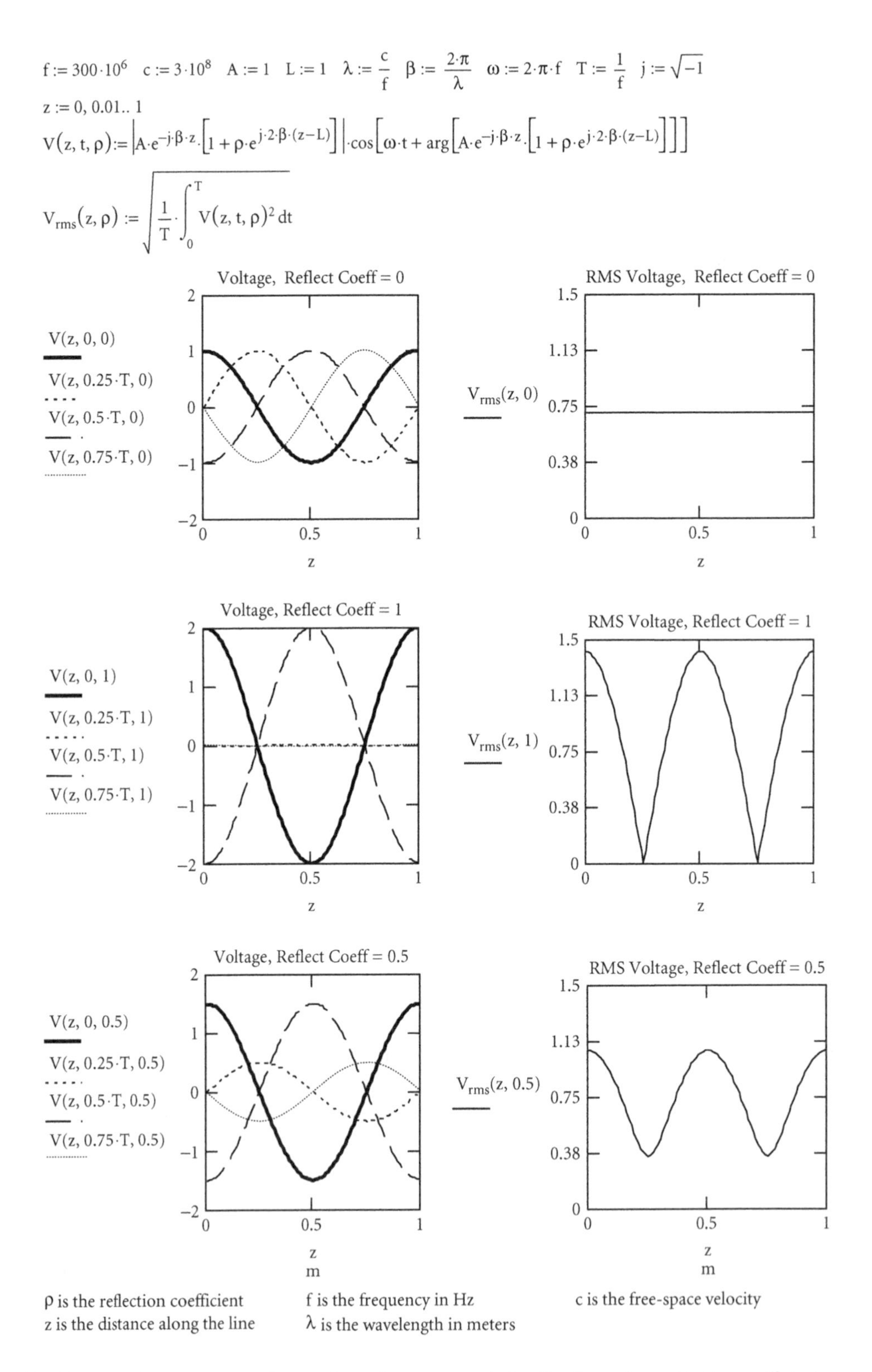

MATHCAD 3.1 Instantaneous and rms voltage along a transmission line for various reflection coefficients.

Using data from the rms plot,

$$\text{VSWR} = \frac{|V|_{max}}{|V|_{min}} \approx \frac{1.1}{0.37} \approx 3$$

Although not specifically shown, the time-average value of the signal is zero at any position z along the line for any reflection coefficient.

In (3.5), $|\rho|$ is the magnitude of the reflection coefficient not its absolute value. It is important to realize this since both the characteristic impedance and load impedance can be complex. For example, if $Z_o = 50\ \Omega$ and $Z_L = 52 - j11\ \Omega$, then

$$\text{VSWR} = \frac{1+|\rho|}{1-|\rho|} = \frac{1+\left|\dfrac{Z_L - Z_o}{Z_L + Z_o}\right|}{1-\left|\dfrac{Z_L - Z_o}{Z_L + Z_o}\right|} = \frac{1+\left|\dfrac{52-j11-50}{52-j11+50}\right|}{1-\left|\dfrac{52-j11-50}{52-j11+50}\right|} \approx \frac{1+|0.11\angle -74^\circ|}{1-|0.11\angle -74^\circ|} \approx \frac{1+0.11}{1-0.11} \approx 1.2$$

When both the load and line impedances are real, the VSWR equation reduces to the ratio of the larger to the smaller impedance:

$$\text{VSWR} = \begin{cases} \dfrac{Z_L}{Z_o} & \text{if } Z_L > Z_o \\ \dfrac{Z_o}{Z_L} & \text{if } Z_o > Z_L \end{cases} \tag{3.7}$$

Again, both the line and load impedances must be purely resistive. As an example, if the load is 300 Ω and the line is 75 Ω, then VSWR = 300/75 = 4. For many applications, a VSWR < 1.5 (or 2) is considered good. It may be, however, too costly to obtain a VSWR of less than 1.5 over the operating frequency range. In some cases, the improvement in performance in designing a system with such a low VSWR may not be noticeable.

The magnitude of the reflection coefficient can be greater than one. This is easily shown. As will be derived later, the characteristic impedance of a transmission line at low frequencies is not entirely real with a value of $\sqrt{L/C}$ but complex:

$$Z_o \approx \sqrt{\frac{R}{2\omega C}} - j\sqrt{\frac{R}{2\omega C}}$$

If the load is inductive with an impedance of

$$Z_L = -j\sqrt{2}\,\text{Im}(Z_o) = j\sqrt{2}\sqrt{\frac{R}{2\omega C}} \tag{3.8}$$

then

$$|\rho| = \left|\frac{Z_L - Z_o}{Z_L + Z_o}\right| = \left|\frac{j\sqrt{2}\sqrt{\dfrac{R}{2\omega C}} - \sqrt{\dfrac{R}{2\omega C}} + j\sqrt{\dfrac{R}{2\omega C}}}{j\sqrt{2}\sqrt{\dfrac{R}{2\omega C}} + \sqrt{\dfrac{R}{2\omega C}} - j\sqrt{\dfrac{R}{2\omega C}}}\right| = \left|\frac{-1+j(\sqrt{2}+1)}{1+j(\sqrt{2}-1)}\right| \approx 2.41 \tag{3.9}$$

It can also be shown that 2.41 is the maximum possible value for the magnitude of the reflection coefficient (for passive loads). At high frequencies, on the other hand, the characteristic impedance has an inductive component:

$$Z_o \approx \sqrt{\frac{\omega L}{2G}} + j\sqrt{\frac{\omega L}{2G}}$$

If the load is capacitive with a value of

$$Z_L \approx -j\sqrt{2}\sqrt{\frac{\omega L}{2G}}$$

then the magnitude of the reflection coefficient is again about 2.41.

The source impedance is a factor in determining the voltage delivered across the input of the line and the current into the line but not the VSWR for sinusoidal steady-state problems. For transient input signals, such as a pulsed voltage waveform, both the source impedance and load impedance determine the size of the reflections along the line. A reflection coefficient is determined at both the source and load for transient problems. The VSWR for a transient situation would require a special definition since the maximum and minimum voltages are not well defined.

For antennas near their resonant frequency, the VSWR can also be expressed in terms of the measured Q of the antenna:

$$\mathrm{VSWR} = \frac{1 + \dfrac{Qn}{\sqrt{(Qn)^2 + 1}}}{1 - \dfrac{Qn}{\sqrt{(Qn)^2 + 1}}} \tag{3.10}$$

where n is the percent difference, expressed as a decimal (e.g., 0.05 corresponding to 5%) between the antenna's actual resonant frequency and the frequency where the Q is measured. Equation (3.10) assumes that the antenna is matched to the transmission line at the resonant frequency of the antenna and that the resistive part of the antenna's impedance does not change much near resonance. (At the resonant frequency, the antenna's impedance is purely resistive.) The variable n should be less than 0.05. As seen in Mathcad 3.2, as the measured Q of the antenna increases, the VSWR at the measurement frequency increases. The Q (not at resonance) is the ratio of the absolute value of the reactance to the resistance of the impedance of the antenna. A larger Q implies a larger reactance and, hence, a greater mismatch from the transmission line's resistive characteristic impedance.

Two other standing wave ratio definitions are occasionally seen. The current standing wave ratio is defined as the ratio of the maximum to the minimum amplitude of the current along the line

$$\mathrm{ISWR} = \frac{|I|_{max}}{|I|_{min}} \tag{3.11}$$

which is also equal to the VSWR. It is usually much more difficult to measure the maximum and minimum value of the current in the conductors than the maximum and minimum values of the voltage across the two conductors. The power standing wave ratio is defined as the square of the VSWR:

$$\mathrm{PSWR} = \frac{|V|_{max}^2}{|V|_{min}^2} \tag{3.12}$$

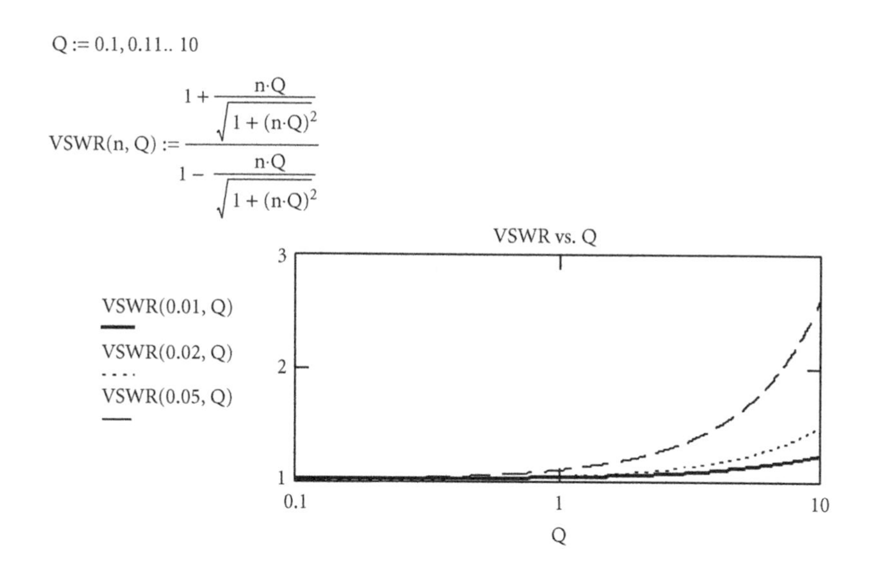

MATHCAD 3.2 Relationship between the measured Q of the antenna and VSWR near resonance.

3.4 The Cost of a VSWR > 1

The impedance of a transmitting antenna is 300 Ω. Instead of using 300 Ω twin-lead to connect the transmitter to the antenna, 75 Ω coax is used. What are the major consequences of this mismatch? [ARRL, '97; Johnson, '50; Moore, '60]

In this case, the transmitter is the source, which is represented by the 5 Ω resistor and sinusoidal signal source as shown in Figure 3.3. Since the load (the 300 Ω antenna) is not properly matched to the coax, there is a mismatch. A VSWR > 1 will exist on the line, which will increase the losses on the transmission line and decrease the power efficiency of the system; however, except at very high frequencies, for very long lines, or at high power levels, these losses are usually negligible. For example, if the coax is one type of RG59, the line loss is 3.4 dB per 100 ft at 144 MHz for VSWR = 1, a perfectly matched system. The VSWR in this example is, however,

$$\text{VSWR} = \frac{1+|\rho|}{1-|\rho|} = \frac{1+\left|\dfrac{Z_L - Z_o}{Z_L + Z_o}\right|}{1-\left|\dfrac{Z_L - Z_o}{Z_L + Z_o}\right|} = \frac{1+\left|\dfrac{300-75}{300+75}\right|}{1-\left|\dfrac{300-75}{300+75}\right|} = 4$$

The *additional* loss due to the mismatch is approximately 1.4 dB per 100 ft of cable. The line loss is greater with an increase in VSWR because the maximum rms value of the current and voltage increase

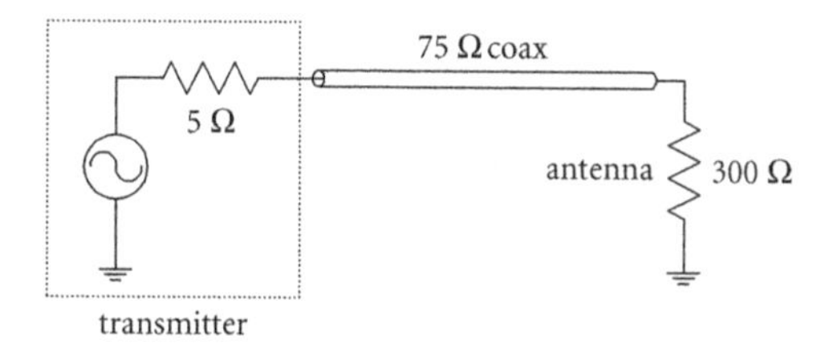

FIGURE 3.3 Transmitter connected to an antenna through a 75 Ω coaxial transmission line.

with VSWR. The ohmic losses in the conductors, I^2R, increase with current, and the dielectric losses in the insulating material between the conductors, V^2G, increase with voltage. If the high VSWR is acceptable, then a low-loss line capable of handling high voltages should be used (e.g., open-air twin-lead line or Heliax coaxial cable).

Probably, the most important consequence of a mismatch is the variation in the load impedance seen by the transmitter. For sinusoidal steady-state signals, the input impedance into an unmatched transmission line is a function of the line length, characteristic impedance of the line, load impedance, and frequency. For a transmitter to deliver effectively its power, the impedance at its output must be within some limited range. The transmitter is usually designed to deliver its optimum power to some specific load. When that load varies, the transmitter's ability to deliver its power decreases. For a lossless line, the maximum and minimum *possible* magnitudes of the input impedance are directly related to the VSWR:

$$\left|Z_{in}\right|_{max} = \frac{V_{max}}{I_{min}} = (\text{VSWR})Z_o, \quad \left|Z_{in}\right|_{min} = \frac{V_{min}}{I_{max}} = \frac{Z_o}{\text{VSWR}} \tag{3.13}$$

However, a large VSWR on a line does not imply that the impedance seen by the transmitter is extreme. The input impedance of a line is, as was previously stated, a function of several parameters including the line length. The VSWR on the line could be high and, yet, the impedance seen by the transmitter could be acceptable over some range of frequencies.

Finally, a large VSWR will increase the radiation of unbalanced transmission lines such as coax since the current levels are greater. This will affect the overall radiation pattern of a balanced antenna system.

3.5 Distinguishing between the Load and Source

A matching network for an HF transceiver is adjusted until the level of the received signal level is maximized. Is this match appropriate for transmitting as well? Explain.

When a system is tuned or adjusted for maximum receiver level, the antenna is the source of energy and the receiver is the load. If the transmission line is not properly matched to the receiver, a mismatch will occur. If the matching network is near the receiver, it can be adjusted until the receiver level is maximized, which may be a $\text{VSWR} = 1$ condition. However, when the system is adjusted for maximum transmission efficiency, the load is the antenna. In this case, for a low VSWR on the line, the antenna should be matched to the transmission line. Note that a low VSWR is not always the most important consideration: the ability of the transmitter to deliver its power into the line must always be considered. It is possible that the transmitter portion of the transceiver is properly matched to the antenna for maximum power efficiency, yet the receiver portion is not optimized for maximum receiving signal level. Being a bidirectional system, there are two loads.

It is a common misconception that changing the length of a line can change the VSWR on a line. If the transmission line only consists of a single, reasonable low-loss line (not multiple lines of different characteristic impedances connected in series), the VSWR is only a function of the load and the line's characteristic impedance. The length of the line is not a factor for reasonable-loss lines.[4] However, changing the length of the line will change the impedance seen by the transmitter, which will affect the loading on the transmitter.

It is also a common misconception that a matching network near a transmitter changes the VSWR on the line: a matching network located near the transmitter has no effect on the VSWR on the line when transmitting. However, a matching network located at the load (i.e., the antenna), would affect the VSWR. Typically, the matching network, often referred to as an antenna tuner, is located at the transmitter

[4]When the line is not low loss, for example, due to excessive line length or lossy conductors, the VSWR changes along the length of the line.

(or transceiver) and is adjusted so that the transmitter portion of the transceiver "sees" an appropriate impedance.

3.6 Transient and Steady-State Input Impedance

The characteristic impedance, Z_o, of a transmission line on a printed circuit board is 40 Ω. If an amplifier at the input of the line is sourcing 0.02 mA, when is the voltage across the input of the transmission line simply $40 \times 0.02 = 0.8$ V?

For budding engineers, it is tempting to model a transmission line with a characteristic impedance of 40 Ω as merely a 40 Ω resistor. For a few situations, the input impedance of the line is purely resistive.

During some transient conditions, such as the transient rise of a digital gate from a Low-to-High state, the momentary impedance seen by the amplifier is Z_o. The characteristic impedance of a line represents the ratio of a forward-traveling voltage waveform to its corresponding forward-traveling current waveform. Under these transient conditions, the transient voltage across the line is $40 \times 0.02 = 0.8$ V. If the load on the transmission line is not equal to Z_o, a reflection will occur at the load. When this reflection arrives at the source, the total voltage across the input of the transmission line is not necessarily 0.8 V.

When the input signal is sinusoidal and the system is at steady state (i.e., all transient voltages and currents have "died off"),[5] then for some special conditions the impedance seen by the amplifier is Z_o. For sinusoidal steady-state conditions, the input impedance of a lossless transmission line is

$$Z_{in} = Z_o \frac{Z_L + jZ_o \tan(\beta l_{th})}{Z_o + jZ_L \tan(\beta l_{th})} \tag{3.14}$$

where l_{th} is the distance to the load, Z_o is the characteristic impedance of the line, Z_L is the load impedance, and $\beta = 2\pi/\lambda$ is the phase constant. Therefore, the impedance seen by the amplifier for sinusoidal steady-state conditions is not generally Z_o. The input impedance of the line is equal to Z_o if $Z_L = Z_o$, corresponding to a matched system:

$$Z_L + jZ_o \tan(\beta l_{th}) = Z_o + jZ_L \tan(\beta l_{th}) \Rightarrow Z_o = Z_L \tag{3.15}$$

When the load impedance is perfectly matched to the transmission line, the impedance seen by the amplifier is $Z_L = Z_o$, and the voltage across the input of the line is $40 \times 0.02 = 0.8$ V.

For dc steady-state conditions, the lossless transmission line is essentially two ideal wires (the series inductive reactance is zero and the shunt capacitive reactance is infinite) and the input impedance is Z_L. If $Z_L = Z_o$, then the voltage across the input of the transmission line is also 0.8 V.

The value of the transient and steady-state impedance looking into various locations of a complex transmission line network (with various resistive loads and interconnecting resistive networks) is given in Table 3.1 for the system shown in Figure 3.4. These impedances determine the reflection and transmission coefficients for transient signals, and the steady-state impedance seen by the source determines the steady-state loading on the source. It is assumed that the three lossless transmission lines of characteristic impedance 65, 50, and 80 Ω are not electrically short, and the resistive circuits are all electrically short. (The transmission lines are shown as two parallel lines.) The lengths of these three transmission lines are not necessarily the same, and their lengths do not affect the given impedances, transient reflection coefficients, and transient transmission coefficients. However, as will be seen in later examples, these lengths will affect the overall transient response. (For sinusoidal steady-state signals, these transmission line lengths do affect the impedances.) The numbers given for the resistances and transmission line characteristic impedances do not necessarily represent real or typical values. They were chosen for numerical convenience and ease of distinction.

[5]After a transient voltage or current has "died off," the resultant waveform is either constant or periodic with time.

TABLE 3.1 Transient Impedance, Steady-State Impedance, Reflection Coefficient, and Transmission Coefficient for the "Circuit" in Figure 3.4

@	$Z_{transient}$ (Ω)	$Z_{steadystate}$ (Ω)	ρ	T
1	$2+(30\|65)$	$2+\left[30\|\left(4+\left\{90\|\left[5+(99\|93)\right]\right\}\right)\right]$	—	—
2	65	$4+\left\{90\|\left[5+(99\|93)\right]\right\}$	—	—
3	$30\|2$	$30\|2$	$\frac{Z_{transient}-65}{Z_{transient}+65}$	$\frac{2Z_{transient}}{Z_{transient}+65}$
4	$4+\left\{90\|\left[5+(50\|80)\right]\right\}$	$4+\left\{90\|\left[5+(99\|93)\right]\right\}$	$\frac{Z_{transient}-65}{Z_{transient}+65}$	$\frac{2Z_{transient}}{Z_{transient}+65}$
5	65	$30\|2$	—	—
6	$50\|80$	$93\|99$	—	—
7	$50\|\left\{5+\left[90\|(4+65)\right]\right\}$	$93\|\left(5+\left\{90\|\left[4+(2\|30)\right]\right\}\right)$	$\frac{Z_{transient}-80}{Z_{transient}+80}$	$\frac{2Z_{transient}}{Z_{transient}+80}$
8	$80\|\left\{5+\left[90\|(4+65)\right]\right\}$	$99\|\left(5+\left\{90\|\left[4+(2\|30)\right]\right\}\right)$	$\frac{Z_{transient}-50}{Z_{transient}+50}$	$\frac{2Z_{transient}}{Z_{transient}+50}$
9	93	93	$\frac{93-50}{93+50}$	$\frac{2(93)}{93+50}$
10	50	$99\|\left(5+\left\{90\|\left[4+(2\|30)\right]\right\}\right)$	—	—
11	99	99	$\frac{99-80}{99+80}$	$\frac{2(99)}{99+80}$
12	80	$93\|\left(5+\left\{90\|\left[4+(2\|30)\right]\right\}\right)$	—	—

When the transient impedances are determined at the various locations, only the discrete resistors and characteristic impedance of the transmission lines connected *directly* to the resistors are seen at these locations. The transmission line characteristic impedances and resistive loads connected "down line" do not affect the transient input impedance. When determining the dc steady-state impedance, the lossless transmission lines can be replaced by perfectly conducting wires, and the circuit analyzed in the standard elementary circuit analysis manner. (The impedance of the voltage source is assumed to be zero.) Therefore, the dc steady-state impedances are affected by the discrete resistances but not the transmission line characteristic impedances.

The voltage reflection and transmission coefficients listed in Table 3.1 are for transient signals only (not sinusoidal steady-state signals). They are determined from the two relationships

$$\rho=\frac{R_L-Z_o}{R_L+Z_o},\quad \mathrm{T}=\frac{2R_L}{R_L+Z_o} \tag{3.16}$$

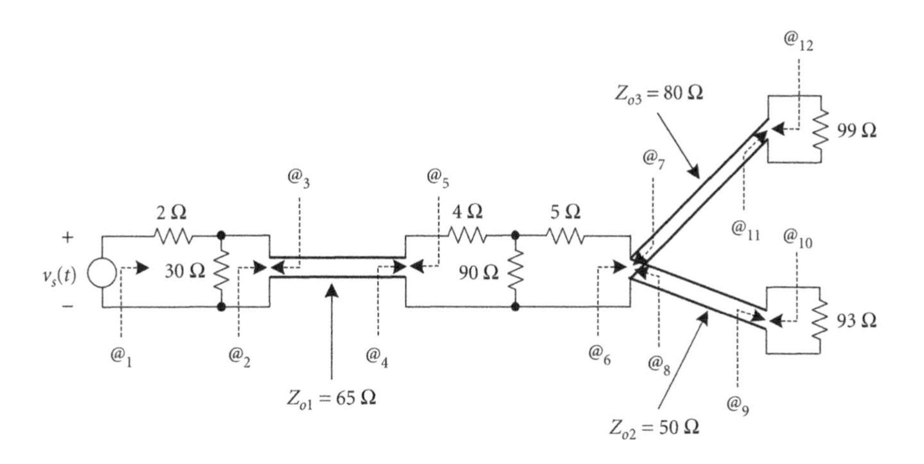

FIGURE 3.4 System consisting of both lumped circuits and distributed circuits (i.e., transmission lines).

where R_L is the load resistance seen by the transmission line for a transient signal traveling in a particular direction and Z_o is the characteristic impedance of the line on which the signal is traveling. The reflection and transmission coefficients are usually defined only for a signal on a transmission line. In Table 3.1, as throughout this book, two vertical lines represent the parallel combination of two elements:

$$R_x \| R_y = \frac{R_x R_y}{R_x + R_y}$$

The results are left unsimplified to show clearly their origin.

3.7 Transient Reflections

Referring to Figure 3.5, the source impedance of a signal source is R_s, and the load impedance of a receiver is R_L. A transmission line of characteristic impedance Z_o connects the source to the receiver. The magnitude of the transmission line propagation function is equal to H (which includes signal attenuation). If the input signal source is a step voltage of amplitude V_o, determine the voltage at the receiver after the first reflection and after the second reflection. By examining the pattern of these reflections, generate a series summation for the voltage level after N reflections. By allowing N to approach infinity, show that the voltage at the receiver after a long time (the steady-state value) is equal to

$$V_\infty = \frac{V_o \dfrac{Z_o}{R_s + Z_o} H(1+\rho_L)}{1 - H^2 \rho_L \rho_s}$$

where ρ_L and ρ_s are the reflection coefficients at the load and source, respectively.

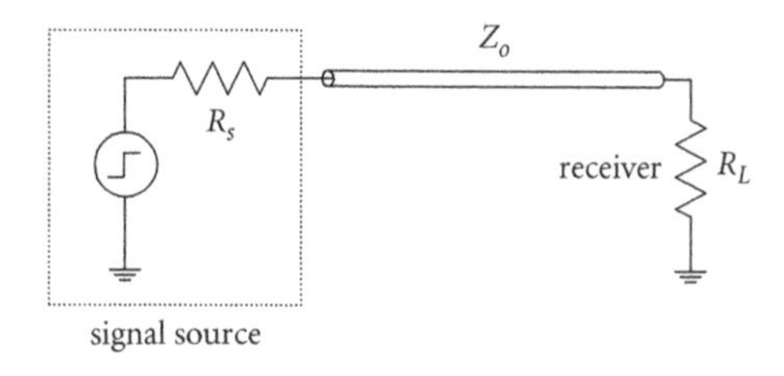

FIGURE 3.5 Source connected to a load via a transmission line.

The total voltage at the load immediately after the first reflection is equal to the sum of the incoming signal and the fraction of this incoming signal reflected off the load (the receiver). The impedance seen by the initial transient signal is equal to Z_o. The load impedance has no affect on this *initial* input impedance: the wave has not reflected off the load and returned to the source to convey any load information. Since the source resistance is equal to R_s, the initial voltage across the input of the line is (using voltage division)

$$V_o \frac{Z_o}{Z_o + R_s} \tag{3.17}$$

The strength of the first incoming signal at the load is obtained by multiplying (3.17) by H, the magnitude of the line propagation function. The reflection coefficient at the load,

$$\rho_L = \frac{R_L - Z_o}{R_L + Z_o} \tag{3.18}$$

multiplied by the initial incoming signal, is the magnitude of the signal reflected off the load:

$$V_o \frac{Z_o}{Z_o + R_s} H \rho_L \tag{3.19}$$

As expected, when $R_L = Z_o$ the reflected signal is zero since ρ_L would be zero. The total amplitude of the signal, at the load, immediately after the first reflection is the sum of the incoming signal and reflected signal:

$$V_o \frac{Z_o}{Z_o + R_s} H + V_o \frac{Z_o}{Z_o + R_s} H \rho_L \tag{3.20}$$

The transmission coefficient may also be used. In this case, the total signal at the load immediately after the first reflection is equal to just the incoming signal multiplied by the transmission coefficient:

$$V_o \frac{Z_o}{Z_o + R_s} HT \quad \text{where } T = \frac{2Z_L}{Z_L + Z_o} \tag{3.21}$$

The *reflected* signal eventually arrives at the input of the transmission line at the source. The reflection coefficient at the source for a signal traveling *toward* the source is (the "load" in this case is the source resistance)

$$\rho_s = \frac{R_s - Z_o}{R_s + Z_o} \tag{3.22}$$

and the amplitude of the reflected signal off the source is

$$V_o \frac{Z_o}{Z_o + R_s} H^2 \rho_L \rho_s \tag{3.23}$$

This represents the signal that is reflected off the source that was previously reflected off the load. The second incoming signal at the load is

$$V_o \frac{Z_o}{Z_o + R_s} H^3 \rho_L \rho_s \tag{3.24}$$

and the magnitude of this reflected signal is

$$V_o \frac{Z_o}{Z_o + R_s} H^3 \rho_L^2 \rho_s \tag{3.25}$$

Therefore, the total signal at the receiver immediately after this second reflection ($N = 2$) is the summation of all four signals:

$$\begin{aligned} &V_o \frac{Z_o}{Z_o + R_s} H + V_o \frac{Z_o}{Z_o + R_s} H\rho_L + V_o \frac{Z_o}{Z_o + R_s} H^3 \rho_L \rho_s + V_o \frac{Z_o}{Z_o + R_s} H^3 \rho_L^2 \rho_s \\ &= V_o \frac{Z_o}{Z_o + R_s} H(1 + \rho_L) + V_o \frac{Z_o}{Z_o + R_s} H^3 \rho_L \rho_s (1 + \rho_L) \end{aligned} \tag{3.26}$$

The grouping of the terms allows the total signal at the receiver after N reflections to be obtained by pattern recognition. (For greater confidence, the $N = 3$ result should be determined.) The series representation for the total signal at the load after N load reflections is therefore

$$V_o \frac{Z_o}{Z_o + R_s} H(1 + \rho_L) \sum_{n=1}^{N} H^{2(n-1)} \rho_L^{n-1} \rho_s^{n-1} \tag{3.27}$$

This equation for the voltage at the load after N reflections can be applied to those situations that involve a single transmission line.

To obtain a compact form for this result for steady-state conditions, the geometric series (for $-1 < r < 1$)

$$s_N = a + ar + ar^2 + \cdots + ar^{N-1} = a \sum_{n=1}^{N} r^{n-1} = \frac{a(1 - r^N)}{1 - r}$$

when the number of terms, N, approaches infinity

$$\lim_{N \to \infty} s_N = \frac{a}{1 - r}$$

can be used. In this case

$$r = H^2 \rho_L \rho_s$$

and

$$\lim_{N \to \infty} s_N = \frac{V_o \dfrac{Z_o}{Z_o + R_s} H(1 + \rho_L)}{1 - H^2 \rho_L \rho_s} \tag{3.28}$$

3.8 Matching at the Receiver and its Cost

Determine the steady-state voltage level at the receiver if the load impedance is set equal to the characteristic impedance, Z_o. Since in practice the receiver impedance cannot usually be set equal to Z_o, how can the total load impedance be matched to the transmission line? Will this method of matching eliminate all reflections on the line? What is the cost of implementing this matching scheme? [Johnson, '93; Montrose, '99]

If the impedance of the receiver (the total load) is set equal to the characteristic impedance of the line, the reflection coefficient at the load is zero:

$$\rho_L = \frac{R_L - Z_o}{R_L + Z_o} = \frac{Z_o - Z_o}{Z_o + Z_o} = 0$$

The steady-state voltage across the receiver is then

$$V_\infty = \frac{V_o \frac{Z_o}{R_s + Z_o} H(1+0)}{1 - H^2 0 \rho_s} = V_o \frac{Z_o}{R_s + Z_o} H \tag{3.29}$$

This voltage is the expected result when the transmission line is matched to its load. The impedance seen by the source for dc steady-state conditions is the load impedance, which is equal to the line's characteristic impedance, Z_o. The voltage across the input of the line is obtained through voltage division.

One method of matching the receiver to the transmission line, so that the reflection coefficient is zero, is to place an appropriate resistor in shunt with the receiver as shown in Figure 3.6. One end of the resistor is connected to the receiver input and the other end is usually connected to the return or reference. The combined parallel impedance of the load and this shunt matching resistor, R_M, should be set equal to Z_o. (This assumes that $R_L > Z_o$.) Ideally, this method should eliminate reflections from the load. However, if the load is not perfectly matched or a signal is induced on the line, this signal traveling toward the source will reflect off the source unless $R_s = Z_o$, where R_s is the resistance of the source (not shown in Figure 3.6). Eventually, all reflections will dissipate.

Assuming the resistive losses of the transmission line are negligible, the steady-state impedance seen by the source for this matching scheme is

$$\frac{R_M R_L}{R_M + R_L} \tag{3.30}$$

Typically, the load impedance, R_L, is much greater than the matching resistor, R_M:

$$\frac{R_M R_L}{R_M + R_L} \approx R_M \quad \text{if } R_M \ll R_L \tag{3.31}$$

Thus, the impedance seen by the source after a long time is much less than without matching. (Without matching, the steady-state impedance seen by the source is R_L.) This increased loading results in greater currents and I^2R line losses. The source may not be capable of delivering this current. This increased current level adds to the power consumption of the circuit, which is probably an important consideration for battery powered systems. If the current level is too great for the driver, the noise margin will also be

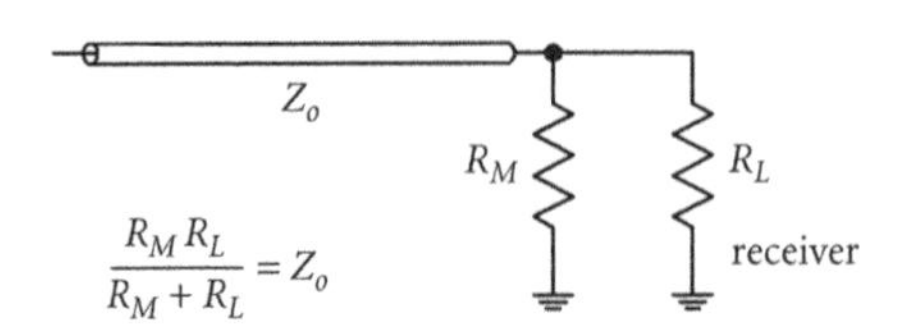

FIGURE 3.6 Matching resistor, R_M, connected across the receiver.

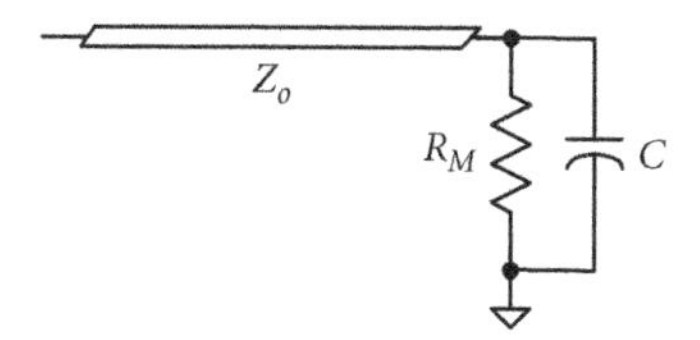

FIGURE 3.7 Modeling the input capacitance of the receiver. The resistance of the receiver is assumed much greater than the matching resistance.

affected.[6] Also, the additional matching resistor adds to the component list and increases the component population density on a board.

This parallel termination does sharpen the response at the receiver (but reduces its amplitude). This can be shown by first including the input capacitance of the receiver as shown in Figure 3.7. The time constant for this *RC* load is given by

$$\tau = \left(Z_o \| R_M\right) C = \frac{Z_o R_M}{Z_o + R_M} C = \frac{Z_o}{2} C \quad \text{if } R_M = Z_o \tag{3.32}$$

where the equivalent resistance seen by the capacitor is the parallel combination of R_M and the transient input impedance of the transmission line. Before the addition of R_M, the time constant is Z_oC (neglecting the input *resistance* of the receiver), which is twice the shunt matching value. The 10–90% rise time for an *RC* circuit to an ideal step function input signal is given by (in seconds)

$$t_{90\%} - t_{10\%} \approx 2.2\, R_{eq} C_{eq} \tag{3.33}$$

The rise time with the shunt resistor is one-half of the value without the shunt resistor:

$$t_{90\%} - t_{10\%} \approx 2.2 \left(\frac{Z_o}{2}\right) C = 1.1 Z_o C \tag{3.34}$$

For this *RC* type circuit, as the rise time decreases, the speed of the response increases.

3.9 Shunt Matching with Distributed Receivers

Referring to Figure 3.8, multiple receivers are connected along one transmission line. A termination shunt resistor is connected across the end of the line. All receivers should be connected between the source and this final shunt resistor. Why? Assume each receiver has a nonzero capacitance. Why should the distance between the tap on the transmission line and each receiver be short?

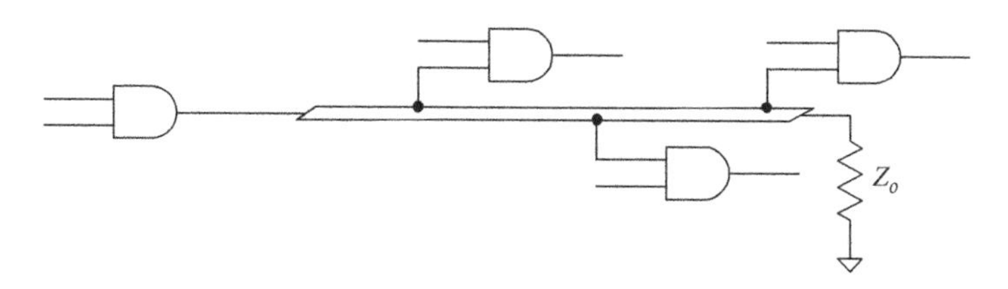

FIGURE 3.8 Multiple receivers connected along the length of a transmission line. The distance from the receivers to the transmission line is electrically short.

[6]For example, if the current drawn by the matching resistor were too great for the driver, the High output voltage would drop in value. Therefore, a noisy signal on the High output could cause the output to appear like a Low.

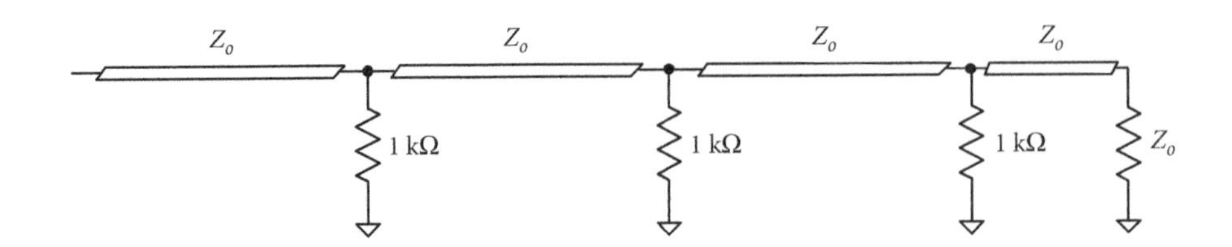

FIGURE 3.9 Identical receivers connected along a transmission line. The final termination is matched to the line.

Ideally, the input impedance of the receivers is infinite. An often used model for a real receiver is a capacitor and resistor in parallel. Connecting the receivers along the line increases the effective capacitance of the transmission line and slows down the input signal (i.e., increases the rise time). Usually the input resistance of the receiver is large compared to the line impedance of the main trunk, and the large resistance of the receiver has negligible loading effect on the line and driver. If the resistance is not large compared to the line impedance, this loading effect can be viewed as a shunt resistance across the transmission line (similar to dielectric resistance).

The length of the lines connecting the receivers to the main trunk should be electrically short; otherwise, these lines will also become transmission lines. Matching may be required at each receiver if these connecting lines are not electrically short. (Even when the lines are electrically short, reflections are still occurring at the mismatch between the receiver impedance and connecting line impedance. These reflections are not usually noticeable or of concern since they quickly return to the source before the source has reached steady state.)

The signal, an electromagnetic wave, travels down the line until the transmission line ends. If the impedance of the line matches the impedances of the final termination, the signal is absorbed by the termination and there are no further reflections. If one or more receivers are in parallel with this matching termination, the overall load impedance will contain a capacitive component. The point is that if the receivers have a significant capacitive component, then a pure resistor for the final distant termination will not be enough to prevent a reflection.

It may be helpful for some readers to examine the model of the line when the three receivers are simply modeled as 1 kΩ resistors as shown in Figure 3.9. When the loads are connected along the line, it is referred to as distributed loading. The reflection coefficient at any of the receiver connections for a transient signal traveling in either direction is approximately zero:

$$\rho_L = \frac{Z_L - Z_o}{Z_L + Z_o} = \frac{\dfrac{Z_o(1\times 10^3)}{Z_o + 1\times 10^3} - Z_o}{\dfrac{Z_o(1\times 10^3)}{Z_o + 1\times 10^3} + Z_o} \approx \frac{Z_o - Z_o}{Z_o + Z_o} = 0 \quad \text{if } Z_o \ll 1\,\text{k}\Omega$$

The effect of these receivers on the signal is negligible since their impedance is large compared to Z_o.

3.10 Microstrip Branching

A microstrip line of impedance Z_o is split into two microstrip lines each of impedance $2Z_o$ as shown in Figure 3.10. How would the lines differ geometrically? Why is this method of impedance matching generally not used? [Johnson, '93; Paul, '92(b); Coombs; Montrose, '99]

When one line is split into two lines, the lines are referred to as bifurcated lines (or T-stubs). Because some designers route lines or traces from a dc perspective, branching lines in this manner is not uncommon. However, if the trace widths (and other impedance related parameters) are identical for all of the traces, then an impedance mismatch will occur at the junction.

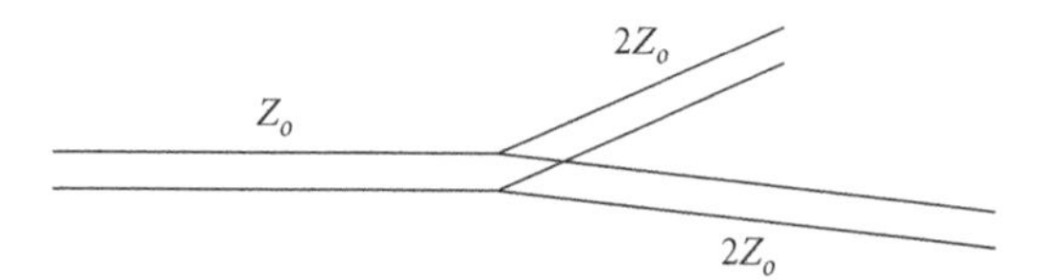

FIGURE 3.10 Transmission line of impedance Z_o splits into two lines, each of impedance $2Z_o$.

To avoid an impedance mismatch, the characteristic impedance of the two lines past the junction can be adjusted to a value of $2Z_o$. (Doubling the characteristic impedance is not necessarily easy.) For transient problems, the reflection coefficient at the junction for a signal traveling toward the two $2Z_o$ cables is equal to

$$\rho_L = \frac{Z_L - Z_o}{Z_L + Z_o} = \frac{\dfrac{2Z_o \times 2Z_o}{2Z_o + 2Z_o} - Z_o}{\dfrac{2Z_o \times 2Z_o}{2Z_o + 2Z_o} + Z_o} = 0$$

For transient signals, the input impedance is not a function of the load for each $2Z_o$ line. To avoid reflections at the end of each of the $2Z_o$ lines, each of the lines can be shunt matched with a $2Z_o$ resistor. Then, the steady-state input impedance would also not be a function of the line length or load.

The characteristic impedance for the microstrip line shown in Figure 3.11 is given by the expressions

$$Z_o = \begin{cases} \dfrac{60}{\sqrt{\varepsilon_{reff}}} \ln\left(\dfrac{8h}{w} + 0.25\dfrac{w}{h}\right) & \text{if } \dfrac{w}{h} \le 1 \\ \dfrac{377}{\sqrt{\varepsilon_{reff}}} \left[\dfrac{w}{h} + 1.393 + 0.667 \ln\left(\dfrac{w}{h} + 1.444\right)\right]^{-1} & \text{if } \dfrac{w}{h} \ge 1 \end{cases}$$

where ε_{reff} is the effective relative permittivity or dielectric constant of the board, w is the land width, and h is the land height above the ground plane. An empirical expression for the effective dielectric constant for the microstrip configuration is

$$\varepsilon_{reff} = 0.475\varepsilon_r + 0.67 \approx \frac{1}{2}(\varepsilon_r + 1)$$

where ε_r is the relative permittivity or dielectric constant of the board. If w = 15 mils, h = 62 mils, and ε_r = 4.7, then the line impedance, Z_o, is about 120 Ω. Referring to Mathcad 3.3, for a 2(120) = 240 Ω line impedance, assuming the PCB material and h are identical, the width, w, of the land must be about 0.54 mils (≈ 0.014 mm)! The width of this 240 Ω line is about a factor of thirty smaller than the 120 Ω line. Such a large difference between two land widths on the same board may add to the cost of production.

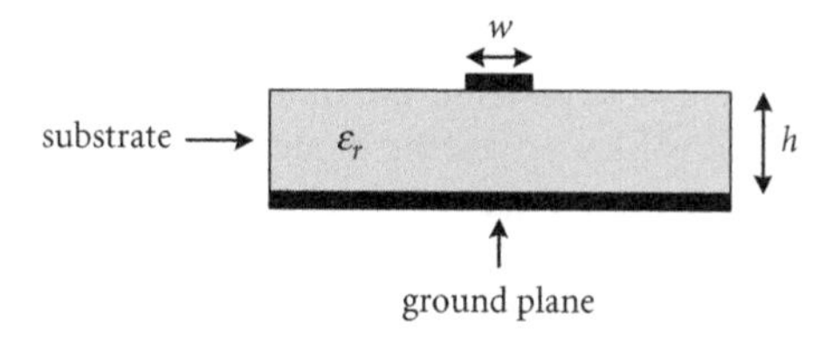

FIGURE 3.11 Cross-sectional view of a microstrip transmission line.

$h := 62 \qquad \varepsilon_r := 4.7$

$w := 0.05, 0.1 .. 2 \cdot h$

$\varepsilon_{eff} := 0.475 \cdot \varepsilon_r + 0.67$

$$Z_{o1}(w) := \frac{60}{\sqrt{\varepsilon_{eff}}} \cdot \ln\left(\frac{8 \cdot h}{w} + 0.25 \cdot \frac{w}{h}\right) \quad Z_{o2}(w) := \frac{377}{\sqrt{\varepsilon_{eff}}} \cdot \left(\frac{w}{h} + 1.393 + 0.677 \cdot \ln\left(\frac{w}{h} + 1.444\right)\right)^{-1}$$

$$Z_o(w) := \text{if}\left(\frac{w}{h} < 1, Z_{o1}(w), Z_{o2}(w)\right)$$

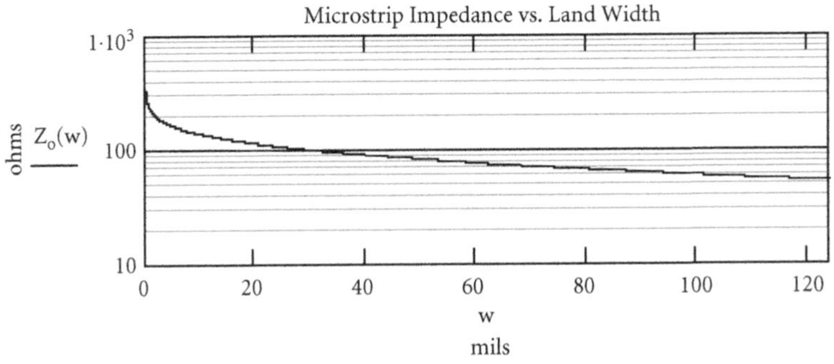

Microstrip Impedance vs. Land Width

$1 \cdot 10^3$

ohms $Z_o(w)$ 100

10

0 20 40 60 80 100 120

w

mils

$Z_o(15) = 123.276$

$w := 1 \qquad \text{Given} \qquad Z_o(w) = 240 \qquad a := \text{Find}(w) \qquad a = 0.544$

The variables are in mils (thousandths of an inch) in this program.

MATHCAD 3.3 Variation of the impedance of a microstrip line with strip width.

When the traces are already small, it may not be possible to decrease significantly their widths. (In practice, because of the close proximity of the traces, the impedance of each of the traces will be affected by nearby traces.)

Typical minimum etched copper widths are currently 4 to 5 mils. If gold is evaporated onto a thin film, 1 to 3 mils can be currently obtained. (For 1 ounce of copper, the thickness of the trace is 1.38 mils.) Land height variation on the same board is not currently a reasonable method for adjusting the line impedance. However, for multilayer boards, land height can vary from layer to layer. There are other factors to consider besides impedance when adjusting the trace width. For example, for a fixed trace thickness, the current-carrying (power-handling) capacity of a trace decreases with decreasing width.

3.11 Shunt Diode Matching

How effective is a diode placed in shunt with a receiver in reducing reflections due to an impedance mismatch? [Rosenstark; Montrose, '99]

A diode (e.g., zener) placed across the receiver, with its anode connected to the common, is intended to turn on when the voltage across the diode exceeds its reverse-biased voltage rating. If the turn-on voltage is selected near the incident voltage level, the diode will turn on when required to prevent overshoot at the receiver. The total voltage across the diode is the sum of the incident signal plus the reflected signal. The diode will prevent the total voltage across it from exceeding its turn-on voltage at the cost of increasing the current through it. Thus, there will be a current surge whenever the diode breaks down.

When the diode is not forward biased (and not in breakdown), the reflection coefficient at the receiver is positive since the total load impedance is (generally) greater than the line impedance. When the diode is reversed biased beyond its breakdown voltage by the peaks of the signal, the dynamic resistance of the diode drops, the diode conducts, and current surges through the diode. The diode is acting as a voltage clipper. The short-duration dynamic resistance may or may not help.

The frequency response of the diode, including the effect of parasitic inductance and capacitance, must also be considered, especially for higher speed input signals. Overshoots greater than the expected turn-on voltage of the diode can pass to the receiver. The analysis of nonlinear loads is very complicated compared to linear loads. Furthermore, this nonlinear device can produce harmonics—a source of noise. One advantage of diodes, however, is that their power consumption is low compared to shunt resistor terminations.

Obviously, two diodes can be used to limit both positive and negative voltage swings during the low-to-high and high-to-low transitions, respectively. One diode can have its anode connected to the line and its cathode connected to the upper supply voltage while the other diode can have its cathode connected to the line and its anode connected to the lower supply voltage (refer to Table 3.2). The diodes are reversed biased (open circuited) during steady state. Again, with the use of diodes there is the problem of current surges when the diodes are turned on, which may be a source of electrical interference.

3.12 Shunt *RC* Matching

How effective is the series *RC* circuit in shunt with a receiver shown in Figure 3.12 in reducing reflections due to an impedance mismatch between the receiver and connecting transmission line? [Montrose, '99; Vo]

Before the mathematical analysis is presented, it is very helpful to analyze qualitatively the response of this *RC* network (or "ac termination") to incident signals. Since the impedance of the capacitor is inversely proportional to the frequency, very high frequencies are not significantly affected by the presence of the capacitor and the load is matched to the line (assuming the input impedance of the receiver is very large). Low-frequency signals are affected by the capacitor, however, and the load impedance seen by the transmission line can be very large resulting in a large mismatch. Probably the greatest advantage of this *RC* matching network over a simple, shunt termination resistor is that its dc steady-state impedance is very large. Hence, its dc power consumption is very low, and the steady-state current loading on the driver is low. Therefore, this circuit can be used in those circuits where a simple shunt resistor would excessively load the driver (e.g., TTL and CMOS circuits). The *RC* matching networks are also used in differential-pair circuits. To keep the two lines balanced, the two resistors must be appropriately matched (or finely adjusted).

The effect of this *RC* matching network on the incident signal will be determined using transform techniques. The transmission coefficient for this load, assuming that the receiver's input impedance is very large compared to the characteristic impedance of the line or impedance of the matching network over the major frequency spectrum of the input signal, is

$$\mathrm{T}(s) = \frac{2Z_L}{Z_L + Z_o} = \frac{2\left(R_M + \dfrac{1}{sC_M}\right)}{R_M + \dfrac{1}{sC_M} + Z_o} = \frac{2R_M\left(s + \dfrac{1}{C_M R_M}\right)}{(R_M + Z_o)\left[s + \dfrac{1}{C_M(R_M + Z_o)}\right]} \tag{3.35}$$

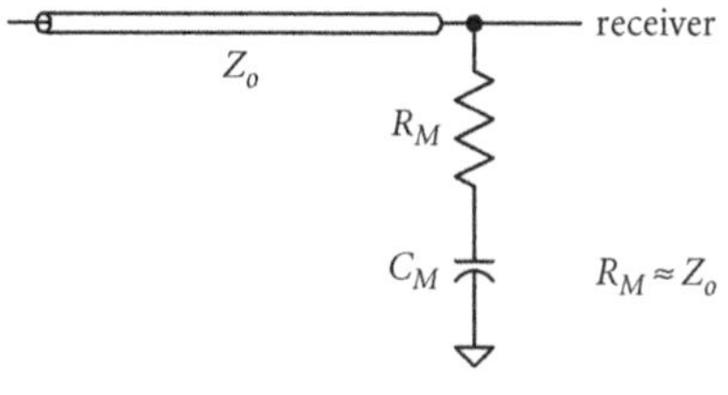

FIGURE 3.12 Series *RC* matching network in shunt with the receiver.

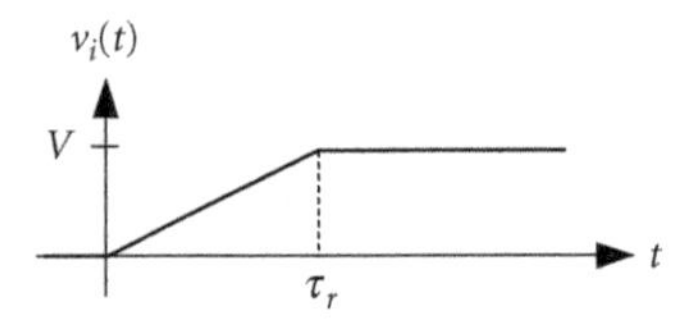

FIGURE 3.13 Transient input signal.

Assume that the input signal is described by (3.36), which is a simple model for a digital signal passing from a Low to a High state:

$$v_i(t) = \frac{V}{\tau_r} t[u(t) - u(t - \tau_r)] + Vu(t - \tau_r) \tag{3.36}$$

This input signal is shown in Figure 3.13. The signal transmitted into the receiver immediately after the first reflection from the receiver will be determined using this input or incident signal. Probably, the least painful method of obtaining this solution is through the use of Laplace transforms. Using the transform pair

$$t[u(t) - u(t-a)] + au(t-a) \Leftrightarrow \frac{1 - e^{-as}}{s^2}$$

the input signal transform is easily obtained:

$$V_i(s) = \frac{V}{\tau_r} \frac{1 - e^{-\tau_r s}}{s^2} \tag{3.37}$$

The output frequency response, corresponding to the transmitted transient signal, is the product of the input signal transform, (3.37), and the transfer function corresponding to the transmission coefficient, (3.35):

$$V_t(s) = V_i(s)\mathrm{T}(s) = \frac{V}{\tau_r} \frac{1 - e^{-\tau_r s}}{s^2} \frac{2R_M\left(s + \dfrac{1}{C_M R_M}\right)}{(R_M + Z_o)\left[s + \dfrac{1}{C_M(R_M + Z_o)}\right]}$$

$$= \frac{V}{\tau_r}\frac{2R_M}{R_M + Z_o}\left\{\frac{1}{s\left[s + \dfrac{1}{C_M(R_M + Z_o)}\right]} + \frac{\dfrac{1}{C_M R_M}}{s^2\left[s + \dfrac{1}{C_M(R_M + Z_o)}\right]}\right\}$$

$$- \frac{V}{\tau_r}\frac{2R_M}{R_M + Z_o}\left\{\frac{1}{s\left[s + \dfrac{1}{C_M(R_M + Z_o)}\right]} + \frac{\dfrac{1}{C_M R_M}}{s^2\left[s + \dfrac{1}{C_M(R_M + Z_o)}\right]}\right\} e^{-\tau_r s}$$

The inverse transform is easily obtained by using the following transform pairs (and the time-shift property):

$$(1 - e^{-bt})u(t) \Leftrightarrow \frac{b}{s(s+b)} \quad \text{and} \quad \left(\frac{at - 1 + e^{-at}}{a}\right)u(t) \Leftrightarrow \frac{a}{s^2(s+a)}$$

The time-domain result is

$$v_t(t) = \frac{2R_M V}{\tau_r}\left\{\begin{array}{l} C_M\left[1-e^{-\frac{t}{C_M(R_M+Z_o)}}\right]+ \\ \dfrac{C_M(R_M+Z_o)}{R_M}\left[\dfrac{t}{C_M(R_M+Z_o)}-1+e^{-\frac{t}{C_M(R_M+Z_o)}}\right] \end{array}\right\}u(t)$$
$$-\frac{2R_M V}{\tau_r}\left\{\begin{array}{l} C_M\left[1-e^{-\frac{t-\tau_r}{C_M(R_M+Z_o)}}\right]+ \\ \dfrac{C_M(R_M+Z_o)}{R_M}\left[\dfrac{t-\tau_r}{C_M(R_M+Z_o)}-1+e^{-\frac{t-\tau_r}{C_M(R_M+Z_o)}}\right] \end{array}\right\}u(t-\tau_r) \quad (3.38)$$

For many aperiodic signals, the highest frequency of interest is inversely proportional to the rise time, τ_r. Therefore, when the rise time of the input signal is very small, the input signal is said to contain significant high-frequency energy. Whether the rise time is "small" or "large," however, the output response is a strong function of the time constant (or equivalent) of the transfer function. In this example, the transfer function is the transmission coefficient function, (3.35). Examining (3.38), the nearest analogue to a time constant is $C_M (R_M + Z_o)$. The transmitted signal will next be approximated near $t = \tau_r$ for large and small time constants, relative to the rise time of the input signal. For small rise-time input signals or for high frequencies

$$C_M(R_M+Z_o) \gg \tau_r \quad (3.39)$$

and (3.38) can be approximated using the standard approximation for the exponential function for small values of its argument:

$$v_t(\tau_r^+) \approx \frac{2R_M V}{\tau_r}\left[\begin{array}{l} C_M\left\{1-\left[1-\dfrac{\tau_r}{C_m(R_M+Z_o)}\right]\right\}+ \\ \dfrac{C_M(R_M+Z_o)}{R_M}\left\{\dfrac{\tau_r}{C_M(R_M+Z_o)}-1+\left[1-\dfrac{\tau_r}{C_M(R_M+Z_o)}\right]\right\} \end{array}\right]$$
$$-\frac{2R_M V}{\tau_r}\left[\begin{array}{l} C_M(1-1)+ \\ \dfrac{C_M(R_M+Z_o)}{R_M}(0-1+1) \end{array}\right] \quad (3.40)$$
$$= \frac{2R_M V}{R_M+Z_o}$$
$$= V \quad \text{if } R_M = Z_o$$

Thus, when the load resistor is matched to the characteristic impedance of the line, the voltage transmitted into the load is equal to V at $t = \tau_r$. This expression does not contain the capacitance, C_M, since at high frequencies the capacitance behaves like a short circuit. Of course, for times much greater than $t = \tau_r$, the input signal is mainly dc and the reactance of the capacitor is large. As previously mentioned, the capacitance is important for reducing the steady-state driver current.

When the time constant of the system is small compared to the rise time of the input signal

$$C_M(R_M + Z_o) \ll \tau_r$$

the reactance of the capacitor is not negligible. Expression (3.38) can also be approximated for this condition. Recalling that e^{-x} is approximately zero for large values of x,

$$\begin{aligned} v_t(\tau_r^+) &\approx \frac{2R_M V}{\tau_r}\left\{ \begin{aligned} &C_M(1-0)+ \\ &\frac{C_M(R_M+Z_o)}{R_M}\left[\frac{\tau_r}{C_M(R_M+Z_o)} - 1 + 0\right] \end{aligned} \right\} \\ &\quad - \frac{2R_M V}{\tau_r}\left[\begin{aligned} &C_M(1-0)+ \\ &\frac{C_M(R_M+Z_o)}{R_M}(0-1+1) \end{aligned} \right] \\ &= \frac{2R_M V}{\tau_r}\left(C_M + \frac{\tau_r}{R_M} - C_M\right) = 2V \end{aligned} \tag{3.41}$$

Therefore, for low-frequency or slow input signals, the series *RC* matching network can double the amplitude of the incident signal.[7] For this reason, the capacitance of the *RC* matching network is selected so that the following condition is satisfied:

$$C_M(R_M + Z_o) \gg \tau_r$$

Typical capacitance values are in the 20–600 pF range. If the trace impedance $R_M \approx Z_o \approx 50\ \Omega$, many faster digital signals can satisfy this inequality:

$$20 \times 10^{-12}(R_M + Z_o) = 2\text{ ns}, \quad 600 \times 10^{-12}(R_M + Z_o) = 60\text{ ns}$$

As is also true for any resistor, the capacitor must be carefully selected so that its resonant frequency is well above the highest frequency of interest of the input signal.

The results in Mathcad 3.4 clearly show the effect of a series *RC* matching network on the input signal. The previous approximations are also plotted on the same graph to show their accuracy. In a real circuit, the capacitance and resistance of the receiver would have some effect on the response. Also, the response would obviously not remain flat but would likely oscillate and eventually fall to the steady-state value of *V*, which is the steady-state value of the input signal. This is not modeled in this example since the incident signal that is reflected from the load and its reflection from the source (and so on) are not included. These multireflections are summed in an identical manner as presented in the transient reflection (and the advanced transient problem) discussion except the Laplace transforms of the reflection and transmission coefficients as well as the input signal are used.

Because the capacitance does have an effect on the signal, there will be some additional propagation delay with the use of the *RC* matching network. Although this delay is not directly seen in the plots, it is indirectly seen through the difference in amplitude between the transmitted signal after 4 ns and the value of *V*. For even the high-*C* case, multiple reflections are required to produce the desired *V*-level at

[7] Well . . . maybe the *RC* network is not actually doubling the signal. At low frequencies, the series *RC* matching network appears like a very large impedance, nearly an open circuit. The reflection coefficient for an open-circuited load is one.

$$Z_o := 75 \quad R_M := Z_o \quad \tau_r := 4\cdot10^{-9} \quad V := 5 \quad C := 200\cdot10^{-12} \qquad C\cdot(R_M+Z_o)=3\times10^{-8}$$

$$t := 0, \frac{1}{100}\cdot\tau_r .. 2\cdot\tau_r \qquad \frac{C}{100}\cdot(R_M+Z_o)=3\times10^{-10}$$

$$\tau(C_M) := C_M\cdot(R_M+Z_o)$$

$$v_t(t, C_M) := \frac{2\cdot R_M\cdot V}{\tau_r}\cdot\left[C_M\cdot\left(1-e^{-\frac{t}{\tau(C_M)}}\right)+\frac{\tau(C_M)}{R_M}\cdot\left(\frac{t}{\tau(C_M)}-1+e^{-\frac{t}{\tau(C_M)}}\right)\right]\cdot\Phi(t)\ ...$$

$$+-\left[\frac{2\cdot R_M\cdot V}{\tau_r}\cdot\left[C_M\cdot\left(1-e^{-\frac{t-\tau_r}{\tau(C_M)}}\right)+\frac{\tau(C_M)}{R_M}\cdot\left(\frac{t-\tau_r}{\tau(C_M)}-1+e^{-\frac{t-\tau_r}{\tau(C_M)}}\right)\right]\cdot\Phi(t-\tau_r)\right]$$

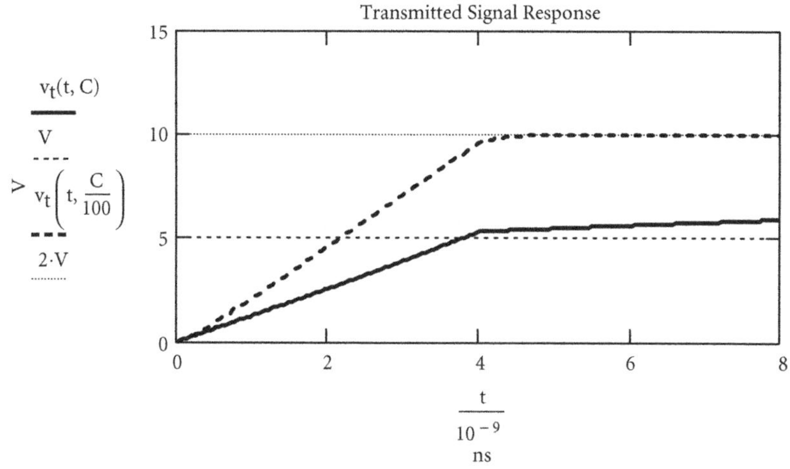

Transmitted Signal Response

Legend: $v_t(t, C)$ (solid); V (dashed); $v_t\left(t, \frac{C}{100}\right)$ (bold dashed); $2\cdot V$ (dotted). Vertical axis: V (0, 5, 10, 15). Horizontal axis: $\frac{t}{10^{-9}}$ ns (0, 2, 4, 6, 8).

MATHCAD 3.4 Effect of a series *RC* matching network on the incident signal.

the receiver. These multiple reflections require time to travel between the load and the receiver and are the source of the delay.

3.13 Matching at the Driver and its Cost

Determine the steady-state voltage level at a receiver if the source impedance is set equal to the characteristic impedance. Since in practice the source impedance cannot usually be set equal to Z_o, how should the total source impedance be matched to the transmission line? Will this method of matching eliminate all reflections of the signal on the line? What is the cost of implementing this matching scheme? [Johnson, '93; Montrose, '99]

If the impedance of the source is set equal to the characteristic impedance of the line, the reflection coefficient at the source is zero:

$$\rho_s = \frac{R_s - Z_o}{R_s + Z_o} = \frac{Z_o - Z_o}{Z_o + Z_o} = 0 \tag{3.42}$$

Using (3.28), the steady-state voltage across the receiver is then

$$V_\infty = \frac{V_o \dfrac{Z_o}{R_s+Z_o} H(1+\rho_L)}{1-H^2\rho_L \times 0} = V_o \frac{Z_o}{Z_o+Z_o} H(1+\rho_L) = \frac{V_o}{2} H(1+\rho_L) \tag{3.43}$$

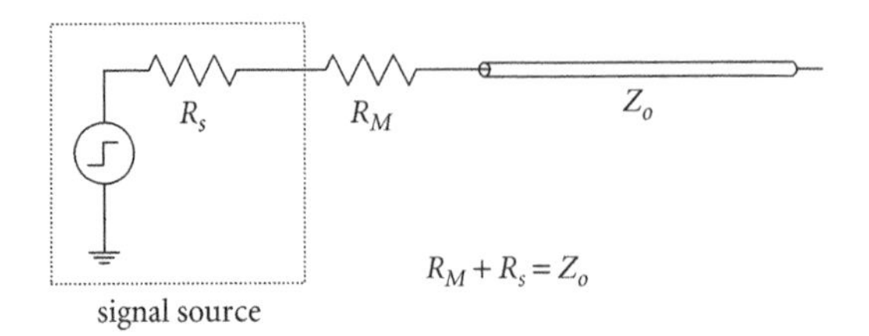

FIGURE 3.14 Matching resistor, R_M, connected in series with the source.

If the receiver impedance is much greater than the line impedance, the load reflection coefficient is approximately one, and the steady-state voltage at the receiver is the same as the source voltage (neglecting line losses). A method of matching the source to the transmission line, so that the reflection coefficient at the source is zero, is to insert an appropriate resistor in series with the source. As shown in Figure 3.14, the combined impedance of the source and this series matching resistor, R_M, should be set equal to Z_o.

By matching at the source, the initial transient voltage across the input of the transmission line is one-half of the source voltage:

$$V_o \frac{Z_o}{Z_o + R_s + R_M} = V_o \frac{Z_o}{Z_o + Z_o} = \frac{V_o}{2}$$

This signal, when it arrives at the load, is reflected off the load. Again, if the reflection coefficient at the load is approximately one, the reflected signal has an amplitude of $V_o/2$, and the total signal level is $(V_o/2) + (V_o/2) = V_o$. When the reflected signal arrives at the source, it is not reflected since the source reflection coefficient is zero. There are no more reflections.

The steady-state impedance seen by the signal source is

$$R_M + R_L$$

where R_L is the load resistance at the end of the transmission line. Without matching, the steady-state impedance seen by the source is R_L. Therefore, loading of the source is usually not a problem with series matching. Although any resistor will absorb power, compared to the shunt matching resistor, the steady-state power consumption of the series matching resistor is low. As with matching at the receiver, the additional matching resistor adds to the component list and circuit density.

Note that inserting a resistor in series with the line, if not planned, is frequently more troublesome than adding a shunt resistor. Also, when there is more than one load with their own connecting lines, series matching is not always feasible (the source reflection coefficient is then a function of the other line impedances). However, if there is only one line between the source and the multiple loads (all located at the end of the line), then series matching is easily implemented and requires only one matching resistor — at the source. When receivers are distributed along the line, the voltage at these receivers is only $V/2$ until the reflected voltage off the end of the line arrives at the location of each of the receivers.

It might be surprising to learn that a series matching resistor can add time delay to a transient input signal. Ideally, at high frequencies, the input impedance into a transmission line seen by a transient signal is just the line's characteristic impedance, Z_o, which is mostly real. However, for low-frequency signals, the characteristic impedance can also contain a capacitive component. This region is referred to as the *RC* region for the transmission line. The characteristic impedance can be approximated for this region as

$$Z_o = \sqrt{\frac{R + j\omega L}{j\omega C}} \approx \sqrt{\frac{R}{2\omega C}} - j\sqrt{\frac{R}{2\omega C}} \quad \text{if } R \gg \omega L \tag{3.44}$$

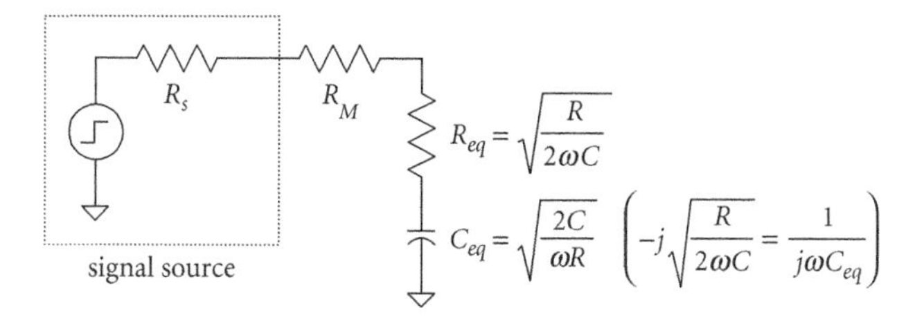

FIGURE 3.15 Modeling the input of the transmission line at low frequencies.

where R and C are the resistance and capacitance per unit length, respectively, of the transmission line leading to the receiver. Referring to Figure 3.15, the time constant for this RC circuit is

$$\tau = R_{eq}\,C_{eq} = \left(R_s + R_M + \sqrt{\frac{R}{2\omega C}}\right)\sqrt{\frac{2C}{\omega R}} \tag{3.45}$$

Of course, the time constant is a transient property, and the impedance used in this expression is a sinusoidal steady-state result. Nevertheless, (3.45) indicates that the "time constant" is larger for lower frequency signals (or the lower frequency spectrum of a transient signal). The rigorous analysis of the response of this RC region of a transmission line to a transient input signal is best handled in the frequency domain using Laplace transforms.

This RC matching network is essentially a filter. As a filter, it can introduce desirable distortion. For example, it can round off any sharp edges, thereby reducing the high-frequency energy of the signal. This will result in lower radiation levels and lower levels of crosstalk.

3.14 Series Matching with Multiple Receivers

How can series matching at the driver be implemented if there are multiple receivers with different transmission lines? [Rosenstark]

With a series termination at the driver and a single transmission line, the initial voltage along the line is one-half of the desired value until the reflected signal returns. Unless this reduced voltage is acceptable for a period of time for receivers along the line, this matching method should not be used. Of course, this lower voltage is not a problem when all the receivers are together near the end of the line.

If multiple lines are connected after the source termination, as shown in Figure 3.16, the reflection coefficient is not zero for the reflected negative-going waves. The transient reflection coefficient for a negative-going reflected transient wave from either line is (i.e., a signal traveling toward the source)

$$\rho = \frac{Z_L - Z_o}{Z_L + Z_o} = \frac{\dfrac{Z_oR_s}{Z_o + R_s} - Z_o}{\dfrac{Z_oR_s}{Z_o + R_s} + Z_o} \approx \frac{\dfrac{Z_o}{2} - Z_o}{\dfrac{Z_o}{2} + Z_o} = -\frac{1}{3} \quad \text{if } R_s = Z_o$$

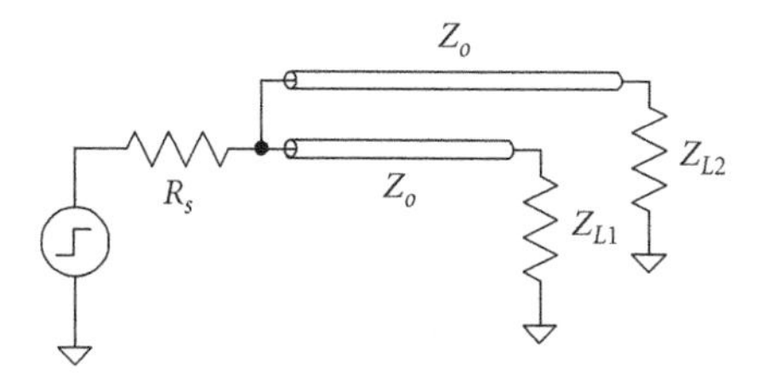

FIGURE 3.16 Two transmission lines connected at the source.

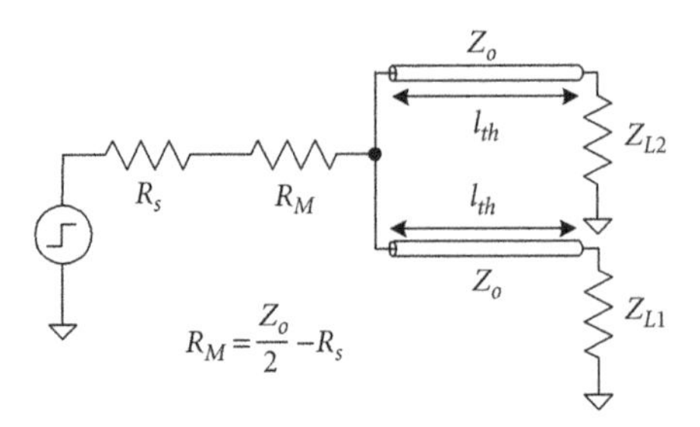

FIGURE 3.17 Matching at the source when the two transmission lines are of identical length.

This system is not properly matched. (An infinite value for R_s would produce a zero reflection coefficient, but the incident signal down each of the lines would be zero. This is considered the trivial case.)

Referring to Figure 3.17, when the length of each of the lines is the same, then a single, resistor series match is possible. If the amplitude of the source is V, then the initial voltage across each of the lines is obtained through voltage division:

$$V\frac{Z_o \| Z_o}{(Z_o \| Z_o) + R_M + R_s} = V\frac{\frac{Z_o}{2}}{\frac{Z_o}{2} + \frac{Z_o}{2} - R_s + R_s} = \frac{V}{2}$$

If both of the loads Z_{L1} and Z_{L2} are very large, the reflection coefficients at both loads are one and the total voltage at the load is then V. The reflected voltage from either load is then $V/2$, as is typical of series matching methods. The negative-going[8] reflection coefficient for either line is

$$\rho = \frac{Z_L - Z_o}{Z_L + Z_o} = \frac{\dfrac{Z_o\left(R_s + \frac{Z_o}{2} - R_s\right)}{Z_o + R_s + \frac{Z_o}{2} - R_s} - Z_o}{\dfrac{Z_o\left(R_s + \frac{Z_o}{2} - R_s\right)}{Z_o + R_s + \frac{Z_o}{2} - R_s} + Z_o} = -\frac{1}{2}$$

while the negative-going transmission coefficient for either line is +1/2. The reflection coefficient at the source for negative-going waves is not zero. However, when the line lengths are identical, the negative-going wave of amplitude $V/2$ on both lines has a reflected value of $-1/2(V/2) = -V/4$ at the source, while the transmitted voltage from the other line has a value of $+1/2(V/2) = +V/4$. Therefore, the sum of this reflected and this transmitted signal on either line is zero. If there were n lines, the value for the series matching resistor would be

$$R_M = \frac{Z_o}{n} - R_s \tag{3.46}$$

For this series matching method when different line lengths are used (having identical characteristic impedances), the negative-going waves from the two lines will not appear at the source at the same time. As a

[8]Negative-going is a colloquial phrase for a signal traveling toward the source.

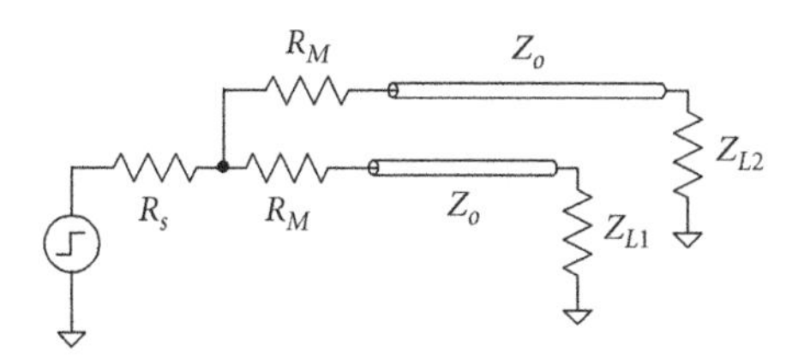

FIGURE 3.18 Separate series matching resistor for each transmission line.

result, the reflected and transmitted signals of $V/4$ magnitude from the first negative-going wave to appear at the source will not be immediately nulled out by the other wave's reflected and transmitted signals.

When a separate series matching resistor is used for each line, a reasonable series match is sometimes possible even for different line lengths. In this case, the value of the series resistor is less than Z_o and is a function of the number of receivers. For the two receiver example shown in Figure 3.18, the reflection coefficient for a negative-going wave on either line is

$$\rho = \frac{Z_L - Z_o}{Z_L + Z_o} = \frac{R_M + [R_s \| (R_M + Z_o)] - Z_o}{R_M + [R_s \| (R_M + Z_o)] + Z_o}$$

The reflection coefficient is zero when

$$R_M + [R_s \| (R_M + Z_o)] = Z_o \quad \text{or} \quad \frac{R_s (R_M + Z_o)}{R_s + R_M + Z_o} = Z_o - R_M \tag{3.47}$$

If the source resistance, R_s, is much less than $R_M + Z_o$, (3.47) reduces to

$$R_M + R_s \approx R_M = Z_o \tag{3.48}$$

When $R_M = Z_o$ and the source resistance is small, the incident input voltage amplitude is V and the voltage across each line is $V/2$, so the transmitted voltage into each high-impedance load is V. For more than two receivers, using a separate series matching resistor of this value (with a small source resistance) can provide a good match.

When the line lengths are identical, then an exact series match is possible. Referring to Figure 3.19, if the source amplitude is V, the initial voltage traveling down each line is obtained using voltage dividers $(R_M = Z_o - 2R_s)$:

$$V \frac{(R_M + Z_o) \| (R_M + Z_o)}{(R_M + Z_o) \| (R_M + Z_o) + R_s} \frac{Z_o}{Z_o + R_M} = V \frac{\frac{R_M + Z_o}{2}}{\frac{R_M + Z_o}{2} + R_s} \frac{Z_o}{Z_o + R_M} = \frac{V}{2}$$

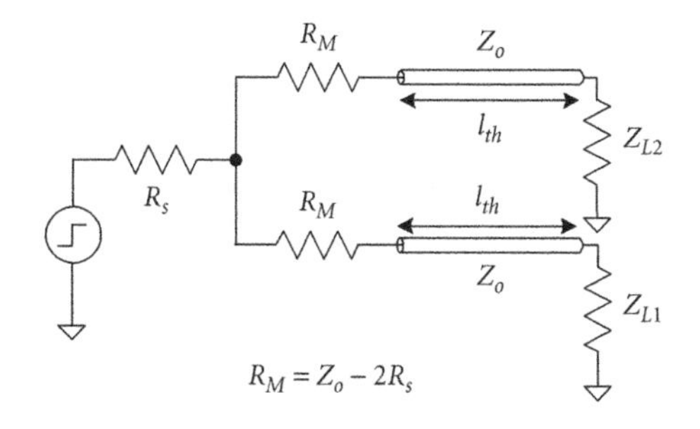

FIGURE 3.19 Exact match when the line lengths are identical.

The reflection coefficient for the negative-going wave is

$$\rho = \frac{Z_L - Z_o}{Z_L + Z_o} = \frac{R_M + \left[(R_M + Z_o) \| R_s\right] - Z_o}{R_M + \left[(R_M + Z_o) \| R_s\right] + Z_o} = \frac{-R_s}{2(Z_o - R_s)}$$

The transmission coefficient is

$$\mathrm{T} = 1 + \rho = \frac{2Z_L}{Z_L + Z_o} = \frac{2Z_o - 3R_s}{2(Z_o - R_s)}$$

If the two line lengths are the same, the total wave traveling back down each line is the sum of the reflected term from the line itself

$$\rho \frac{V}{2} = \frac{-R_s V}{4(Z_o - R_s)}$$

and the transmitted term from the neighboring line, obtained using voltage dividers:

$$\mathrm{T} \frac{V}{2} \frac{(R_M + Z_o) \| R_s}{\left[(R_M + Z_o) \| R_s\right] + R_M} \frac{Z_o}{Z_o + R_M} = \frac{R_s V}{4(Z_o - R_s)}$$

Notice that the sum of these two signals is zero. Therefore, when the two line lengths are identical, this series termination method provides a proper match. If there were n lines, the value for each of the series-matching resistors would be

$$R_M = Z_o - nR_s \tag{3.49}$$

For this series matching method when the line lengths are different, the transmitted and reflected signals would not cancel at the same time. However, the magnitude of the transmitted or reflected signal will usually be less than with the use of a single matching resistor since

$$\left|\frac{V}{4}\right| > \left|\frac{R_s V}{4(Z_o - R_s)}\right| \quad \text{if } |Z_o - R_s| > R_s \tag{3.50}$$

3.15 Effects of Nonzero Source and Load Reflection Coefficients

If the electrical properties and length of a transmission line cannot be changed, why are the amplitudes of the signal reflections reduced by decreasing the source and load reflection coefficients? When is the magnitude of the product of these two coefficients always less than one?

Especially after a system is in place, the physical and electrical properties of any transmission lines are usually not (easily) changeable. If reflections are to be reduced, then the source and load impedances must be changed. Reducing the reflection coefficients at the source and load will reduce the magnitude of any reflections off the source and load, respectively. This will reduce "ringing" and the time required to reach the desired steady-state value.

The reflection coefficient at the load is equal to

$$\rho_L = \frac{R_L - Z_o}{R_L + Z_o}$$

A signal of amplitude A incident to (or "striking") the load will have a reflection with a magnitude of $A|\rho_L|$. Reducing the reflection coefficient at the load reduces the magnitude of this reflected signal. The reflection coefficient at the source is equal to

$$\rho_s = \frac{R_s - Z_o}{R_s + Z_o}$$

A signal of amplitude B incident to the source will have a reflection with a magnitude of $B|\rho_s|$. Reducing the reflection coefficient at the source reduces the magnitude of this reflected signal. The magnitude of the product of the load and source reflection coefficients is equal to

$$|\rho_L \rho_s| = \left| \frac{R_L - Z_o}{R_L + Z_o} \frac{R_s - Z_o}{R_s + Z_o} \right|$$

For any real, finite nonzero values for the load, source, and characteristic impedance, both the load and source reflection coefficients are less than one (since their denominators are greater than their respective numerators). The product of two numbers less than one is also less than one. Of course, if the source or load is matched, the corresponding reflection coefficient is equal to zero.

3.16 Signal Bounce as a Function of Time

Assume that the nth signal that arrives at the load can be described by the following expression (where $n = 0$ corresponds to the first signal to appear at the load):

$$V_n = \frac{V_o Z_o}{R_s + Z_o} H\left(\rho_L \rho_s H^2\right)^n (\rho_L + 1)$$

Assuming a lossless transmission line, show that the *percent difference* between the signal at the load after $(2n+1)T$ (where T is the one-way propagation delay of the cable) and its steady-state value is

$$\Delta\% = 100 |\rho_L \rho_s|^{n+1}$$

[Johnson, '93]

Using (3.27), the percent difference between the final steady-state value and total signal amplitude after N reflections is $(N = n + 1)$

$$\Delta\% = 100 \left| \frac{\dfrac{V_o \dfrac{Z_o}{R_s + Z_o} H(1+\rho_L)}{1 - H^2 \rho_L \rho_s} - \sum_{n=1}^{N} \dfrac{V_o Z_o}{R_s + Z_o} H(\rho_L \rho_s H^2)^{n-1} (\rho_L + 1)}{\dfrac{V_o \dfrac{Z_o}{R_s + Z_o} H(1+\rho_L)}{1 - H^2 \rho_L \rho_s}} \right| \tag{3.51}$$

or

$$\Delta\% = 100 \left| 1 - (1 - H^2 \rho_L \rho_s) \sum_{n=1}^{N} (\rho_L \rho_s H^2)^{n-1} \right| \tag{3.52}$$

Using the geometric series

$$s_M = a + ar + ar^2 + \cdots + ar^{N-1} = a\sum_{n=1}^{N} r^{n-1} = \frac{a(1-r^N)}{1-r}$$

the summation can be rewritten as

$$\begin{aligned} \Delta\% &= 100\left|1-(1-H^2\rho_L\rho_s)\frac{1-(\rho_L\rho_s H^2)^N}{1-\rho_L\rho_s H^2}\right| = 100\left|\rho_L\rho_s H^2\right|^N \\ &= 100\left|\rho_L\rho_s\right|^N = 100\left|\rho_L\rho_s\right|^{n+1} \end{aligned} \tag{3.53}$$

If the line is lossless, H is one.

3.17 Settling Time

Referring to (3.53), as n approaches infinity, what does $\Delta\%$ approach? Then, show that the *additional* time (compared to a perfectly matched system) required for the signal to settle down to $\Delta\%$ of its steady-state value is approximately

$$T\left[\frac{2\ln\left(\frac{\Delta\%}{100}\right)-\ln\left|\rho_L\rho_s\right|}{\ln\left|\rho_L\rho_s\right|}\right]$$

Check this result slightly after $t = T$ if the characteristic impedance of the lossless line is 52 Ω, source resistance is 10 Ω, and load impedance is 100 Ω. [Johnson, '93]

Since the magnitudes of both the load and source reflection coefficients are less than or equal to one for real impedances, their product is less than or equal to one. Since any real system is not perfectly matched, the product of these coefficients is less than one. Any number less than one raised to a large n power is small. As expected, the magnitude of the signal approaches its steady-state value after a long time. Solving for n in (3.53) by first dividing both sides by 100 results in

$$\frac{\Delta\%}{100} = \left|\rho_L\rho_s\right|^{n+1}$$

where n is an integer ($n = 0$ corresponds to the first reflection at the load). Then, taking the natural logarithm of both sides

$$\ln\left(\frac{\Delta\%}{100}\right) = \ln\left(\left|\rho_L\rho_s\right|^{n+1}\right) = (n+1)\ln\left|\rho_L\rho_s\right|$$

and solving for $n + 1$,

$$n+1 = \frac{\ln\left(\frac{\Delta\%}{100}\right)}{\ln\left|\rho_L\rho_s\right|} \tag{3.54}$$

This expression should not be used when either reflection coefficient is zero since the natural logarithm is not finite with an argument of zero. The percent difference equation is valid for $t = (2n+1)T$, where T is the one-way propagation delay of the cable. Solving for n

$$n = \frac{t}{2T} - \frac{1}{2} \tag{3.55}$$

and substituting (3.55) into (3.54) yields

$$\frac{t}{2T} - \frac{1}{2} + 1 = \frac{t+T}{2T} = \frac{\ln\left(\frac{\Delta\%}{100}\right)}{\ln|\rho_L \rho_s|}$$

Solving for t

$$t = \frac{2T\ln\left(\frac{\Delta\%}{100}\right)}{\ln|\rho_L \rho_s|} - T \tag{3.56}$$

For a perfectly matched system, there is no increase in the settling time due to reflections: the time required to travel down the line is merely the one-way propagation delay, T. Expression (3.56) can be written as

$$\frac{2T\ln\left(\frac{\Delta\%}{100}\right)}{\ln|\rho_L \rho_s|} - T = T\left[\frac{2\ln\left(\frac{\Delta\%}{100}\right) - \ln|\rho_L \rho_s|}{\ln|\rho_L \rho_s|}\right] \tag{3.57}$$

The line is assumed lossless in this equation. As a check, for the given impedances,

$$\rho_s \approx -0.68, \quad \rho_L \approx 0.32$$

and the voltage at the load immediately after the first reflection is

$$V_0 = \frac{V_o Z_o}{R_s + Z_o}(\rho_L + 1) = \frac{(5)\,52}{10+52}(0.316+1) \approx 5.5\text{ V}$$

The steady-state voltage at the receiver is

$$V_\infty = 5\frac{100}{100+10} \approx 4.6\text{ V}$$

since the impedance seen by the source after a long time is the load impedance, 100 Ω. At $t = T$,

$$|\Delta\%| = \left|100\frac{4.55 - 5.52}{4.55}\right| \approx 21\%$$

Using the derived equation, the time corresponding to 21% overshoot is

$$t = T\left[\frac{2\ln\left(\frac{\Delta\%}{100}\right) - \ln|\rho_L\rho_s|}{\ln|\rho_L\rho_s|}\right] \approx T\left[\frac{2\ln\left(\frac{21}{100}\right) - \ln|(0.32)0.68|}{\ln|(0.32)0.68|}\right] \approx T$$

In general, the nearest odd-multiple of T should provide an approximate result for the time.

3.18 Settling Time vs. Reflection Coefficient

Using (3.57), plot the time required for a Low-to-High transition to settle to 1% of its final value vs. the load reflection coefficient. Assume that the characteristic impedance of the lossless line is 52 Ω and the source resistance is 10 Ω. Allow the load impedance to vary from 1 Ω to 10 kΩ, and $T = 1$ s. According to the plot, for what value of the reflection coefficient is the settling time a minimum? Is this the expected value?

The source reflection coefficient is about −0.677, and the expression for the time assuming $T = 1$ s is

$$\left[\frac{2\ln\left(\frac{\Delta\%}{100}\right) - \ln|\rho_L\rho_s|}{\ln|\rho_L\rho_s|}\right] = \left[\frac{2\ln(0.01) - \ln|\rho_L 0.677|}{\ln|\rho_L 0.677|}\right] \quad \text{where } \rho_L = \frac{Z_L - 52}{Z_L + 52}$$

The load impedance varies from 1 Ω to 10 kΩ, so the reflection coefficient at the load varies from about −0.96 to 0.99. The plot is provided in Mathcad 3.5. As expected, the settling time is a minimum when the load is matched to the line (i.e., the load reflection coefficient is equal to zero). In this case, the settling time is zero since there are no reflections. For open-circuited and short-circuited loads, the settling time is over 20*T*. (This example assumes no losses. When a signal travels over large distances, the losses of the line should be included.)

$$t_1(\rho_{L1}) := \frac{2\cdot\ln(0.01) - \ln(|\rho_{L1}\cdot 0.677|)}{\ln(|\rho_{L1}\cdot 0.677|)} \qquad t_2(\rho_{L2}) := \frac{2\cdot\ln(0.01) - \ln(|\rho_{L2}\cdot 0.677|)}{\ln(|\rho_{L2}\cdot 0.677|)}$$

$$\rho_{L1} := -1, -0.999 .. -0.001 \qquad \rho_{L2} := 0.001, 0.01 .. 1$$

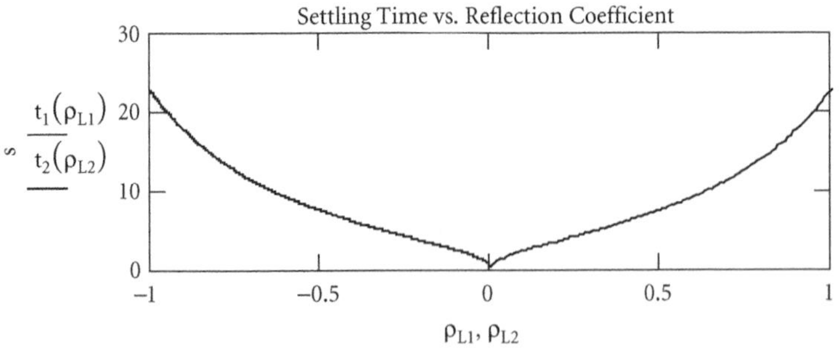

MATHCAD 3.5 Settling time vs. the reflection coefficient.

3.19 Receiver Voltage when Rise Time = Line Delay

Assume the rise time (the time required for the signal to rise from 10–90% of its final value) of a 0-to-5 V ramp source (low-to-high transition) is 5 ns. Furthermore, assume that a 1 meter RG/8U type cable (L = 0.078 μH/ft, C = 29 pF/ft) connects the 10 Ω signal source impedance to a 15 kΩ receiver load impedance (TTL devices). Plot the output voltage at the load from 0 to 100 ns. Compare the rise time to the round-trip propagation delay.

The characteristic impedance of this lossless cable is

$$Z_o = \sqrt{\frac{L}{C}} = \sqrt{\frac{0.078\times 10^{-6}}{29\times 10^{-12}}} \approx 52\ \Omega$$

The one-way propagation delay for this line is obtained from the line length and velocity of propagation. Since the line is lossless, the velocity of propagation for any frequency signal on the line is

$$v = \frac{1}{\sqrt{LC}} = \frac{1}{\sqrt{(0.078\times 10^{-6})(29\times 10^{-12})}} \approx 6.7\times 10^{8}\ \text{ft/s} \approx 2.0\times 10^{8}\ \text{m/s}$$

As a check, this velocity is less than the speed of light, $c = 3\times 10^{8}$ m/s. The one-way propagation delay is

$$T = \frac{l_{th}}{v} \approx \frac{1}{2.0\times 10^{8}} \approx 5\ \text{ns}$$

All frequency components on this lossless line travel at the same speed down the line. For most lossy lines, the velocity is a function of the frequency. (A special line where losses are present and yet the velocity is independent of frequency is when $R/L = G/C$.) The voltage at the load due to the nth reflection is obtained from (3.27):

$$V_n = \frac{V_o Z_o}{R_s + Z_o} H(\rho_L \rho_s H^2)^n (\rho_L + 1)$$

where $n = 0$ corresponds to the first signal to appear at the load. Rather than using a ramp function to represent the low-to-high transition, an exponential-sinusoidal function is used. It is a smooth function that can model overshoot. The solution is given in Mathcad 3.6. In this example, the rise time is approximately the same as the one-way propagation delay. Notice the overshoot and oscillation at the receiver.

An alternative method of representing the input signal involves the error function given in (3.58):

$$erf(x) = \frac{2}{\sqrt{\pi}} \int_0^x e^{-\lambda^2} d\lambda \tag{3.58}$$

The function given in (3.59) varies from exactly 0 (at $t = 0$) to about V in about τ seconds:

$$v_s(t) = V\left[erf\left(\frac{t}{k\tau}\right)\right]u(t) \tag{3.59}$$

where $u(t)$ is the unit step function (= Φ(t) in Mathcad 3.7). The coefficient k is varied to obtain the desired rise time. Although no overshoot or oscillation is modeled with this function, only one parameter,

$$f := 31.1\cdot 10^{6}$$

$$v_s(t) := 5\cdot\left(1 - e^{-3\cdot 10^{8}\cdot|t|}\cdot\cos(2\cdot\pi\cdot f\cdot t)\right)\cdot\Phi(t)$$

$$t := 0, 0.5\cdot 10^{-9} .. 3\cdot 10^{-8}$$

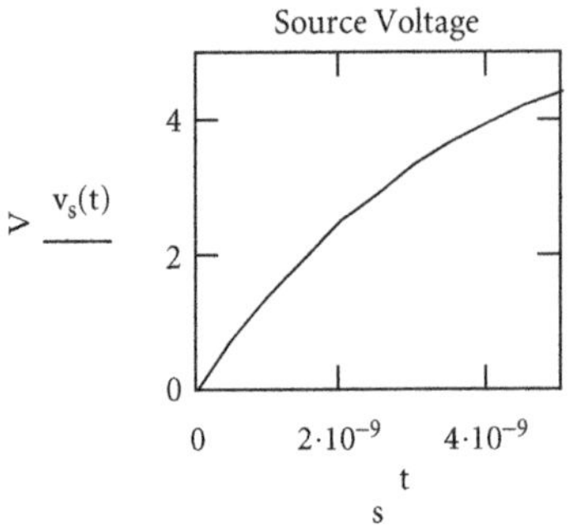

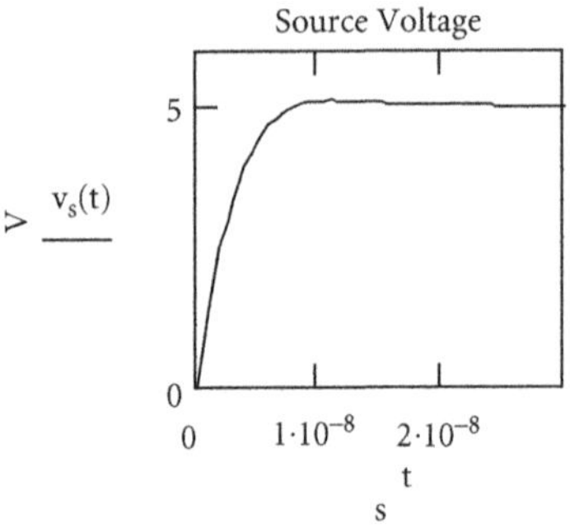

Guess $t := 1\cdot 10^{-9}$ Given $v_s(t) = 4.5$ $a := \mathrm{Find}(t)$ $a = 5.362\times 10^{-9}$

Given $v_s(t) = 0.5$ $b := \mathrm{Find}(t)$ $b = 3.437\times 10^{-10}$

$$a - b = 5.018\times 10^{-9}$$

$Z_o := 52$ $R_s := 10$ $R_L := 15\cdot 10^{3}$ $T := 5\cdot 10^{-9}$

$\rho_L := \dfrac{R_L - Z_o}{R_L + Z_o}$ $\rho_L = 0.993$ $\rho_s := \dfrac{R_s - Z_o}{R_s + Z_o}$ $\rho_s = -0.667$

$$t := 0, 0.5\cdot 10^{-9} .. 10\cdot 10^{-8}$$

$$v_L(t) := \sum_{n=1}^{12} v_s[t - (2\cdot n - 1)\cdot T]\cdot\frac{Z_o}{R_s + Z_o}\cdot(\rho_L\cdot\rho_s)^{n-1}\cdot(\rho_L + 1)$$

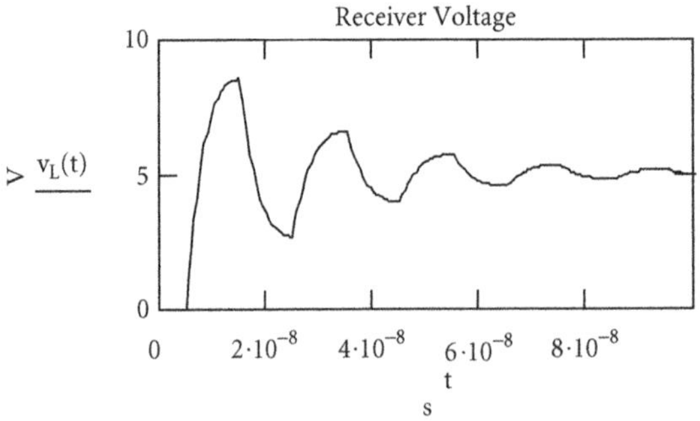

The variable f is the overshoot oscillation frequency.
Note that the 10–90% rise time is equal to 5 ns.

MATHCAD 3.6 Receiver voltage when the rise time of the input signal is equal to the one-way propagation delay of the transmission line.

$\tau := 5 \cdot 10^{-9}$ $\quad V := 5$ $\quad k := 0.94$

$t := 0, 0.1 \cdot 10^{-9} .. 5 \cdot \tau$

$v_s(t) := V \cdot \text{erf}\left(\frac{t}{k \cdot \tau}\right) \cdot \Phi(t)$

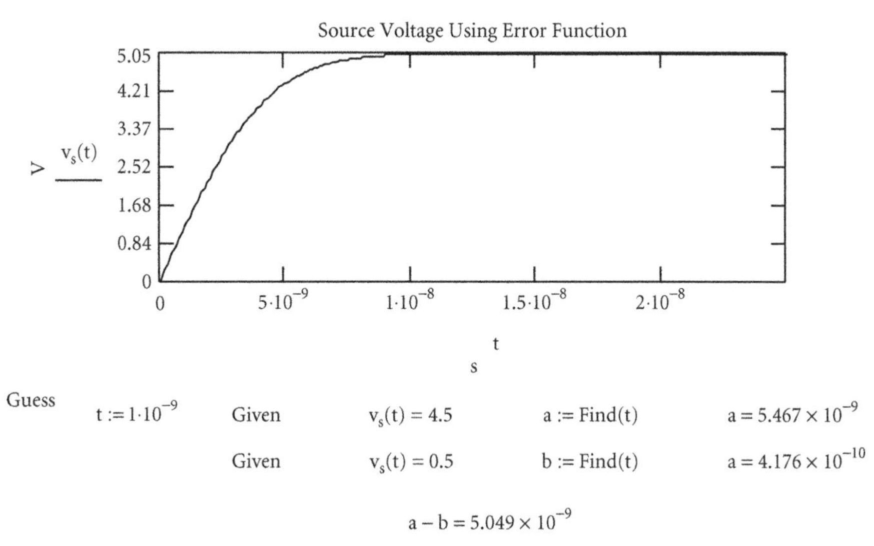

Note that the 10–90% rise time is 5 ns.

MATHCAD 3.7 Error function representation of the transient input signal.

k, must be adjusted to set the rise time. The input signal can also be modeled as

$$v_s(t) = \left[\frac{V}{2} + \frac{V}{2} erf\left(\frac{t-\tau}{k\tau}\right)\right] u(t) \tag{3.60}$$

This function has a slow initial rise near $t = 0$. It gradually rises from about 0 to V. Equation (3.60) is plotted in Mathcad 3.8. The coefficient k is 0.56 for this particular case.

3.20 Receiver Voltage when Rise Time ≪ Line Delay

Assume the rise time (the time required for the signal to rise from 10–90% of its final value) of a 0-to-5 V ramp source (low-to-high transition) is 5 ns. Furthermore, assume that a 3 meter RG/8U type cable (L = 0.078 μH/ft, C = 29 pF/ft) connects the 10 Ω signal source impedance to a 15 kΩ receiver load impedance (TTL devices). Plot the output voltage at the load from 0 to 300 ns. Compare the rise time to the round-trip propagation delay.

All the parameters for this example are identical to the previous case except that the one-way propagation delay is greater:

$$T = \frac{l_{th}}{v} \approx \frac{3}{2.0 \times 10^8} = 15\,\text{ns}$$

The solution is given in Mathcad 3.9.

$\tau := 2.3273 \cdot 10^{-9}$ $\quad V := 5 \quad k := 0.56$

$t := 0, 0.01 \cdot \tau .. 3 \cdot \tau$

$$v_s(t) := \left(\frac{V}{2} + \frac{V}{2} \cdot \mathrm{erf}\left(\frac{t - \tau}{k \cdot \tau}\right)\right) \cdot \Phi(t)$$

Source Using Modified Error Function

5.05, 4.21, 3.37, 2.52, 1.68, 0.84, 0

$> \frac{v_s(t)}{}$

0, $2 \cdot 10^{-9}$, $4 \cdot 10^{-9}$, $6 \cdot 10^{-9}$

$\frac{t}{s}$

Guess $\quad t := 3 \cdot 10^{-9}$ $\quad$ Given $\quad v_s(t) = 0.9 \cdot V$ $\quad a := \mathrm{Find}(t)$ $\quad a = 3.508 \times 10^{-9}$

Given $\quad v_s(t) = 0.1 \cdot V$ $\quad b := \mathrm{Find}(t)$ $\quad b = 1.146 \times 10^{-9}$

$a - b = 2.362 \times 10^{-9}$

Note that the 10–90% rise time is 2.4 ns.

MATHCAD 3.8 Error function (with time translation and dc shift) representation of the transient input signal.

$f := 31.1 \cdot 10^6$

$$v_s(t) := 5 \cdot \left(1 - e^{-3 \cdot 10^8 \cdot |t|} \cdot \cos(2 \cdot \pi \cdot f \cdot t)\right) \cdot \Phi(t)$$

$Z_o := 52 \quad R_s := 10 \quad R_L := 15 \cdot 10^3 \quad T := 15 \cdot 10^{-9}$

$\rho_L := \frac{R_L - Z_o}{R_L + Z_o} \quad \rho_L = 0.993 \quad \rho_s := \frac{R_s - Z_o}{R_s + Z_o} \quad \rho_s := -0.677$

$t := 0, 0.5 \cdot 10^{-9} .. 30 \cdot 10^{-8}$

$$v_L(t) := \sum_{n=1}^{12} v_s[t - (2 \cdot n - 1) \cdot T] \cdot \frac{Z_o}{R_s + Z_o} \cdot (\rho_L \cdot \rho_s)^{n-1} \cdot (\rho_L + 1)$$

Receiver Voltage

10, 5, 0

$> \frac{v_L(t)}{}$

0, $5 \cdot 10^{-8}$, $1 \cdot 10^{-7}$, $1.5 \cdot 10^{-7}$, $2 \cdot 10^{-7}$, $2.5 \cdot 10^{-7}$

$\frac{t}{s}$

MATHCAD 3.9 Receiver voltage when the rise time of the input signal is less than the one-way propagation delay of the transmission line.

In this example, the rise time is less than the one-way propagation delay.[9] Again, notice the overshoot and oscillation at the receiver. The highest frequency of interest for a 5 ns rise-time signal is approximately $1/(\pi \times 5\text{ ns}) \approx 64\text{ MHz}$, corresponding to a wavelength of

$$\lambda = \frac{v}{f} = \frac{2.0\times10^8}{64\times10^6} \approx 3.1\text{ m}$$

Thus, this 3 m long transmission line is not electrically short at 64 MHz. This coaxial cable is acting like a transmission line.

3.21 Receiver Voltage when Rise Time ≫ Line Delay

Assume the rise time (the time required for the signal to rise from 10–90% of its final value) of a 0-to-5 V ramp source (low-to-high transition) is 5 ns. Furthermore, assume that a 0.1 meter RG/8U type cable (L = 0.078 μH/ft, C = 29 pF/ft) connects the 10 Ω signal source impedance to a 15 kΩ receiver load impedance (TTL devices). Plot the output voltage at the load from 0 to 30 ns. Compare the rise time to the round-trip propagation delay.

All the parameters for this example are identical to those of the previous cases except the one-way propagation delay is less:

$$T = \frac{l_{th}}{v} = \frac{0.1}{2.0\times10^8} \approx 0.5\text{ ns}$$

$$f := 31.1\cdot10^6$$

$$v_s(t) := 5\cdot\left(1 - e^{-3\cdot10^8\cdot|t|}\cdot\cos(2\cdot\pi\cdot f\cdot t)\right)\cdot\Phi(t)$$

$$Z_o := 52 \quad R_s := 10 \quad R_L := 15\cdot10^3 \quad T := 0.5\cdot10^{-9}$$

$$\rho_L := \frac{R_L - Z_o}{R_L + Z_o} \quad \rho_L = 0.993 \quad \rho_s := \frac{R_s - Z_o}{R_s + Z_o} \quad \rho_s = -0.677$$

$$t := 0, 0.5\cdot10^{-9} .. 3\cdot10^{-8}$$

$$v_L(t) := \sum_{n=1}^{32} v_s[t - (2\cdot n - 1)\cdot T]\cdot\frac{Z_o}{R_s + Z_o}\cdot(\rho_L\cdot\rho_s)^{n-1}\cdot(\rho_L + 1)$$

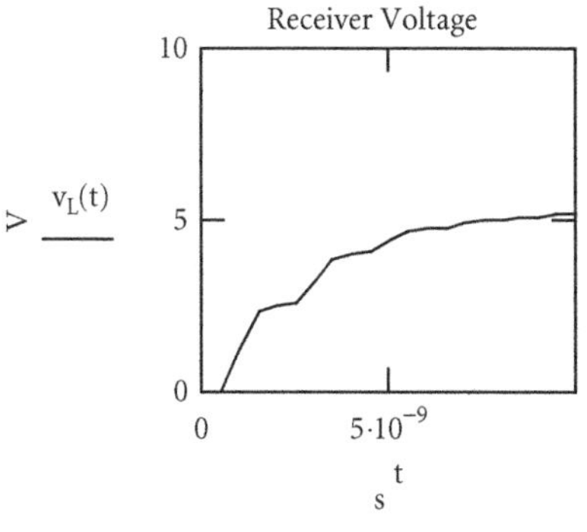

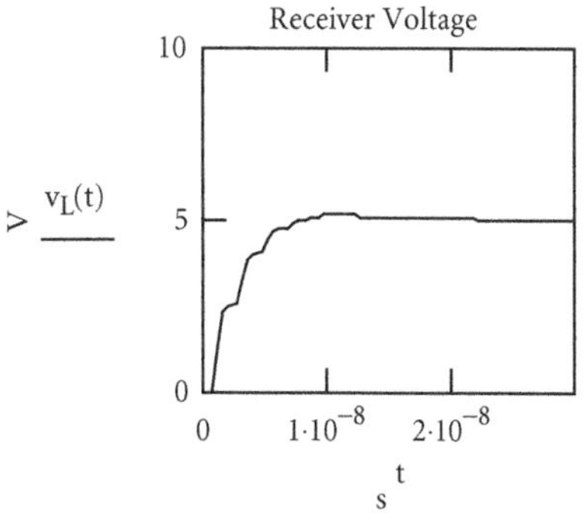

MATHCAD 3.10 Receiver voltage when the rise time of the input signal is much greater than the one-way propagation delay of the transmission line.

[9]Although the rise time is not *much* less than the line delay, the oscillation in the receiver voltage is still clearly present.

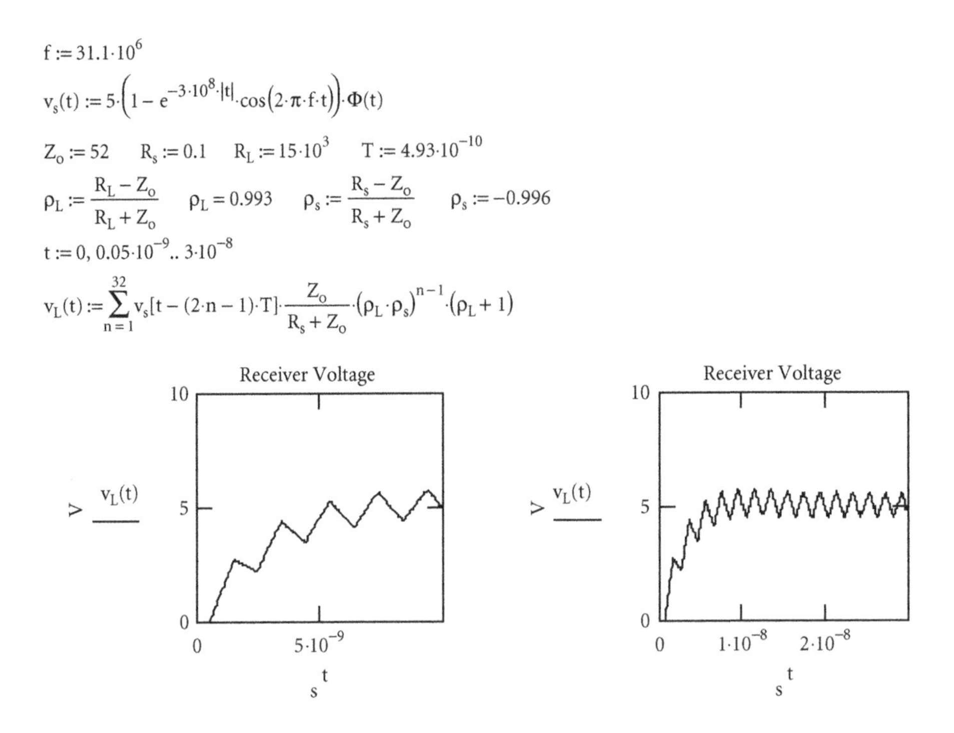

MATHCAD 3.11 Receiver voltage when the rise time of the input signal is much greater than the one-way propagation delay of the transmission line. In this case, however, the source resistance is very small compared to the characteristic impedance of the line.

The plot of the load voltage vs. time is given in Mathcad 3.10. In this case, the one-way delay of the line is one-tenth of the rise time of the input signal. Sometimes, line terminations for matching purposes are recommended when the one-way delay is equal to or greater than one-sixth of the rise time.

In this example, the rise time is much greater than the one-way propagation delay. As seen in Mathcad 3.10, there is little overshoot or oscillation. This 0.1 m long line is electrically short and is not acting like a transmission line in the normal use of the term: it is acting more like two lossless conductors. For lines that are electrically short, the source will still be rising in voltage by the time the signal arrives at the load. When the signal reflects off the receiver and returns to the driver, the input signal could still be rising. Although this line is short, reflections still occur at both the source and receiver.

It is interesting to note that if the source resistance is very small compared to the line impedance, small amplitude oscillations are clearly present at the high-impedance receiver, yet the *average* value of the signal approaches its steady-state value quickly. Because the magnitude of the product of the source and load reflection coefficients is nearly one, however, the time required for the amplitude of this small oscillation to decay significantly is very large. In Mathcad 3.11, these oscillations are clearly seen.

3.22 Advanced Transient Problem

For the transmission line circuit given in Figure 3.20, determine the values of the voltages at several different locations for a transient input signal. Allow the time, t, to vary from $t = 0$ to a time slightly after a single reflection has occurred at the most distant load.

In Table 3.1, the transient input impedances and the reflection and transmission coefficients at each of the cable-resistor junctions are listed. This information is used in this example. The one-way propagation

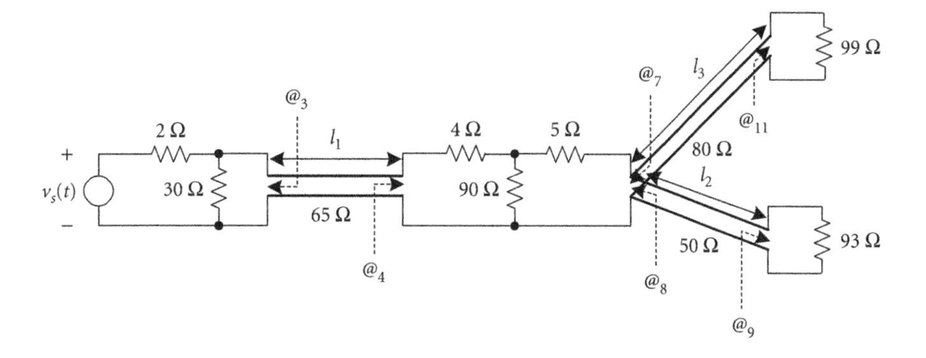

FIGURE 3.20 Circuit consisting of lumped and distributed elements.

times are given by

$$T_1 = \frac{l_1}{v_1}, \quad T_2 = \frac{l_2}{v_2}, \quad T_3 = \frac{l_3}{v_3}$$

where v_1 is the velocity of a signal on the lossless 65 Ω line, v_2 is the velocity of a signal on the lossless 50 Ω line, and v_3 is the velocity of a signal on the lossless 80 Ω line. It will be assumed that the lengths are such that

$$T_1 = T_2 = \frac{T_3}{2}$$

and all three transmission lines are electrically long. Although determining the voltages for advanced transient problems may be tedious, it is not intellectually challenging once the transmission and reflection coefficients have been determined. Voltage division is also commonly used in the analysis.

If the transient voltage source turns on at $t = 0$ and very quickly reaches a maximum amplitude of A, the voltages along the transmission line circuit are as shown in Figure 3.21 for times slightly greater than $t = 0$. Obviously, most of the voltages are zero since nonzero time is required for the source signal to travel down the transmission lines. The ⌐▶ symbol in these figures indicates the direction of the transient signal.

After $t = 0$, the transient signal of amplitude $A/1.1$ travels down the 65 Ω transmission line. For the time slightly greater than T_1, the total voltage at each of the junctions is as shown in Figure 3.22. Because the T-shaped resistor network is electrically short, any voltage across its input results in an instantaneous

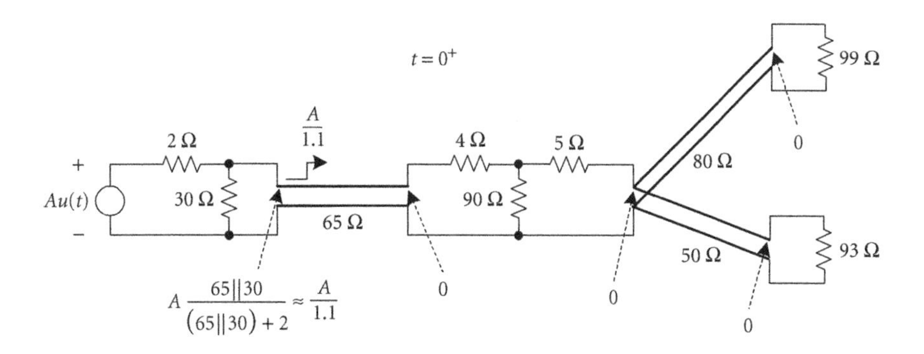

FIGURE 3.21 Voltages at a time slightly greater than $t = 0$.

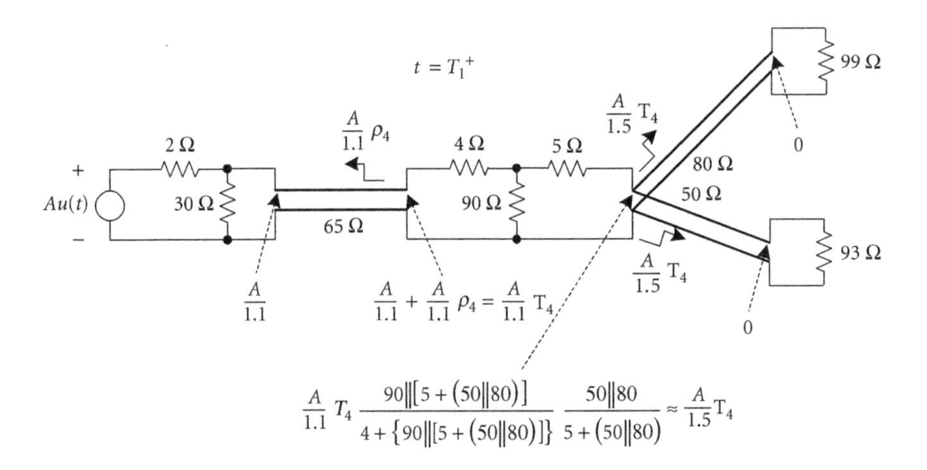

FIGURE 3.22 Voltages at a time slightly greater than $t = T_1$.

voltage across its output. The voltage across both inputs of the 50 Ω and 80 Ω transmission lines are identical since the lines are in parallel. Voltage division was used twice to obtain the given one-line solution. Also shown above the 65 Ω line in Figure 3.22 is the magnitude of the reflected signal traveling back toward the source.

For times greater than T_1, signals are present on all three transmission lines. On the 65 Ω line, two signals are present traveling in different directions. On the 50 Ω and 80 Ω lines, one signal is traveling on each line toward their respective loads. For the time slightly greater than $2T_1$, the voltages are as shown in Figure 3.23. To reduce the clutter, *not* all of the signals traveling on the transmission lines are shown.

The last finite time to be analyzed is for t slightly greater than $3T_1$ (Figure 3.24). Since both the reflected signal from the source junction and the reflected signal from the 93 Ω load are present simultaneously at the T resistor network, there are many signals present within this T network. Even the best students can become daunted. (For future times, the reflected signal from the 99 Ω load will also be present, further increasing the complexity of the solution.) For circuits of this complexity, unless symbolic expressions are required, it is recommended that a simulation package be used to determine the values of the voltages as a function of time. (These expressions were checked using SPICE.)

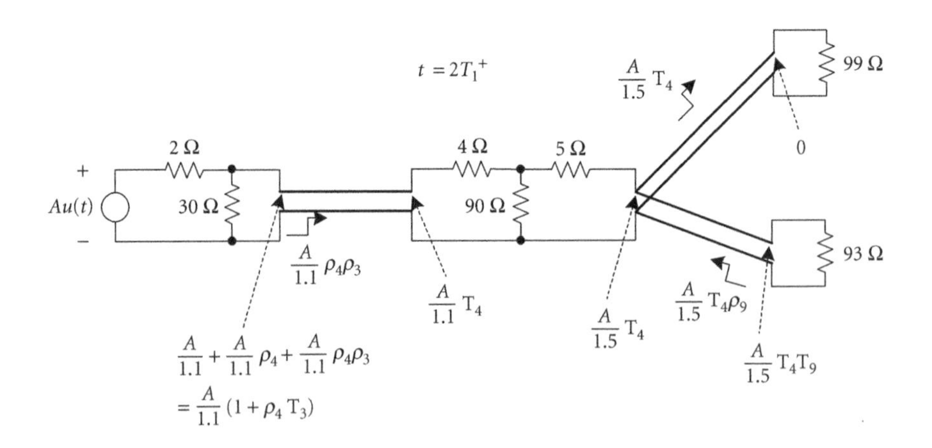

FIGURE 3.23 Voltages at a time slightly greater than $t = 2T_1$.

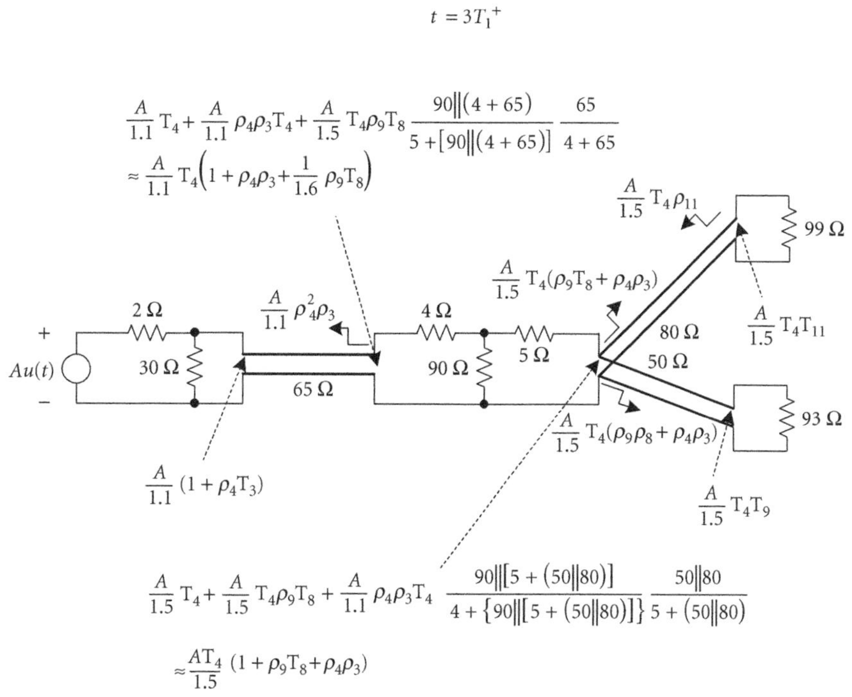

$t = 3T_1^+$

$$\frac{A}{1.1}T_4 + \frac{A}{1.1}\rho_4\rho_3 T_4 + \frac{A}{1.5}T_4\rho_9 T_8 \frac{90\|(4+65)}{5+[90\|(4+65)]}\frac{65}{4+65} \approx \frac{A}{1.1}T_4\left(1+\rho_4\rho_3+\frac{1}{1.6}\rho_9 T_8\right)$$

$\frac{A}{1.5}T_4\rho_{11}$ 99 Ω

$\frac{A}{1.5}T_4(\rho_9 T_8 + \rho_4\rho_3)$

2 Ω $\frac{A}{1.1}\rho_4^2\rho_3$ 4 Ω 80 Ω $\frac{A}{1.5}T_4T_{11}$

+ $Au(t)$ − 30 Ω 65 Ω 90 Ω 5 Ω 50 Ω

$\frac{A}{1.5}T_4(\rho_9\rho_8 + \rho_4\rho_3)$ 93 Ω

$\frac{A}{1.1}(1+\rho_4 T_3)$ $\frac{A}{1.5}T_4T_9$

$$\frac{A}{1.5}T_4 + \frac{A}{1.5}T_4\rho_9 T_8 + \frac{A}{1.1}\rho_4\rho_3 T_4 \frac{90\|[5+(50\|80)]}{4+\{90\|[5+(50\|80)]\}}\frac{50\|80}{5+(50\|80)} \approx \frac{AT_4}{1.5}(1+\rho_9 T_8 + \rho_4\rho_3)$$

FIGURE 3.24 Voltages at a time slightly greater than $t = 3T_1$.

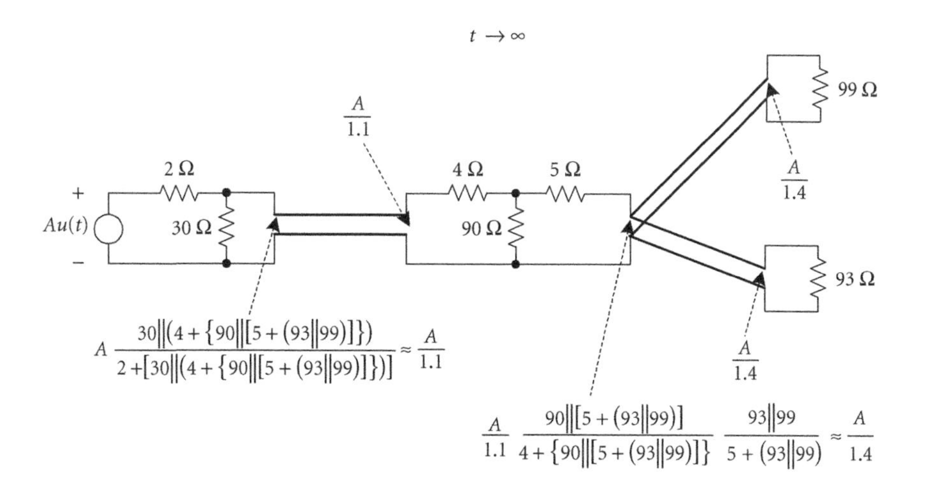

$$A\frac{30\|(4+\{90\|[5+(93\|99)]\})}{2+[30\|(4+\{90\|[5+(93\|99)]\})]} \approx \frac{A}{1.1}$$

$$\frac{A}{1.1}\frac{90\|[5+(93\|99)]}{4+\{90\|[5+(93\|99)]\}}\frac{93\|99}{5+(93\|99)} \approx \frac{A}{1.4}$$

FIGURE 3.25 Voltages after a very long time.

Finally, the steady-state voltages are as shown in Figure 3.25. The lossless transmission lines are replaced by perfectly conducting wires and the circuit analyzed in the standard manner. Voltage division was used to obtain these results.

In Mathcad 3.12, the voltage at the output of the 65 Ω line is plotted from $t = 0$ to slightly past $3T_1$. The values for the reflection and transmission coefficients were obtained from Table 3.1. The rise time

$\tau := 5 \cdot 10^{-9} \quad V := 5 \quad k := 0.94$

$\rho_3 := -0.94 \quad \rho_4 := -0.37 \quad T_4 := 0.63 \quad T_8 := 0.725 \quad \rho_9 := 0.30 \quad T_{d1} := 20 \cdot 10^{-9}$

$t := 0, 0.1 \cdot 10^{-9} .. 20 \cdot \tau$

$$v_s(t) := V \cdot \text{erf}\left(\frac{t}{k \cdot \tau}\right) \cdot \Phi(t)$$

$$v_{o65}(t) := v_s(t - T_{d1}) \cdot \frac{T_4}{1.1} + v_s(t - 3 \cdot T_{d1}) \cdot \frac{T_4}{1.1} \cdot \left(\rho_4 \cdot \rho_3 + \frac{\rho_9 \cdot T_8}{1.6}\right)$$

Output Voltage of 65 Ω Line

> $v_{o65}(t)$

4, 2, 0

0, $2 \cdot 10^{-8}$, $4 \cdot 10^{-8}$, $6 \cdot 10^{-8}$, $8 \cdot 10^{-8}$

$\frac{t}{s}$

MATHCAD 3.12 Voltage at the output of the 65 Ω line from $t = 0$ to a time slightly greater than $t = 3T_1$.

of the input signal is about 5 ns, and the lengths of the transmission lines are such that

$$T_1 = T_2 = \frac{T_3}{2} = 20 \text{ ns}$$

3.23 Ringing in Lumped Circuits

Briefly explain why an impedance mismatch in a lumped circuit (such as an electrically-short transmission line) can cause ringing and, hence, signal distortion.

If the load is not matched to the characteristic impedance of the connecting line, an incident signal will reflect off the load. If the source is also not matched to the characteristic impedance, this reflected signal from the load, when it eventually arrives at the source, will reflect off the source. This reflected signal from the source, when it arrives at the load, will then reflect off the load. These reflections off the source and load continue until the amplitude of the last reflection is negligible. The summation of all the signals at the load will appear as ringing with an oscillation frequency and decay rate. Even if a line is electrically short, the line still behaves as a transmission line in the sense that it has a nonzero inductance and capacitance. A lumped circuit is by definition electrically short: the conductors connecting the elements are electrically-short transmission lines.

If an impedance mismatch is present between the line and load, the signal will reflect off the load. For lumped circuits, the signal that results from the sum of all of the incident and reflected signals is referred to as ringing. The Q of the circuit and the energy content of the input signal determines whether this ringing is significant.

3.24 More Shunt Matching

If a driver is connected to a receiver via a transmission line and the receiver input impedance is not equal to the characteristic impedance of the connecting line, a mismatch will exist. In theory, a termination resistor of impedance equal to the characteristic impedance of the line can be placed in shunt with the receiver to match the line to the receiver. Explain when this is a valid technique. When the driver consists of certain digital devices, however, this terminating resistor can heavily load down the driver in steady-state operation; in other words, the driver cannot properly supply the necessary current. Determine the minimum value of line impedance that can be matched in this manner for an AND gate of each of the following digital types: TTL, CMOS, and ECL.

If the impedance of the receiver, Z_R, is equal to the characteristic impedance of the line, Z_o, the reflection coefficient at the load is zero. Usually, however, the receiver input impedance is not equal to the line impedance. If the input impedance of the receiver is large compared to the line impedance, a matching resistor, R_M, placed across (in shunt with) the receiver input with a value equal to the line impedance will reasonably match the receiver to the line:

$$Z_L = \frac{Z_R R_M}{Z_R + R_M} \approx \frac{Z_R R_M}{Z_R} = R_M = Z_o \quad \text{if } Z_R \gg Z_o,\ Z_R \gg R_M$$

In steady-state operation, the transients have disappeared, and the impedance seen by the driver is merely the total load at the end of the lossless transmission line. The transmission line appears as an ideal set of wires connecting the driver to the receiver. Although not specifically shown in Figure 3.26, the driver and transmission line have a common return with the matching resistor.

For a 74LS08, 2-input AND chip, typical values for the High and Low output voltages and currents are, for an operating voltage of $V_{CC} = 5.25$ V,

$$V_{OH} = 3.4\text{ V},\quad I_{OH} = -0.4\text{ mA}$$
$$V_{OL} = 0.35\text{ V},\quad I_{OL} = 8\text{ mA}$$

The current levels associated with the High and Low states are not the same for TTL devices. Although students will occasionally tie the output of TTL devices such as this directly to LED's, many LED's require more than 0.4 mA.[10] This can make the device more susceptible to EMI by heavily loading down the device and changing its intended operating range (increasing its noise susceptibility). For example, for the Low state of TTL, as the load current increases, the Low output voltage increases. Eventually, the receiving gate will no longer interpret the voltage as Low. Also, if the output LED is High asserted, only 0.4 mA is available to drive the LED (the negative value of −0.4 mA implies that the device is sourcing current or positive current is passing out of the device). If LED's must be driven directly with TTL, it is best if the LED's are Low asserted with the smallest possible number of other gates tied to the same output (or a high output-current bus driver used). The minimum value for the matching resistor and line impedance is

$$Z_{o,min} = \max\left(\left|\frac{V_{OH}}{I_{OH}}\right| = 8.5\text{ k}\Omega, \left|\frac{V_{OL}}{I_{OL}}\right| = 44\ \Omega\right) = 8.5\text{ k}\Omega \tag{3.61}$$

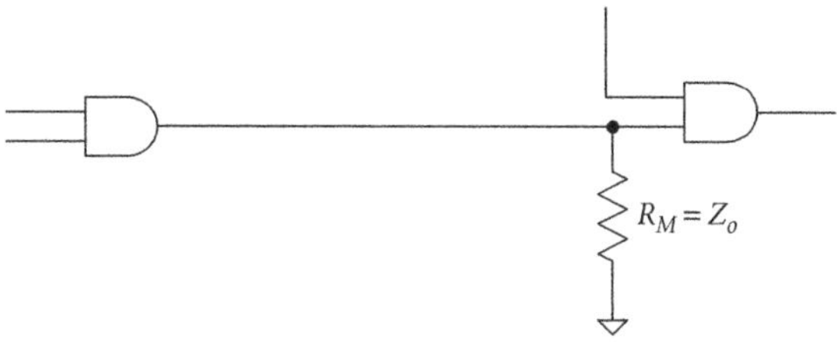

$R_M = Z_o$

FIGURE 3.26 In dc steady-state conditions, the ideal transmission line connecting the driver to the receiver appears like a wire with no resistance.

[10]Generally, green LEDs are less efficient than red LEDs. For example, 4.4 mA was required to turn on barely one particular green LED at a dc voltage of 1.9 V.

If the steady-state load seen by the driver is less than 8.5 kΩ, the current magnitude will exceed 0.4 mA.

For a 74HC08, 2-input CMOS AND chip, typical values for the High and Low output voltages and currents are, for an operating voltage of $V_{CC} = 6$ V,

$$V_{OH} = 5.7\ \text{V},\quad I_{OH} = -5.2\ \text{mA}$$
$$V_{OL} = 0.2\ \text{V},\quad I_{OL} = 5.2\ \text{mA}$$

Notice that the current levels associated with the High and Low states are the same for this CMOS device. The minimum value for the matching resistor and line impedance is

$$Z_{o,min} = \max\left(\left|\frac{V_{OH}}{I_{OH}}\right| = 1.1\,\text{k}\Omega, \left|\frac{V_{OL}}{I_{OL}}\right| = 38\,\Omega\right) = 1.1\,\text{k}\Omega \tag{3.62}$$

For a 10H104, 2-input ECL AND chip, typical values for the High and Low output voltages and currents are, for an operating voltage of $V_{CC} = -5.2$ V,

$$V_{OH} = -0.81\ \text{V},\quad I_{OH} = -24\ \text{mA}$$
$$V_{OL} = -1.63\ \text{V},\quad I_{OL} = -7.4\ \text{mA}$$

The output current is a function of the output pull-up resistor and voltage (a 50 Ω pull-up resistor connected to a −2 V supply was used in this case). The current levels associated with the High and Low states are different for ECL devices but in the same direction. The minimum value for the matching resistor and line impedance is

$$Z_{o,min} = \max\left(\left|\frac{V_{OH}}{I_{OH}}\right| = 34\,\Omega, \left|\frac{V_{OL}}{I_{OL}}\right| = 220\,\Omega\right) = 220\,\Omega \tag{3.63}$$

3.25 Shunt Matching with a Split Termination for a TTL System

A split termination consists of a pull-down resistor of value R_1 and pull-up resistor of value R_2 at the receiver. Resistor R_1 is pulled down to the lower supply voltage (most negative voltage) and R_2 is pulled up to the higher supply voltage (most positive voltage). Determine the Thévenin equivalent of this split termination, and use this information to determine the values for R_1 and R_2 to match the receiver to the transmission line of characteristic impedance Z_o. For a 65 Ω line, determine the values for R_1 and R_2 for a TTL NAND gate. Use the current levels for both the High and Low states and the maximum possible supply voltages. Note: the High and Low states for a TTL driver source and sink current, respectively. [Johnson, '93; Montrose, '99]

The matching network at the receiver, known as a Thévenin match, is shown in Figure 3.27. The Thévenin resistance of this matching network is obtained by turning-off the supply voltages (resulting in zero volts between the supply voltages V_{CC} and V_{EE}) and then determining the net resistance of the resultant network. The net or equivalent resistance is R_1 in parallel with R_2 (assuming that the input impedance of the receiver is much greater than either resistance):

$$R_{th} = \frac{R_1 R_2}{R_1 + R_2} \tag{3.64}$$

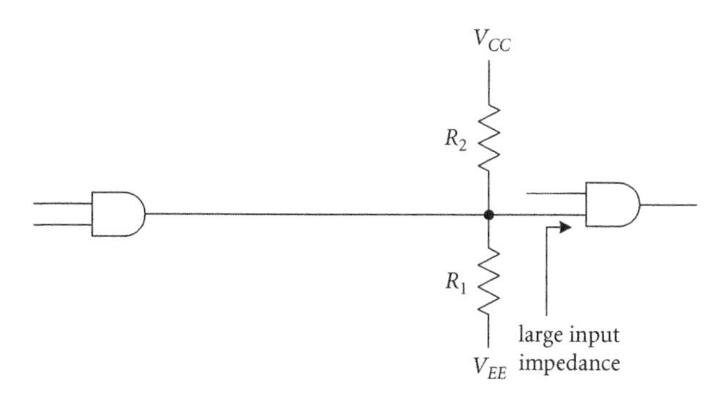

FIGURE 3.27 Thévenin matching network.

The Thévenin voltage is the voltage across the open-circuited output of the network. The "output" is the voltage at the node between R_1 and R_2. Using voltage division,

$$V_{th} = \frac{R_2 V_{EE} + R_1 V_{CC}}{R_1 + R_2} \tag{3.65}$$

Thus, the Thévenin matching network allows the line voltage to be set to some optimal value between the lowest and highest voltage. To avoid ringing, it is important to match the characteristic impedance Z_o of the line to the load. This matching is accomplished by setting Z_o equal to the equivalent resistance of the matching network:

$$R_{th} = \frac{R_1 R_2}{R_1 + R_2} = Z_o$$

However, the driver must be capable of supplying the necessary current drawn by this matching network. It is sometimes possible through the careful selection of the matching resistors to supply some of this current via the matching network supply voltage. The equivalent steady-state circuit is shown in Figure 3.28.

For a 74LS08, 2-input AND chip, typical values for the High and Low output voltages and currents are, for an operating voltage of $V_{CC} = 5.25$ V,

$$V_{OH} = 3.4 \text{ V}, \quad I_{OH} = -0.4 \text{ mA}$$
$$V_{OL} = 0.35 \text{ V}, \quad I_{OL} = 8 \text{ mA}$$

The first equation that must be satisfied for a 65 Ω line involves the High-level voltage and current:

$$I = \frac{V_s - V_{th}}{R_{th}} = \frac{3.4 - V_{th}}{65} < 0.4 \text{ mA} \quad \Rightarrow V_{th} > 3.37 \text{ V}$$

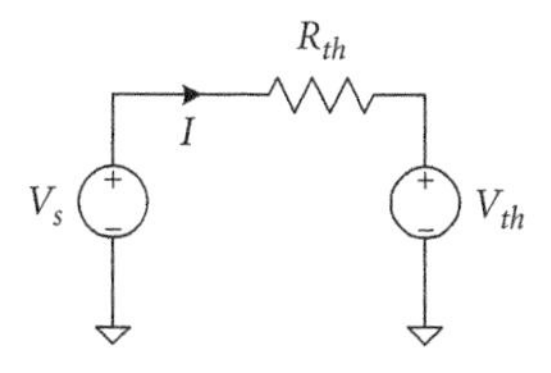

FIGURE 3.28 Equivalent circuit of a Thévenin matching network.

The voltage V_{th} must also be less than 3.4 V if the current, I, is to remain positive. The second equation involves the Low-level voltage and current:

$$-I = \frac{V_{th} - V_s}{R_{th}} = \frac{V_{th} - 0.35}{65} < 8\,\text{mA} \quad \Rightarrow V_{th} < 0.87\text{ V}$$

Of course, V_{th} must also be greater than 0.35 V if the current, I, is to remain negative. Obviously, the Thévenin voltage cannot be both less than 0.87 V and greater than 3.37 V. Thus, no combination of R_1 and R_2 will prevent the driver from being loaded down and yet still match the receiver to the 65 Ω transmission line. The smallest possible characteristic impedance for the transmission line that will not load the driver is easily obtained:

$$\frac{3.4 - V_{th}}{R_{th}} = 0.4\text{ mA} \quad \Rightarrow 3.4 - 0.4 \times 10^{-3} R_{th} = V_{th}$$

$$\frac{V_{th} - 0.35}{R_{th}} = 8\text{ mA} \quad \Rightarrow 0.35 + 8 \times 10^{-3} R_{th} = V_{th}$$

$$\therefore R_{th} \approx 360\,\Omega \leq Z_o \tag{3.66}$$

Referring to Equation (3.61), 360 Ω is an improvement over 8.5 kΩ, but 360 Ω is still not a reasonable impedance for most PCB transmission lines. These TTL devices are not capable of driving an end termination.

Notice that when the two resistors are equal in value, the "idle"[11] voltage is midway between (or the average of) the two supply voltages:

$$V_{int} = \frac{V_{EE} + V_{CC}}{2} \quad \text{if } R_1 = R_2 \tag{3.67}$$

The current drawn for a High output state is

$$I = \frac{V_{OH} - V_{th}}{R_{th}} \tag{3.68}$$

while the current drawn for a Low output state is

$$I = \frac{V_{OL} - V_{th}}{R_{th}} \tag{3.69}$$

Therefore, the current requirements for the two states are similar when $R_1 = R_2$. When $R_1 > R_2$, the Thévenin voltage

$$V_{th} = \frac{R_2 V_{EE} + R_1 V_{CC}}{R_1 + R_2}$$

or the idle voltage is greater than the average value, and the Low-state current requirements are greater than the High-state current requirements. When $R_2 > R_1$, the idle voltage is less than the average value, and the High-state current requirements are greater than the Low-state current requirements.

[11]Idle corresponds to the time of no signal activity.

3.26 Shunt Matching with a Split Termination for a CMOS System

Repeat the previous analysis for a CMOS AND gate. Note: the High and Low states for a CMOS driver sources and sinks current, respectively. Why are CMOS gates usually unaffected by reflections unless the interconnections are very long?

The analysis is similar to that presented previously. For a 74HC08, 2-input AND chip, typical values for the High and Low output voltages and currents are, for an operating voltage of $V_{CC} = 6$ V,

$$V_{OH} = 5.7\text{ V},\quad I_{OH} = -5.2\text{ mA}$$
$$V_{OL} = 0.2\text{ V},\quad I_{OL} = 5.2\text{ mA}$$

The first equation that must be satisfied involves the High-level voltage and current:

$$I = \frac{V_s - V_{th}}{R_{th}} = \frac{5.7 - V_{th}}{65} < 5.2\text{ mA} \quad \Rightarrow \quad V_{th} > 5.4\text{ V}$$

Of course, V_{th} must also be less than 5.7 V if the current, I, is to remain positive. The second equation involves the Low-level voltage and current:

$$-I = \frac{V_{th} - V_s}{R_{th}} = \frac{V_{th} - 0.2}{65} < 5.2\text{ mA} \quad \Rightarrow \quad V_{th} < 0.54\text{ V}$$

The voltage V_{th} must also be greater than 0.2 V if the current, I, is to remain negative. Obviously, the Thévenin voltage cannot be less than 0.54 V and greater than 5.4 V simultaneously. Thus, no combination of R_1 and R_2 will prevent the driver from being loaded down and yet still match the receiver to the 65 Ω transmission line. The smallest possible characteristic impedance for the transmission line that will not load the driver is easily obtained:

$$\frac{5.7 - V_{th}}{R_{th}} = 5.2\text{ mA} \qquad \Rightarrow 5.7 - 5.2\times 10^{-3} R_{th} = V_{th}$$

$$\frac{V_{th} - 0.2}{R_{th}} = 5.2\text{ mA} \qquad \Rightarrow 0.2 + 5.2\times 10^{-3} R_{th} = V_{th}$$

$$\therefore R_{th} \approx 530\,\Omega \le Z_o \tag{3.70}$$

Referring to Equation (3.62), 530 Ω is an improvement over 1.1 kΩ, but 530 Ω is still not a reasonable impedance for most PCB transmission lines. These CMOS devices are not capable of driving an end termination. However, for CMOS gates, the reflections are usually not important because CMOS gates are slow; that is, the rise/fall times are long (30–50 ns).[12]

3.27 Shunt Matching with a Split Termination for an ECL System

Repeat the previous analysis for an ECL AND gate, but assume that the characteristic impedance of the transmission line is 120 Ω instead of 65 Ω. Note: both the High and Low states for an ECL driver source current. (Check: the rule-of-thumb given in databooks is $R_1 = 2.6Z_o$, $R_2 = 1.6Z_o$.) [Blood; Wakerly]

[12]Whoops! This statement is no longer true. CMOS technology has improved drastically since the mid-90's. Rise times have decreased to 0.5 ns and less.

For ECL gates, reflections are especially important. Typically, the undershoot should not exceed 10% and the overshoot should not exceed 35% of the logic swing. If the signal overshoots too much, the transistors may go into saturation, slowing the logic. If the logic undershoots too much, the noise margin may be exceeded resulting in a false Low trigger. For a 10H104, 2-input AND chip, typical values for the High and Low output voltages and currents are, for an operating voltage of $V_{EE} = -5.2$ V,

$$V_{OH} = -0.81\,\text{V}, \quad I_{OH} = -24\,\text{mA}$$
$$V_{OL} = -1.63\,\text{V}, \quad I_{OL} = -7.4\,\text{mA}$$

The output current is a function of the output pull-up resistor and voltage (a 50 Ω pull-up resistor connected to a –2 V supply was used in this case). The analysis is similar to that presented previously. The first equation that must be satisfied involves the High-level voltage and current:

$$I = \frac{V_s - V_{th}}{R_{th}} = \frac{-0.81 - V_{th}}{120} < 24\,\text{mA} \quad \Rightarrow V_{th} > -3.7\,\text{V}$$

Of course, V_{th} must also be less than –0.81 V ($V_{th} < -0.81$ V) if the current, I, is to remain positive. The second equation involves the Low-level voltage and current:

$$I = \frac{V_s - V_{th}}{R_{th}} = \frac{-1.63 - V_{th}}{120} < 7.4\,\text{mA} \quad \Rightarrow V_{th} > -2.5\,\text{V}$$

The Thévenin voltage, V_{th}, must also be less than –1.63 V ($V_{th} < -1.63$ V) if the current, I, is to remain positive. Combining all these inequalities yields

$$-2.5 < V_{th} < -1.63\,\text{V}$$

Selecting $V_{th} = -2$ V and realizing that $R_{th} = 120$ Ω, the two equations required to solve for the unknown matching resistor values are ($V_{CC} = 0$ V and $V_{EE} = -5.2$ V)

$$V_{th} = -2 = \frac{R_2 V_{EE} + R_1 V_{CC}}{R_1 + R_2} = \frac{-5.2 R_2}{R_1 + R_2} \tag{3.71}$$

$$R_{th} = 120 = \frac{R_1 R_2}{R_1 + R_2} \tag{3.72}$$

There are two equations and two unknowns. Solving for the two unknowns, $R_1 = 312$ Ω and $R_2 =$ 195 Ω. The rule-of-thumb yields $R_1 \approx 310$ Ω and $R_2 \approx 190$ Ω. (Close common values for these two resistors are 200 Ω and 300 Ω.) The disadvantage of this matching scheme is the power consumed by the resistors.

The smallest possible characteristic impedance for the transmission line that will not load the driver is easily obtained:

$$\frac{-0.81 - V_{th}}{R_{th}} = 24\,\text{mA} \quad \Rightarrow -0.81 - 24 \times 10^{-3} R_{th} = V_{th}$$

$$\frac{-1.63 - V_{th}}{R_{th}} = 7.4\,\text{mA} \quad \Rightarrow -1.63 - 7.4 \times 10^{-3} R_{th} = V_{th}$$

$$\therefore R_{th} \approx 49\,\Omega \le Z_o$$

Referring to Equation (3.63), 49 Ω is an improvement over 220 Ω. Furthermore, this is a reasonable range for a line impedance on a PCB.

The derivation of the rule-of-thumb is simple. Assuming that $V_{th} = -2$ V (the smallest necessary voltage) and $R_{th} = Z_o$ is greater than or equal to 50 Ω (smallest possible impedance),

$$-2 = \frac{-5.2R_2}{R_1 + R_2},\ 50 = \frac{R_1 R_2}{R_1 + R_2} \quad \Rightarrow R_2 \approx 1.6Z_o \text{ and } R_1 = 1.6R_2 \approx 2.6Z_o \tag{3.73}$$

This combination of resistors will provide the necessary match to the transmission line of impedance Z_o, and the driver will be capable of supplying the necessary current as long as this line impedance is greater than or equal to 50 Ω.

3.28 Split-Termination Equivalent

Referring to the previous TTL split-termination discussion, when many matching terminations are expected, a single termination resistor connected to some intermediate voltage between the upper and lower supply voltages can be utilized at each receiver instead of a split termination. Determine the value of this single termination resistor. If the lower supply voltage is V_{EE} and the upper supply voltage is V_{CC}, show that this intermediate voltage is

$$V_{int} = \frac{R_2 V_{EE} + R_1 V_{CC}}{R_1 + R_2}$$

This single resistor is known as the Thévenin resistance of the matching network, and the single intermediate voltage is the Thévenin voltage of the matching network. From (3.64), the resistance value is

$$R_{th} = \frac{R_1 R_2}{R_1 + R_2} = Z_o$$

The Thévenin voltage can be obtained by determining the voltage at the transmission line output (at the junction of the two matching resistors) through superposition. Referring to Figure 3.27, when the lower supply is turned off, the voltage at the resistor junction due to V_{CC} is, through voltage division,

$$V_{CC} \frac{R_1}{R_1 + R_2}$$

When the upper supply is turned off, the voltage at the junction due to V_{EE} is, through voltage division,

$$V_{EE} \frac{R_2}{R_1 + R_2}$$

For linear systems, superposition can be applied. The total voltage due to both supplies is the superposition or sum of the voltages due to each supply acting alone:

$$V_{CC} \frac{R_1}{R_1 + R_2} + V_{EE} \frac{R_2}{R_1 + R_2} = \frac{R_2 V_{EE} + R_1 V_{CC}}{R_1 + R_2} \tag{3.74}$$

This value is also the voltage on the transmission line during idle moments. As long as the driver current requirements are satisfied, the resistor values can be adjusted so that the voltage at the receiver is either a Low or High during idle moments.

3.29 Experimentally Determining the Line Impedance

How can the split-termination setup be used to match a transmission line to a receiver? Assume a step generator is available, and the resistors R_1 and R_2 are variable. If a step voltage of magnitude V is applied at the input of a transmission line of characteristic impedance Z_o, determine the value of the voltage appearing at the receiver at T_D and at the driver at $2T_D$ (T_D is the one-way time delay of the line) as a function of R_1, R_2, and Z_o.

The Thévenin resistance of the load, assuming the impedance of the receiver is much greater than either R_1 or R_2, is

$$R_{th} = \frac{R_1 R_2}{R_1 + R_2}$$

and the reflection coefficient at the load is

$$\rho_L = \frac{Z_L - Z_o}{Z_L + Z_o} = \frac{\dfrac{R_1 R_2}{R_1 + R_2} - Z_o}{\dfrac{R_1 R_2}{R_1 + R_2} + Z_o}$$

The voltage applied to the input of the transmission line is V. The voltage reflected from the load is therefore $\rho_L V$. If the source, or driver, is matched to the transmission line, the signal reflected off the load is not reflected off the source, and there are no further reflections. A matching pad can be used at the driver to perform this matching function. The total voltage at the driver after the reflected signal returns to the source is the sum of the forward and backward traveling waves:

$$V + V\rho_L = V + V\frac{\dfrac{R_1 R_2}{R_1 + R_2} - Z_o}{\dfrac{R_1 R_2}{R_1 + R_2} + Z_o} = V\left[\frac{2R_1 R_2}{R_1 R_2 + (R_1 + R_2)Z_o}\right] \tag{3.75}$$

Equation (3.75) is the total voltage at the receiver immediately after $t = T_D$ and at the driver immediately after $t = 2T_D$. Resistors R_1 and R_2 can be varied until the reflected signal is zero (ideally), the reflection coefficient is zero, and the load is matched to the transmission line. The voltage across the input of the transmission line can be monitored to verify that the reflected signal is zero: the voltage at the input should not increase in value after the step voltage is applied.

A faster and more systematic method of obtaining the resistor values is through a measurement of the signal amplitude at the input of the line after the reflected signal returns to the input. In Figure 3.29,

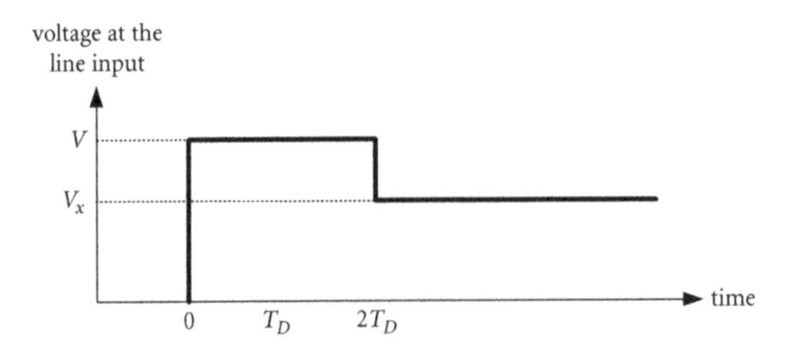

FIGURE 3.29 Voltage at the input of a transmission line with a mismatched load. The source is assumed matched to the line.

this steady-state voltage amplitude is denoted as V_x. The amplitude was given in (3.75):

$$V_x = V\left[\frac{2R_1R_2}{R_1R_2 + (R_1 + R_2)Z_o}\right] \Rightarrow Z_o = \frac{R_1R_2}{R_1 + R_2}\left(\frac{2V - V_x}{V_x}\right) \tag{3.76}$$

If the values of the two resistors are known, as well as the incident and final voltage levels, the characteristic impedance of the line can be determined from this expression. Once the line impedance Z_o is known, the resistor values can be adjusted to satisfy the following impedance matching equation:

$$Z_o = \frac{R_1R_2}{R_1 + R_2} \tag{3.77}$$

Of course, the loading effect of these resistors on the driver must also be considered.

3.30 Series Matching and Dynamic Output Resistance

If a receiver is not matched to its connecting transmission line of impedance Z_o, determine why a series resistor of resistance Z_o between the driver and transmission line can be used instead of a shunt resistor at the receiver. What is an advantage and a disadvantage of matching in this manner? In practice, this series resistance is less than Z_o because of the nonzero resistance of the driver output. Determine typical series matching resistor values for the High and Low states of a TTL, a CMOS, and an ECL gate. [Johnson, '93; Blood]

A series matching resistor at the driver, a source termination, was previously analyzed. Although the reflection coefficient at the load, the receiver, is not zero, the total voltage at the receiver is the proper voltage level if the load reflection coefficient, ρ, is nearly one. One-half the source voltage, V, is sent down the line (because of the voltage division between the source termination and line impedance), and $V/2 + \rho V/2 = V$. The reflected signal from the load, when it arrives at the source, is not rereflected because the reflection coefficient at the source for this negative-going wave is zero.

An advantage of matching in this manner is the lower steady-state current required from the driver. With shunt matching at the receiver, the matching resistor will load down the driver in steady state unless the matching resistor is "properly biased." The impedance seen by the initial transient signal for a series terminated line is the sum of R_M, the series matching resistor, and Z_o, the line impedance. With shunt matching at the receiver, the impedance seen by the initial transient signal is Z_o, the cable impedance.

If the input signal changes quickly, this difference in the steady-state current between the series and shunt matching may not be a factor since a steady-state value may not be reached before the input changes state. For slowly varying signals, the power dissipated in a series matching resistor is less than that dissipated in a shunt matching resistor.

Another advantage of series matching is that a better match is more likely obtainable. The output impedance of most gates is mainly resistive (the small resistance dominates over the shunt capacitive reactance). The input impedance of most gates, however, is resistive and capacitive (capacitance of the emitter-base junction or the capacitance of the gate-source junction). This capacitance is not removed with resistive matching networks. If the transmission line connecting the driver and receiver is not electrically long, then the series matching resistor will provide some damping of the signal if ringing is present. Also, if the characteristic impedance of the line is not known or varying (e.g., straight wire connections), then a series resistor is also helpful. In this case, the small-valued resistor can dampen the oscillations by lowering the Q of the circuit.

One of the disadvantages of the source termination is that the initial voltage traveling down the line is one-half of the desired voltage V. Other gates connected to the same transmission line (before the final load) will first detect a voltage of $V/2$ and then $V/2 + V/2 = V$ at some later time. This $V/2$ may be interpreted as a Low voltage. With shunt matching, gates can be connected along the line since the initial voltage traveling down the line is V.

Another disadvantage of series matching is the greater time constant at the load. The impedance seen at the load looking toward the source (or for that matter anywhere along the line) is equal to Z_o for a transient signal. Modeling the receiver as a capacitance, C, the RC time constant for the transients is about Z_oC for series matching at the driver. However, for shunt matching at the load, the total impedance seen at the load is the Z_o of the transmission line in parallel with the Z_o of the matching resistor. The total impedance is therefore $Z_o/2$, and the time constant for the transients is $Z_oC/2$ for shunt matching at the receiver. The consequence of the larger time constant can be clearly seen by imagining that the initial transient signal is a zero-to-five volt unit step. The capacitance at the receiver will prevent the signal from rising instantaneously from zero to five volts (the voltage across a capacitor cannot change instantaneously). After a few time constants, the signal will be approximately five volts. For series matching, the time required for the load signal to rise to five volts is twice that of shunt matching since the time constant is twice as large.

Ideally, the output impedance of a gate is zero. Actually, the output impedance is nonlinear and sometimes a function of the output current state (High or Low) and logic family. The dynamic resistance of the output is often obtained by determining or approximating the slope of the v-i output curve at some operating point. Once this output resistance is determined, the value of the matching resistor is reduced by this amount so that the sum of the series resistor and output resistance of the gate is equal to the characteristic impedance of the transmission line.

Sometimes the concepts of static and dynamic resistance are not clearly understood. Obviously, for a resistor, a linear element, the ratio of the voltage across to the current through the resistor is equal to R, its resistance:

$$R = \frac{v}{i} \tag{3.78}$$

This linear relationship between the current and voltage is shown in Figure 3.30. Whether the voltage across the resistor is a dc signal, a small-amplitude ac signal, a large-amplitude ac signal, a small-amplitude ac signal with a dc offset, or a large-amplitude ac signal with a dc offset, the resistance is the same. However, for nonlinear devices such as diodes and many logic devices, the resistance is a function of the voltage across and current through the device. The static resistance of a diode, for example, is the ratio of the dc voltage across to the dc current through the diode. This resistance is *not* necessarily equal to the slope of the v-i curve (the derivative of the voltage with respect to the current) for the diode at this dc operating point. However, the dynamic resistance of a diode for a small-amplitude ac signal is the slope of the v-i curve at its dc operating point (V_{dc}):

$$r_{ac} = \left.\frac{dv}{di}\right|_{V_{dc}} \tag{3.79}$$

The nonlinear relationship between the current and voltage for a diode is shown in Figure 3.31. For the large voltage swings associated with many transient conditions, the dynamic resistance clearly changes

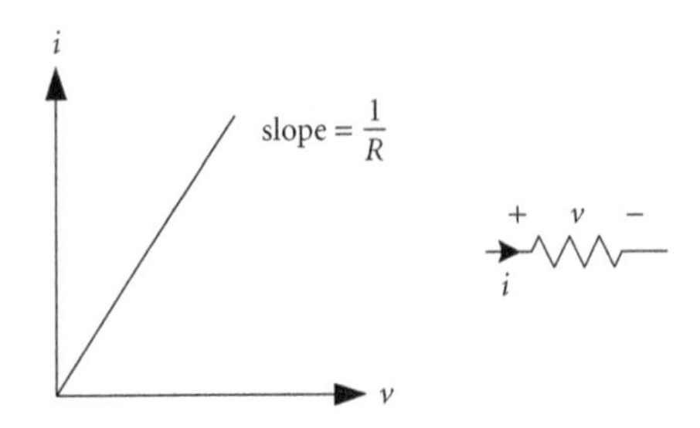

FIGURE 3.30 Current-voltage relationship for a linear resistor.

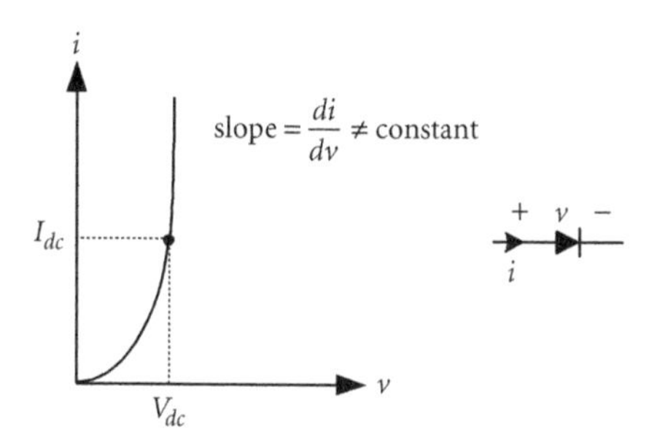

FIGURE 3.31 Current-voltage curve for a nonlinear diode.

with the voltage, and it is not necessarily just the slope at one particular operating point. For example, the slope of the *i*-*v* curve for a diode changes rather dramatically from 0 to 1 V. The resistance is not formally defined for these large voltage swings for a nonlinear device. Sometimes, an average dynamic resistance, the dynamic resistance at some central operating point, or the maximum or minimum dynamic resistance over the range of interest is used.

For TTL logic, the slope of the *v*-*i* curve is a function of the output state and series type. For the gates of one LS series, the dynamic resistance is approximately

$$\left|\frac{\Delta V_{out}}{\Delta I_{out}}\right| \approx \left|\frac{2.5-3}{15\times 10^{-3}-9\times 10^{-3}}\right| \approx 83\,\Omega \quad \text{for the High State}$$

$$\left|\frac{\Delta V_{out}}{\Delta I_{out}}\right| \approx \left|\frac{0.36-0.5}{7\times 10^{-3}-16\times 10^{-3}}\right| \approx 16\,\Omega \quad \text{for the Low State}$$

When determining the slope of the *v*-*i* curve for the Low state, it is important that the voltage level correspond to typical Low-level values because the slope is not constant from 0 to 3 V. For example, the *v*-*i* slope in the 0.6 to 1.6 voltage range is much greater than 16 Ω; however, the noise margin for this Low voltage state range is not as great, and this logic should not normally be operated in this range. Since the output resistance is a function of the state level, a compromise is necessary when selecting a series termination resistor.

Another method of determining an approximate dynamic output resistance utilizes the output current and voltage corresponding to a typical High and Low state. For 74HC logic with V_{CC} = 4.5 V,

$$V_{OH} = 3.94\text{ V},\quad I_{OH} = -4.0\text{ mA}$$

$$V_{OL} = 0.33\text{ V},\quad I_{OL} = 4.0\text{ mA}$$

Since the *v*-*i* curve passes through V_{CC} at the maximum voltage,

$$\left|\frac{\Delta V_{out}}{\Delta I_{out}}\right| \approx \left|\frac{4.5-3.94}{4\times 10^{-3}}\right| = 140\,\Omega \quad \text{for the High State}$$

and since the *v*-*i* curve passes through the origin at the minimum voltage,

$$\left|\frac{\Delta V_{out}}{\Delta I_{out}}\right| \approx \left|\frac{0.33}{4\times 10^{-3}}\right| \approx 83\,\Omega \quad \text{for the Low State}$$

The advantage of this approach is that the determination of the slope from the *v-i* graphs is not required. Again, note that there is not just one dynamic resistance for these gates. For example, the High-state resistance can be 45 Ω and the Low-state resistance can be 37 Ω.

For ECL logic the slopes of the *v-i* curves for the High and Low states are approximately the same for a wide load range:

$$\left|\frac{\Delta V_{out}}{\Delta I_{out}}\right| \approx \left|\frac{-1-(-0.8)}{40\times10^{-3}-4\times10^{-3}}\right| \approx 6\ \Omega \quad \text{for the High State}$$

$$\left|\frac{\Delta V_{out}}{\Delta I_{out}}\right| \approx \left|\frac{-1.9-(-1.8)}{30\times10^{-3}-12\times10^{-3}}\right| \approx 6\ \Omega \quad \text{for the Low State}$$

A constant slope for both states is expected since ECL logic operates the transistors in the active region, resulting in fast speeds. (When a transistor is driven into saturation, as with some TTL devices, the devices are often slower to respond. It takes more time for the transistor to turn off when in saturation.) For current levels less than 4 mA, the dynamic resistance is greater than 6 Ω since the slope of the *v-i* curve is greater in this range.

3.31 Driver Current for Series and Shunt Matching

If a series matching resistor is used, sketch the source drive current if the output voltage rises from V_L to V_H and the time is less than $2T_D$ (T_D is the one-way time delay of the line). Determine the source drive current for this series matching scheme after $2T_D$. Repeat for shunt resistor matching at the receiver. Assume the line is electrically long and the output impedance of the gate does not affect the current.

The impedance seen by the output of the driver for times less than $2T_D$ is

$$R_s + Z_o \tag{3.80}$$

where R_s is the series matching resistor and Z_o is the characteristic impedance of the line. The driver does not see the effect of the load impedance until the incident signal, reflected off the receiver, arrives at the driver. This negative-going reflected wave is not rereflected at the source since the reflection coefficient at the source is zero. Steady-state conditions are reached at $2T_D$, and the impedance seen by the source is the sum of the series matching resistance and impedance of the receiver, a large value, ideally infinite. The source current drops to zero after $2T_D$, as shown in Figure 3.32, if the receiver input impedance is infinite.

For shunt matching, the initial impedance seen by the driver is Z_o. Because the reflection coefficient at the load is zero, there is no reflected negative-going signal. Steady-state conditions are reached at the source as soon as the source voltage has reached V_H as shown in Figure 3.33. This example illustrates one of the differences between series and shunt matching. The power absorbed by the shunt matching resistor, I^2R, is generally greater than the power absorbed by the series matching resistor (the current

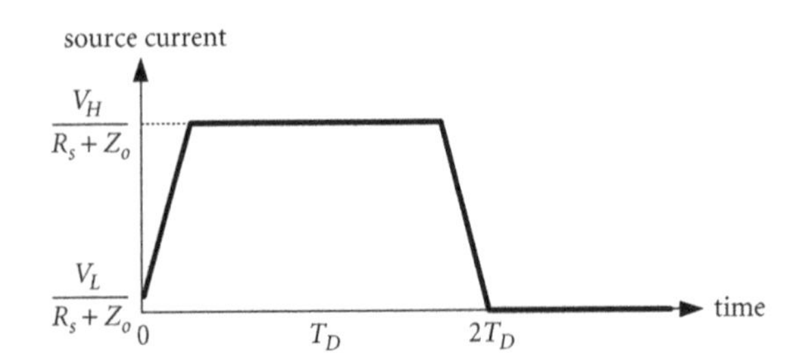

FIGURE 3.32 Source drive current for a series matching resistor.

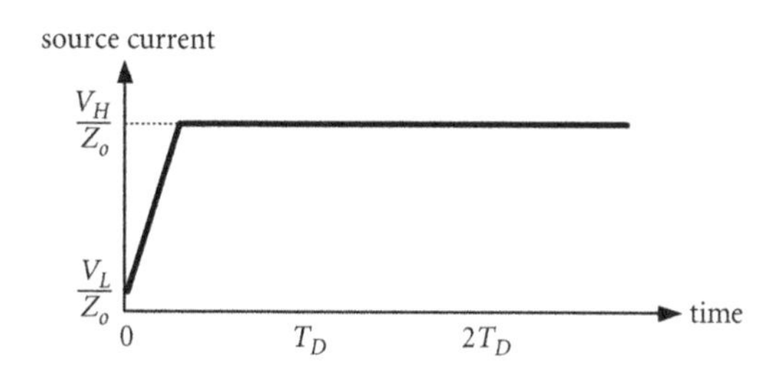

FIGURE 3.33 Source drive current for a shunt matching resistor.

drops to zero for series matching). If the source signal is slowly varying between High and Low, then this difference may be significant. However, if the source signal is rapidly changing between High and Low, the steady-state current level difference may not be significant. (As previously discussed, a capacitor can be placed in series with the shunt resistor to reduce the power absorbed.)

3.32 Summary of Matching Methods

In table form, summarize the advantages and disadvantages of the various matching methods. What are some of the consequences of not matching? [Johnson, '93; Johnson, '03; Montrose, '99; Vo]

Many methods for matching have been discussed. When a transmission line is not properly terminated or matched, reflections will occur. A convenient summary of the advantages and disadvantages of these many methods is provided in Table 3.2. Although not previously discussed, both-end termination can

TABLE 3.2 Advantages and Disadvantages of Different Matching Methods

Matching Method	Advantages	Disadvantages
Series resistor termination driver R Z_o receiver	Low power consumption Low dc drive current Lower high-frequency emissions Multiple fanout of lines from common source possible Ideal for point-to-point connections	Delay added Difficult to implement if not planned for Rounding of input signal Half of expected voltage appears along line for a period of time — not appropriate for distributed loading Reflection from load
Shunt/Parallel resistor termination driver Z_o R receiver	Easy to add after circuit is built assuming receiver accessible Distributed loading along line is ok No reflection from load Multiple loads at end of line are ok Undistorted wave along line	High power consumption High dc drive current levels Reduces noise margin May require current driver Reflections from source can occur for induced signals on the line
Shunt/ac *RC* termination driver Z_o R C receiver	Low dc power consumption Distributed loading along line is ok Very low reflections from load	Medium power consumption High-frequency response maybe affected Two elements

(*Continued*)

TABLE 3.2 (Continued)

Matching Method	Advantages	Disadvantages
Shunt diode termination V_{CC}, D_1, Z_o, driver, D_2 receiver, V_{EE}	Low power consumption Limits overshoot	High-frequency response maybe affected Two elements Some reflections still occur
Shunt split termination V_{CC}, R_2, Z_o, driver, R_1 receiver, V_{EE}	Distributed loading along line is ok No reflection from load Multiple loads along line are ok Line voltage set without driver Undistorted wave along line	Medium-high power consumption High drive current levels Reduces noise margin Two elements May require current driver Difficult in mixed-logic systems

be employed. By matching at both the source (or driver) and load (or receiver), reflections are attenuated at a greater rate. However, the voltage along the line remains at half its "normal" value since the load is also matched. (With only series matching at the source, the half-value signal is doubled after it reflects off the unmatched high-impedance load.) The discussions have emphasized matching for single-ended drivers and receivers, where one end of the transmission line is at ground or reference potential. However, very similar methods can be employed for differential drivers and receivers.

Sometimes reflections from an unmatched or unterminated line can be tolerated, especially if the line is short and data rate is low. There are some advantages in not matching. The most obvious is that an unterminated line does not require additional matching components. Also, terminated lines can add additional loading to the driver, which can decrease the noise margin of the system and increase the power delivered by the driver.

When a system requires matching and it is not properly terminated, there are several potential consequences. The amplitude of the signal at the receiver can reach twice the incident signal amplitude, which can destroy or damage the receiver. The amplitude at the receiver can be much different from the expected value, which can affect the noise margin of the system and generate false signal levels. Finally, the signal at the receiver and source can contain excessive ringing, which can increase emissions from the circuit.

3.33 Relationship Between Sinusoidal Input and Output Voltage

Frequently a line is said to be electrically short if its length, l_{th}, is less than one-tenth of the wavelength of the highest frequency of interest. Using the transmission line equations for a lossless line, determine the ratio of the magnitude of the input voltage to the output voltage of a transmission line of real characteristic impedance Z_o connected to a real load Z_L. Plot this ratio as a function of the electrical length (l_{th}/λ) as l_{th}/λ varies from 0 to 1. Assume Z_L/Z_o is equal to 0.5, 1, 2, and 1,000. From these graphs, deduce why the one-tenth wavelength guideline is used. [Paul, '92(b)]

If the load on a transmission line is located at $z = l_{th}$ and the source is located at $z = 0$, then the general expression for the voltage along a *lossless* line in the frequency domain is

$$V_s(z) = V^+ e^{-j\beta z}[1+\rho(z)] = V^+ e^{-j\beta z}\left[1+\rho_L e^{j2\beta(z-l_{th})}\right] \tag{3.81}$$

where

$$\rho(z)=\rho_L e^{j2\beta(z-l_{th})} \quad \text{where } \rho_L=\frac{Z_L-Z_o}{Z_L+Z_o}=\frac{\dfrac{Z_L}{Z_o}-1}{\dfrac{Z_L}{Z_o}+1} \tag{3.82}$$

and β is the phase constant for the line. These expressions are in the frequency domain. The amplitude of this phasor represents the amplitude of the sinusoidal signal, and the phase of this phasor represents the phase angle of the sinusoidal signal in the time domain. The voltage along a transmission line varies in time in a sinusoidal fashion.

The voltage along a line consists of a positive-going and negative-going signal. These signals are clearly present in (3.81) if it is expanded:

$$V_s(z)=V^+e^{-j\beta z}+V^+\rho_L e^{j\beta z}e^{-j2\beta l_{th}}=V^+e^{-j\beta z}+V^-e^{j\beta z} \tag{3.83}$$

where

$$\frac{V^-}{V^+}=\rho_L e^{-j2\beta l_{th}} \tag{3.84}$$

(The variables V^+ and V^- are not generally entirely real.) The first term in (3.83) is the signal traveling in the positive z direction, while the second term is the signal traveling in the negative z direction. The voltage at the source, $z = 0$, is

$$V_s(0)=V^+\left(1+\rho_L e^{-j2\beta l_{th}}\right) \tag{3.85}$$

and the voltage at the load, $z = l_{th}$, is

$$V_s(l_{th})=V^+e^{-j\beta l_{th}}(1+\rho_L) \tag{3.86}$$

Thus, the ratio of the input voltage to the output voltage is

$$\frac{V_s(0)}{V_s(l_{th})}=\frac{V^+\left(1+\rho_L e^{-j2\beta l_{th}}\right)}{V^+e^{-j\beta l_{th}}(1+\rho_L)}=\frac{1+\rho_L\left[\cos(2\beta l_{th})-j\sin(2\beta l_{th})\right]}{(1+\rho_L)\left[\cos(\beta l_{th})-j\sin(\beta l_{th})\right]} \tag{3.87}$$

Substituting the phase constant, $\beta=2\pi/\lambda$, the desired input to output ratio is obtained:

$$\frac{V_s(0)}{V_s(l_{th})}=\frac{1+\rho_L\left[\cos\left(4\pi\frac{l_{th}}{\lambda}\right)-j\sin\left(4\pi\frac{l_{th}}{\lambda}\right)\right]}{(1+\rho_L)\left[\cos\left(2\pi\frac{l_{th}}{\lambda}\right)-j\sin\left(2\pi\frac{l_{th}}{\lambda}\right)\right]} \tag{3.88}$$

Equation (3.88) is plotted in Mathcad 3.13. Radians must be used in the trigonometric arguments.

$\rho_L(x) := \dfrac{x-1}{x+1} \qquad j := \sqrt{-1} \qquad \lambda := 1$

$\rho_L(0.5) = -0.333 \qquad \rho_L(1) = 0 \qquad \rho_L(2) := 0.333 \qquad \rho_L(1000) := 0.998$

$l_{th} := 0, 0.005 .. 1$

$$H(\rho_L, l_{th}) := \frac{1 + \rho_L \cdot \left(\cos\left(4\cdot\pi\cdot\frac{l_{th}}{\lambda}\right) - j\cdot\sin\left(4\cdot\pi\cdot\frac{l_{th}}{\lambda}\right)\right)}{(1+\rho_L)\cdot\left(\cos\left(2\cdot\pi\cdot\frac{l_{th}}{\lambda}\right) - j\cdot\sin\left(2\cdot\pi\cdot\frac{l_{th}}{\lambda}\right)\right)}$$

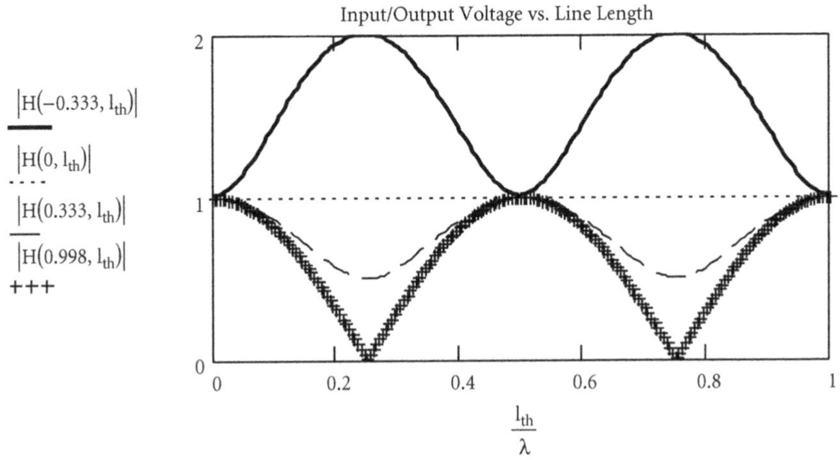

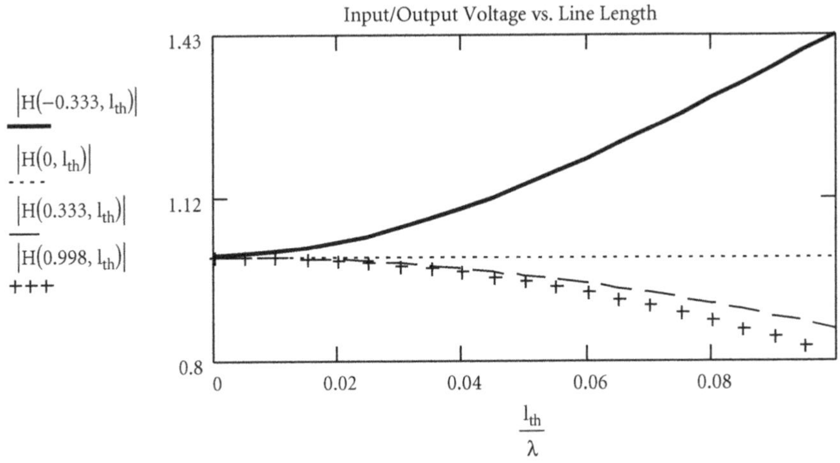

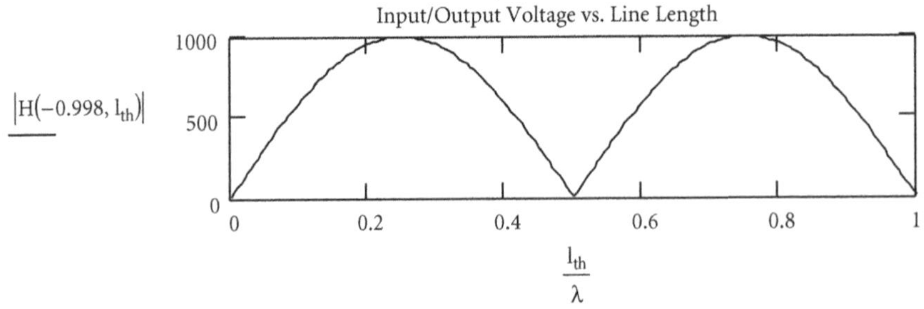

MATHCAD 3.13 Magnitude of the ratio of the input to the output voltage for a transmission line with various reflection coefficients.

The voltage along a transmission line varies in a sinusoidal fashion with a spatial period of λ. The *magnitude* of the voltage, which has a period of $\lambda/2$, is plotted in Mathcad 3.13. For an electrically-short line, a line less than one-tenth of a wavelength of the highest frequency of interest, the sinusoidal variation of the voltage is not yet apparent. For an electrically-short line, a linear approximation for the voltage is often used.

So as not to mislead the reader into believing that the voltage variation relative to the input is always small for lossless short lines, the last plot illustrates the case of a large, negative reflection coefficient (when the load resistance is much less than the characteristic impedance of the line). Even at one-tenth of a wavelength, the voltage difference between the source voltage and load voltage is quite large, but a full sinusoidal variation is still not present at one-tenth of a wavelength. Electrically-short lines do not exhibit sinusoidal voltage or current variations. For electrically-short lines, the load impedance should be close to the line impedance (reflection coefficient close to zero) if the voltage variation from the input to output is to remain small.

For situations involving large mismatches, the variation in voltage and current is large for electrically-short *lossless* lines. However, for any practical line, losses are present. As the VSWR on the line increases, the losses become more important. These losses will dampen this voltage variation.

Since students frequently have difficulty determining the actual expression for the voltage along a transmission line from the given expression

$$V_s(z) = V^+e^{-j\beta z}[1+\rho(z)] = V^+e^{-j\beta z}\left[1+\rho_L e^{j2\beta(z-l_{th})}\right]$$

a few helpful suggestions will be provided. First, if the voltage across the input of a transmission line, $z = 0$, is known, it is *not* equal to V^+ but

$$V_s(0) = V^+\left(1+\rho_L e^{-j2\beta l_{th}}\right)$$

However, V^+ can be determined from the voltage across the input of the line:

$$V^+ = \frac{V_s(0)}{1+\rho_L e^{-j2\beta l_{th}}} \tag{3.89}$$

The variable V^+ is not necessarily entirely real. If a real voltage source is connected across the input of the line, then the voltage across the input of the line is not necessarily equal to the source's voltage unless the source's impedance is zero (i.e., ideal voltage source). Voltage division between the sinusoidal steady-state input impedance of the line and the source impedance can be used to determine the voltage across the input of the line. If the voltage across the output, $z = l_{th}$, of the line is known instead, then V^+can be determined from this voltage:

$$V_s(l_{th}) = V^+e^{-j\beta l_{th}}(1+\rho_L) \quad \Rightarrow V^+ = \frac{V_s(l_{th})e^{j\beta l_{th}}}{1+\rho_L} \tag{3.90}$$

If the current is known through the load impedance connected across the output of the line, then the product of this current (phasor) and this load impedance is this output voltage. Finally, if the voltages across both the input and output of the line are known, then V^+ and the phase constant, β, can both be determined by solving simultaneously for these two variables in the equations given in (3.91):

$$V_s(0) = V^+\left(1+\rho_L e^{-j2\beta l_{th}}\right) \quad \text{and} \quad V_s(l_{th}) = V^+e^{-j\beta l_{th}}(1+\rho_L) \tag{3.91}$$

3.34 The Sinusoidal Current Expression

When is the current along a transmission line equal to the voltage divided by the line's characteristic impedance?

Unfortunately, the expression for the current along a transmission line is not usually equal to the voltage expression divided by the line's characteristic impedance:

$$I_s(z) \neq \frac{V_s(z)}{Z_o} = \frac{V^+ e^{-j\beta z}\,[1+\rho(z)]}{Z_o}$$

The load is located at $z = l_{th}$ and the source is located at $z = 0$. However, if the reflection coefficient is zero and the line is matched, then the current is equal to the voltage divided by the line impedance:

$$I_s(z) = \frac{V_s(z)}{Z_o} = \frac{V^+ e^{-j\beta z}}{Z_o} \quad \text{if } \rho_L = 0 \tag{3.92}$$

When the line is matched to its load, the reflected voltage and current are zero. In this case, the voltage and current on the line consist of a single wave traveling in the direction of the load. When the line is not matched to its load, the voltage and current on the line consist of two signals traveling in the direction of the load *and* source. For this reason, the current expression along the line is (in the frequency domain)

$$I_s(z) = \frac{V^+ e^{-j\beta z}[1-\rho(z)]}{Z_o} \tag{3.93}$$

where the spatial reflection coefficient is given as

$$\rho(z) = \rho_L e^{j2\beta(z-l_{th})} \quad \text{where } \rho_L = \frac{Z_L - Z_o}{Z_L + Z_o} = \frac{\frac{Z_L}{Z_o} - 1}{\frac{Z_L}{Z_o} + 1} \tag{3.94}$$

The current and voltage along a transmission line vary in a sinusoidal fashion. However, the current and voltage are not necessarily in phase. The reason for the minus sign preceding $\rho(z)$ can be quickly understood:

$$\begin{aligned} I_s(z) &= \frac{V^+ e^{-j\beta z}\left[1 - \rho_L e^{j2\beta(z-l_{th})}\right]}{Z_o} = \frac{V^+ e^{-j\beta z} - V^+ \rho_L e^{j\beta z} e^{-j2\beta l_{th}}}{Z_o} \\ &= \frac{V^+ e^{-j\beta z}}{Z_o} - \frac{V^- e^{j\beta z}}{Z_o} \end{aligned} \tag{3.95}$$

where

$$\frac{V^-}{V^+} = \rho_L e^{-j2\beta l_{th}} \tag{3.96}$$

The first term in (3.95) is the signal that is traveling in the positive z direction toward the load. The second term is the portion of the current that is traveling in the negative z direction toward the source. This negative-going term is zero if the load is matched to the line. The minus sign is required in this expression since the power for the reflected signal, traveling in the negative z direction, must also be in the negative z direction: the power for the reflected signal is in the direction of the source. This real power for the reflected signal is

$$\begin{aligned} P_{avg,V^-} &= \frac{1}{2}\mathrm{Re}(V_s I_s^*) = \frac{1}{2}\mathrm{Re}\left[V^- e^{j\beta z}\left(-\frac{V^- e^{j\beta z}}{Z_o}\right)^*\right] \\ &= \frac{1}{2}\mathrm{Re}\left[-V^- e^{j\beta z}\frac{(V^-)^* e^{-j\beta z}}{Z_o}\right] = -\frac{|V^-|^2}{2Z_o} \end{aligned} \tag{3.97}$$

where I_s^* is the complex conjugate of the current phasor. It was assumed in this expression that the characteristic impedance of the line, Z_o, is purely real. If it is complex, then its complex conjugate must be taken. This slightly complicates the final expression since the real portion of the terms inside Re( ) must be determined. The reflected power is negative as required for this reflected signal. The power for the forward-traveling component of the signal is positive:

$$\begin{aligned} P_{avg,V^+} &= \frac{1}{2}\mathrm{Re}\left(V_s I_s^*\right) = \frac{1}{2}\mathrm{Re}\left[V^+ e^{-j\beta z}\left(\frac{V^+ e^{-j\beta z}}{Z_o}\right)^*\right] \\ &= \frac{1}{2}\mathrm{Re}\left[V^+ e^{-j\beta z}\frac{(V^+)^* e^{j\beta z}}{Z_o}\right] = \frac{|V^+|^2}{2Z_o} \end{aligned} \tag{3.98}$$

In Mathcad 3.14, the current along a lossless 50 Ω transmission line is plotted for various loads and times. The line is one wavelength in length. Note that the current, as with the voltage, repeats every wavelength. However, the voltage and current are not necessarily in phase. Also, when the load is matched to the line, a zero reflection coefficient condition, the current waveform consists of a single wave traveling in the positive z direction. There is no reflected signal from the load. Another obvious check on the solution is that the current is zero at the open-circuited load for all time.

As previously stated, the voltage and current are not necessarily in phase along a transmission line:

$$V_s(z) = V^+ e^{-j\beta z}\left[1 + \rho_L e^{j2\beta(z-l_{th})}\right], \quad I_s(z) = \frac{V^+ e^{-j\beta z}\left[1 - \rho_L e^{j2\beta(z-l_{th})}\right]}{Z_o}, \quad \rho_L = \frac{Z_L - Z_o}{Z_L + Z_o}$$

However, when the line is lossless and the load is matched to the line (so the load is also purely resistive),

$$V_s(z) = V^+ e^{-j\beta z}, \quad I_s(z) = \frac{V^+ e^{-j\beta z}}{Z_o} \quad \text{if } Z_L = Z_o, \ \rho_L = 0$$

and the voltage and current are in phase:

$$\frac{V_s(z)}{I_s(z)} = Z_o$$

As will be seen shortly, when the load is either a short circuit or an open circuit, the voltage and current are 90° out of phase.

$$f := 30\cdot 10^{6} \quad A := 50 \quad L := 0.25\cdot 10^{-6} \quad C := 100\cdot 10^{-12} \quad \omega := 2\cdot\pi\cdot f \quad T := \frac{1}{f} \quad j := \sqrt{-1}$$

$$Z_o := \sqrt{\frac{L}{C}} \quad Z_o = 50 \quad v := \frac{1}{\sqrt{L\cdot C}} \quad v = 2\times 10^{8} \quad \lambda := \frac{v}{f} \quad \beta := \frac{2\cdot\pi}{\lambda} \quad l_{th} := \lambda \quad l_{th} = 6.667$$

$$z := 0, 0.01\cdot l_{th} .. l_{th}$$

$$I(z, t, \rho_L) := \left|\frac{A}{Z_o}\cdot e^{-j\cdot\beta\cdot z}\cdot\left[1 - \rho_L\cdot e^{j\cdot 2\cdot\beta\cdot(z - l_{th})}\right]\right|\cdot\cos\left[\omega\cdot t + \arg\left[\frac{A}{Z_o}\cdot e^{-j\cdot\beta\cdot z}\cdot\left[1 - \rho_L\cdot e^{j\cdot 2\cdot\beta\cdot(z - l_{th})}\right]\right]\right]$$

Current for Various Loads

A: I(z, 0, 0), I(z, 0, 0.5), I(z, 0, 1), I(z, 0, −1)

z, m: 0, 1.67, 3.33, 5, 6.67

Current for Matched Load

A: I(z, 0, 0), I(z, 0.15·T, 0), I(z, 0.3·T, 0), I(z, 0.45·T, 0)

z, m: 0, 1.67, 3.33, 5, 6.67

Current for Open Load

A: I(z, 0, 1), I(z, 0.15·T, 1), I(z, 0.3·T, 1), I(z, 0.45·T, 1)

z, m: 0, 1.67, 3.33, 5, 6.67

MATHCAD 3.14 Current along a transmission line for various times and loads. The *spatial* period of the current is λ.

3.35 The Sinusoidal Input Impedance

Why does the sinusoidal input impedance along a transmission line repeat every one-half of a wavelength instead of one wavelength? What is the input impedance of a quarter wavelength line?

The impedance looking into a transmission line is equal to the voltage across the line divided by the current into the line:

$$Z_{in} = \frac{V_s(0)}{I_s(0)} = \frac{V^+e^{-j\beta(0)}\left[1+\rho_L e^{j2\beta(0-l_{th})}\right]}{\dfrac{V^+e^{-j\beta(0)}\left[1-\rho_L e^{j2\beta(0-l_{th})}\right]}{Z_o}} = Z_o\frac{1+\rho_L e^{-j2\beta l_{th}}}{1-\rho_L e^{-j2\beta l_{th}}} = Z_o\frac{e^{j\beta l_{th}}+\rho_L e^{-j\beta l_{th}}}{e^{j\beta l_{th}}-\rho_L e^{-j\beta l_{th}}}$$

$$= Z_o\frac{\left[\cos(\beta l_{th})+j\sin(\beta l_{th})\right]+\rho_L\left[\cos(\beta l_{th})-j\sin(\beta l_{th})\right]}{\left[\cos(\beta l_{th})+j\sin(\beta l_{th})\right]-\rho_L\left[\cos(\beta l_{th})-j\sin(\beta l_{th})\right]}$$

$$= Z_o \frac{[1 + j\tan(\beta l_{th})] + \rho_L[1 - j\tan(\beta l_{th})]}{[1 + j\tan(\beta l_{th})] - \rho_L[1 - j\tan(\beta l_{th})]}$$

$$= Z_o \frac{[1 + j\tan(\beta l_{th})] + \frac{Z_L - Z_o}{Z_L + Z_o}[1 - j\tan(\beta l_{th})]}{[1 + j\tan(\beta l_{th})] - \frac{Z_L - Z_o}{Z_L + Z_o}[1 - j\tan(\beta l_{th})]}$$

Simplifying the last expression, the commonly used equation for the input impedance of a lossless transmission line is obtained:

$$Z_{in} = Z_o \frac{Z_L + jZ_o \tan(\beta l_{th})}{Z_o + jZ_L \tan(\beta l_{th})} = Z_o \frac{Z_L + jZ_o \tan\left(\frac{2\pi}{\lambda} l_{th}\right)}{Z_o + jZ_L \tan\left(\frac{2\pi}{\lambda} l_{th}\right)} \tag{3.99}$$

Radians must be used when determining the tangent of βl_{th}. The source is located at $z = 0$, and the load is located at $z = l_{th}$. The input impedance is a function of the tangent of $2\pi l_{th}/\lambda$. The cosine of $2\pi l_{th}/\lambda$ repeats every $l_{th} = \lambda$. The sine of $2\pi l_{th}/\lambda$ also repeats every $l_{th} = \lambda$. However, the tangent of $2\pi l_{th}/\lambda$, which is equal to the ratio of $\sin(\beta l_{th})$ to $\cos(\beta l_{th})$, repeats every $l_{th} = \lambda/2$. Mathematically, this is why the input impedance repeats every one-half of a wavelength. The input impedance is the ratio of the voltage to the current. Although the voltage and current repeat every one wavelength, the ratio of the voltage to the current repeats every one-half of a wavelength. For example, imagine that the voltage along a line varies from 1 to −1 V, and the current along the same line varies from 5 to −5 mA. When the voltage across the line is equal to 1 V and current into the line is equal to 5 mA, the input impedance is 200 Ω (assuming the current and voltage are in phase). However, at one-half of a wavelength into the line, the voltage is −1 V, current is −5 mA, and input impedance is again 200 Ω.

Two special cases for the input impedance will be briefly examined. When the load corresponds to a short,

$$Z_{in} = jZ_o \tan(\beta l_{th}) \quad \text{if } Z_L = 0$$

The input impedance is purely imaginary (or zero) and, hence, the voltage and current are 90° out of phase. The input impedance is purely reactive or imaginary since there are no losses in the line. The only possible source for a real component of the input impedance of a lossless line is the load, which is a short in this case. The lossless line is composed entirely of *L*'s and *C*'s. These statements are also true when the load corresponds to an open circuit:

$$Z_{in} = -j \frac{Z_o}{\tan(\beta l_{th})} \quad \text{if } Z_L = \infty$$

Transmission lines can also be used as impedance transformers or matching devices. For example, when a transmission line is $\lambda/4$ in length,

$$Z_{in} = Z_o \frac{Z_L + jZ_o \tan\left(\frac{2\pi}{\lambda}\frac{\lambda}{4}\right)}{Z_o + jZ_L \tan\left(\frac{2\pi}{\lambda}\frac{\lambda}{4}\right)} = Z_o \frac{jZ_o}{jZ_L} = \frac{Z_o^2}{Z_L} \Rightarrow Z_o = \sqrt{Z_{in} Z_L} \quad \text{if } l_{th} = \frac{\lambda}{4} \tag{3.100}$$

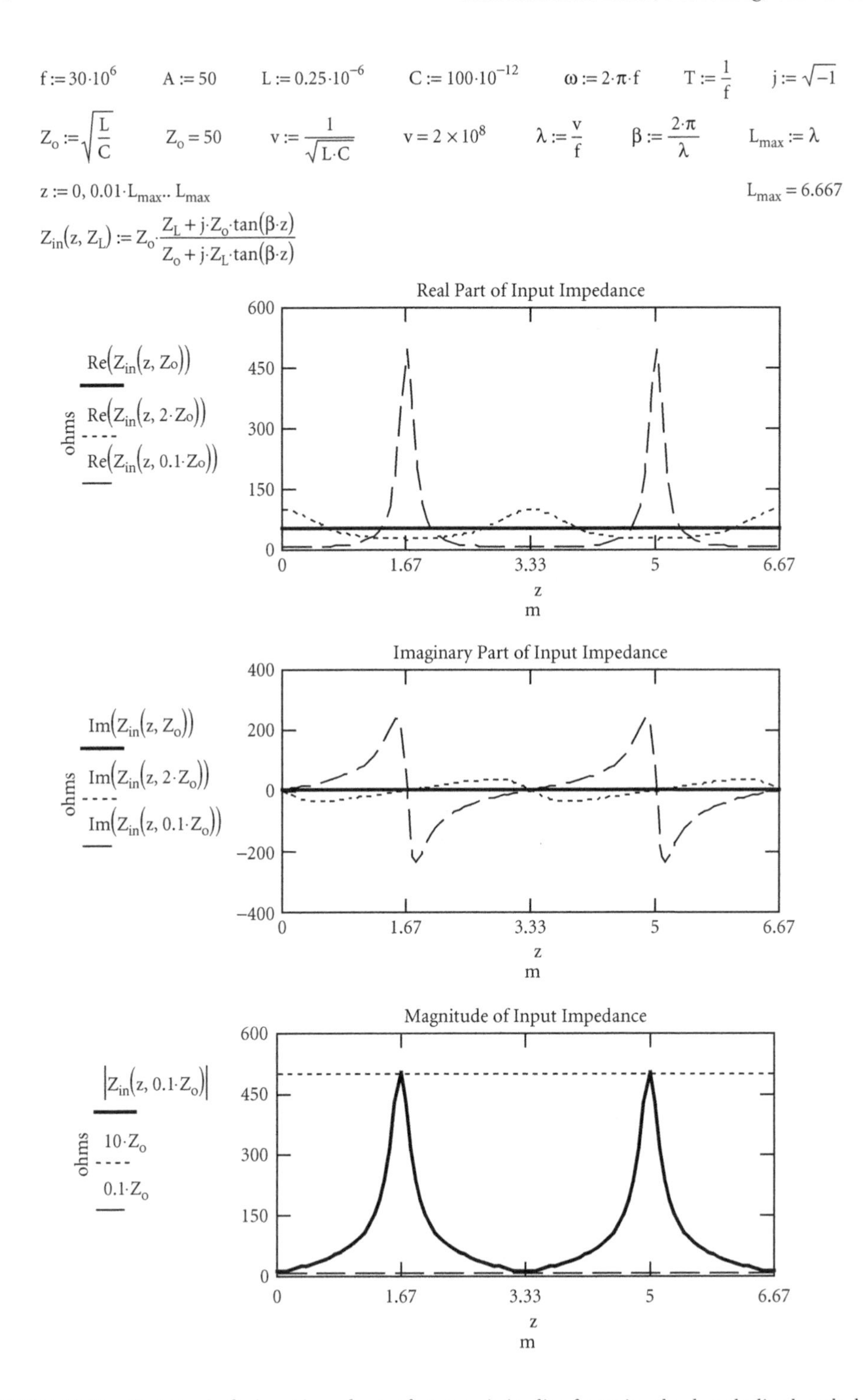

MATHCAD 3.15 Variation in the input impedance of a transmission line for various loads as the line length changes.

To obtain an input impedance of Z_{in}, a $\lambda/4$ long, lossless transmission line should have a characteristic impedance equal to the geometric mean of the load impedance and desired input impedance. For example, at 30 MHz, the free-space wavelength is 10 m. Thus, an open-air twin-lead line of length 2.5 m with a characteristic impedance of about 170 Ω can transform a 300 Ω load to 100 Ω. The line can also be odd integer multiples of a quarter wavelength. The matching is exact only at those frequencies where the line length is an odd integer multiple of a quarter wavelength. As the frequency deviates from these specific values, the input impedance deviates from the desired Z_{in}. In other words, this matching transmission line transformer has a finite and narrow bandwidth (i.e., frequency sensitive). By using two or more quarter-wave lines in series, the bandwidth can be improved. Referring to the previous example, one line of length 2.5 m and impedance of about 240 Ω can transform the impedance from 300 Ω to 200 Ω. Then, a second 2.5 m line with an impedance of about 140 Ω could transform the 200 Ω down to 100 Ω. As the number of sections increase to a very large number, the length of the line becomes several wavelengths long. Transmission-line matching techniques can also be applied when using coatings on metals and other materials to reduce electromagnetic reflections.

The input impedance of a lossless 50 Ω line for various load impedances as the line length varies from zero to one wavelength (≈ 6.7 m) is given in Mathcad 3.15. As stated previously, the input impedance does indeed repeat every one-half of a wavelength. When the load is equal to the line impedance (the matched condition), the input impedance is constant and equal to the line impedance. Notice that for the two nonmatching conditions shown, both the real and imaginary parts of the input impedance vary dramatically along the line. In addition, even though the load impedance is entirely real in both cases, the input impedance is inductive in nature (corresponding to a positive imaginary term) for certain positions along the line and capacitive in nature (corresponding to a negative imaginary term) for certain other positions along the line. A lossless transmission line can be modeled as a large number of *LC* circuits. These *LC* circuits are the source of these imaginary terms. Finally, recall that the maximum and minimum *possible* input impedances for a lossless line are

$$|Z_{in}|_{max} = \frac{V_{max}}{I_{min}} = (\text{VSWR})\,Z_o, \quad |Z_{in}|_{min} = \frac{V_{min}}{I_{max}} = \frac{Z_o}{\text{VSWR}}$$

When $Z_L = 2Z_o$, the VSWR = 2, and when $Z_L = 0.1Z_o$, the VSWR = 10. The range on the magnitude of the input impedance when $Z_L = 0.1Z_o$ is confirmed on the last plot in Mathcad 3.15.

3.36 Coaxial Cable Branching

Seventy-five ohm CATV coax is connected to the 75 Ω input of a living room TV. Instead of using a commercial splitter, additional 75 Ω coax is used to connect a bedroom TV's 75 Ω input to the CATV line. This connection to the CATV signal, via the 75 Ω coax, is directly made at the living room TV as is shown in Figure 3.34. Determine the reflection coefficient seen by the incoming CATV line at the living room TV. What is the major consequence of this mismatch? If the cable losses are neglected, will the signal level at both TVs be approximately the same? Explain.

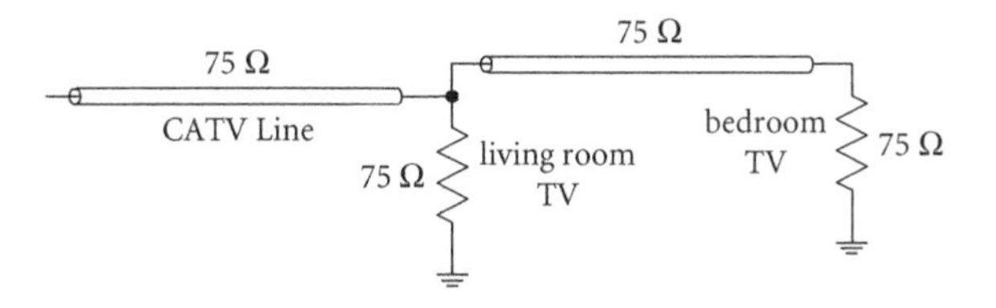

FIGURE 3.34 Splitting of a CATV line without regard to matching.

This analysis assumes sinusoidal steady-state conditions. The reflection coefficient at the living room TV is equal to

$$\rho_L = \frac{Z_L - Z_o}{Z_L + Z_o} = \frac{\dfrac{75\times 75}{75+75} - 75}{\dfrac{75\times 75}{75+75} + 75} \approx -0.33$$

The total load seen by the CATV line at the Living Room TV (Z_L) is equal to the 75 Ω input of the living room TV in parallel with the input impedance of the line leading to the bedroom TV. Since this line to the bedroom TV is matched to the 75 Ω input of the bedroom TV, the line's input impedance is also 75 Ω.

There are several consequences of the nonzero reflection coefficient. A VSWR > 1 along the CATV line will increase the losses along this line and, hence, decrease the power level at the living room TV. This may result in a few dB of additional loss per 100 ft of line. Also, the rms voltage level at the living room TV and bedroom TV will be a function of the length of the incoming line. (If both lines are electrically small, the voltages should be about the same.)

The impedance seen by the CATV source will no longer be the expected value of 75 Ω, and the source may be loaded down by this line-length dependent impedance. If the cable resistive losses are neglected, the signal level at both TVs will be approximately the same: the living room TV and bedroom cable are in parallel, and the voltage across parallel elements is the same. Because their impedances are identical and the bedroom cable TV is matched to the TV impedance, the current and power levels are also the same at both TVs.

Referring to Figure 3.34, suppose that the incoming CATV cable is split before it reaches either TV using separate 150 Ω cables as is shown in Figure 3.35 (assuming that a coaxial cable of this characteristic impedance can be obtained). Determine the reflection coefficient at this cable junction as seen by the CATV line. Determine the reflection coefficient seen by each 150 Ω cable at each TV. How should the 150 Ω cables be matched to the TV inputs?

To determine the reflection coefficient at the junction between the 75 Ω and 150 Ω lines, the input impedance into the 150 Ω lines is required. Once these input impedances are determined, their parallel combination is the total load seen by the CATV line. Unfortunately, the length of each 150 Ω line and frequency of operation are required to determine the input impedance for each of these lines at the junction. Once this information is known, the input impedance for each line may be determined from the following equation for sinusoidal steady-state signals:

$$Z_{in} = Z_o \frac{Z_L + jZ_o \tan(\beta l_{th})}{Z_o + jZ_L \tan(\beta l_{th})}$$

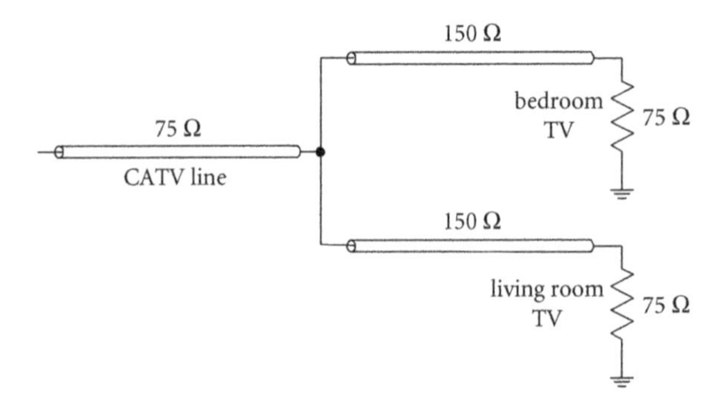

FIGURE 3.35 An attempt to provide some matching at the junction of the CATV line and the lines leading to the TVs.

where l_{th} is the length of the line and $\beta = 2\pi/\lambda$ is the phase constant for the line. The reflection coefficient at each TV, which is not a function of the line length, is

$$\rho_L = \frac{Z_L - Z_o}{Z_L + Z_o} = \frac{75-150}{75+150} \approx -0.33$$

If the load impedance at each TV is adjusted so it is equal to 150 Ω, the input impedance at the junction for each of the 150 Ω lines is equal to 150 Ω for all frequencies and line lengths. This constant input impedance is one advantage of matching the load to the line. In this case, the reflection coefficient *at the junction* is zero:

$$\rho_L = \frac{Z_L - Z_o}{Z_L + Z_o} = \frac{\dfrac{150\times150}{150+150} - 75}{\dfrac{150\times150}{150+150} + 75} = 0$$

When the cables are matched to their loads, the VSWR on the CATV line (and the 150 Ω lines) is one, and the system is matched everywhere. Then, the voltage levels at both TVs are the same regardless of the line length if the line losses are negligible.

One method of matching the 75 Ω input of the TV to the 150 Ω line is by inserting a 75 Ω resistor in series with the line at the TV. This primitive method, of course, splits the voltage evenly between this series resistor and TV. Also, one-half of the power is wasted in this series resistor. The impedance can be stepped up from 75 Ω to 150 Ω without this large resistive loss by using an impedance-matching transformer or an *LC* matching network. In each case, the bandwidth and losses associated with the matching device should be considered.

Finally, note that 150 Ω is not a typical characteristic impedance for a coaxial cable (although it is available). For coax, characteristic impedances equal to 35 Ω, 48 Ω, 50 Ω, 52 Ω, 53.5 Ω, 73 Ω, 75 Ω, 93 Ω, 95 Ω, and 125 Ω can be obtained (this list is *not* comprehensive).

3.37 "Y" Splitter for "Hair-Ball" Networks

A "Y" splitter connects an incoming 75 Ω cable to two 75 Ω cables. A "Y" splitter consists of three resistors, each of value $Z_o/3$, connected in a "Y" arrangement as shown in Figure 3.36. The impedance of each of the three transmission lines is Z_o. One transmission line is connected to each end of the "Y." Determine the reflection coefficient seen by each of the three cables at the "Y." Determine the signal loss in dB of this splitter. [Johnson, '93]

If the lines are properly matched at their loads or only the transient reflection coefficients are desired, the equivalent circuit seen by any one of the transmission lines because of the symmetry is given in Figure 3.37.

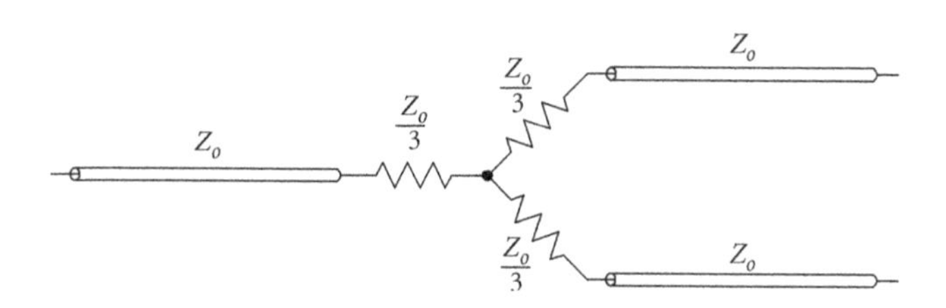

FIGURE 3.36 "Hair-ball" matching network for three transmission lines.

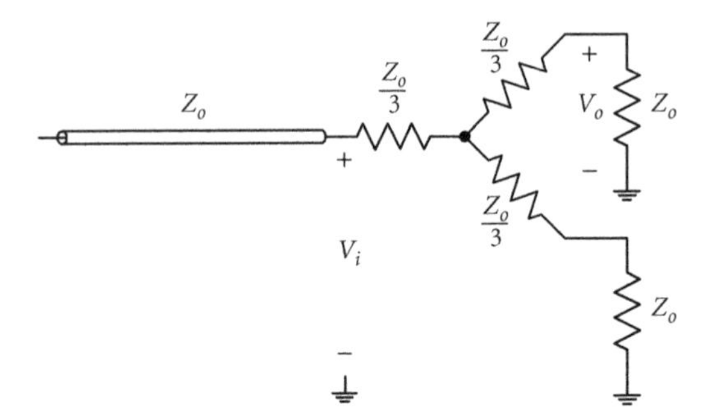

FIGURE 3.37 Equivalent circuit for the "hair-ball" network and connecting lines if (1) the lines are all matched at their loads or (2) only the transient coefficients are desired.

The reflection coefficient is zero, and the system is matched from any direction:

$$\rho_L = \frac{Z_L - Z_o}{Z_L + Z_o} = \frac{\left[\frac{Z_o}{3} + \frac{\left(\frac{Z_o}{3}+Z_o\right)\left(\frac{Z_o}{3}+Z_o\right)}{\left(\frac{Z_o}{3}+Z_o\right)+\left(\frac{Z_o}{3}+Z_o\right)}\right] - Z_o}{\left[\frac{Z_o}{3} + \frac{\left(\frac{Z_o}{3}+Z_o\right)\left(\frac{Z_o}{3}+Z_o\right)}{\left(\frac{Z_o}{3}+Z_o\right)+\left(\frac{Z_o}{3}+Z_o\right)}\right] + Z_o} = \frac{\left(\frac{Z_o}{3}+\frac{2Z_o}{3}\right) - Z_o}{\left(\frac{Z_o}{3}+\frac{2Z_o}{3}\right) + Z_o} = 0$$

The cost associated with this matching scheme is the attenuation due to the resistors. The ratio of the voltage at the input of either of the two *output* lines to the voltage at the output of the *input* line is obtained through voltage division:

$$\frac{V_o}{V_i} = \frac{Z_o}{Z_o + \frac{Z_o}{3}} \frac{\frac{\left(Z_o + \frac{Z_o}{3}\right)}{2}}{\frac{\left(Z_o + \frac{Z_o}{3}\right)}{2} + \frac{Z_o}{3}} = \frac{1}{2}$$

The loss is rather large. The loss in dB of this "hair-ball" network is

$$loss_{dB} = 20 \log 2 \approx 6 \text{ dB}$$

It can be shown that for N cables meeting at a junction, the resistance of each splitter resistor is

$$Z_o\left(\frac{N-2}{N}\right) \tag{3.101}$$

and the loss of the splitter in dB is

$$loss_{dB} = 20 \log (N-1) \tag{3.102}$$

These equations agree with the previous results when $N = 3$.

3.38 Stub Tuning

Explain how short-circuited or open-circuited transmission lines can be used to tune a mismatched transmission line. [Seshadri; Anderson, '85]

Stub tuning is a method of matching a load (Z_L) to its transmission line and, thus, eliminating the reflections on the line from the source to the location of the stub. A stub line is a length of transmission line that is usually either short circuited or open circuited[13] at its far end ($Z_{Lstub} = 0$ or ∞). The stub line is connected in parallel with the transmission line leading to the load. As shown in Figure 3.38, usually the characteristic impedance of the stub's line is equal to the characteristic impedance of the line leading to the load.[14]

When a load is not matched to its transmission line, the VSWR is greater than one, and the input impedance seen by the source varies with the length of the line and frequency of operation. Obviously, the impedance looking into the line toward the load will also vary with the load impedance. This impedance will generally have both a real and an imaginary part. By inserting a stub line of the appropriate length, l_{stub}, and appropriate position from the load, l_1, the parallel combination of the input impedance of the line and stub can be made equal to the characteristic impedance at the stub location:

$$Z_{intotal} = Z_o = Z_{inl} \| Z_{instub}$$

$$\text{where } Z_{inl} = Z_o \frac{Z_L + jZ_o \tan(\beta l_1)}{Z_o + jZ_L \tan(\beta l_1)}, \quad Z_{instub} = Z_o \frac{Z_{Lstub} + jZ_o \tan(\beta l_{stub})}{Z_o + jZ_{Lstub} \tan(\beta l_{stub})} \tag{3.103}$$

The lines are assumed lossless. A simple numerical example will be helpful. Assume that for a short-circuited stub ($Z_{Lstub} = 0$)

$$Z_o = 50\ \Omega,\ C = 100\ \text{pF/m},\ f = 30\ \text{MHz},\ Z_L = 120 + j200\ \Omega$$

$$l_1 = 1.53\ \text{m},\ l_{stub} = 0.372\ \text{m}$$

Then,

$$Z_{inl} \approx 5.9 - j16\ \Omega,\ Z_{instub} \approx j18\ \Omega \ \Rightarrow Z_{intotal} \approx \frac{(5.9 - j16)\,j18}{5.9 - j16 + j18} \approx 50\ \Omega$$

From the source to the stub's position, the VSWR is one (for a properly designed stub) and *this portion* of the line is matched. However, the VSWR on the stub line is obviously quite large (since Z_{Lstub} is either

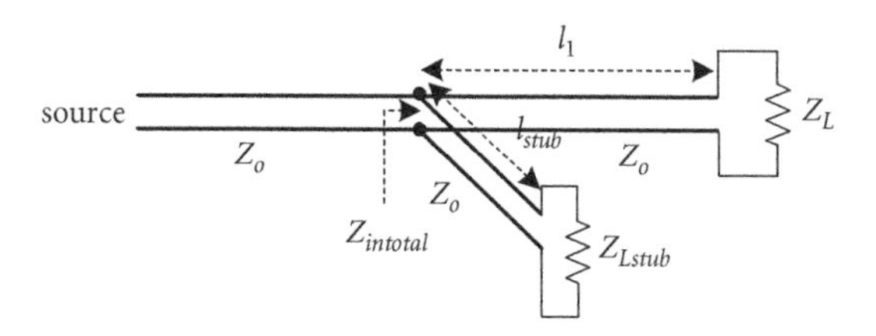

FIGURE 3.38 Single-stub tuner.

[13]The terms short stub and open stub are commonly used.

[14]To obtain a match, it is not necessary that the characteristic impedance of the stub line be equal to the characteristic impedance of the line between the source and load.

$f := 30 \cdot 10^6 \quad L := 0.25 \cdot 10^{-6} \quad C := 100 \cdot 10^{-12} \quad \omega := 2 \cdot \pi \cdot f \quad T := \frac{1}{f} \quad j := \sqrt{-1}$

$Z_o := \sqrt{\frac{L}{C}} \quad Z_o = 50 \quad v := \frac{1}{\sqrt{L \cdot C}} \quad v = 2 \times 10^8 \quad \lambda := \frac{v}{f} \quad \beta := \frac{2 \cdot \pi}{\lambda} \quad \lambda = 6.667$

$$Z_{in1}(l_{th1}, Z_L) := Z_o \cdot \frac{Z_L + j \cdot Z_o \cdot \tan(\beta \cdot l_{th1})}{Z_o + j \cdot Z_L \cdot \tan(\beta \cdot l_{th1})}$$

$$Z_{instub}(l_{thstub}, Z_{Lstub}) := Z_o \cdot \frac{Z_{Lstub} + j \cdot Z_o \cdot \tan(\beta \cdot l_{thstub})}{Z_o + j \cdot Z_{Lstub} \cdot \tan(\beta \cdot l_{thstub})}$$

$$Z_{intotal}(l_{th1}, Z_L, l_{thstub}, Z_{Lstub}) := \frac{Z_{in1}(l_{th1}, Z_L) \cdot Z_{instub}(l_{thstub}, Z_{Lstub})}{Z_{in1}(l_{th1}, Z_L) + Z_{instub}(l_{thstub}, Z_{Lstub})}$$

$Z_L := 50 + j \cdot 20 \quad Z_{Lstub} := 0$

Guess $l_{th1} := 1 \quad l_{thstub} := 1$

Given

$\mathrm{Im}(Z_{intotal}(l_{th1}, Z_L, l_{thstub}, Z_{Lstub})) = 0 \qquad \mathrm{Re}(Z_{intotal}(l_{th1}, Z_L, l_{thstub}, Z_{Lstub})) = Z_o$

$SOLN := \mathrm{Minerr}(l_{th1}, l_{thstub}) \quad SOLN_0 = 1.667 \quad SOLN_1 = 1.263$

$$Z_{intotal}(SOLN_0, Z_L, SOLN_1, Z_{Lstub}) = 50$$

$Z_L := 120 + j \cdot 200 \quad Z_{Lstub} := 0$

Guess $l_{th1} := 1 \quad l_{thstub} := 1$

Given

$\mathrm{Im}(Z_{intotal}(l_{th1}, Z_L, l_{thstub}, Z_{Lstub})) = 0 \qquad \mathrm{Re}(Z_{intotal}(l_{th1}, Z_L, l_{thstub}, Z_{Lstub})) = Z_o$

$SOLN := \mathrm{Minerr}(l_{th1}, l_{thstub}) \quad SOLN_0 = 1.527 \quad SOLN_1 = 0.372$

$$Z_{intotal}(SOLN_0, Z_L, SOLN_1, Z_{Lstub}) = 50$$

MATHCAD 3.16 Solving for the position and length of a single-stub tuner for two different loads.

a short circuit or an open circuit), and the VSWR is still greater than one on the main line between the stub location and load.

The set of nonlinear equations in (3.103) is solved in Mathcad 3.16 for two different loads so that Z_{in1} is equal to Z_o. The characteristic impedance is assumed real (resistive). In these examples, a short-circuited stub is used. Because there are multiple solutions to this set of nonlinear equations, initial starting values (referred to as "Guess" in the program) are required. The final solution is checked in both cases where $l_1 = l_{th1} = SOLN_0$ and $l_{stub} = l_{thstub} = SOLN_1$.

The mystery of how a stub matches is solved once the variation of the input impedance with length on a short-circuited or an open-circuited line is understood. A short-circuited or an open-circuited ideal stub can generate *any* reactance desired at a particular frequency by varying the length of the stub. In Mathcad 3.17, the input reactance of a shorted and opened stub are shown vs. stub length. For a stub with a short-circuited load, the reactance is positive (corresponding to inductance) up to a length of $\lambda/4$. For a stub with an open-circuited load, the reactance is negative (corresponding to capacitance) up to a length of $\lambda/4$. A short-circuited or an open-circuited stub constructed of low-loss transmission line will have zero or a very small input resistance. The input impedance will be almost entirely imaginary or reactive. This reactance is used to cancel out the reactance at the location of the stub due to the mismatched load on the main line. Although the short-circuited or open-circuited stub can generate any reactance, the sensitivity of the reactance to line length will be quite large for large values of reactance.

The previous set of nonlinear equations has been solved analytically for short-circuited and open-circuited stubs. The results, which are given in Table 3.3, are written in terms of the normalized admittance

$f := 20 \cdot 10^{6}$ $\quad L := 0.25 \cdot 10^{-6}$ $\quad C := 100 \cdot 10^{-12}$ $\quad \omega := 2 \cdot \pi \cdot f$ $\quad T := \frac{1}{f}$ $\quad j := \sqrt{-1}$

$Z_o := \sqrt{\frac{L}{C}}$ $\quad Z_o = 50$ $\quad v := \frac{1}{\sqrt{L \cdot C}}$ $\quad v = 2 \times 10^{8}$ $\quad \lambda := \frac{v}{f}$ $\quad \beta := \frac{2 \cdot \pi}{\lambda}$

$l_{thstub} := 0, 0.0001 \cdot \lambda .. \lambda$

$$Z_{instub}(Z_{Lstub}, l_{thstub}) := Z_o \cdot \frac{Z_{Lstub} + j \cdot Z_o \cdot \tan(\beta \cdot l_{thstub})}{Z_o + j \cdot Z_{Lstub} \cdot \tan(\beta \cdot l_{thstub})}$$

Impedance of Shorted and Opened Stubs

ohms: $Im(Z_{instub}(0, l_{thstub}))$ —— ; $Im(Z_{instub}(10^{9}, l_{thstub}))$ - - - -

1000, 0, −1000

0 1 2 3 4 5 6 7 8 9 10

l_{thstub}

m

$\frac{\lambda}{4} = 2.5$ $\quad \frac{\lambda}{2} = 5$ $\quad \frac{3 \cdot \lambda}{4} = 7.5$ $\quad \lambda = 10$

MATHCAD 3.17 Variation of the input impedance of a short-circuited and an open-circuited lossless transmission line.

TABLE 3.3 Solutions for the Position and Length of Short-Circuited and Open-Circuited Stubs [Seshadri]

Z_{Lstub}	One Solution	Another Solution
0 Ω	$l_1 = \frac{1}{\beta}\tan^{-1}\left\{\frac{1-g_L}{b_L + \sqrt{g_L[(1-g_L)^2 + b_L^2]}}\right\}$	$l_1 = \frac{1}{\beta}\tan^{-1}\left\{\frac{1-g_L}{b_L - \sqrt{g_L[(1-g_L)^2 + b_L^2]}}\right\}$
	$l_{stub} = \frac{1}{\beta}\tan^{-1}\left[\sqrt{\frac{g_L}{(1-g_L)^2 + b_L^2}}\right]$	$l_{stub} = \frac{\pi}{\beta} - \frac{1}{\beta}\tan^{-1}\left[\sqrt{\frac{g_L}{(1-g_L)^2 + b_L^2}}\right]$
∞ Ω	$l_1 = \frac{1}{\beta}\tan^{-1}\left\{\frac{1-g_L}{b_L + \sqrt{g_L[(1-g_L)^2 + b_L^2]}}\right\}$	$l_1 = \frac{1}{\beta}\tan^{-1}\left\{\frac{1-g_L}{b_L - \sqrt{g_L[(1-g_L)^2 + b_L^2]}}\right\}$
	$l_{stub} = \frac{\pi}{\beta} - \frac{1}{\beta}\tan^{-1}\left[\sqrt{\frac{(1-g_L)^2 + b_L^2}{g_L}}\right]$	$l_{stub} = \frac{1}{\beta}\tan^{-1}\left[\sqrt{\frac{(1-g_L)^2 + b_L^2}{g_L}}\right]$

$$\beta = \frac{2\pi}{\lambda} = \frac{2\pi f}{v} = 2\pi f\sqrt{LC} \quad \text{lossless lines}$$

(normalized relative to the characteristic impedance of the line) as

$$y_L = \frac{1}{z_L} = \frac{1}{\dfrac{Z_L}{Z_o}} = \frac{Z_o}{Z_L} = g_L + jb_L \tag{3.104}$$

These equations are nonlinear, and there are two possible solutions for stubs located within $\lambda/2$ of the load. Radians should be used when determining the inverse tangents. The previous numerical case will now be checked using these expressions. If

$$Z_o = 50\ \Omega,\quad C = 100\ \text{pF/m},\quad f = 30\ \text{MHz},\quad Z_L = 120 + j200\ \Omega,\quad \beta = 0.942\ \text{rad/m}$$

then the normalized admittance is

$$y_L = \frac{50}{120 + j200} \approx 0.11 - j0.18 = g_L + jb_L$$

Therefore, for a short-circuited stub, one solution is

$$l_1 \approx \frac{1}{0.942}\tan^{-1}\left\{\frac{1-0.11}{-0.18+\sqrt{0.11\left[(1-0.11)^2+(-0.18)^2\right]}}\right\} \approx 1.5\ \text{m}$$

$$l_{stub} \approx \frac{1}{0.942}\tan^{-1}\left[\sqrt{\frac{0.11}{(1-0.11)^2+(-0.18)^2}}\right] \approx 0.37\ \text{m}$$

These results are the same as given in Mathcad 3.16. The second solution is

$$l_1 \approx \frac{1}{0.942}\tan^{-1}\left\{\frac{1-0.11}{-0.18-\sqrt{0.11\left[(1-0.11)^2+(-0.18)^2\right]}}\right\} \approx 2.2\ \text{m}$$

$$l_{stub} \approx \frac{\pi}{0.942} - \frac{1}{0.942}\tan^{-1}\left[\sqrt{\frac{0.11}{(1-0.11)^2+(-0.18)^2}}\right] \approx 3.0\ \text{m}$$

Care must be exercised when determining the arctangent of a negative number in the second (or third) quadrant. (By changing the initial "Guess" values, these results can also be obtained using the method given in Mathcad 3.16.) The advantage of using these equations is that they quickly provide the two possible closest solutions to the load without providing any initial "Guess" values.

Like other matching networks, this transmission-line based matching network has a limited bandwidth. That is, the network may provide an excellent match at and very near a particular design frequency, but as the frequency varies away from this particular frequency, the match deteriorates. The VSWR is a measure of the quality of a match, and it is defined for this stub match as

$$\text{VSWR} = \frac{1+|\rho|}{1-|\rho|} \quad \text{where } \rho = \frac{Z_L - Z_o}{Z_L + Z_o} = \frac{Z_{intotal} - Z_o}{Z_{intotal} + Z_o} \tag{3.105}$$

In Mathcad 3.18, the VSWR is plotted vs. frequency for the results obtained in Mathcad 3.16. If the bandwidth is based on those frequencies corresponding to a VSWR of 1.5, then the bandwidth is about 30.7 MHz − 29.3 MHz = 1.4 MHz, corresponding to about 5% bandwidth. For frequencies below 29.3 MHz and above 30.7 MHz the VSWR provided by this stub is greater than 1.5. This implies that the impedance seen by the source is not necessarily equal to Z_o. Most sources are designed to operate over a specific range of loads. When the load is not within this expected range, excessive loading and frequency "pulling" can occur. The bandwidth and response will vary with the stub's position and length. Single-stub tuners are considered low-bandwidth matching devices.

$L := 0.25\cdot10^{-6}$ $C := 100\cdot10^{-12}$ $Z_o := \sqrt{\frac{L}{C}}$ $Z_o = 50$ $v := \frac{1}{\sqrt{L\cdot C}}$ $v := 2\times10^{8}$ $j := \sqrt{-1}$

$Z_L := 120 + j\cdot200$ $l_{th1} := 1.527$ $l_{thstub} := 0.372$ $Z_{Lstub} := 0$

$f := 28\cdot10^{6}, 28.1\cdot10^{6} .. 32\cdot10^{6}$

$$\beta(f) := \frac{2\cdot\pi\cdot f}{v}$$

$$Z_{in1}(f) := Z_o\cdot\frac{Z_L + j\cdot Z_o\cdot\tan(\beta(f)\cdot l_{th1})}{Z_o + j\cdot Z_L\cdot\tan(\beta(f)\cdot l_{th1})} \qquad Z_{instub}(f) := Z_o\cdot\frac{Z_{Lstub} + j\cdot Z_o\cdot\tan(\beta(f)\cdot l_{thstub})}{Z_o + j\cdot Z_{Lstub}\cdot\tan(\beta(f)\cdot l_{thstub})}$$

$$Z_{intotal}(f) := \frac{Z_{in1}(f)\cdot Z_{instub}(f)}{Z_{in1}(f) + Z_{instub}(f)} \qquad VSWR(f) := \frac{1 + \left|\frac{Z_{intotal}(f) - Z_o}{Z_{intotal}(f) + Z_o}\right|}{1 - \left|\frac{Z_{intotal}(f) - Z_o}{Z_{intotal}(f) + Z_o}\right|}$$

VSWR Vs. Frequency

VSWR(f) ——
1.5 - - - -

3
2.5
2
1.5
1
28 29 30 31 32

$\frac{f}{10^6}$
MHz

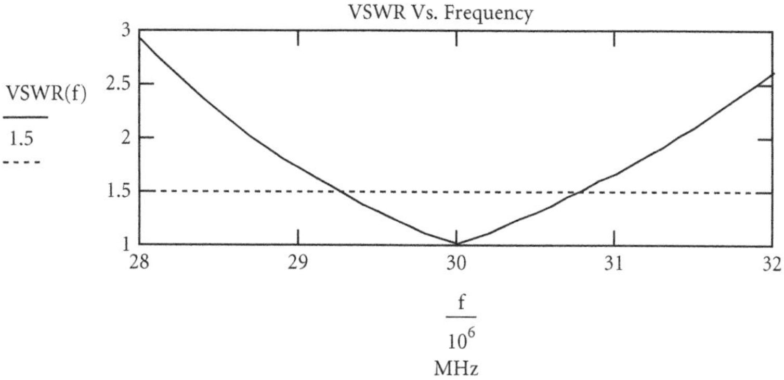

MATHCAD 3.18 Variation of the VSWR with frequency for a single-stub tuner. The best match is at 30 MHz.

Again, the VSWR on the stub line itself is large since its load (e.g., short or open circuit) is obviously not matched to its line impedance. When the VSWR is large, serious line currents and losses can occur in the stub line. Although many possible stub positions and lengths will satisfy the matching requirements, shorter stub line lengths will allow for lower losses. For this reason, the stub length is often selected to be less than $\lambda/2$. When the frequency of operation is very high (e.g., GHz range), to obtain reasonably sized stubs, the line length may have to be greater than $\lambda/2$. Also, at these higher frequencies, adjustment of the stub length for fine tuning may be difficult.

Although the specific examples in this discussion have used short-circuited stubs, open-circuited stubs can also be used. However, because it is often easier to construct a short-circuited stub and the short-circuited stub will radiate less power for some line types (e.g., coax), short-circuited stubs are more common. An inexpensive method of short circuiting coax is capping its end with aluminum foil (rather than just placing a single wire between the inner and outer conductors). In commercial coaxial stubs, adjustable short-circuiting plungers are used. Open circuits are more difficult to construct since a true, or an ideal, open circuit has infinite impedance at all frequencies, and an open circuit does not absorb any power. A real open circuit has nonzero parasitic capacitance and finite impedance, and there will be some power loss due to radiation.

If the load impedance changes, it is usually inconvenient to use a single-stub matching line: the stub has to be attached to a new location along the line, and its length must be adjusted. To control both the resistive and reactive portions of the impedance, both the stub's position and length must be adjustable. By using two stubs as shown in Figure 3.39, however, only the length of the stubs needs to be adjusted as the load changes. The location of the stubs along the line does not have to be adjusted as the load changes. (The two variables in this case are the two lengths of the stubs.) Although an approximation for the position and length of the double stubs can be calculated, it is often simpler to determine the best match

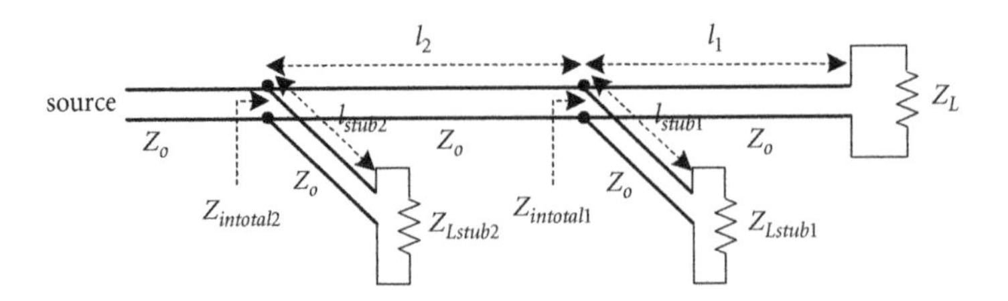

FIGURE 3.39 Double-stub tuner.

(i.e., lowest VSWR) by trial and error. Often the stubs are fixed at some close spacing such as $l_2 = 3\lambda/8$. When they are at a fixed location along a line, sometimes the desired VSWR cannot be obtained and additional cable (e.g., $\lambda/4$ in length) must be inserted between the double-stub tuner and load.

In Mathcad 3.19, a double-stub tuner is designed for the following parameters:

$$Z_o = 50\,\Omega,\quad C = 100\,\text{pF/m},\quad f = 30\,\text{MHz},\quad Z_L = 120 + j200\,\Omega,\quad l_1 = \frac{\lambda}{3},\quad l_2 = \frac{3\lambda}{8}$$

The lengths of the two stubs are

$$l_{stub1} \approx 2.2 = \frac{\lambda}{3},\quad l_{stub2} \approx 0.47 \approx \frac{\lambda}{14}$$

$f := 30 \cdot 10^6 \qquad L := 0.25 \cdot 10^{-6} \qquad C := 100 \cdot 10^{-12} \qquad \omega := 2 \cdot \pi \cdot f \qquad j := \sqrt{-1}$

$Z_o := \sqrt{\frac{L}{C}} \qquad Z_o = 50 \qquad v := \frac{1}{\sqrt{L \cdot C}} \qquad v = 2 \times 10^8 \qquad \lambda := \frac{v}{f} \qquad \beta := \frac{2 \cdot \pi}{\lambda} \qquad \lambda = 6.667$

$$Z_{in1}(l_{th1}, Z_L) := Z_o \cdot \frac{Z_L + j \cdot Z_o \cdot \tan(\beta \cdot l_{th1})}{Z_o + j \cdot Z_L \cdot \tan(\beta \cdot l_{th1})} \qquad Z_{instub1}(l_{thstub1}, Z_{Lstub1}) := Z_o \cdot \frac{Z_{Lstub1} + j \cdot Z_o \cdot \tan(\beta \cdot l_{thstub1})}{Z_o + j \cdot Z_{Lstub1} \cdot \tan(\beta \cdot l_{thstub1})}$$

$$Z_{instub2}(l_{thstub2}, Z_{Lstub2}) := Z_o \cdot \frac{Z_{Lstub2} + j \cdot Z_o \cdot \tan(\beta \cdot l_{thstub2})}{Z_o + j \cdot Z_{Lstub2} \cdot \tan(\beta \cdot l_{thstub2})}$$

$$Z_{intotal1}(l_{th1}, Z_L, l_{thstub1}, Z_{Lstub1}) := \frac{Z_{in1}(l_{th1}, Z_L) \cdot Z_{instub1}(l_{thstub1}, Z_{Lstub1})}{Z_{in1}(l_{th1}, Z_L) + Z_{instub1}(l_{thstub1}, Z_{Lstub1})}$$

$$Z_{in2}(l_{th1}, Z_L, l_{thstub1}, Z_{Lstub1}, l_{th2}) := Z_o \cdot \frac{Z_{intotal1}(l_{th1}, Z_L, l_{thstub1}, Z_{Lstub1}) + j \cdot Z_o \cdot \tan(\beta \cdot l_{th2})}{Z_o + j \cdot Z_{intotal1}(l_{th1}, Z_L, l_{thstub1}, Z_{Lstub1}) \cdot \tan(\beta \cdot l_{th2})}$$

$$Z_{intotal2}(l_{th1}, Z_L, l_{thstub1}, Z_{Lstub1}, l_{th2}, Z_{Lstub2}, l_{thstub2}) := \left(\frac{1}{Z_{instub2}(l_{thstub2}, Z_{Lstub2})} \ldots + \frac{1}{Z_{in2}(l_{th1}, Z_L, l_{thstub1}, Z_{Lstub1}, l_{th2})} \right)^{-1}$$

$Z_L := 120 + j \cdot 200 \qquad l_{th1} := \frac{\lambda}{3} \qquad l_{th2} := \frac{3 \cdot \lambda}{8} \qquad Z_{Lstub1} := 0 \qquad Z_{Lstub2} := 0$

$l_{thstub1} := 0.5 \qquad l_{thstub2} := 1$

Given

$\mathrm{Im}(Z_{intotal2}(l_{th1}, Z_L, l_{thstub1}, Z_{Lstub1}, l_{th2}, Z_{Lstub2}, l_{thstub2})) = 0$

$\mathrm{Re}(Z_{intotal2}(l_{th1}, Z_L, l_{thstub1}, Z_{Lstub1}, l_{th2}, Z_{Lstub2}, l_{thstub2})) = Z_o$

$SOLN := \mathrm{Minerr}(l_{thstub1}, l_{thstub2}) \qquad SOLN_0 = 2.217 \qquad SOLN_1 = 0.466$

$Z_{intotal2}(l_{th1}, Z_L, SOLN_0, Z_{Lstub1}, l_{th2}, Z_{Lstub2}, SOLN_1) = 50$

MATHCAD 3.19 Solving for the lengths of a double-stub tuner. The positions of the stubs were specified.

Increasing the distance l_1 simplifies the design process, especially for larger impedance mismatches. Recall that the input impedance of a transmission line repeats every $\lambda/2$. Thus, if the load impedance is very large compared to the characteristic impedance of the line, as the distance from the load increases (but before the distance of $\lambda/2$ is reached), the real part of the input impedance has the possibility of becoming closer to the characteristic impedance of the line (and the input reactance closer to zero). Therefore, the stub tuners would not have to "work as hard." Triple-stub tuners are even more difficult to design analytically, but they can provide additional benefits such as reduced sensitivity to stub-to-stub spacing.

3.39 Inductive Loading

A 50 Ω lossless transmission line (with C = 100 pF/m) operating at 30 MHz has a length of 8 m. Its output is open circuited. If a 0.1 μH inductor (L_{ex}) is inserted in series with one of the conductors midway down the line, as is shown in Figure 3.40, determine the expressions for the current everywhere along the line. Assume the source has an amplitude of 1 V with a resistance of 2 Ω. How would the analysis be different if the voltage along the line is desired? [Collin, '85]

Inductive loading is used in telephone systems (to reduce distortion and attenuation) and in antennas (to increase the current level, increase the radiation resistance, and cancel capacitive reactance). The expression for the current, in the frequency domain, along a lossless transmission line is from (3.93)

$$I_s(z) = \frac{V^+ e^{-j\beta z}\left[1 - \left(\dfrac{Z_L - Z_o}{Z_L + Z_o}\right) e^{j2\beta(z - l_{th})}\right]}{Z_o}$$

This general form for the current will be used for both halves of the transmission line. Before delving into the details, the β will be determined. Since the line is lossless, the velocity of propagation for any frequency signal on the line is

$$v = \frac{1}{\sqrt{LC}} = \frac{1}{\sqrt{(0.25\times10^{-6})(100\times10^{-12})}} = 2.0\times10^{8} \text{ m/s}$$

which is less than the speed of light, $c = 3\times10^{8}$ m/s. (The inductance per unit length of the transmission line, L, is not the same as L_{ex}, the external loading inductance.) The wavelength for signals within the transmission line is

$$\lambda = \frac{v}{f} = \frac{2.0\times10^{8}}{30\times10^{6}} \approx 6.7 \text{ m}$$

Therefore, the transmission line is 1.2λ in length. The phase constant is

$$\beta = \frac{2\pi}{\lambda} \approx 0.94 \text{ rad/m}$$

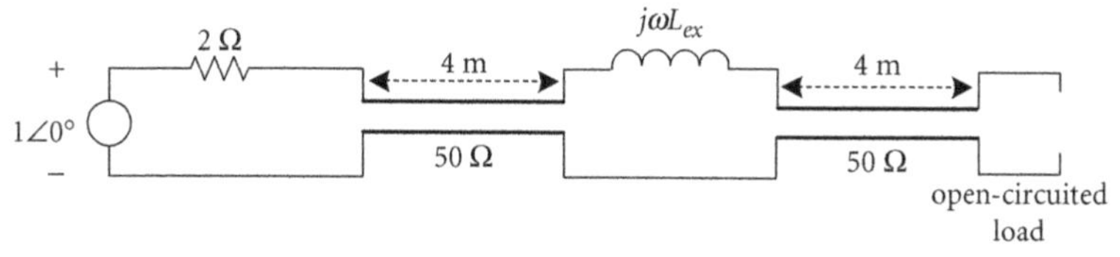

FIGURE 3.40 Series inductive loading midway along a transmission line.

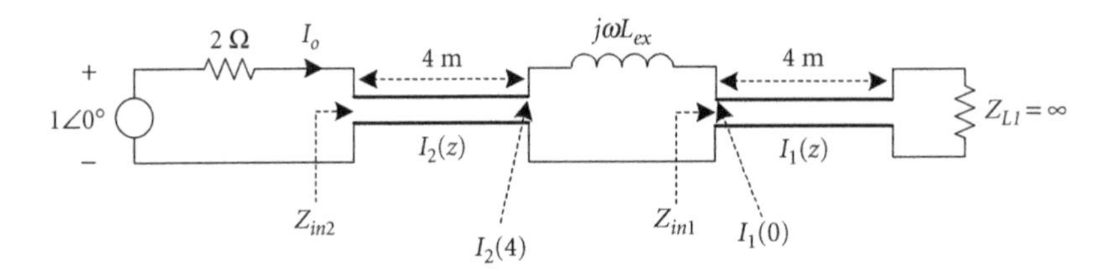

FIGURE 3.41 Current and impedance variables used in the analysis.

When determining the current (or voltage) along transmission lines, it is recommended that the impedance seen by the source be initially determined. Once this input impedance is determined, the initial current into (or voltage across) the transmission line connected to the source can be determined. This leads to the value for V^+.

Generally, when working with a circuit involving more than one transmission line, the input impedances looking into the transmission lines farthest from the source are determined first. These input impedances are then used, along with any connecting circuitry, as the loads for the next set of transmission lines closest to the source. This process is repeated until the input impedance at the source is obtained. This process is clearly illustrated in this example.

The two lines are connected together via the loading inductor. Referring to Figure 3.41, the impedance seen by this loading inductor looking toward the open-circuited load is (using radians when taking the tangent of βl_{th})

$$Z_{in1} = Z_o \frac{Z_{L1} + jZ_o \tan(\beta l_{th})}{Z_o + jZ_{L1} \tan(\beta l_{th})} = 50\frac{\infty + j50\tan(4\beta)}{50 + j\infty\tan(4\beta)} = \frac{50}{j\tan(4\beta)} \approx -j70\,\Omega$$

The total load seen by the transmission line closest to the source is then this input impedance in series with the impedance of the inductor L_{ex}:

$$Z_{L2} = Z_{in1} + j\omega L_{ex} = -j69 + j2\pi(30\times10^6)(0.1\times10^{-6}) \approx -j50\,\Omega$$

(If two transmission lines were connected to this one inductor, the parallel combination of the input impedance of both transmission lines would be calculated first and then this result added in series with the inductor's impedance.) The impedance seen by the source is then

$$Z_{in2} = 50\frac{Z_{L2} + j50\tan(4\beta)}{50 + jZ_{L2}\tan(4\beta)} \approx -j8\,\Omega$$

Now that the line input impedance is known, the current from the source can be determined as

$$I_o = I_{2s}(0) = \frac{1\angle 0^\circ}{2 + Z_{in2}} \approx 0.03 + j0.12\text{ A}$$

and equated to the current expression for the line closest to the source evaluated at $z = 0$:

$$I_{2s}(0) = \frac{V_2^+ e^{-j\beta(0)}\left[1 - \left(\frac{Z_{L2} - 50}{Z_{L2} + 50}\right)e^{j2\beta(0-4)}\right]}{50} \approx 0.03 + j0.12\text{ A}$$

Notice that the load used in this expression is the total load seen by the transmission line closest to the source, not the open-circuited load. Substituting the value obtained for Z_{L2} and solving for the variable V_2^+,

$$V_2^+ \approx 1.3 + j2.9 \text{ V}$$

Therefore, the unsimplified expression for the current in the line closest to the source is

$$I_{2s}(z) \approx \frac{(1.3 + j2.9)e^{-j0.94z}\left[1 - \left(\frac{-j50 - 50}{-j50 + 50}\right)e^{j2(0.94)(z-4)}\right]}{50}$$

The general expression for the current on the line closest to the open-circuited load is

$$I_{1s}(z) = \frac{V_1^+ e^{-j\beta(z-4)}\left[1 - \left(\frac{Z_{L1} - Z_o}{Z_{L1} + Z_o}\right)e^{j2\beta(z-8)}\right]}{Z_o}$$

The variable z has been shifted by 4 m because the $z = 0$ reference is at the source, not at the inductor junction. Since the current into the inductor must be equal to the current out of the inductor, the current is continuous at the junction. The current expression previously obtained can then be used to determine the initial current into the transmission line closest to the final load:

$$I_{2s}(4) \approx \frac{(1.3 + j2.9)e^{-j0.94(4)}\left[1 - \left(\frac{-j50 - 50}{-j50 + 50}\right)e^{j2(0.94)(4-4)}\right]}{50} \approx -0.02 - j0.09 \text{ A}$$

$$= I_{1s}(4) = \frac{V_1^+ e^{-j\beta(4-4)}\left[1 - \left(\frac{Z_{L1} - Z_o}{Z_{L1} + Z_o}\right)e^{j2\beta(4-8)}\right]}{Z_o}$$

Substituting the infinite value for Z_{L1} and solving for the variable V_1^+,

$$V_1^+ = -3.5 - j1.5 \text{ V}$$

The unsimplified expression for the current is then

$$I_{1s}(z) \approx \frac{(-3.5 - j1.5)e^{-j0.94(z-4)}\left[1 - e^{j2(0.94)(z-8)}\right]}{50}$$

The approach is quite similar when determining the voltage along the lines. There are a few minor differences. First, the voltage expression (3.81) along a lossless line is used:

$$V_s(z) = V^+ e^{-j\beta z}\left[1 + \left(\frac{Z_L - Z_o}{Z_L + Z_o}\right)e^{j2\beta(z - l_{th})}\right]$$

$$f := 30\cdot10^{6} \quad L := 0.25\cdot10^{-6} \quad C := 100\cdot10^{-12} \quad \omega := 2\cdot\pi\cdot f \quad T := \frac{1}{f} \quad j := \sqrt{-1}$$

$$Z_o := \sqrt{\frac{L}{C}} \quad Z_o = 50 \quad v := \frac{1}{\sqrt{L\cdot C}} \quad \lambda := \frac{v}{f} \quad \beta := \frac{2\cdot\pi}{\lambda} \quad V_s := 1 \quad R_s := 2 \quad l_{th} := 1.2\cdot\lambda$$

$$z := 0, \frac{l_{th}}{100} .. l_{th}$$

$$Z_{in1}(Z_{L1}) := Z_o\cdot\frac{Z_{L1} + j\cdot Z_o\cdot\tan\left(\beta\cdot\frac{l_{th}}{2}\right)}{Z_o + j\cdot Z_{L1}\cdot\tan\left(\beta\cdot\frac{l_{th}}{2}\right)} \qquad Z_{L2}(L_{ex}, Z_{L1}) := Z_{in1}(Z_{L1}) + j\cdot\omega\cdot L_{ex}$$

$$Z_{in2}(L_{ex}, Z_{L1}) := Z_o\cdot\frac{Z_{L2}(L_{ex}, Z_{L1}) + j\cdot Z_o\cdot\tan\left(\beta\cdot\frac{l_{th}}{2}\right)}{Z_o + j\cdot Z_{L2}(L_{ex}, Z_{L1})\cdot\tan\left(\beta\cdot\frac{l_{th}}{2}\right)} \qquad I_o(L_{ex}, Z_{L1}) := \frac{V_s}{R_s + Z_{in2}(L_{ex}, Z_{L1})}$$

$$V_{pos2}(L_{ex}, Z_{L1}) := \frac{I_o(L_{ex}, Z_{L1})\cdot Z_o}{1 - \left(\frac{Z_{L2}(L_{ex}, Z_{L1}) - Z_o}{Z_{L2}(L_{ex}, Z_{L1}) + Z_o}\right)\cdot e^{-j\cdot 2\cdot\beta\cdot\frac{l_{th}}{2}}}$$

$$I_2(z, L_{ex}, Z_{L1}) := \frac{V_{pos2}(L_{ex}, Z_{L1})\cdot e^{-j\cdot\beta\cdot z}\cdot\left[1 - \left(\frac{Z_{L2}(L_{ex}, Z_{L1}) - Z_o}{Z_{L2}(L_{ex}, Z_{L1}) + Z_o}\right)\cdot e^{j\cdot 2\cdot\beta\cdot\left(z - \frac{l_{th}}{2}\right)}\right]}{Z_o}$$

$$V_{pos1}(L_{ex}, Z_{L1}) := \frac{I_2\left(\frac{l_{th}}{2}, L_{ex}, Z_{L1}\right)\cdot Z_o}{1 - \left(\frac{Z_{L1} - Z_o}{Z_{L1} + Z_o}\right)\cdot e^{-j\cdot 2\cdot\beta\cdot\frac{l_{th}}{2}}}$$

$$I_1(z, L_{ex}, Z_{L1}) := \frac{V_{pos1}(L_{ex}, Z_{L1})\cdot e^{-j\cdot\beta\cdot\left(z - \frac{l_{th}}{2}\right)}\cdot\left[1 - \left(\frac{Z_{L1} - Z_o}{Z_{L1} + Z_o}\right)\cdot e^{j\cdot 2\cdot\beta\cdot(z - l_{th})}\right]}{Z_o}$$

$$I(z, L_{ex}, Z_{L1}) := \text{if}\left(z < \frac{l_{th}}{2}, I_2(z, L_{ex}, Z_{L1}), I_1(z, L_{ex}, Z_{L1})\right)$$

MATHCAD 3.20 Current along the transmission line with no series loading inductance.

Second, the initial voltage across the input of the line closest to the source should be determined instead of the current into the line:

$$V_o = V_{2s}(0) = 1\angle 0^\circ \frac{Z_{in2}}{2 + Z_{in2}}$$

Finally, the voltage is discontinuous between the two lines because of the voltage drop across the coil:

$$V_{1s}(4) = V_{2s}(4)\frac{Z_{in1}}{Z_{in1} + j\omega L_{ex}}$$

If the coil were not present, both the current and voltage would be continuous between the two lines.

In Mathcad 3.20, the magnitude of the current is plotted for various loads, setting the loading inductance equal to zero. These plots are used as a check on the solutions. As the load increases, the variation in the current magnitude increases. Recall that the VSWR increases with the impedance mismatch, and the VSWR is a measure of the current (and voltage) variation.

In Mathcad 3.21, the current is plotted for two values of the series loading inductance. The current is, and must be, zero at the load when the load is an open circuit. For these plots, the line length is electrically long and equal to 1.2λ. For a certain range of values, the loading inductance can increase the average current magnitude over both lines. Because the line is not electrically short, the total inductance of the line, minus the loading inductance, is not clearly defined. However, if the frequency of the source is decreased so that the total line length is electrically short, then the total inductance of the line is (8 m)(0.25 μH) = 2 μH.

In Mathcad 3.22, the line length is electrically short. Again, for open-circuited loads, the current must go to zero at the open circuit. Again, the average current magnitude can be increased for a range of loading inductance. Finally, the current magnitude is approximately uniform (or at least linear) over each of the two electrically-short lines.

3.40 Low-Loss Lines and Short Lines

Briefly explain why the propagation transfer function for sinusoidal signals on a matched transmission line is of the form

$$H = e^{-z\sqrt{(R+j\omega L)j\omega C}}$$

If the line losses are low but not negligible, determine *H*. What is *H* representing in this case? If the line losses are low but not negligible and the line is short, determine *H*. For a short low-loss line, determine the amplitude of the voltage at the receiver (or load) if the input voltage amplitude is *A*. Define "short." Define "low loss."

When the differential equations for the voltage and current on a matched transmission line are solved, the solutions are of the form

$$Ae^{-\alpha z}\cos(\omega t - \beta z) \tag{3.106}$$

where z is the position along the line, α is the attenuation coefficient of the cable, β is the phase constant of the cable, and $\omega = 2\pi f$. For a line matched to its load, there are no reflected or negative-going signals. This general form for the steady-state voltage and current includes line loss via the exponential term and includes line delay via the βz term. This equation can also be written in the frequency domain or phasor form as

$$Ae^{-\alpha z}e^{-j\beta z} = Ae^{-(\alpha + j\beta)z} \tag{3.107}$$

$$f := 30 \cdot 10^{6} \quad L := 0.25 \cdot 10^{-6} \quad C := 100 \cdot 10^{-12} \quad \omega := 2 \cdot \pi \cdot f \quad T := \frac{1}{f} \quad j := \sqrt{-1}$$

$$Z_o := \sqrt{\frac{L}{C}} \quad Z_o = 50 \quad v := \frac{1}{\sqrt{L \cdot C}} \quad \lambda := \frac{v}{f} \quad \beta := \frac{2 \cdot \pi}{\lambda} \quad V_s := 1 \quad R_s := 2 \quad l_{th} := 1.2 \cdot \lambda$$

$$z := 0, \frac{l_{th}}{100} .. \, l_{th}$$

$$Z_{in1}(Z_{L1}) := Z_o \cdot \frac{Z_{L1} + j \cdot Z_o \cdot \tan\left(\beta \cdot \frac{l_{th}}{2}\right)}{Z_o + j \cdot Z_{L1} \cdot \tan\left(\beta \cdot \frac{l_{th}}{2}\right)} \qquad Z_{L2}(L_{ex}, Z_{L1}) := Z_{in1}(Z_{L1}) + j \cdot \omega \cdot L_{ex}$$

$$Z_{in2}(L_{ex}, Z_{L1}) := Z_o \cdot \frac{Z_{L2}(L_{ex}, Z_{L1}) + j \cdot Z_o \cdot \tan\left(\beta \cdot \frac{l_{th}}{2}\right)}{Z_o + j \cdot Z_{L2}(L_{ex}, Z_{L1}) \cdot \tan\left(\beta \cdot \frac{l_{th}}{2}\right)} \qquad I_o(L_{ex}, Z_{L1}) := \frac{V_s}{R_s + Z_{in2}(L_{ex}, Z_{L1})}$$

$$V_{pos2}(L_{ex}, Z_{L1}) := \frac{I_o(L_{ex}, Z_{L1}) \cdot Z_o}{1 - \left(\frac{Z_{L2}(L_{ex}, Z_{L1}) - Z_o}{Z_{L2}(L_{ex}, Z_{L1}) + Z_o}\right) \cdot e^{-j \cdot 2 \cdot \beta \cdot \frac{l_{th}}{2}}}$$

$$I_2(z, L_{ex}, Z_{L1}) := \frac{V_{pos2}(L_{ex}, Z_{L1}) \cdot e^{-j \cdot \beta \cdot z} \cdot \left[1 - \left(\frac{Z_{L2}(L_{ex}, Z_{L1}) - Z_o}{Z_{L2}(L_{ex}, Z_{L1}) + Z_o}\right) \cdot e^{j \cdot 2 \cdot \beta \cdot \left(z - \frac{l_{th}}{2}\right)}\right]}{Z_o}$$

$$V_{pos1}(L_{ex}, Z_{L1}) := \frac{I_2\left(\frac{l_{th}}{2}, L_{ex}, Z_{L1}\right) \cdot Z_o}{1 - \left(\frac{Z_{L1} - Z_o}{Z_{L1} + Z_o}\right) \cdot e^{-j \cdot 2 \cdot \beta \cdot \frac{l_{th}}{2}}}$$

$$I_1(z, L_{ex}, Z_{L1}) := \frac{V_{pos1}(L_{ex}, Z_{L1}) \cdot e^{-j \cdot \beta \cdot \left(z - \frac{l_{th}}{2}\right)} \cdot \left[1 - \left(\frac{Z_{L1} - Z_o}{Z_{L1} + Z_o}\right) \cdot e^{j \cdot 2 \cdot \beta \cdot (z - l_{th})}\right]}{Z_o}$$

$$I(z, L_{ex}, Z_{L1}) := \text{if}\left(z < \frac{l_{th}}{2}, I_2(z, L_{ex}, Z_{L1}), I_1(z, L_{ex}, Z_{L1})\right)$$

Inductive-Loaded Open-Circuited Line

$|I(z, 0, 10^9)|$, $|I(z, 10^{-7}, 10^9)|$, $|I(z, 10^{-6}, 10^9)|$ — A

0.15, 0.1, 0.05, 0

0, 2, 4, 6, 8

z, m

MATHCAD 3.21 Current along the transmission line when the series loading inductance is present.

$f := 30\cdot 10^6 \quad L := 0.25\cdot 10^{-6} \quad C := 100\cdot 10^{-12} \quad \omega := 2\cdot\pi\cdot f \quad T := \frac{1}{f} \quad j := \sqrt{-1}$

$Z_o := \sqrt{\frac{L}{C}} \quad Z_o = 50 \quad v := \frac{1}{\sqrt{L\cdot C}} \quad \lambda := \frac{v}{f} \quad \beta := \frac{2\cdot\pi}{\lambda} \quad V_s := 1 \quad R_s := 2 \quad l_{th} := \frac{\lambda}{10}$

$z := 0, \frac{l_{th}}{100} .. l_{th}$

$$Z_{in1}(Z_{L1}) := Z_o\cdot\frac{Z_{L1} + j\cdot Z_o\cdot\tan\left(\beta\cdot\frac{l_{th}}{2}\right)}{Z_o + j\cdot Z_{L1}\cdot\tan\left(\beta\cdot\frac{l_{th}}{2}\right)} \qquad Z_{L2}(L_{ex}, Z_{L1}) := Z_{in1}(Z_{L1}) + j\cdot\omega\cdot L_{ex}$$

$$Z_{in2}(L_{ex}, Z_{L1}) := Z_o\cdot\frac{Z_{L2}(L_{ex}, Z_{L1}) + j\cdot Z_o\cdot\tan\left(\beta\cdot\frac{l_{th}}{2}\right)}{Z_o + j\cdot Z_{L2}(L_{ex}, Z_{L1})\cdot\tan\left(\beta\cdot\frac{l_{th}}{2}\right)} \qquad I_o(L_{ex}, Z_{L1}) := \frac{V_s}{R_s + Z_{in2}(L_{ex}, Z_{L1})}$$

$$V_{pos2}(L_{ex}, Z_{L1}) := \frac{I_o(L_{ex}, Z_{L1})\cdot Z_o}{1 - \left(\frac{Z_{L2}(L_{ex}, Z_{L1}) - Z_o}{Z_{L2}(L_{ex}, Z_{L1}) + Z_o}\right)\cdot e^{-j\cdot 2\cdot\beta\cdot\frac{l_{th}}{2}}}$$

$$I_2(z, L_{ex}, Z_{L1}) := \frac{V_{pos2}(L_{ex}, Z_{L1})\cdot e^{-j\cdot\beta\cdot z}\cdot\left[1 - \left(\frac{Z_{L2}(L_{ex}, Z_{L1}) - Z_o}{Z_{L2}(L_{ex}, Z_{L1}) + Z_o}\right)\cdot e^{j\cdot 2\cdot\beta\cdot\left(z - \frac{l_{th}}{2}\right)}\right]}{Z_o}$$

$$V_{pos1}(L_{ex}, Z_{L1}) := \frac{I_2\left(\frac{l_{th}}{2}, L_{ex}, Z_{L1}\right)\cdot Z_o}{1 - \left(\frac{Z_{L1} - Z_o}{Z_{L1} + Z_o}\right)\cdot e^{-j\cdot 2\cdot\beta\cdot\frac{l_{th}}{2}}}$$

$$I_1(z, L_{ex}, Z_{L1}) := \frac{V_{pos1}(L_{ex}, Z_{L1})\cdot e^{-j\cdot\beta\cdot\left(z - \frac{l_{th}}{2}\right)}\cdot\left[1 - \left(\frac{Z_{L1} - Z_o}{Z_{L1} + Z_o}\right)\cdot e^{j\cdot 2\cdot\beta\cdot(z - l_{th})}\right]}{Z_o}$$

$$I(z, L_{ex}, Z_{L1}) := \text{if}\left(z < \frac{l_{th}}{2}, I_2(z, L_{ex}, Z_{L1}), I_1(z, L_{ex}, Z_{L1})\right)$$

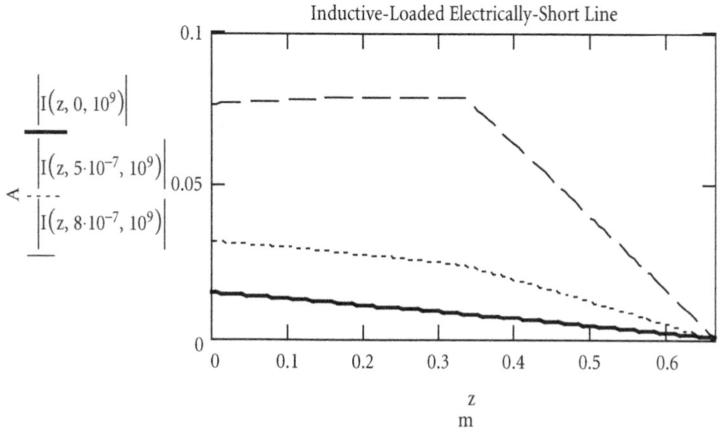

MATHCAD 3.22 Current along an electrically-short transmission line when a series loading inductance is present.

Transfer functions are a frequency-domain concept. To convert a frequency-domain function back to the "real" time domain is straightforward:

$$\text{Re}\left[Ae^{-(\alpha+j\beta)z}e^{j\omega t}\right] = \text{Re}\left[Ae^{-\alpha z}e^{j(\omega t-\beta z)}\right]$$

where Re( ) is the real nonimaginary portion of the function. Euler's formula

$$e^{jx} = \cos x + j\sin x \quad \text{where } j = \sqrt{-1}$$

is used to rewrite the complex exponential term, yielding

$$\text{Re}\{Ae^{-\alpha z}[\cos(\omega t - \beta z) + j\sin(\omega t - \beta z)]\} = Ae^{-\alpha z}\cos(\omega t - \beta z) \tag{3.108}$$

which is the original time-domain expression.

For a transmission line, the propagation constant, γ, is defined as

$$\gamma = \alpha + j\beta = \sqrt{(R + j\omega L)(G + j\omega C)} = \sqrt{(R + j\omega L)j\omega C} \quad \text{if } G = 0 \tag{3.109}$$

Inserting (3.109) into (3.107) yields the expected form for the propagation transfer function:

$$e^{-(\alpha+j\beta)z} = e^{-z\sqrt{(R+j\omega L)j\omega C}} \tag{3.110}$$

(The constant A was set to one, which assumes the matched line has no "gain.") Now, if the conductor losses are zero, then

$$H = e^{-z\sqrt{(j\omega L)j\omega C}} = e^{-z\sqrt{-\omega^2 LC}} = e^{-z\omega\sqrt{-1}\sqrt{LC}} = e^{-j\omega\sqrt{LC}z} \quad \text{if } R = 0 \tag{3.111}$$

Converting this expression back to the "real" time domain yields the impulse function (i.e, the time-domain version of the transfer function) for the line:

$$\begin{aligned} h(t) &= \text{Re}\left[e^{-j\omega\sqrt{LC}z}e^{j\omega t}\right] = \text{Re}\left[e^{j(\omega t-\omega\sqrt{LC}z)}\right] \\ &= \cos\left(\omega t - \omega\sqrt{LC}z\right) = \cos(\omega t - \beta z) \quad \text{if } R = 0 \end{aligned}$$

where $\beta = \omega\sqrt{LC}$ is the phase constant for a lossless line. As expected, there is no exponential term and no losses or signal decay, but there is a phase delay term, $\omega\sqrt{LC}z$. Time is required for the signal to travel to the end of the line. The velocity of the signal on a lossless line can be obtained from this expression:

$$h(t) = \cos\left[\omega\left(t - \frac{z}{\frac{1}{\sqrt{LC}}}\right)\right] \quad \text{if } R = 0$$

The velocity is therefore

$$v = \frac{1}{\sqrt{LC}} \quad \text{if } R = 0$$

Note that the velocity is not a function of the frequency. The velocity of a signal on a lossless transmission line is not a function of the frequency.

For low losses, the resistance, R, is small compared to the inductive reactance, ωL. The propagation constant can then be approximated as

$$\gamma=\sqrt{(R+j\omega L)\,j\omega C}=\sqrt{-\omega^2LC+j\omega CR}=j\sqrt{LC}\sqrt{\omega^2-\frac{j\omega R}{L}}$$
$$=j\omega\sqrt{LC}\sqrt{1+\frac{R}{j\omega L}}\approx j\omega\sqrt{LC}\left(1+\frac{R}{2j\omega L}\right)=\frac{R}{2}\sqrt{\frac{C}{L}}+j\omega\sqrt{LC}\quad \text{if } R\ll\omega L \tag{3.112}$$

where the first two terms of the binomial expansion

$$(1+x)^n=1+nx+\frac{n(n-1)}{2}x^2+\frac{n(n-1)(n-2)}{6}x^3+\cdots$$

were used to approximate the complex quantity under the square root. In this case, $n = 1/2$ and $x = R/(j\omega L)$. For this approximation to be good, x must be much less than one so that the x^2, x^3, and higher order terms can be neglected; in other words, the resistance, R, must be small compared to the inductive reactance ωL. The expression for the impulse function for low-loss lines is

$$h(t)\approx \text{Re}\left[e^{-\left(\frac{R}{2}\sqrt{\frac{C}{L}}+j\omega\sqrt{LC}\right)z}\right]=e^{-\frac{R}{2}\sqrt{\frac{C}{L}}z}\cos\left(\omega t-\omega\sqrt{LC}z\right)\quad \text{if } R\ll\omega L \tag{3.113}$$

Unlike the no-loss situation, the low-loss case involves an exponential decay term. Of course, there is still a time delay present on low-loss lines. The signal is attenuated as it passes through the cable. It is sometimes helpful to rewrite the attenuation constant, α, as a function of the characteristic impedance of a no-loss line:

$$\alpha=\frac{R}{2}\sqrt{\frac{C}{L}}=\frac{R}{2Z_o} \tag{3.114}$$

The line resistance per meter, R, is usually much less than the characteristic impedance of the line.

Now, if the line is “short,” the expression for $h(t)$ can be further simplified by recalling the series definition for the exponential function:

$$e^x=1+x+\frac{x^2}{2}+\cdots$$

If x is much less than one, then a good approximation for the exponential is $1+x$. For short low-loss lines, z is small, α is small, αz is small (much less than one), and

$$h(t)\approx\left[1-\left(\frac{R}{2}\sqrt{\frac{C}{L}}\right)z\right]\cos\left(\omega t-\omega\sqrt{LC}z\right)\quad \text{if } R\ll\omega L \tag{3.115}$$

The amplitude of the voltage at the receiver for a short, low-loss matched line is

$$A\left[1-\left(\frac{R}{2}\sqrt{\frac{C}{L}}\right)z\right] \tag{3.116}$$

As expected, the amplitude at the receiver decreases with z, the distance the signal travels, and decreases with R, the resistance per unit length of the line.

3.41 Inductive Line

A resistive load R_L is connected to a lossless transmission line of characteristic impedance Z_o. Assuming the line is electrically small, show that for small values of R_L the input impedance is

$$Z_{in} = R_L + j\omega L_T\left(1 - \frac{R_L^2 C_T}{L_T}\right)$$

The parameters L_T and C_T represent the *total* inductance and capacitance of the transmission line, respectively. Determine the relationship between the load resistance and characteristic impedance of the line so that the input reactance is inductive. [Adler]

When a line is connected to a low-impedance load, it is common to assume that its input reactance is inductive and that the system is mainly magnetic in nature. For sinusoidal steady-state conditions, the input impedance of a lossless transmission line is

$$Z_{in} = Z_o \frac{Z_L + jZ_o \tan(\beta l_{th})}{Z_o + jZ_L \tan(\beta l_{th})}$$

where l_{th} is the distance to the load, Z_o is the characteristic impedance of the line, Z_L is the load impedance, and $\beta = 2\pi/\lambda$ is the phase constant. If the line is electrically short, then

$$\beta l_{th} = \frac{2\pi}{\lambda} l_{th} \tag{3.117}$$

should be small since the length of the line, l_{th}, is a small fraction of a wavelength, λ. The series definitions for sine and cosine are helpful in determining an approximate expression for tangent x:

$$\sin x = x - \frac{x^3}{3!} + \frac{x^5}{5!} - \cdots, \quad \cos x = 1 - \frac{x^2}{2!} + \frac{x^4}{4!} - \cdots$$

The series definition for tangent of x is therefore

$$\tan x = \frac{\sin x}{\cos x} = \frac{x - \dfrac{x^3}{3!} + \dfrac{x^5}{5!} - \cdots}{1 - \dfrac{x^2}{2!} + \dfrac{x^4}{4!} - \cdots}$$

If x is small (i.e., much less than one), then the high-order terms of x are negligible compared to x and one:

$$\tan x \approx \frac{x}{1} = x \quad \text{for small } x$$

This allows the tangent term in the Z_{in} expression to be approximated as

$$\tan(\beta l_{th}) \approx \beta l_{th} = \omega \sqrt{\frac{L_T}{l_{th}} \frac{C_T}{l_{th}}} l_{th} = \omega \sqrt{L_T C_T} \tag{3.118}$$

Again, L_T and C_T correspond to the total inductance and capacitance, respectively, of the electrically-short lossless line. The definition for the phase constant for a lossless line,

$$\beta = \omega\sqrt{LC}$$

is a function of the inductance, L, and capacitance, C, *per unit length*. If the line is not electrically short, the significance of total inductance and capacitance for the transmission line is not clear.

The input impedance for an electrically-short line terminated with a load resistance, R_L, is

$$Z_{in} \approx Z_o \frac{R_L + jZ_o\omega\sqrt{L_T C_T}}{Z_o + jR_L\omega\sqrt{L_T C_T}} = Z_o \frac{R_L + jZ_o\omega\sqrt{L_T C_T}}{Z_o + jR_L\omega\sqrt{L_T C_T}} \frac{Z_o - jR_L\omega\sqrt{L_T C_T}}{Z_o - jR_L\omega\sqrt{L_T C_T}}$$

$$= Z_o \frac{R_L Z_o + Z_o R_L \omega^2 L_T C_T + jZ_o^2\omega\sqrt{L_T C_T} - jR_L^2\omega\sqrt{L_T C_T}}{Z_o^2 + R_L^2\omega^2 L_T C_T}$$

$$\approx Z_o \frac{R_L Z_o + jZ_o^2\omega\sqrt{L_T C_T}}{Z_o^2} = R_L + jZ_o\omega\sqrt{L_T C_T}$$

or

$$Z_{in} \approx R_L + j\omega L_T \tag{3.119}$$

since $Z_o = \sqrt{L_T / C_T}$. The last approximation assumed that the characteristic impedance of the cable was much greater than the load resistance (i.e., "small" values of R_L)

$$Z_o \gg R_L \tag{3.120}$$

and that the line was electrically short:

$$\beta l_{th} = \omega\sqrt{L_T C_T} \ll 1 \tag{3.121}$$

The input impedance consists of a resistive term and an inductive term (the imaginary part of the impedance is positive, which represents an inductive reactance, since $X_L = \omega L_T$ while $X_C = -1/(\omega C_T)$).

This input impedance approximation, (3.119), although consisting of both a resistive and an inductive term, is not the same expression as given in the problem statement. Another approximation can be obtained by first rewriting the input impedance approximation as

$$Z_{in} \approx Z_o \frac{R_L + jZ_o\omega\sqrt{L_T C_T}}{Z_o + jR_L\omega\sqrt{L_T C_T}} = jZ_o\omega\sqrt{L_T C_T}\, \frac{1 + \dfrac{R_L}{jZ_o\omega\sqrt{L_T C_T}}}{1 + \dfrac{jR_L\omega\sqrt{L_T C_T}}{Z_o}}$$

$$= jZ_o\omega\sqrt{L_T C_T}\left(1 + \frac{R_L}{jZ_o\omega\sqrt{L_T C_T}}\right)\left(1 + \frac{jR_L\omega\sqrt{L_T C_T}}{Z_o}\right)^{-1} \tag{3.122}$$

For small values of x, the binomial expansion states that $(1+x)^n \approx 1+nx.$ Since $n = -1$ in this case, (3.122) can be approximated as

$$Z_{in} \approx jZ_o\omega\sqrt{L_T C_T}\left(1+\frac{R_L}{jZ_o\omega\sqrt{L_T C_T}}\right)\left(1-\frac{jR_L\omega\sqrt{L_T C_T}}{Z_o}\right)$$

$$= R_L(1+\omega^2 L_T C_T) + j\omega\sqrt{L_T C_T}\left(Z_o - \frac{R_L^2}{Z_o}\right)$$

Substituting in the definition for the characteristic impedance for a lossless line

$$Z_o = \sqrt{\frac{L}{C}} = \sqrt{\frac{L_T}{C_T}}$$

results in the desired expression:

$$Z_{in} \approx R_L(1+\omega^2 L_T C_T) + j\omega L_T\left(1-\frac{R_L^2 C_T}{L_T}\right) \approx R_L + j\omega L_T\left(1-\frac{R_L^2 C_T}{L_T}\right) \qquad (3.123)$$

In this approximation, it is necessary that

$$1-\frac{R_L^2 C_T}{L_T} > 0 \quad \text{or} \quad \frac{L_T}{C_T} = Z_o^2 > R_L^2 \qquad (3.124)$$

if the reactive portion of the impedance is inductive. This inequality is satisfied for low-impedance loads. The real part of both approximations (3.119) and (3.123) are identical. The imaginary part of (3.123), however, involves the load resistance, while (3.119) does not.

For low-distortion short lines, the reactive term in Z_{in} is generally small. If the capacitance of the cable is increased, the input reactance is decreased. This can reduce cable-related distortion since the reactive term is frequency dependent.

Provide an equivalent circuit for (3.123). If the characteristic impedance of the line is 50 Ω, the line is 1 m in length, and the highest frequency of interest is 20 MHz, determine the range of loads so that the input reactance is inductive.

An equivalent circuit from (3.123) is obtained by inspection. It is provided in Figure 3.42. Since the characteristic impedance of the line is 50 Ω,

$$50 = \sqrt{\frac{L}{C}} = \sqrt{\frac{L_T}{C_T}} \Rightarrow \frac{C_T}{L_T} = \frac{1}{2{,}500}$$

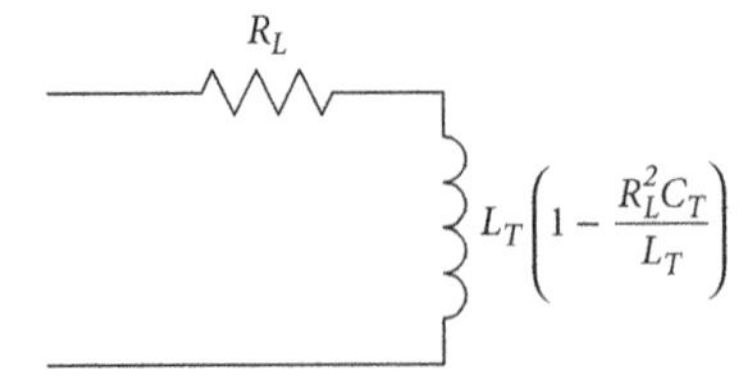

FIGURE 3.42 Equivalent circuit for the input impedance of an electrically-small transmission line connected to a low-impedance load.

For the reactive portion of this model to be inductive, its reactance must be positive:

$$\omega L_T\left(1-\frac{R_L^2C_T}{L_T}\right)>0 \;\Rightarrow 1>\frac{R_L^2C_T}{L_T}=\frac{R_L^2}{2{,}500} \;\Rightarrow R_L<50\,\Omega$$

Actually, this expression should only be used when

$$Z_o \gg R_L$$

and when the line is electrically short:

$$\beta l_{th}=\omega\sqrt{L_TC_T}\ll 1$$

Therefore, the load impedance should be less than about 5 Ω. The free-space wavelength corresponding to 20 MHz is

$$\lambda_o=\frac{3\times 10^8}{20\times 10^6}=15\,\text{m}$$

Although the velocity of a wave in a 50 Ω cable is less than the speed of light, a 1 m long line should be electrically short at 20 MHz.

3.42 Capacitive Line

A resistive load R_L is connected to a lossless transmission line of characteristic impedance Z_o. Assuming the line is electrically small, show for large values of R_L that the input admittance is

$$Y_{in}=G_L+j\omega C_T\left(1-\frac{G_L^2L_T}{C_T}\right)$$

where G_L is the load conductance. The parameters L_T and C_T represent the total inductance and capacitance of the transmission line, respectively. Determine the relationship between the load resistance and characteristic impedance of the line so that the input reactance is capacitive. [Adler]

The derivation of this result is similar to the low-impedance load version given previously. For this reason, many of the steps will be omitted. When a line is connected to a high-impedance load, it is common to assume that its input reactance is capacitive and that the system is mainly electric in nature. For sinusoidal steady-state conditions, the input admittance of a lossless transmission line is

$$Y_{in}=\frac{1}{Z_{in}}=\frac{1}{Z_o}\frac{Z_o+jZ_L\tan(\beta l_{th})}{Z_L+jZ_o\tan(\beta l_{th})}$$

where l_{th} is the distance to the load, Z_o is the characteristic impedance of the line, Z_L is the load impedance, and $\beta=2\pi/\lambda$ is the phase constant. If the line is electrically short, then

$$\beta l_{th}=\frac{2\pi}{\lambda}l_{th}=\omega\sqrt{L_TC_T}\ll 1$$

The variables L_T and C_T correspond to the total inductance and capacitance, respectively, of the electrically-short line. The input admittance for an electrically-short line terminated with a load resistance, R_L, is

$$Y_{in} \approx \frac{1}{Z_o}\frac{Z_o + jR_L\omega\sqrt{L_TC_T}}{R_L + jZ_o\omega\sqrt{L_TC_T}} = \frac{1}{Z_oR_L}\frac{Z_o + jR_L\omega\sqrt{L_TC_T}}{1+\dfrac{jZ_o\omega\sqrt{L_TC_T}}{R_L}}$$

$$= \frac{1}{Z_oR_L}\left(Z_o + jR_L\omega\sqrt{L_TC_T}\right)\left(1+\frac{jZ_o\omega\sqrt{L_TC_T}}{R_L}\right)^{-1}$$

$$\approx \frac{1}{Z_oR_L}\left(Z_o + jR_L\omega\sqrt{L_TC_T}\right)\left(1-\frac{jZ_o\omega\sqrt{L_TC_T}}{R_L}\right)$$

$$= \frac{1}{R_L}(1+\omega^2L_TC_T) + j\omega\sqrt{L_TC_T}\left(\frac{1}{Z_o} - \frac{Z_o}{R_L^2}\right)$$

or

$$Y_{in} \approx \frac{1}{R_L} + j\omega\sqrt{L_TC_T}\left(\frac{1}{Z_o} - \frac{Z_o}{R_L^2}\right) \tag{3.125}$$

The last approximations assumed that the characteristic impedance of the cable was much less than the load resistance ("large" values of R_L)

$$Z_o \ll R_L \tag{3.126}$$

and again that the line was electrically short:

$$\beta l_{th} = \omega\sqrt{L_TC_T} \ll 1 \tag{3.127}$$

Substituting in the definition for the characteristic impedance for a lossless line

$$Z_o = \sqrt{\frac{L}{C}} = \sqrt{\frac{L_T}{C_T}}$$

results in the expression

$$Y_{in} = \frac{1}{R_L} + j\omega C_T\left(1 - \frac{L_T}{R_L^2C_T}\right) \tag{3.128}$$

or $(G_L = 1/R_L)$

$$Y_{in} = G_L + j\omega C_T\left(1 - \frac{G_L^2L_T}{C_T}\right) \tag{3.129}$$

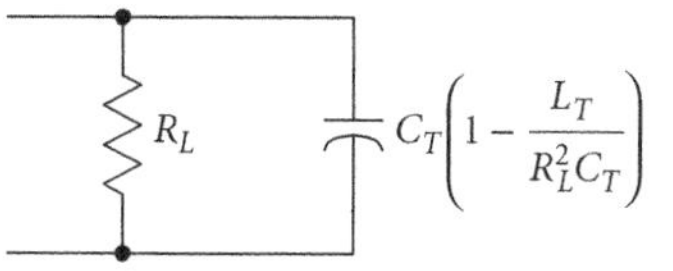

FIGURE 3.43 Equivalent circuit for the input impedance of an electrically-small transmission line connected to a high-impedance load.

In this approximation, it is necessary that

$$1-\frac{G_L^2 L_T}{C_T}>0 \quad \text{or} \quad \frac{L_T}{C_T}=Z_o^2<\frac{1}{G_L^2}=R_L^2 \tag{3.130}$$

if the imaginary portion of the admittance is capacitive. This inequality is satisfied for high-impedance loads. The input admittance consists of a resistive term and capacitive term (the imaginary part of the admittance is positive, which represents a capacitive reactance, since $Y_L=-1/(\omega L_T)$ while $Y_C=\omega C_T$).

Provide an equivalent circuit for (3.129). If the characteristic impedance of the line is 50 Ω, the line is 1 m in length, and the highest frequency of interest is 20 MHz, determine the range of load resistors so that the input reactance is capacitive.

An equivalent circuit from the expression for the input admittance of a high-impedance load is obtained by inspection ($G_L=1/R_L$). It is provided in Figure 3.43. Since Z_o = 50 Ω,

$$50=\sqrt{\frac{L}{C}}=\sqrt{\frac{L_T}{C_T}} \quad \Rightarrow \frac{L_T}{C_T}=2,500$$

To be capacitive,

$$\omega C_T\left(1-\frac{G_L^2 L_T}{C_T}\right)>0 \quad \Rightarrow 1>\frac{G_L^2 L_T}{C_T}=\frac{2,500}{R_L^2} \quad \Rightarrow R_L>50\,\Omega$$

Actually, this expression should only be used when

$$Z_o \ll R_L$$

and when the line is electrically short. Therefore, the load impedance should be greater than about 500 Ω. A 1 m long line should be electrically short at 20 MHz.

3.43 The Lossy Expressions for Sinusoidal Steady-State

Provide the sinusoidal steady-state expressions for the input impedance, voltage, and current along a lossy transmission line. How do they differ in form from the lossless expressions?

If a line is not excessively long (e.g., 100 ft), a line is not intentionally made lossy (e.g., ferrite-loaded cables), or a line's loss is not critical in the specific application, then the lossless equations are often adequate. The lossy equations are actually quite similar in form to the lossless equations. Although the lossy versions involve more manipulation of complex numbers, when a numerical package is used, the extra effort is frequently negligible. The equations for the voltage, current, and input impedance for a lossy transmission line are, respectively,

$$V_s(z) = V^+ e^{-\gamma z}\left[1+\rho(z)\right] = V^+ e^{-\gamma z}\left[1+\rho_L e^{2\gamma(z-l_{th})}\right] \tag{3.131}$$

$$I_s(z) = \frac{V^+}{Z_o} e^{-\gamma z}\left[1-\rho(z)\right] = \frac{V^+}{Z_o} e^{-\gamma z}\left[1-\rho_L e^{2\gamma(z-l_{th})}\right] \tag{3.132}$$

$$Z_{in} = Z_o \frac{Z_L + Z_o \tanh(\gamma l_{th})}{Z_o + Z_L \tanh(\gamma l_{th})} \tag{3.133}$$

where

$$\gamma = \alpha + j\beta = \sqrt{(R+j\omega L)(G+j\omega C)} \tag{3.134}$$

The load on the transmission line is located at $z = l_{th}$, and the source is located at $z = 0$. With the addition of loss, the voltage and current of both the incident and reflected voltage and current waveforms are

$f := 30\cdot 10^6$ $L := 0.25\cdot 10^{-6}$ $C := 100\cdot 10^{-12}$ $R := 1$ $G := 0$ $\omega := 2\cdot\pi\cdot f$ $T := \frac{1}{f}$ $j := \sqrt{-1}$

$Z_o := \sqrt{\frac{L}{C}}$ $Z_o = 50$ $v := \frac{1}{\sqrt{L\cdot C}}$ $\lambda := \frac{v}{f}$ $V_s := 1$ $R_s := 2$ $l_{th} := 10\cdot\lambda$ $l_{th} = 66.667$

$\gamma := \sqrt{(R + j\cdot\omega\cdot L)\cdot(G + j\cdot\omega\cdot C)}$

$z := 0, \frac{l_{th}}{100} .. l_{th}$

$Z_{in}(Z_L) := Z_o\cdot\frac{Z_L + Z_o\cdot\tanh(\gamma\cdot l_{th})}{Z_o + Z_L\cdot\tanh(\gamma\cdot l_{th})}$ $V_o(Z_L) := V_s\cdot\frac{Z_{in}(Z_L)}{Z_{in}(Z_L) + R_s}$

$V_{pos}(Z_L) := \frac{V_o(Z_L)}{1 + \left(\frac{Z_L - Z_o}{Z_L + Z_o}\right)\cdot e^{2\cdot\gamma\cdot(-l_{th})}}$ $V(z, Z_L) := V_{pos}(Z_L)\cdot e^{-\gamma\cdot z}\cdot\left[1 + \left(\frac{Z_L - Z_o}{Z_L + Z_o}\right)\cdot e^{2\cdot\gamma\cdot(z - l_{th})}\right]$

MATHCAD 3.23 Voltage along a lossy transmission line for two different loads.

attenuated along the line. Also, note that the input impedance is a function of the hyperbolic tangent:

$$\tanh z = \frac{\sinh z}{\cosh z} = \frac{\frac{e^z - e^{-z}}{2}}{\frac{e^z + e^{-z}}{2}} = \frac{e^z - e^{-z}}{e^z + e^{-z}}$$

The hyperbolic tangent is not equal to the well-known ordinary tangent.

Although the VSWR can be defined as a function of distance along a lossy line, the VSWR has the most use and meaning for low-loss transmission lines. The VSWR for a lossy line is greatest at the load and is smallest at the source or generator. For a very lossy line, the reflected signal from the load is quite small by the time it arrives at the source. Thus, the VSWR is essentially determined by the initial signal from the source, and there is little effect from the reflected signal to cause a variation in the voltage or current magnitude at the source. At the load, however, interaction between the incident and reflected signals can be significant.

In Mathcad 3.23, the magnitude of the voltage along a lossy line is plotted. When the load is matched to the line, there is still voltage variation along the line due to the line losses, but the voltage magnitude decreases in a monotonic fashion (i.e., no peaks or valleys). When the load is not matched to the line, there is a sinusoidal-like variation in the voltage magnitude. The VSWR, which is the ratio of the maximum amplitude to the minimum amplitude of the voltage or current, varies along the line. The VSWR can be defined using an adjacent peak and valley. Then, the VSWR has a maximum value near the load and a minimum value near the source or generator (at $z = 0$). When using matching networks, it is thus important for long lossy lines that the VSWR be examined at the load.

3.44 Telephone Lines and the "*RC*" Region

Assume that telephone wires are #24 AWG[15] with $R = 50$ mΩ/ft, $C = 12$ pF/ft, and $L = 120$ nH/ft. Plot the magnitude and phase angle of the characteristic impedance from 250 Hz to 2.7 kHz. Is the telephone wire mainly capacitive or inductive over this voice band range? Where does the resistive portion of the impedance exceed the reactive portion of the impedance? Where does phase distortion occur? No phase distortion occurs when the phase constant, β, is proportional to the frequency, ω, since the velocity is then independent of the frequency. For example, no phase distortion occurs if

$$v = \frac{\omega}{\beta} = \frac{\omega}{\omega\sqrt{\mu\varepsilon}} = \frac{1}{\sqrt{\mu\varepsilon}}$$

Why is the telephone twisted pair used in this "*RC*" region? [Johnson, '50; Johnson, '93; Kimbark; Glasford]

Although many telephone lines are twisted pair, untwisted wires are also used. The general expression for the characteristic impedance of a lossy line is

$$Z_o = \sqrt{\frac{R + j\omega L}{G + j\omega C}}$$

If the dielectric losses are negligible (typically true for frequencies below the microwave range, when the dielectric is not wet, and if corona is not present), $G = 0$, and

$$Z_o = \sqrt{\frac{R + j\omega L}{j\omega C}} \quad \text{if } G = 0$$

[15]Other wire sizes are used, such as #22 AWG.

The phase constant, β, is equal to the imaginary part of the propagation constant:

$$\beta = \mathrm{Im}(\gamma) = \mathrm{Im}\left[\sqrt{(R + j\omega L)\, j\omega C}\right]$$

The analysis in Mathcad 3.24 assumes that the resistance does not change with frequency (i.e., skin effect is negligible).

Since the phase angle of the impedance is negative over this entire frequency range, the telephone line is capacitive over the voice range, which is why this range of the line impedance is referred to as the "*RC*" region.[16] The resistive portion of the impedance (the real portion) is approximately 3 to 20 Ω greater than the magnitude of the capacitive reactance of the impedance (the imaginary portion) over this range. The resistive portion is approximately 600 Ω near 900 Hz. The "*RC*" nature of the line can also be shown analytically. If $G = 0$, the expression for the characteristic impedance can be separated into a real and an imaginary term:

$$Z_o = \sqrt{\frac{R + j\omega L}{j\omega C}} = \sqrt{\frac{L}{C}}\sqrt{\frac{1}{2} + \frac{1}{2}\sqrt{1 + \left(\frac{R}{\omega L}\right)^2}} - j\sqrt{\frac{L}{C}}\sqrt{-\frac{1}{2} + \frac{1}{2}\sqrt{1 + \left(\frac{R}{\omega L}\right)^2}} \tag{3.135}$$

As a simple check, at frequencies where $\omega L \gg R$, this expression reduces to

$$Z_o \approx \sqrt{\frac{L}{C}}\sqrt{\frac{1}{2} + \frac{1}{2}} - j\sqrt{\frac{L}{C}}\sqrt{-\frac{1}{2} + \frac{1}{2}} = \sqrt{\frac{L}{C}}$$

The characteristic impedance is entirely real at high frequencies, which allows for a real matching termination impedance. Notice that at low frequencies where $\omega L << R$, the reactive portion approaches the resistive portion in (3.135):

$$\begin{aligned} Z_o &\approx \sqrt{\frac{L}{C}}\sqrt{\frac{1}{2} + \frac{1}{2}\sqrt{\left(\frac{R}{\omega L}\right)^2}} - j\sqrt{\frac{L}{C}}\sqrt{-\frac{1}{2} + \frac{1}{2}\sqrt{\left(\frac{R}{\omega L}\right)^2}} \\ &\approx \sqrt{\frac{L}{C}}\sqrt{\frac{1}{2}\sqrt{\left(\frac{R}{\omega L}\right)^2}} - j\sqrt{\frac{L}{C}}\sqrt{\frac{1}{2}\sqrt{\left(\frac{R}{\omega L}\right)^2}} \approx \sqrt{\frac{R}{2\omega C}} - j\sqrt{\frac{R}{2\omega C}} \end{aligned} \tag{3.136}$$

In other words, the phase angle of the impedance approaches −45°. The reactive portion of the impedance can sometimes be ignored when matching the load or termination impedance to the characteristic impedance. However, for very long lines, the capacitive portion of the impedance can be significant and additional components may be required to prevent distortion of these low-frequency signals.

The characteristic impedance of the telephone line is not constant over this frequency range, and distortion is expected. It is best to examine the phase constant for the line, however, to determine if phase distortion is expected. As seen on the graph, the phase constant, the imaginary portion of γ, is not linearly proportional to the frequency, and some phase distortion is expected. (For a line to be distortionless, two conditions must be satisfied: the attenuation constant, α, should be independent of frequency, and the phase constant, β, should be linearly proportional to frequency.)

An analytical approximation for the phase velocity, v, for this "*RC*" range can be obtained from the definition of β and by assuming that the resistance, R, is much greater than the inductive reactance,

[16]When operated in this region, the line is also referred to as a charge-diffusion transmission line.

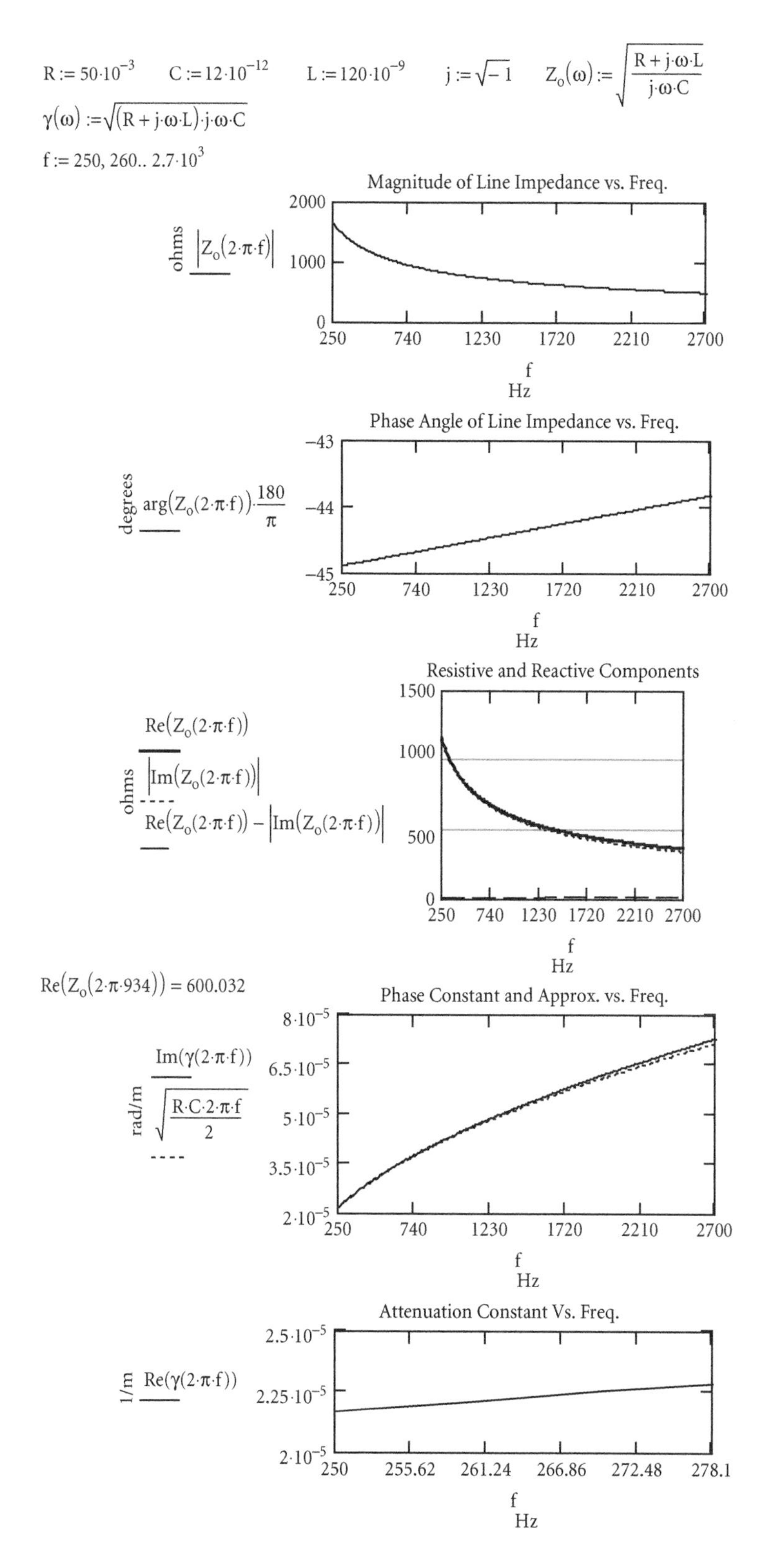

MATHCAD 3.24 Impedance, phase constant, and attenuation constant vs. frequency for a transmission line operating in the "*RC*" region.

ωL:

$$\beta = \text{Im}\left[\sqrt{(R+j\omega L)\,j\omega C}\right] \approx \text{Im}\left(\sqrt{Rj\omega C}\right) = \text{Im}\left(\sqrt{RC\omega\angle 90^\circ}\right)$$
$$= \text{Im}\left(\sqrt{RC\omega}\angle 45^\circ\right) = \text{Im}\left[\sqrt{RC\omega}\left(\frac{1}{\sqrt{2}} + j\frac{1}{\sqrt{2}}\right)\right] = \sqrt{\frac{RC\omega}{2}} \tag{3.137}$$

(At the highest frequency of interest, 2.7 kHz, the resistance per foot is 50 mΩ and the inductive reactance per foot is 2 mΩ.) This approximation for the phase constant is very close to the exact result as seen in the Mathcad solution. An approximation for the phase velocity follows:

$$v = \frac{\omega}{\beta} \approx \sqrt{\frac{2\omega}{RC}} \tag{3.138}$$

Since the velocity of the signal is *dependent* on the frequency of the signal, distortion will occur for any signal containing more than one frequency component. (This distortion can be corrected.) Telephone lines are used in this "*RC*" region because the attenuation of the line, which is a function of the real part of γ, is very low. Thus, the losses due to the line are low. This occurs at frequencies such that the inductive reactance can be ignored:

$$R \gg \omega L \quad \text{or} \quad \frac{R}{L} \gg \omega = 2\pi f$$

The inductive reactance is low because the frequency is low and inductance is small (the wires are close together). For this telephone wire, R/L = 420 krad/s (≈ 66 kHz).

On some low-frequency lines (e.g., telephone lines and audio cables), the inductance of the line is intentionally increased by inserting inductors at uniform intervals along the line or wrapping the conductors with high-permeability magnetic tape. When the inductance is intentionally increased in this manner, it is referred to as inductive loading. By increasing the inductance of the line, the line will tend to have less distortion and attenuation over a limited frequency range. Since the velocity on a line is

$$v = \frac{\omega}{\beta} = \frac{\omega}{\text{Im}\left[\sqrt{(R+j\omega L)\,j\omega C}\right]}$$

when the inductive reactance is large compared to the resistance, the velocity reduces to

$$v = \frac{\omega}{\beta} = \frac{\omega}{\text{Im}\left[\sqrt{j\omega L\left(\frac{R}{j\omega L}+1\right)j\omega C}\right]} \approx \frac{\omega}{\text{Im}\left[j\omega\sqrt{LC}\left(1+\frac{1}{2}\frac{R}{j\omega L}+\frac{1}{8}\frac{R^2}{\omega^2 L^2}\right)\right]}$$
$$= \frac{\omega}{\omega\sqrt{LC}\left(1+\frac{1}{8}\frac{R^2}{\omega^2L^2}\right)} = \frac{1}{\sqrt{LC}\left(1+\frac{1}{8}\frac{R^2}{\omega^2L^2}\right)} \quad \text{if } \omega L \gg R \tag{3.139}$$

This dependency of the velocity on frequency decreases as the inductance increases. Furthermore, the attenuation factor

$$\alpha = \text{Re}(\gamma) = \text{Re}\left[\sqrt{(R+j\omega L)\,j\omega C}\right] \approx \text{Re}\left[j\omega\sqrt{LC}\left(1+\frac{1}{2}\frac{R}{j\omega L}+\frac{1}{8}\frac{R^2}{\omega^2 L^2}\right)\right]$$
$$= \omega\sqrt{LC}\left(\frac{1}{2}\frac{R}{\omega L}\right) = \frac{R}{2}\sqrt{\frac{C}{L}} \tag{3.140}$$

decreases as L increases. (The binomial expansion was used to obtain these approximations.) It should not be surprising that increasing L decreases the attenuation. Increasing L increases the high-frequency characteristic impedance; therefore, for a given input power to the line, the current and, hence, the I^2R losses will be less. (Of course, the voltage across the conductors will be greater, but the dielectric losses, which are a function of G, should be small.) Obviously, increasing the inductance, increases the inductive reactance. When the inductive reactance is much greater than the resistance, the resistance can be ignored, and the transmission line can be modeled with L's and C's: the characteristic impedance and signal velocity are then frequency independent.

There are several costs associated with inductive loading. The most obvious is the additional cost of either the coils or magnetic tape. A cost specific to magnetic tape is the nonlinearities associated with the magnetic material. A cost specific to lumped-inductive loading (or loading coils) is the additional high-frequency attenuation: the inductors are acting as low-pass filters.[17] As the spacing between the inductors decreases, the inductors appear more like part of the line and this effect is lessened. Ideally, the spacing between the inductors should be electrically small if the line is to appear "smooth." Otherwise, the inductors appear as loads along the transmission line. Finally, an additional cost of inductive loading is the increased signal delay:

$$\frac{1}{\sqrt{LC}\left(1+\dfrac{1}{8}\dfrac{R^2}{\omega^2L^2}\right)}<\sqrt{\frac{2\omega}{RC}} \quad \text{or} \quad 1<\frac{2\omega L}{R}+\frac{R}{2\omega L}+\frac{R^3}{32\omega^3L^3}$$

Since $2\omega L/R$ and $R/(2\omega L)$ are inverses of each other, their sum is greater than one. Over very long distances, the additional delay may be important. For delay lines, this additional delay is desirable. Inductive loading can be used in exotic speaker cable where low distortion is desired and in cables where space is at a premium (e.g., in a submarine).

3.45 Transmission Line Parameter Expressions

Tabulate the various exact and approximate expressions given for Z_o, α, β, and v for transmission lines at low, mid, and high frequencies. [Miner; Ramo]

It is convenient to have the various expressions for the characteristic impedance, Z_o, the attenuation coefficient, α, the phase constant, β, and the (phase) velocity, v, in one location. The exact expressions are given in Table 3.4. When R is in Ω/m, L is in H/m, C is in F/m, and G is in Ω^{-1}/m, then Z_o has units of Ω, α has units of nepers/m, β has units of radians/m, and v has units of m/s. Although L is a function of frequency, often the frequency dependence of R is only important at low frequencies and the frequency dependence of G is only important at very high frequencies. If Z_o and γ are experimentally determined, then $R+j\omega L$ and $G+j\omega C$ of the transmission line can be readily determined from (3.141):

$$R+j\omega L=\gamma Z_o \quad \text{and} \quad G+j\omega C=\frac{\gamma}{Z_o} \tag{3.141}$$

since

$$\sqrt{\frac{R+j\omega L}{G+j\omega C}}=\sqrt{\frac{\gamma Z_o}{\dfrac{\gamma}{Z_o}}}=Z_o \quad \text{and} \quad \sqrt{(R+j\omega L)(G+j\omega C)}=\sqrt{\gamma Z_o\left(\frac{\gamma}{Z_o}\right)}=\gamma$$

[17]High-frequency data lines (such as DSL's or digital subscriber lines) are adversely affected by the use of loading coils.

TABLE 3.4 Exact Expressions for the Characteristic Impedance, Propagation Constant, and Velocity

Characteristic Impedance

$$Z_o = R_o + jX_o = \sqrt{\frac{R+j\omega L}{G+j\omega C}} = \sqrt{\frac{(RG+\omega^2 LC)+j\omega(LG-RC)}{G^2+(\omega C)^2}}$$

$$= \frac{[(RG+\omega^2 LC)^2+(\omega LG-\omega RC)^2]^{\frac{1}{4}}}{\sqrt{G^2+(\omega C)^2}}\left\{\cos\left[\frac{1}{2}\tan^{-1}\left(\frac{\omega LG-\omega RC}{RG+\omega^2 LC}\right)\right] + j\sin\left[\frac{1}{2}\tan^{-1}\left(\frac{\omega LG-\omega RC}{RG+\omega^2 LC}\right)\right]\right\}$$

Propagation Constant

$$\gamma = \alpha + j\beta = \sqrt{(R+j\omega L)(G+j\omega C)} = \sqrt{(RG-\omega^2 LC)+j\omega(RC+LG)}$$

$$= [(RG-\omega^2 LC)^2+\omega^2(RC+LG)^2]^{\frac{1}{4}}\cos\left[\frac{1}{2}\tan^{-1}\left(\frac{\omega RC+\omega LG}{RG-\omega^2 LC}\right)\right]$$

$$+ j[(RG-\omega^2 LC)^2+\omega^2(RC+LG)^2]^{\frac{1}{4}}\sin\left[\frac{1}{2}\tan^{-1}\left(\frac{\omega RC+\omega LG}{RG-\omega^2 LC}\right)\right]$$

$$= \sqrt{\frac{[(RG-\omega^2 LC)^2+\omega^2(RC+LG)^2]^{\frac{1}{2}}+RG-\omega^2 LC}{2}}$$

$$+ j\sqrt{\frac{[(RG-\omega^2 LC)^2+\omega^2(RC+LG)^2]^{\frac{1}{2}}-RG+\omega^2 LC}{2}}$$

(assuming α and β are >0)

Velocity

$$v = \frac{\omega}{\beta} = \frac{\omega}{[(RG-\omega^2 LC)^2+\omega^2(RC+LG)^2]^{\frac{1}{4}}\sin\left[\frac{1}{2}\tan^{-1}\left(\frac{\omega RC+\omega LG}{RG-\omega^2 LC}\right)\right]}$$

$$= \frac{\omega}{\sqrt{\frac{[(RG-\omega^2 LC)^2+\omega^2(RC+LG)^2]^{\frac{1}{2}}-RG+\omega^2 LC}{2}}}$$

(assuming $\beta > 0$)

To determine the approximations listed in Table 3.5, Table 3.6, and Table 3.7, the binomial expansion approximation was used

$$(1+x)^n \approx 1+nx+\frac{n(n-1)}{2}x^2 \quad \text{if } x \ll 1$$

the half-angle formulas (where the sign is a function of the quadrant of the angle) were used

$$\cos\left(\frac{\theta}{2}\right) = \pm\sqrt{\frac{1+\cos\theta}{2}}, \quad \sin\left(\frac{\theta}{2}\right) = \pm\sqrt{\frac{1-\cos\theta}{2}}$$

TABLE 3.5 Approximate Low-Frequency Expressions for the Real and Imaginary Components of the Characteristic Impedance, Real and Imaginary Components of the Propagation Constant, and Velocity

	$\omega \approx 0$ dc	$R \gg \omega L$, $G \gg \omega C$ low frequency, high loss, attenuator region, low distortion	$R \gg \omega L$, $G \ll \omega C$ low frequency, RC region
R_o	$\sqrt{\frac{R}{G}}$	$\sqrt{\frac{R}{G}}\left(1+\frac{\omega^2 LC}{4RG}-\frac{3\omega^2C^2}{8G^2}+\frac{\omega^2L^2}{8R^2}\right)$ $\approx\sqrt{\frac{R}{G}}$	$\sqrt{\frac{R}{2\omega C}}\left(1-\frac{GL}{4RC}+\frac{G}{2\omega C}+\frac{\omega L}{2R}\right)$ $\approx\sqrt{\frac{R}{2\omega C}}$
X_o	0	$\sqrt{\frac{R}{G}}\left(\frac{\omega L}{2R}-\frac{\omega C}{2G}\right)\approx 0$	$\sqrt{\frac{R}{2\omega C}}\left(-1-\frac{GL}{4RC}+\frac{G}{2\omega C}+\frac{\omega L}{2R}\right)$ $\approx-\sqrt{\frac{R}{2\omega C}}$
α	$\sqrt{RG}$	$\sqrt{RG}\left(1-\frac{\omega^2 LC}{4RG}+\frac{\omega^2C^2}{8G^2}+\frac{\omega^2L^2}{8R^2}\right)$ $\approx\sqrt{RG}$	$\sqrt{\frac{\omega RC}{2}}\left(1+\frac{GL}{4RC}+\frac{G}{2\omega C}-\frac{\omega L}{2R}\right)$ $\approx\sqrt{\frac{\omega RC}{2}}$
β	0	$\sqrt{RG}\left(\frac{\omega C}{2G}+\frac{\omega L}{2R}\right)\approx 0$	$\sqrt{\frac{\omega RC}{2}}\left(1+\frac{GL}{4RC}-\frac{G}{2\omega C}+\frac{\omega L}{2R}\right)$ $\approx\sqrt{\frac{\omega RC}{2}}$
v		$\frac{2\sqrt{RG}}{RC+GL}$	$\frac{\sqrt{2\omega}}{\sqrt{RC}\left(1+\frac{GL}{4RC}-\frac{G}{2\omega C}+\frac{\omega L}{2R}\right)}$ $\approx\sqrt{\frac{2\omega}{RC}}$

the relationship between the rectangular and polar form of a complex number was used

$$A+jB=\sqrt{A^2+B^2}\angle\tan^{-1}\left(\frac{B}{A}\right)$$

the formula for the square root of a complex number when in polar form was used

$$\sqrt{C\angle\theta}=\sqrt{C}\angle\left(\frac{\theta}{2}\right)$$

TABLE 3.6 Approximate High-Frequency Expressions for the Real and Imaginary Components of the Characteristic Impedance, Real and Imaginary Components of the Propagation Constant, and Velocity

	$R \ll \omega L$, $G \gg \omega C$, high frequency, LG region	$R \ll \omega L$, $G \ll \omega C$, high frequency, low loss, low distortion
R_o	$\sqrt{\frac{\omega L}{2G}}\left(1-\frac{RC}{4GL}+\frac{\omega C}{2G}+\frac{R}{2\omega L}\right)$ $\approx\sqrt{\frac{\omega L}{2G}}$	$\sqrt{\frac{L}{C}}\left(1+\frac{RG}{4\omega^2 LC}-\frac{3G^2}{8\omega^2 C^2}+\frac{R^2}{8\omega^2 L^2}\right)$ $\approx\sqrt{\frac{L}{C}}$
X_o	$\sqrt{\frac{\omega L}{2G}}\left(1-\frac{RC}{4GL}-\frac{\omega C}{2G}-\frac{R}{2\omega L}\right)$ $\approx\sqrt{\frac{\omega L}{2G}}$	$\sqrt{\frac{L}{C}}\left(\frac{G}{2\omega C}-\frac{R}{2\omega L}\right)\approx 0$
α	$\sqrt{\frac{\omega GL}{2}}\left(1+\frac{RC}{4GL}-\frac{\omega C}{2G}+\frac{R}{2\omega L}\right)$ $\approx\sqrt{\frac{\omega GL}{2}}$	$\frac{R}{2}\sqrt{\frac{C}{L}}+\frac{G}{2}\sqrt{\frac{L}{C}}$
β	$\sqrt{\frac{\omega GL}{2}}\left(1+\frac{RC}{4GL}+\frac{\omega C}{2G}-\frac{R}{2\omega L}\right)$ $\approx\sqrt{\frac{\omega GL}{2}}$	$\omega\sqrt{LC}\left(1-\frac{RG}{4\omega^2 LC}+\frac{G^2}{8\omega^2 C^2}+\frac{R^2}{8\omega^2 L^2}\right)$ $\approx\omega\sqrt{LC}$
v	$\frac{\sqrt{2\omega}}{\sqrt{GL}\left(1+\frac{RC}{4GL}+\frac{\omega C}{2G}-\frac{R}{2\omega L}\right)}$ $\approx\sqrt{\frac{2\omega}{GL}}$	$\frac{1}{\sqrt{LC}\left(1-\frac{RG}{4\omega^2 LC}+\frac{G^2}{8\omega^2 C^2}+\frac{R^2}{8\omega^2 L^2}\right)}$ $\approx\frac{1}{\sqrt{LC}}$

and simple complex algebra[18] such as

$$\frac{1}{A+jB}=\frac{1}{A+jB}\,\frac{A-jB}{A-jB}=\frac{A-jB}{A^2+B^2}$$

was used.

[18]Yes, the phrase "simple complex algebra" is an oxymoron.

TABLE 3.7 Approximate Mid-Frequency Expressions for the Real and Imaginary Components of the Characteristic Impedance, Real and Imaginary Components of the Propagation Constant, and Velocity

	$R \approx \omega L$ $G \approx 0$ mid frequency, high R loss	$R \approx 0$ $G \approx \omega C$ mid frequency, high G loss
R_o	$\sqrt{\frac{L}{C}}\sqrt{\frac{\sqrt{1+\left(\frac{R}{\omega L}\right)^2}+1}{2}}$	$\sqrt{\frac{\frac{L}{C}}{\sqrt{1+\left(\frac{G}{\omega C}\right)^2}}}\sqrt{\frac{\sqrt{1+\left(\frac{G}{\omega C}\right)^2}+1}{2}}$
X_o	$-\sqrt{\frac{L}{C}}\sqrt{\frac{\sqrt{1+\left(\frac{R}{\omega L}\right)^2}-1}{2}}$	$\sqrt{\frac{\frac{L}{C}}{\sqrt{1+\left(\frac{G}{\omega C}\right)^2}}}\sqrt{\frac{\sqrt{1+\left(\frac{G}{\omega C}\right)^2}-1}{2}}$
α	$\omega\sqrt{LC}\sqrt{\frac{\sqrt{1+\left(\frac{R}{\omega L}\right)^2}-1}{2}}$	$\omega\sqrt{LC}\sqrt{\frac{\sqrt{1+\left(\frac{G}{\omega C}\right)^2}-1}{2}}$
β	$\omega\sqrt{LC}\sqrt{\frac{\sqrt{1+\left(\frac{R}{\omega L}\right)^2}+1}{2}}$	$\omega\sqrt{LC}\sqrt{\frac{\sqrt{1+\left(\frac{G}{\omega C}\right)^2}+1}{2}}$
v	$\frac{1}{\sqrt{LC}\sqrt{\frac{\sqrt{1+\left(\frac{R}{\omega L}\right)^2}+1}{2}}}$	$\frac{1}{\sqrt{LC}\sqrt{\frac{\sqrt{1+\left(\frac{G}{\omega C}\right)^2}+1}{2}}}$

3.46 S Parameters

What are *S* parameters, and why are they useful? Which of the *S* parameters correspond to the reflection and transmission coefficients? [Dworsky; Vendelin]

As is emphasized throughout this book, at higher frequencies where conductors are not electrically short, it is essential that the conductors not be modeled as a lumped circuit. It is usually necessary to introduce and determine reflection and transmission coefficients. Fortunately, lower cost network analyzers are now available to measure conveniently and quickly these and many other parameters of interest to engineers. These powerful instruments, however, measure dimensionless complex quantities referred to as *S* or scattering parameters. It is the purpose of this discussion to explain briefly the relationship between these *S* parameters and the more familiar concepts such as reflection coefficient, transmission coefficient, and input impedance.

The use of the Thévenin equivalent to describe a single-port network is common in undergraduate programs (a port can be any pair of input or output terminals). Although the Thévenin equivalent is not usually the same as what is inside the network,[19] from the outside it has the same current-voltage response. Most networks contain more than one port. To describe or model a multiport network, such as simple audio low-pass filter or a complex microwave amplifier, without worrying about the details of the "black box," various models are available. These models use parameters to describe the relationship between the input and output variables of interest, such as voltage and current. Probably most electrical engineers are familiar with the use of *h* parameters, at lower frequencies, to model a transistor. At high frequencies, *S* parameters are probably the simplest to work with and to measure. Unlike many other parameters that can be used to model or describe a network, *S* parameters do not require that the input and output of the network be short circuited and open circuited. Besides the obvious, possible negative reaction of a network to being short circuited or open circuited, at very high frequencies it is difficult to produce a true open or short circuit. The impedance of an open circuit may be strongly capacitive in nature (and not infinite), while the impedance of a short circuit may be strongly inductive in nature (and not zero). To further complicate matters, as the frequency of the measurement changes, the impedance of the open or short circuit will also vary. With *S* parameters, the source and load impedance is usually assumed equal to a resistance of 50 Ω. If the resonant frequency of this 50 Ω resistor is well above the highest testing frequency, its impedance should be mostly resistive and independent of frequency.

Although the definition for the *S* parameters can just be provided, it is much more edifying to derive them starting from the transmission line equations. Imagine that a sinusoidal signal source with an internal impedance of Z_o is connected to a single-port network via a transmission line of characteristic impedance Z_o as shown in Figure 3.44. The input impedance of the network is not necessarily equal to Z_o so there may be both a forward or an incident wave as well as a backward or reflected wave. The sinusoidal steady-state voltage and current along the transmission line is then given by (3.83) and (3.95), modified for the possibility of a lossy line:

$$V_s(z) = V^+ e^{-\gamma z} + V^- e^{\gamma z} \tag{3.142}$$

$$I_s(z) = \frac{V^+ e^{-\gamma z}}{Z_o} - \frac{V^- e^{\gamma z}}{Z_o} \tag{3.143}$$

where $\gamma = \alpha + j\beta = \sqrt{(R + j\omega L)(G + j\omega C)}$ is the propagation constant for the line. Solving for the incident and reflected terms,

$$V^+ e^{-\gamma z} = \frac{1}{2}[V_s(z) + Z_o I_s(z)]$$

$$V^- e^{\gamma z} = \frac{1}{2}[V_s(z) - Z_o I_s(z)]$$

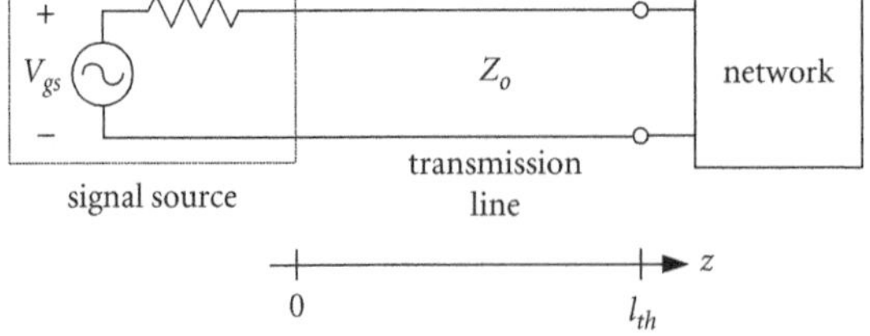

FIGURE 3.44 Total voltage on a transmission line is the sum of an incident and a reflected wave.

[19]For example, the power dissipated by the Thévenin equivalent is not necessarily equal to the power dissipated by the actual network.

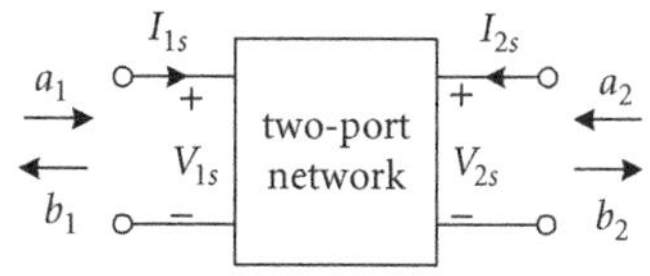

FIGURE 3.45 Incident and reflected parameters for a two-port network.

The incident and reflective parameters are defined, respectively, as these terms divided by $\sqrt{Z_o}$:

$$a = \frac{V^+e^{-\gamma z}}{\sqrt{Z_o}} = \frac{1}{2}\left[\frac{V_s(z)}{\sqrt{Z_o}} + \sqrt{Z_o}I_s(z)\right] \tag{3.144}$$

$$b = \frac{V^-e^{\gamma z}}{\sqrt{Z_o}} = \frac{1}{2}\left[\frac{V_s(z)}{\sqrt{Z_o}} - \sqrt{Z_o}I_s(z)\right] \tag{3.145}$$

By dividing by the normalizing factor $\sqrt{Z_o}$, both a^2 and b^2 have units of power.[20] These complex parameters possess both an amplitude and a phase and can be evaluated along the line at the position $z = l_{th}$.

For any number of ports of interest, a pair of a and b parameters can be defined. This is analogous to defining a voltage across and current into each port of interest. A two-port network is shown in Figure 3.45 with the incident, a, and reflected, b, parameters defined at each port. Notice that the incident parameters are defined entering the ports and the reflected parameters are defined leaving the ports (so the direction of the z axis for each of the ports is now irrelevant). Also, the characteristic impedance of each transmission line connected to each port is assumed equal to Z_o. In this discussion, the a and b parameters are defined as

$$a_1 = \frac{1}{2}\left(\frac{V_{1s}}{\sqrt{Z_o}} + \sqrt{Z_o}I_{1s}\right),\quad b_1 = \frac{1}{2}\left(\frac{V_{1s}}{\sqrt{Z_o}} - \sqrt{Z_o}I_{1s}\right) \tag{3.146}$$

$$a_2 = \frac{1}{2}\left(\frac{V_{2s}}{\sqrt{Z_o}} + \sqrt{Z_o}I_{2s}\right),\quad b_2 = \frac{1}{2}\left(\frac{V_{2s}}{\sqrt{Z_o}} - \sqrt{Z_o}I_{2s}\right) \tag{3.147}$$

Finally, the S parameters can be defined. The relationship between the incident and reflected waves at these ports is given in terms of the S variables:

$$b_1 = S_{11}a_1 + S_{12}a_2 \tag{3.148}$$

$$b_2 = S_{21}a_1 + S_{22}a_2 \tag{3.149}$$

The reflected wave from each port is a function of the incident waves into each port.

The classic definitions for each of the S parameters are obtained by setting either a_1 or a_2 equal to zero. However, this does not imply short circuiting (or open circuiting), but it does imply matching. To begin with, S_{11} is obtained from (3.148) as

$$S_{11} = \left.\frac{b_1}{a_1}\right|_{a_2=0} = \rho_1 \left(\begin{array}{l}\text{reflection coefficient at port 1 with}\\ \text{port 2 terminated with an impedance of } Z_o\end{array}\right) \tag{3.150}$$

[20]Other normalizing factors can be used and are seen in the literature.

If the load connected to port 2 is equal to Z_o, then a_2 is zero since the load is matched to the line. The parameter S_{11} is the ratio of the reflected to incident wave at port 1, which is equal to the voltage reflection coefficient at this port or ρ_1. Similarly, if the voltage source is located at port 2, then

$$S_{22} = \left.\frac{b_2}{a_2}\right|_{a_1=0} = \rho_2 \left(\begin{array}{l}\text{reflection coefficient at port 2 with} \\ \text{port 1 terminated with an impedance of } Z_o\end{array}\right) \tag{3.151}$$

A commonly used S parameter is S_{21}:

$$S_{21} = \left.\frac{b_2}{a_1}\right|_{a_2=0} = \mathrm{T}_1 \left(\begin{array}{l}\text{transmission coefficient from port 1 to port 2} \\ \text{with port 2 terminated with an impedance of } Z_o\end{array}\right) \tag{3.152}$$

The last parameter is

$$S_{12} = \left.\frac{b_1}{a_2}\right|_{a_1=0} = \mathrm{T}_2 \left(\begin{array}{l}\text{transmission coefficient from port 2 to port 1} \\ \text{with port 1 terminated with an impedance of } Z_o\end{array}\right) \tag{3.153}$$

Once these S parameters are measured or determined, other properties of interest can be determined. For example, the input impedance at port 1 can be determined from S_{11} using (3.146):

$$S_{11} = \frac{b_1}{a_1} = \frac{\frac{1}{2}\left(\frac{V_{1s}}{\sqrt{Z_o}} - \sqrt{Z_o} I_{1s}\right)}{\frac{1}{2}\left(\frac{V_{1s}}{\sqrt{Z_o}} + \sqrt{Z_o} I_{1s}\right)} = \frac{\frac{V_{1s}}{I_{1s}} - Z_o}{\frac{V_{1s}}{I_{1s}} + Z_o} = \frac{Z_{in1} - Z_o}{Z_{in1} + Z_o} \Rightarrow Z_{in1} = Z_o \frac{1 + S_{11}}{1 - S_{11}} \tag{3.154}$$

Again, this assumes that the termination at port 2 is equal to Z_o. When the reflection coefficient is equal to 1, corresponding to an open circuit, $S_{11} = 1$ and $Z_{in1} = \infty$. When the reflection coefficient is equal to -1, corresponding to a short circuit, $S_{11} = -1$ and $Z_{in1} = 0$. Similarly, the input impedance at port 2 can be determined from S_{22}:

$$S_{22} = \frac{b_2}{a_2} = \frac{\frac{1}{2}\left(\frac{V_{2s}}{\sqrt{Z_o}} - \sqrt{Z_o} I_{2s}\right)}{\frac{1}{2}\left(\frac{V_{2s}}{\sqrt{Z_o}} + \sqrt{Z_o} I_{2s}\right)} = \frac{\frac{V_{2s}}{I_{2s}} - Z_o}{\frac{V_{2s}}{I_{2s}} + Z_o} = \frac{Z_{in2} - Z_o}{Z_{in2} + Z_o} \Rightarrow Z_{in2} = Z_o \frac{1 + S_{22}}{1 - S_{22}} \tag{3.155}$$

A voltage transfer function is obtained from S_{21}. If a sinusoidal voltage source, V_{gs}, with an internal impedance of Z_o is connected directly to port one, then

$$-V_{gs} + I_{1s} Z_o + V_{1s} = 0 \quad \Rightarrow V_{1s} = V_{gs} - I_{1s} Z_o$$

Also, if the port 2 termination is Z_o, then

$$V_{2s} = -I_{2s} Z_o$$

Therefore,

$$S_{21} = \frac{b_2}{a_1} = \frac{\frac{1}{2}\left(\frac{V_{2s}}{\sqrt{Z_o}} - \sqrt{Z_o} I_{2s}\right)}{\frac{1}{2}\left(\frac{V_{1s}}{\sqrt{Z_o}} + \sqrt{Z_o} I_{1s}\right)} = \frac{\frac{V_{2s}}{\sqrt{Z_o}} + \sqrt{Z_o}\frac{V_{2s}}{Z_o}}{\frac{V_{gs} - I_{1s}Z_o}{\sqrt{Z_o}} + \sqrt{Z_o} I_{1s}} = \frac{2V_{2s}}{V_{gs}} \tag{3.156}$$

The parameter S_{21} is the ratio of twice the output voltage to the generator's open-circuit voltage. The traditional voltage transfer function, defined as the ratio of the total voltage across the output to the *total* voltage across the input of a network is also easily obtained once two of the S parameters are known:

$$A_v = \frac{b_2}{a_1 + b_1} = \frac{\frac{b_2}{a_1}}{1 + \frac{b_1}{a_1}} = \frac{S_{21}}{1 + S_{11}} \tag{3.157}$$

Again, this assumes that the output port 2 is terminated in an impedance of Z_o so that the total voltage across the output is just $b_2 + a_2 = b_2 + 0 = b_2$. When the input impedance of port 1 is matched to the transmission line, $S_{11} = 0$ and $A_v = S_{21}$.

Incident and reflected powers are easily written in terms of the *a*'s and *b*'s. Working with (3.144) and (3.145), it is apparent why the *a*'s and *b*'s are referred to as power waves[21]:

$$\frac{1}{2}a_1 a_1^* = \frac{1}{2}|a_1|^2 = \frac{V_1^+}{2\sqrt{Z_o}}\left(\frac{V_1^+}{\sqrt{Z_o}}\right)^* = \frac{|V_1^+|^2}{2|Z_o|} = \text{average power incident to port 1} \tag{3.158}$$

$$\frac{1}{2}b_1 b_1^* = \frac{1}{2}|b_1|^2 = \frac{V_1^-}{2\sqrt{Z_o}}\left(\frac{V_1^-}{\sqrt{Z_o}}\right)^* = \frac{|V_1^-|^2}{2|Z_o|} = \text{average power reflected from port 1} \tag{3.159}$$

$$\frac{1}{2}a_2 a_2^* = \frac{1}{2}|a_2|^2 = \frac{V_2^+}{2\sqrt{Z_o}}\left(\frac{V_2^+}{\sqrt{Z_o}}\right)^* = \frac{|V_2^+|^2}{2|Z_o|} = \text{average power incident to port 2} \tag{3.160}$$

$$\frac{1}{2}b_2 b_2^* = \frac{1}{2}|b_2|^2 = \frac{V_2^-}{2\sqrt{Z_o}}\left(\frac{V_2^-}{\sqrt{Z_o}}\right)^* = \frac{|V_2^-|^2}{2|Z_o|} = \text{average power reflected from port 2} \tag{3.161}$$

The actual power transmitted into port 1 is the difference between the incident and reflected powers:

$$\begin{aligned}\frac{1}{2}|a_1|^2 - \frac{1}{2}|b_1|^2 &= \frac{|V_1^+|^2}{2|Z_o|} - \frac{|V_1^-|^2}{2|Z_o|} = \frac{|V_1^+|^2}{2|Z_o|} - \frac{|\rho_1 V_1^+|^2}{2|Z_o|} \\ &= \frac{|V_1^+|^2}{2|Z_o|}\left(1 - |\rho_1|^2\right) = \text{power transmitted into port 1}\end{aligned} \tag{3.162}$$

[21]The asterisk superscript in these expressions represents complex conjugation.

Equation (3.162) is the classical result for the power transmitted into the load of a transmission line. When the magnitude of the reflection coefficient is one, corresponding to a short circuit or an open circuit, no power is absorbed by the load. Similarly, the power transmitted into port 2 is

$$\frac{1}{2}|a_2|^2 - \frac{1}{2}|b_2|^2 = \frac{|V_2^+|^2}{2|Z_o|} - \frac{|V_2^-|^2}{2|Z_o|} = \frac{|V_2^+|^2}{2|Z_o|} - \frac{|\rho_2 V_2^+|^2}{2|Z_o|}$$
$$= \frac{|V_2^+|^2}{2|Z_o|}\left(1-|\rho_2|^2\right) = \text{power transmitted into port 2} \tag{3.163}$$

It is also fairly obvious that

$$|S_{11}|^2 = \frac{|b_1|^2}{|a_1|^2} = \frac{\text{power reflected from port 1}}{\text{power incident to port 1}} \tag{3.164}$$

$$|S_{22}|^2 = \frac{|b_2|^2}{|a_2|^2} = \frac{\text{power reflected from port 2}}{\text{power incident to port 2}} \tag{3.165}$$

$$|S_{21}|^2 = \frac{|b_2|^2}{|a_1|^2} = \frac{\text{power reflected from port 2}}{\text{power incident to port 1}} \tag{3.166}$$

$$|S_{12}|^2 = \frac{|b_1|^2}{|a_2|^2} = \frac{\text{power reflected from port 1}}{\text{power incident to port 2}} \tag{3.167}$$

Relationship (3.166) is also referred to as the transducer power gain and is the ratio of the power delivered to the load to the power available from the source. In dB, the insertion loss is defined as

$$\text{IL}_{\text{dB}} = -10\log|S_{21}|^2 = -20\log|S_{21}| \tag{3.168}$$

A simple but important application of S parameters will now be presented. Assume that the source and load impedance connected to the amplifier shown in Figure 3.46 are matched to the line and equal to 50 Ω. For this amplifier, $S_{11} = 0.9$ and $S_{21} = -9.5$ are obtained using a network analyzer. Immediately, the input resistance is obtained from (3.154)

$$R_{in} = Z_o \frac{1+S_{11}}{1-S_{11}} = 50\frac{1+0.9}{1-0.9} = 950\,\Omega$$

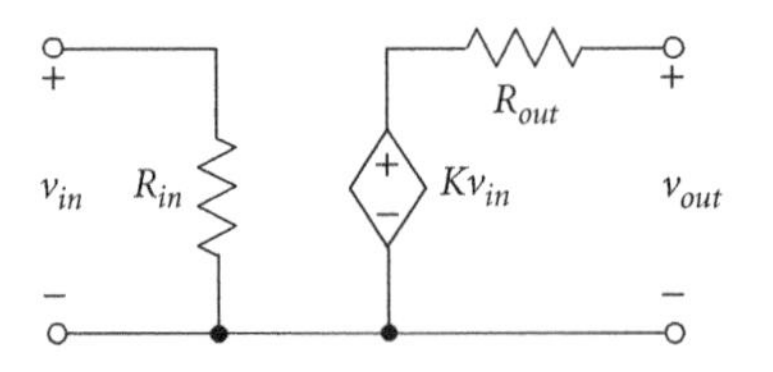

FIGURE 3.46 Simple model for an amplifier.

while the voltage gain is obtained from (3.157):

$$A_v = \frac{S_{21}}{1+S_{11}} = \frac{-9.5}{1+0.9} = -5$$

Again, this gain is assuming that the load across this amplifier is equal to 50 Ω. The relationship between this voltage gain and the two remaining unknown parameters is determined using voltage division:

$$v_{out} = Kv_{in}\frac{50}{50+R_{out}} \quad \Rightarrow \frac{v_{out}}{v_{in}} = A_v = -5 = K\frac{50}{50+R_{out}} \tag{3.169}$$

The measurement of the output port parameter S_{22} as 0.95 allows both K and R_{out} to be determined. In practice, to obtain this parameter the amplifier would be reversed (or flipped) in the test jig (so that the input port is now the output port and vice versa). Therefore,

$$R_{out} = Z_o\frac{1+S_{22}}{1-S_{22}} = 50\frac{1+0.95}{1-0.95} = 1.95\,\text{k}\Omega$$

Substituting into (3.169), the constant K is determined to be −200.

When using S parameters, the two basic assumptions in their derivation should not be forgotten. First, the transmission lines connected to each port should have the same characteristic impedance. Second, each transmission line should be terminated in its characteristic impedance. There are many situations where a network model is desired for loads not equal to 50 Ω. For example, when evaluating the performance of filters, often the actual in-situ load (and source) impedance is not 50 Ω. In this case, a resistive matching pad can be used to generate the desired load. The pad is used between the output of the network and the 50 Ω load. The loss of the pad should be considered in the analysis.

3.47 Using the Sinusoidal Reflection Coefficient for Transient Problems

The concepts of reflection and transmission coefficients are applicable for sinusoidal steady-state analysis for real or complex loads and characteristic line impedances. However, for transient problems, the load must be resistive and the characteristic line impedance lossless if the reflection and transmission coefficients are used in a simple manner. For loads that contain capacitors and/or inductors and line impedances that are lossy, transforms are used or differential equations set up and then solved. Why?

When using the reflection coefficient for transient analysis, it is assumed that the transient wave instantaneously reflects off the load. For loads containing inductors and capacitors, a time constant is associated with the charging and discharging of the load. For transmission lines modeled with losses, there is also a time constant associated with the charging and discharging of the line. This time constant implies that the voltage does not instantaneously change at the load. A differential equation must be solved (or transforms used) since the inductance and capacitance are defined by the two differential equations

$$v = L\frac{di}{dt}, \quad i = C\frac{dv}{dt}$$

For sinusoidal steady-state analysis, these differential equations reduce to

$$V_s = Lj\omega I_s, \quad I_s = Cj\omega V_s$$

The Fourier transform of the transient signal can be used to determine the frequency range where most of the signal's energy exists. From this information, the highest frequency of interest can be determined for the transient signal. Once the highest frequency of interest is determined, sinusoidal steady-state analysis can be performed at this highest frequency to provide some insight.

3.48 Effect of Receiver Capacitance on Transient Behavior

A lossless line of characteristic impedance Z_o is properly terminated. Somewhere along the line, a capacitor, C, is connected across the line (i.e., between the two conductors) as shown in Figure 3.47. Verify that the reflection and transmission coefficients seen by the line at this load are, respectively,

$$\rho(\omega)=\frac{-j\omega CZ_o}{2+j\omega CZ_o},\quad \mathrm{T}(\omega)=\frac{1}{1+\dfrac{j\omega CZ_o}{2}}$$

Since the load is matched to the line, the impedance seen by the capacitor looking toward the load is Z_o. The load seen by the first section of the line (to the left of the capacitor junction) is $1/(j\omega C)$, the impedance of the capacitor, in parallel with this Z_o. The reflection and transmission coefficients in sinusoidal steady state are then

$$\rho(\omega)=\frac{Z_L-Z_o}{Z_L+Z_o}=\frac{\dfrac{Z_o\dfrac{1}{j\omega C}}{Z_o+\dfrac{1}{j\omega C}}-Z_o}{\dfrac{Z_o\dfrac{1}{j\omega C}}{Z_o+\dfrac{1}{j\omega C}}+Z_o}=\frac{-j\omega CZ_o}{2+j\omega CZ_o} \tag{3.170}$$

$$\mathrm{T}(\omega)=1+\rho(\omega)=\frac{1}{1+\dfrac{j\omega CZ_o}{2}} \tag{3.171}$$

In theory, the effect of the capacitive load can be eliminated (or at least reduced) by locally changing one or more characteristics of the transmission line. For a microstrip transmission line, a simple way of reducing the line's capacitance is by reducing the strip's width. The length of the adjusted segment of the transmission line is a function of the desired change in capacitance. So that the adjusted segment does not appear like a different transmission line, the length of the segment should appear electrically small.

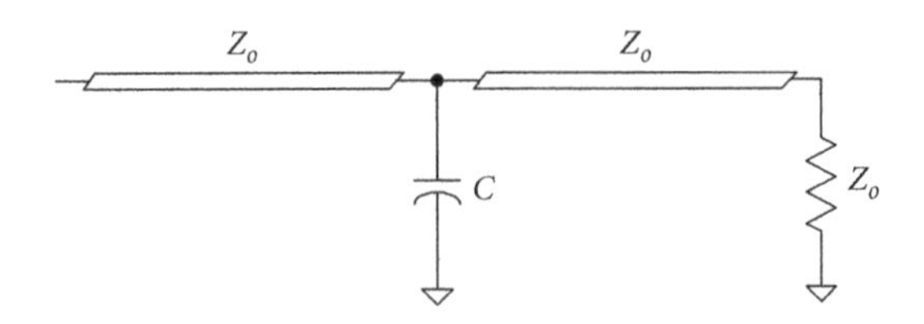

FIGURE 3.47 Capacitive load between two transmission lines.

3.49 Complete Reflection due to Excessive Capacitance

Referring to Figure 3.47, above what frequency is the signal almost completely reflected from the capacitive load? Well below this frequency, what mathematical operation is the capacitor performing? Determine the minimum rise time of a transient signal that will pass the capacitive load relatively unaffected.

The magnitude of the reflection coefficient is approximately one when

$$|\rho(\omega)| = \left|\frac{-j\omega CZ_o}{2+j\omega CZ_o}\right| \approx 1 \;\; \text{if } \omega CZ_o >> 2 \;\; \Rightarrow \omega >> \frac{2}{CZ_o} \tag{3.172}$$

As expected, as the capacitance increases, the frequency where complete reflection occurs decreases: the impedance of the capacitor decreases with an increase in C or increase in frequency. Almost the entire signal is reflected off the capacitor when the magnitude of the reflection coefficient is near one. At low frequencies,

$$\rho(\omega) \approx \frac{-j\omega CZ_o}{2} \;\; \text{if } \omega CZ_o << 2 \tag{3.173}$$

the reflection coefficient is entirely imaginary. This implies differentiation in the time domain as expected for a capacitor:

$$i = C\frac{dv}{dt} \;\; (I_s = Cj\omega V_s)$$

The transmission coefficient is nearly one when

$$\mathrm{T}(\omega) = \frac{1}{1+\dfrac{j\omega CZ_o}{2}} \approx 1 \;\; \text{if } \frac{\omega CZ_o}{2} << 1 \;\; \Rightarrow \omega << \frac{2}{CZ_o} \tag{3.174}$$

Frequencies that satisfy this inequality pass by the capacitor relatively unaffected. For a transient signal of rise time τ_r, the highest frequency of interest is approximately $1/(\pi\tau_r)$ (Hz). Therefore, the rise time of the signal must be much greater than CZ_o if the capacitor is to have negligible effect:

$$\omega = 2\pi f_{max} = 2\pi\frac{1}{\pi\tau_r} = \frac{2}{\tau_r} << \frac{2}{CZ_o} \;\; \Rightarrow \tau_r >> CZ_o \tag{3.175}$$

Again, the rise time of the transient must be much larger than CZ_o for the capacitor to appear like an open circuit to the transient. If the capacitor is representing an HCT100 CMOS gate and the line impedance is 65 Ω, then

$$\tau_r >> CZ_o = 3.5\times10^{-12}(65) \approx 0.23 \text{ ns}$$

For an HCT CMOS gate, the rise time is 4.7 ns, which is much greater than 0.23 ns.

3.50 Amplitude of Mismatch "Blimp" from Receiver Capacitance

Show that the magnitude of the "blimp" reflected from a capacitor at low frequencies (large rise-time transients) is approximately

$$\frac{CZ_oV}{2\tau_r}$$

where V is the magnitude of the incident transient signal.

To solve this problem rigorously for nearly any input signal, the transform of the input transient, $V(\omega)$, should be multiplied by the transfer function

$$\rho(\omega) = \frac{-j\omega CZ_o}{2 + j\omega CZ_o} \tag{3.176}$$

and then the inverse transform of this product determined. This time-domain result represents the output voltage, the reflected time signal. The maximum value of this function represents the magnitude of the "blimp." If certain assumptions are made concerning the input signal, much of this work can be avoided.

First, assume that the input signal is a trapezoidal function with an amplitude of V and a rise time of τ_r. Second, assume that the rise time is large or the highest frequency of interest is relatively small. Then, the reflection coefficient "transfer" function can be approximated as

$$\rho(\omega) \approx \frac{-j\omega CZ_o}{2} \quad \text{if } \omega CZ_o << 2 \tag{3.177}$$

As previously stated, for a transient signal of rise time τ_r, the highest frequency of interest is approximately $1/(\pi\tau_r)$ (Hz). Therefore, the approximation for the reflection coefficient is valid when

$$\frac{2}{\tau_r} CZ_o << 2 \quad \text{or} \quad CZ_o << \tau_r \tag{3.178}$$

Under this condition, the "blimp" is proportional to the time derivative of the transient signal since the reflection coefficient is proportional to $j\omega$. This is a precious transform property. If the transform of the trapezoidal pulse is obtained and then multiplied by this simpler expression for $\rho(\omega)$, the inverse transform of the product is the time-domain output signal. This same result is also obtained without determining any transforms by merely differentiating the original time-based input signal with respect to time (and multiplying by the $CZ_o/2$ factor). For a trapezoidal function, the maximum value of its derivative is the slope of the trapezoid during the rise or fall of the function:

$$\frac{V}{\tau_r}$$

Therefore, the maximum amplitude of the reflected signal is

$$|V\rho(\omega)| \approx \frac{V}{\tau_r} \frac{CZ_o}{2} \tag{3.179}$$

If the capacitor is representing an HCT100 CMOS gate and the line impedance is 65 Ω, then the amplitude of the reflected pulse is about one-fortieth of the incident voltage:

$$|V\rho(\omega)| \approx \frac{VCZ_o}{2\tau_r} = \frac{V(3.5\times10^{-12})65}{2(4.7\times10^{-9})} \approx \frac{V}{40}$$

Occasionally, individuals will merely take the highest frequency of interest, in this case $1/(\pi\tau_r)$, and substitute it into the transfer function. Then, the magnitude of this result is multiplied by the amplitude of the input signal, and this result is claimed to be the maximum amplitude of the reflected signal. Using (3.177),

$$\left| V \frac{-j\frac{2\pi}{\pi\tau_r} CZ_o}{2} \right| = \frac{V}{\tau_r} CZ_o \tag{3.180}$$

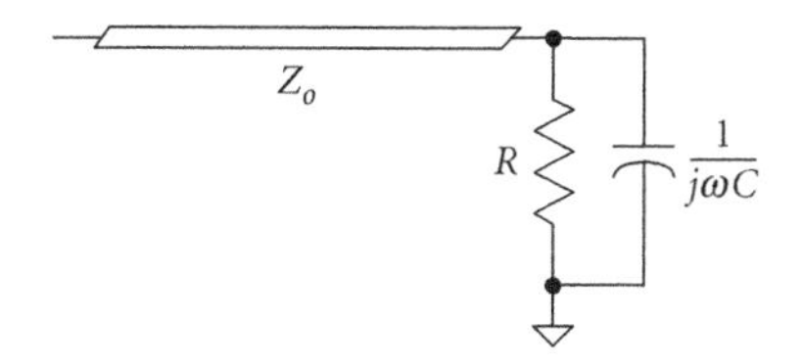

FIGURE 3.48 Transmission line connected to an *RC* load.

Although in this particular example, (3.179) and (3.180) are quite close, this approach is **not** recommended. One fault with this method is that the magnitude of the transform, multiplied by the "amplitude" of the input signal, evaluated at the highest frequency of interest is not necessarily the maximum amplitude of the output time signal. Furthermore, the time signal of the output or reflected pulse is the sum of many frequency components, not just the highest frequency of interest.

Determine the reflection and transmission coefficients for a transmission line of impedance Z_o connected to a parallel *RC* circuit (a simple model for a receiver input) as shown in Figure 3.48. Using reasonable values, determine the effect of *C* on a transient 74AS00 TTL signal.

The reflection coefficient is

$$\rho(\omega) = \frac{Z_L - Z_o}{Z_L + Z_o} = \frac{\dfrac{R\dfrac{1}{j\omega C}}{R + \dfrac{1}{j\omega C}} - Z_o}{\dfrac{R\dfrac{1}{j\omega C}}{R + \dfrac{1}{j\omega C}} + Z_o} = \frac{(R - Z_o) - j\omega RCZ_o}{(R + Z_o) + j\omega RCZ_o} \tag{3.181}$$

and the transmission coefficient is

$$\mathrm{T}(\omega) = 1 + \rho(\omega) = \frac{\dfrac{2R}{R + Z_o}}{1 + \dfrac{j\omega RCZ_o}{R + Z_o}} \tag{3.182}$$

The total signal at the receiver is the incident signal, *V*, multiplied by this transmission coefficient. As expected, at low frequencies the total voltage across the receiver involves voltage division between the load resistance and line impedance:

$$\mathrm{T}(\omega) \approx \frac{\dfrac{2R}{R + Z_o}}{1} = \frac{2R}{R + Z_o} \quad \text{if} \left|\frac{j\omega RCZ_o}{R + Z_o}\right| << 1 \tag{3.183}$$

The capacitance has a negligible effect on the signal when

$$\left|\frac{j\omega RCZ_o}{R + Z_o}\right| << 1 \qquad \Rightarrow \omega << \frac{R + Z_o}{RZ_o}\frac{1}{C} = \frac{1}{\left(R \| Z_o\right)C} \tag{3.184}$$

The time constant for this load is the parallel combination of *R* and Z_o, which is the net resistance seen by the capacitor, multiplied by *C*. For a 74AS00 TTL device connected to a 65 Ω transmission line, which is matched to the resistive portion of the load ($R = Z_o = 65\ \Omega$),

$$C = 3\ \text{pF}, \quad \tau_r = 1.7\ \text{ns}$$

$$\omega = 2\pi f_{max} = \frac{2\pi}{\pi\tau_r} = \frac{2}{\tau_r} << \frac{1}{\left(R \| Z_o\right)C} \quad \Rightarrow \tau_r >> 2\left(R \| Z_o\right)C \approx 0.2\ \text{ns}$$

The rise time of the TTL device satisfies this inequality, and the input capacitance of the receiver should not have a significant effect on the transient input signal. When the capacitance does have an effect, it will slow down the signal. To reduce this effect, the net resistance at the receiver (R in parallel with Z_o) should be as small as possible without excessively loading the driver.

3.51 When Not to Match!

Under what circumstances should the driver *not* be matched to the receiver? [Whitlock, '98]

The importance of matching has been strongly emphasized. However, in some circumstances, matching is not desirable. When the line connecting a driver to a receiver is electrically short at all frequencies of interest, then sometimes the load is "voltage matched" to the driver. Voltage matching is not the same as impedance matching the load to the characteristic impedance of the transmission line.

Frequently, a driver has a low output impedance (i.e., a low Thévenin impedance at its output), such as 1 Ω. Rarely is the input impedance of an amplifier, or a receiver, selected to be low and equal to the driver's output impedance. Usually, the input impedance is high, such as 100 kΩ. Connecting a driver with a low output impedance to a receiver with a high input impedance is referred to as "voltage matching." The largest signal voltage is transferred to the input of an amplifier, or a receiver, when its input impedance is much greater than the driver's output impedance. When the high-frequency signals on the line connecting the driver to the receiver can be neglected, voltage matching is common.

A receiver with a low input impedance would be a good impedance match to a driver with a low output impedance. A conjugate impedance match, in theory, would provide the maximum power transfer. In practice, though, systems are not designed for maximum power transfer. A low input impedance receiver will adversely affect the performance of the system. First, a low input impedance receiver could excessively load down the driver (with extreme current demands). Second, the voltage across the input of the receiver under a matched scenario would only be one-half of the driver's voltage, which will affect the maximum voltage swing.

In some applications, the impedance of the driver is increased or decreased (e.g., via a transformer). This transformation or adjustment of the driver's output impedance might be to satisfy a signal-to-noise requirement of the receiver or amplifier. Some amplifier data sheets provide information relating the signal-to-noise vs. the impedance seen by the amplifier's *input*. Often, there is an optimal expected input impedance for the amplifier to produce the greatest signal-to-noise ratio. Adjustment of the input impedance seen by an amplifier to optimize the signal-to-noise ratio is referred to as noise matching.

Some products have the ability to select between several different input impedances. Many modern oscilloscopes, for example, have two possible input impedances: 50 Ω and 1 MΩ. When connected to a 50 Ω cable, the lower input impedance is selected for high-frequency signals, where the cable is not electrically short, while the higher input impedance is used for lower frequency signals, where the cable is electrically short. Some video devices have both a 75 Ω input and "high-Z" input. The high input impedance, or "high-Z" input, is used when the video device is connected along the cable connecting the driver (e.g., camera) and receiver (e.g., monitor). This tap on the line is referred to as "looping-thru" or "bridging" onto a line. This high input impedance has little effect or loading on a 75 Ω cable connecting the driver to the receiver. When the video device is used as the final termination on an electrically-long line, then the 75 Ω input can be used to reduce reflections and provide impedance matching.

4

Passive Contact Probes

An ideal probe has no effect on the circuit under test. Of course, a real probe does affect the circuit under test. Especially at high frequencies and high circuit impedances, the probe must be carefully selected. When selecting a probe to perform a measurement, the probe's input impedance, bandwidth, gain, and susceptibility to noise should be known. The probes discussed in this chapter make physical contact with the circuit under test.[1]

4.1 Low-Impedance Passive Probe

A low-impedance passive probe consists of a 50 Ω coaxial cable with a series input resistor of R. Assume that the input impedance at the scope end is also 50 Ω. What are the advantages of this probe? What is the probe gain? If a 10:1 probe is desired with an input impedance of 500 Ω, what value should be selected for R? What are the disadvantages of this probe? [Smith, '93; Tektronix]

When an oscilloscope probe is connected to a circuit, it is common for inexperienced engineers to assume that the probe has little or no effect on the circuit itself. A good probe would extract little energy from the circuit under test and introduce negligible distortion. Unfortunately, this is not always the case, especially if high frequencies are involved or the probe is contacting a high-impedance point in the circuit.

A simple low-impedance passive probe, also referred to as a resistive-input, resistive-divider, or Z_o probe, connected to a scope is shown in Figure 4.1. Two obvious advantages of this low-impedance probe are its low cost and ease of construction. Another advantage is that the loading effect on the circuit can be reduced by merely increasing the value of R. Since the impedance looking into the cable is 50 Ω (the cable is matched to the scope), the input impedance of the probe for the circuit under test is

$$Z_{probe} = R + 50\,\Omega \tag{4.1}$$

Note that this impedance is independent of the frequency. The probe response is flat, and it does not distort the signal, which is another advantage. Of course, when the nonideal characteristics of the resistor are modeled, the impedance of the resistor becomes frequency dependent, and the probe response is no longer flat.

The probe gain is simply obtained through voltage division:

$$V_{ys} = V_{xs}\frac{50}{50 + R} \tag{4.2}$$

This voltage also appears across the scope since the transmission line impedance is equal to the scope input impedance. If a 10:1 probe is desired (a probe that will output one-tenth of the circuit voltage), R should be set to 450 Ω:

$$\frac{V_{ys}}{V_{xs}} = \frac{1}{10} = \frac{50}{50+R} \quad \Rightarrow R = 450\,\Omega$$

[1]The reading of the unique and practical book entitled *High Frequency Measurements and Noise in Electronic Circuits* written by Douglas C. Smith inspired the author to write this chapter.

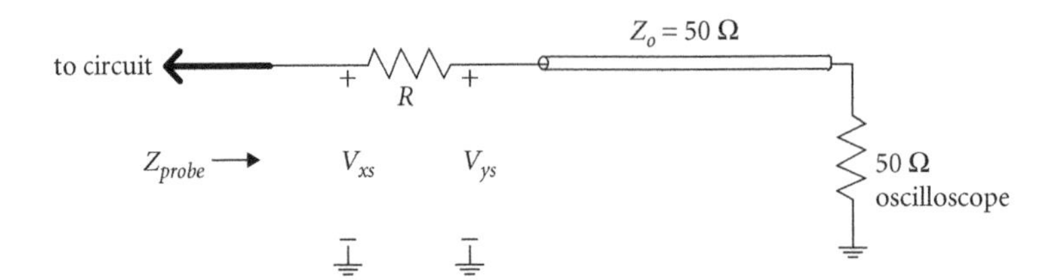

FIGURE 4.1 Low-impedance passive probe connected to a scope through a 50 Ω transmission line.

In this case, the input impedance of the probe is 500 Ω. As the probe step-down voltage ratio increases, the input impedance increases, which implies less loading. For example, if R is equal to 1 kΩ, then $Z_{probe} = 1{,}050\ \Omega$ and $V_{ys}/V_{xs} = 1/21$.

One limitation of this probe is the need for a 50 Ω scope input. Many older scopes have only a high-impedance input (e.g., 1 MΩ) and, thus, a mismatch occurs between the 50 Ω cable and scope. An external 50 Ω resistor in shunt with the input of the scope would be required to match the scope to the cable. (Modern scopes often have a selectable input impedance of either 50 Ω or 1 MΩ.) A second limitation can arise at high frequencies where the capacitance and inductance of the external resistor, R, are no longer negligible.

4.2 Improved Model of the Low-Impedance Passive Probe

For the probe given in Figure 4.1, assume that the resistor R has a parasitic parallel capacitance of C_p. At what frequency does the probe gain begin to change? Does the gain increase or decrease? How does the input impedance of this passive probe change with frequency? What are the input impedance's maximum and minimum values?

A real resistor has parasitic capacitance (and inductance), often modeled in parallel with the resistor. The improved model for the probe is shown in Figure 4.2. The gain of this probe is modified by this parasitic capacitance. In the frequency domain, the gain is (for sinusoidal steady-state conditions)

$$\frac{V_{ys}}{V_{xs}} = \frac{50}{50 + \dfrac{R\dfrac{1}{j\omega C_p}}{R + \dfrac{1}{j\omega C_p}}} = \frac{50}{50 + \dfrac{R}{1 + j\omega R C_p}} \tag{4.3}$$

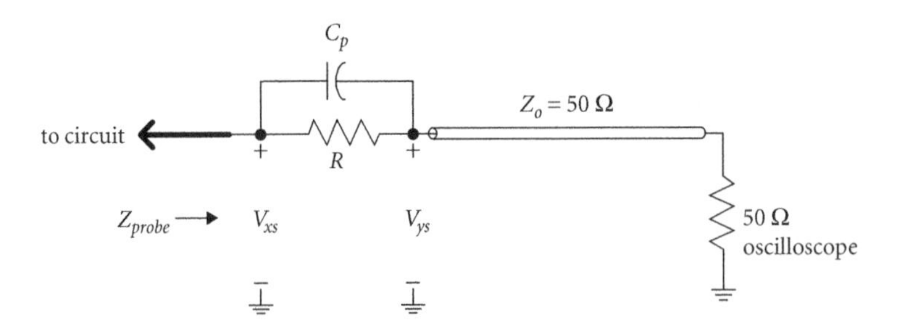

FIGURE 4.2 Improved model of the low-impedance probe in Figure 4.1.

At low frequencies,

$$\frac{V_{ys}}{V_{xs}} \approx \frac{50}{50+R} \quad \text{if } 1 \gg \omega RC_p \text{ or } \omega \ll \frac{1}{RC_p} \tag{4.4}$$

and the gain is not affected by the capacitance (the gain is flat). At higher frequencies,

$$\frac{V_{ys}}{V_{xs}} \approx \frac{50}{50+\dfrac{1}{j\omega C_p}} = \frac{j\omega(50C_p)}{1+j\omega(50C_p)} \quad \text{if } 1 \ll \omega RC_p \text{ or } \omega \gg \frac{1}{RC_p} \tag{4.5}$$

and the gain is a function of the capacitance and frequency. At even higher frequencies, the gain approaches unity:

$$\frac{V_{ys}}{V_{xs}} \approx \frac{j\omega C_p 50}{j\omega C_p 50} = 1 \quad \text{if } 1 \ll \omega RC_p \text{ and } 1 \ll \omega(50C_p) \tag{4.6}$$

This result is reasonable since the reactance of the capacitor is very small at high frequencies, and the capacitor effectively short circuits the resistor R. (Of course, at very high frequencies the resistor's parasitic inductance must be included in the model.)

When the gain expression, (4.3), is written in standard Bode magnitude plotting form

$$\frac{V_{ys}}{V_{xs}} = \frac{50}{50+R} \frac{1+\dfrac{j\omega}{\dfrac{1}{RC_p}}}{1+\dfrac{j\omega}{\dfrac{50+R}{50RC_p}}} \tag{4.7}$$

it is clear that the two cutoff or break frequencies are

$$\omega_c\text{'s} = \frac{1}{RC_p}, \frac{50+R}{50RC_p} \tag{4.8}$$

Since

$$\frac{1}{RC_p} < \frac{R+50}{50RC_p}$$

the lowest break frequency is $1/(RC_p)$. The gain is indeed flat (i.e., constant with frequency) up to the first break frequency, $1/(RC_p)$. Then, the gain increases at 20 dB/decade. After the second, higher break frequency, $(R+50)/(50RC_p)$, the gain is again flat.

The input impedance of the probe is

$$Z_{probe} = \frac{R\dfrac{1}{j\omega C_p}}{R+\dfrac{1}{j\omega C_p}} + 50 = \frac{R}{1+j\omega RC_p} + 50 = (R+50)\frac{1+\dfrac{j\omega}{\dfrac{R+50}{50RC_p}}}{1+\dfrac{j\omega}{\dfrac{1}{RC_p}}}$$

Written in this standard form, the two break frequencies of the input impedance are clearly visible. These break frequencies are identical to (4.8). The impedance is flat until $1/(RC_p)$ where it decreases at 20 dB/decade. Well above the second break frequency, $(R+50)/(50RC_p)$, the impedance is constant with frequency. The maximum impedance occurs at low frequencies

$$Z_{probe,max} \approx R+50 \quad \text{if } \omega \ll \frac{1}{RC_p} \tag{4.9}$$

while the minimum impedance occurs at high frequencies:

$$Z_{probe,min} \approx 50 \quad \text{if } \omega \gg \frac{R+50}{50RC_p} \tag{4.10}$$

4.3 Operating Range of the Low-Impedance Passive Probe

In reference to the low-impedance passive probe shown in Figure 4.2, what is a typical upper usable frequency? Why should this probe not be used to probe high-Q LC circuits operated near resonance?

It was determined previously that the first break frequency for the probe is

$$\omega_c = \frac{1}{RC_p} \quad \text{or} \quad f_c = \frac{1}{2\pi RC_p}$$

Near and above this frequency, the response of the probe changes with frequency. Ideally, a probe's frequency characteristics should be flat, and the probe should not change the characteristics of the signal being probed. For a 10:1 probe, $R = 450\ \Omega$. A typical parasitic capacitance is 1 pF.[2] This corresponds to a break frequency of about 350 MHz. At these higher frequencies, the parasitic inductance of the resistor probably should be considered. When measuring a 0.5 ns rise-time signal, this break frequency would probably be too low since the highest frequency of interest is about $1/(\pi \times 0.5\ \text{ns}) \approx 640$ MHz.

When $R = 450\ \Omega$, the input impedance of the probe before the first break frequency is about 500 Ω. Greater input impedances can be obtained by increasing R at the expense of a greater voltage step-down ratio. In high-Q circuits, the impedance of the circuit may become extremely large compared to 500 Ω, especially near resonance, and the probe may heavily load down the circuit under test. For low-impedance circuit measurements, such as in ground-noise measurements, this probe should be adequate.

4.4 Improved Model of the Cable and Scope

In reference to Figure 4.2, assume that the scope input can be modeled as a 50 Ω resistor in parallel with a 2 pF capacitor, and the parasitic capacitance of the probe resistor $R = 450\ \Omega$ is about 2 pF. The capacitance of the 50 Ω cable is approximately 29.5 pF/ft, and the cable is 3 ft long. Plot the magnitude of the input impedance of this low-impedance passive probe vs. frequency. [Smith, '93]

Before a transmission line can be modeled as a lumped circuit, its electrical length must be determined. If the line is electrically small, then the lumped model is appropriate. The frequency corresponding to a

[2]The operating frequency of the resistive probe can be increased by reducing this (already small) parasitic capacitance.

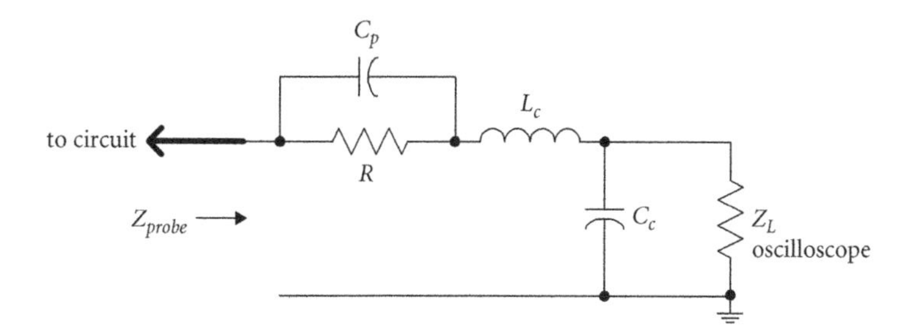

FIGURE 4.3 Modeling the cable between the probe and oscilloscope as a lumped *LC* circuit.

wavelength of 3 ft is

$$f = \frac{v}{\lambda} \approx \frac{v}{3 \times 0.305} \text{ Hz}$$

where v is the velocity of propagation in m/s for a signal traveling in the cable. Since the capacitance is given as 29.5 pF/ft ≈ 97 pF/m and the characteristic impedance of the line is 50 Ω, the inductance per meter is about 0.24 μH/m:

$$Z_o = \sqrt{\frac{L}{C}} \quad \Rightarrow L = CZ_o^2 \approx 0.24\,\mu\text{H/m}$$

The velocity of a signal on the cable is then

$$v = \frac{1}{\sqrt{LC}} \approx 2.1 \times 10^8 \text{ m/s}$$

which is obviously less than the speed of light $(c = 3 \times 10^8 \text{ m/s})$. Using this velocity, the frequency where this 3 ft cable is one wavelength in length is $f \approx$ 230 MHz. Therefore, for frequencies less than about 23 MHz, this cable can be considered electrically small and can be modeled as a lumped circuit.[3]

The model of the situation is given in Figure 4.3 where L_c is the total inductance and C_c is the total capacitance of the cable. The input impedance is

$$Z_{probe} = \frac{R\dfrac{1}{j\omega C_p}}{R + \dfrac{1}{j\omega C_p}} + j\omega L_c + \frac{Z_L \dfrac{1}{j\omega C_c}}{Z_L + \dfrac{1}{j\omega C_c}} \tag{4.11}$$

In this case, the complex load impedance of the scope, Z_L, consists of a 50 Ω resistor (R_s) in parallel with a 2 pF capacitor (C_s):

$$Z_L = R_s \Big\| \frac{1}{j\omega C_s} = \frac{R_s}{1 + j\omega R_s C_s}$$

[3]Of course, if the frequency is too low, then the cable is in the "*RC*" region. In this region, the characteristic impedance of the cable is not merely $\sqrt{L/C}$, and the resistance of the cable is not negligible compared to the inductive reactance of the cable.

$C_p := 2 \cdot 10^{-12}$ $R := 450$ $L_c := 0.242 \cdot 10^{-6} \cdot 0.915$ $C_c := 96.7 \cdot 10^{-12} \cdot 0.915$ $j := \sqrt{-1}$

$x := 50, 50.1..\,90$

$C_s := 2 \cdot 10^{-12}$ $R_s := 50$

$$\omega(x) := \left(x + 1 - 10 \cdot \text{floor}\left(\frac{x}{10}\right)\right) \cdot 10^{\text{floor}\left(\frac{x}{10}\right)}$$

$$Z_L(\omega) := \frac{R_s \cdot \frac{1}{j \cdot \omega \cdot C_s}}{R_s + \frac{1}{j \cdot \omega \cdot C_s}} \qquad Z_{probe}(\omega) := \frac{R \cdot \frac{1}{j \cdot \omega \cdot C_p}}{R + \frac{1}{j \cdot \omega \cdot C_p}} + j \cdot \omega \cdot L_c + \frac{Z_L(\omega) \cdot \frac{1}{j \cdot \omega \cdot C_c}}{Z_L(\omega) + \frac{1}{j \cdot \omega \cdot C_c}}$$

$$G_{probe}(\omega) := \frac{\dfrac{Z_L(\omega) \cdot \frac{1}{j \cdot \omega \cdot C_c}}{Z_L(\omega) + \frac{1}{j \cdot \omega \cdot C_c}}}{Z_{probe}(\omega)}$$

Input Impedance of Low-Impedance Probe

ohms: $|Z_{probe}(\omega(x))|$ (solid) and $\frac{|Z_{probe}(\omega(0))|}{\sqrt{2}}$ (dashed); vertical axis 300 to 550 (300, 350, 400, 450, 500, 550); horizontal axis $\frac{\omega(x)}{2 \cdot \pi}$ Hz: $1 \cdot 10^4$, $1 \cdot 10^5$, $1 \cdot 10^6$, $1 \cdot 10^7$, $1 \cdot 10^8$

Gain of Low-Impedance Probe

dB: $20 \cdot \log(|G_{probe}(\omega(x))|)$ (solid) and $20 \cdot \log(|G_{probe}(\omega(0))|) - 3$ (dashed); vertical axis −30 to −15 (−15, −20, −25, −30); horizontal axis $\frac{\omega(x)}{2 \cdot \pi}$ Hz: $1 \cdot 10^4$, $1 \cdot 10^5$, $1 \cdot 10^6$, $1 \cdot 10^7$, $1 \cdot 10^8$

MATHCAD 4.1 Input impedance magnitude and gain of the low-impedance probe connected to an oscilloscope via a 50 Ω cable.

The input impedance and gain response are provided in Mathcad 4.1. For the magnitude of the impedance, the cutoff frequency for the probe is about 100 MHz corresponding to about $1/\sqrt{2}$ of the maximum value of the response. (Of course, the cable is not strictly electrically short at this frequency, so this result should be used with caution.) The magnitude of the input impedance of the probe is flat up to 10 MHz with a value of about 500 Ω. The gain of the system, the ratio of the voltage at the scope to the voltage at the probe, is also plotted (in dB). The 3 dB break frequency of the gain response is about 40 MHz. If the inductive term is neglected and (4.11) written in standard form

$$Z_{probe} \approx (R+R_s)\frac{1+\dfrac{j\omega}{\dfrac{R+R_s}{RR_s(C_s+C_c+C_p)}}}{\left(1+\dfrac{j\omega}{\dfrac{1}{RC_p}}\right)\left(1+\dfrac{j\omega}{\dfrac{1}{R_s(C_s+C_c)}}\right)} \tag{4.12}$$

then the break frequencies are determined by inspection:

$$\omega_c\text{'s} = \frac{1}{R_s(C_s+C_c)}, \frac{R+R_s}{RR_s(C_s+C_c+C_p)}, \frac{1}{RC_p} \tag{4.13}$$

The lowest break frequency is about 35 MHz when the cable inductance is neglected. Therefore, the cable inductance can be neglected up to 23 MHz. (The impedance of the inductor at 35 MHz is about 50 Ω, which is much less than R.)

4.5 High-Impedance Passive Probe

Passive high-impedance oscilloscope probes are popular (e.g., 10X probe). Unfortunately, they can lead to significant error. Why are they so popular? Model the probe tip as a 9 MΩ resistor in parallel with a 10 pF capacitor. Assume that this probe is connected via a cable to a scope, and the input of the scope is modeled as a 1 MΩ resistor in parallel with a 20 pF capacitor. Why not connect the probe to a scope with a 50 Ω input as with the low-impedance probe? The characteristic impedance of the cable connecting the probe tip to the scope is given as 170 Ω. What is the reflection coefficient at the scope? Are reflections expected from the scope and, if so, will these reflections be troublesome? Sometimes, the center conductor of the connecting cable is constructed of a nickel-chromium alloy that is very resistive or lossy (around 40–120 Ω/ft). Why? [Wood; Smith, '93]

High-impedance (or Hi-Z) probes are used in virtually every laboratory. The main reasons these probes are used are their *expected* (or perceived) high input impedance (and negligible loading of the circuit under test), their simple scale-conversion relationship,[4] and their ability to measure higher voltages.

Referring to Figure 4.4, the input impedance at V_{ys} (looking into the transmission line) would be equal to 170 Ω if it had a 170 Ω load. There would be no reason to connect the 170 Ω cable to the 50 Ω input

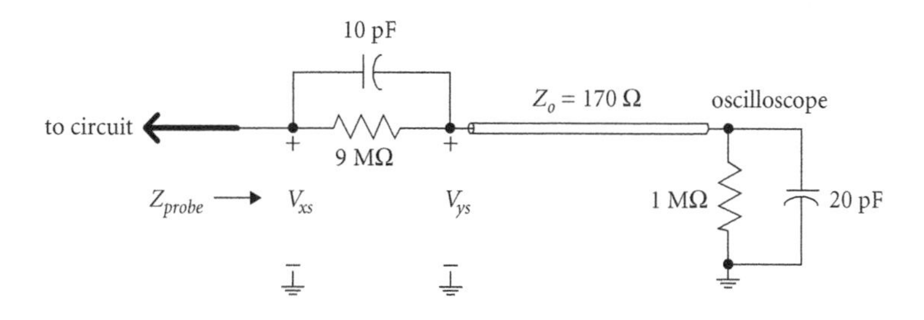

FIGURE 4.4 High-impedance passive probe connected to a scope through a 170 Ω transmission line.

[4]Some modern scopes can recognize their own brand of high-impedance probe and automatically compensate for this loss in gain.

of the scope (if this input was available). If this system is to be matched at the scope, the total load impedance should be equal to 170 Ω. This would eliminate high-frequency reflections from the scope.

The reflection coefficient at the scope is

$$\rho=\frac{Z_L-Z_o}{Z_L+Z_o}=\frac{\dfrac{1\times10^6\dfrac{1}{j\omega(20\times10^{-12})}}{1\times10^6+\dfrac{1}{j\omega(20\times10^{-12})}}-170}{\dfrac{1\times10^6\dfrac{1}{j\omega(20\times10^{-12})}}{1\times10^6+\dfrac{1}{j\omega(20\times10^{-12})}}+170}=\frac{10^6-170-j\omega(3.4\times10^{-3})}{10^6+170+j\omega(3.4\times10^{-3})}$$

$$=\frac{(10^6-170)\left(1-\dfrac{j\omega}{\dfrac{10^6-170}{3.4\times10^{-3}}}\right)}{(10^6+170)\left(1+\dfrac{j\omega}{\dfrac{10^6+170}{3.4\times10^{-3}}}\right)}\quad\text{where}\ \frac{\dfrac{10^6\pm170}{3.4\times10^{-3}}}{2\pi}\approx47\ \text{MHz}$$

The reflection coefficient is approximately one up to the cutoff frequency of 47 MHz. (Actually, it is also approximately one for frequencies greater than 47 MHz.) As will be seen shortly, at this frequency, the transmission line is not electrically short. Furthermore, the models for the probe and scope may have to be improved at this frequency.

A reflection coefficient of zero implies a matched load. A reflection coefficient of one implies a poorly matched system. Reflections are expected for this probe. These reflections will be troublesome if the line is not electrically short. Although a lossy transmission line will reduce the signal strength at the scope, it will also dampen reflections. This will reduce the Q of the system and the amplitude of any ringing on the line. The resistance of the cable can be increased by using conductors with a lower conductivity and smaller cross-sectional dimensions (i.e., a smaller cable). Compared to a 50 Ω cable, a 170 Ω cable will have a smaller capacitance per unit length (if the inductance per unit length is kept approximately constant).

Since they are not commonly available, the measured parameters of a real lossy cable will be given. For one lossy coaxial cable with a #41 AWG copper-nickel alloy inner conductor, 15 mil foam polyethylene dielectric, and silver-plated copper spiral outer conductor, the measured parameters at 20 MHz are

$$R=50\ \Omega/\text{ft},\quad L=93\ \text{nH/ft},\quad G=4.2\ \mu\text{S/ft},\quad C=11\ \text{pF/ft} \tag{4.14}$$

At $\omega=2\pi(20\times10^6)$, the characteristic impedance of this cable is

$$Z_o=\sqrt{\frac{R+j\omega L}{G+j\omega C}}\approx150-j120\ \Omega\approx190\angle-38^\circ\ \Omega \tag{4.15}$$

Typically, at higher frequencies, the characteristic impedance of a "normal" cable is mostly real. (Neglecting the ohmic and dielectric resistances, the nominal characteristic impedance of this cable is about 92 Ω.) As is apparent from (4.15), the inner conductor's high resistance has a strong influence on the cable's impedance (at least at the measured frequency). Assuming that the cable's parameters do not change with frequency, the phase angle of Z_o is still greater than 2° at 1 GHz.

4.6 Input Impedance of a High-Impedance Passive Probe

Assume that the cable capacitance for the high-impedance passive probe is approximately 9 pF/ft. For a 6 ft 170 Ω cable, at what frequencies is the lumped-circuit model accurate? Plot the magnitude of the input impedance of this high-impedance passive probe vs. frequency. What is the upper usable frequency of this probe?

The lumped model representation for the line is valid if the line is electrically short. To determine the electrical length of the cable, the velocity of propagation inside the cable is required. Since the capacitance is given as 9 pF/ft ≈ 30 pF/m and the high-frequency characteristic impedance of the line is 170 Ω, the inductance per meter is 0.87 μH/m:

$$Z_o = \sqrt{\frac{L}{C}} \quad \Rightarrow L = CZ_o^2 \approx 0.87\ \mu\text{H/m}$$

The velocity of propagation in the cable is therefore

$$v = \frac{1}{\sqrt{LC}} \approx 2.0 \times 10^8 \text{ m/s} \tag{4.16}$$

which is less than the speed of light. The frequency where this 6 ft cable is one wavelength in length is about 110 MHz. Therefore, for frequencies less than about 11 MHz, this cable can be modeled as a lumped circuit. The R of the cable is negligible compared to the resistance of the series probe resistor.

The previous low-frequency model for the probe, cable, and scope input is given in Figure 4.5. The input impedance is

$$Z_{probe} = \frac{R\dfrac{1}{j\omega C_p}}{R + \dfrac{1}{j\omega C_p}} + j\omega L_c + \frac{Z_L \dfrac{1}{j\omega C_c}}{Z_L + \dfrac{1}{j\omega C_c}} \tag{4.17}$$

In Mathcad 4.2, both the magnitude of the input impedance and gain in dB of the system are plotted. The plots indicate that the cutoff frequency for the system is about 1 kHz. After this cutoff frequency, both the input impedance and gain are clearly a function of frequency. A conservative, upper usable frequency for this probe would be 1 kHz. If the probe is used beyond this frequency, signal distortion can occur. (Between about 10 kHz and 1 MHz, the gain response is again flat.) This distortion is definitely a disadvantage of the "high-impedance" probe. It is also interesting to note that when the cable inductance is eliminated (i.e., set to zero), the cutoff frequency remains about 1 kHz. (The reactance of the inductor is small compared to the other impedances in the system at this frequency.)

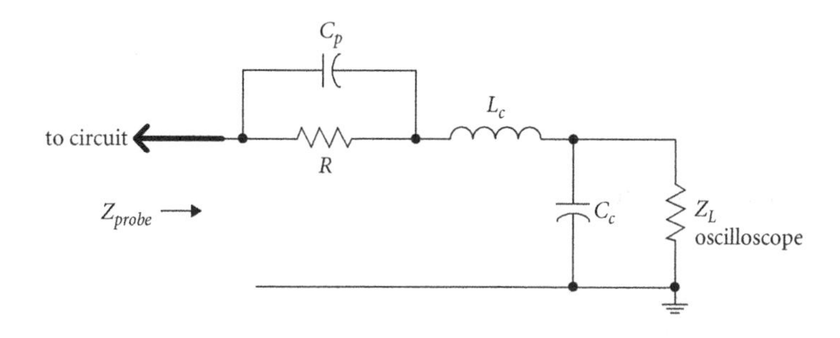

FIGURE 4.5 Lumped-circuit representation of the probe, cable, and scope input.

$C_p := 10 \cdot 10^{-12}$ $\quad R := 9 \cdot 10^6$ $\quad L_c := 0.87 \cdot 10^{-6} \cdot 1.83$ $\quad C_c := 30 \cdot 10^{-12} \cdot 1.83$ $\quad j := \sqrt{-1}$

$x := 10, 10.1 .. 80$ $\quad C_s := 20 \cdot 10^{-12}$ $\quad R_s := 10^6$

$$\omega(x) := \left(x + 1 - 10 \cdot \text{floor}\left(\frac{x}{10}\right)\right) \cdot 10^{\text{floor}\left(\frac{x}{10}\right)}$$

$$Z_L(\omega) := \frac{R_s \cdot \frac{1}{j \cdot \omega \cdot C_s}}{R_s + \frac{1}{j \cdot \omega \cdot C_s}} \qquad Z_{probe}(\omega) := \frac{R \cdot \frac{1}{j \cdot \omega \cdot C_p}}{R + \frac{1}{j \cdot \omega \cdot C_p}} + j \cdot \omega \cdot L_c + \frac{Z_L(\omega) \cdot \frac{1}{j \cdot \omega \cdot C_c}}{Z_L(\omega) + \frac{1}{j \cdot \omega \cdot C_c}}$$

$$G(\omega) := \frac{\dfrac{Z_L(\omega) \cdot \frac{1}{j \cdot \omega \cdot C_c}}{Z_L(\omega) + \frac{1}{j \cdot \omega \cdot C_c}}}{Z_{probe}(\omega)}$$

MATHCAD 4.2 Input impedance magnitude and gain of the high-impedance probe connected to an oscilloscope via a 170 Ω cable.

4.7 High-Impedance Probe Compensator

For passive high-impedance probes, the probe resistance and capacitance can sometimes be adjusted to flatten the frequency response, referred to as probe compensation. Assume that the input resistance of the scope is R_x and the combined capacitance of the scope, the cable, and another adjustable capacitor is C_x. By deriving the expression for the ratio of the voltage across the scope to the input voltage to the

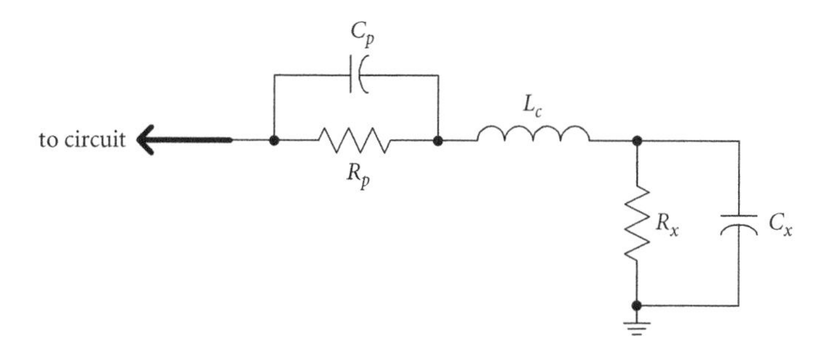

FIGURE 4.6 Modeling the probe, cable, and oscilloscope.

probe, explain why probe compensation is used. Obtain an approximate expression for this ratio for low frequencies. Obtain an approximate expression for this ratio for high frequencies. These probes should be adjusted even if a different channel of the same scope is used. Why?

The gain for the circuit given in Figure 4.6, the ratio of the voltage across the scope to the voltage at the input to the probe, is

$$G(\omega)=\frac{\dfrac{R_x\dfrac{1}{j\omega C_x}}{R_x+\dfrac{1}{j\omega C_x}}}{\dfrac{R_x\dfrac{1}{j\omega C_x}}{R_x+\dfrac{1}{j\omega C_x}}+j\omega L_c+\dfrac{R_p\dfrac{1}{j\omega C_p}}{R_p+\dfrac{1}{j\omega C_p}}}\approx\frac{R_x}{R_x+R_p\left(\dfrac{j\omega R_xC_x+1}{j\omega R_pC_p+1}\right)} \tag{4.18}$$

or in standard Bode form

$$G(\omega)=\frac{R_x}{R_x+R_p}\frac{1+\dfrac{j\omega}{\dfrac{1}{R_pC_p}}}{1+\dfrac{j\omega}{\dfrac{R_x+R_p}{R_xR_p(C_p+C_x)}}} \tag{4.19}$$

(To simplify the expression, the inductance of the cable was neglected. Over the frequency range of interest, it is reasonable to neglect this inductance since its reactance is small.) The purpose of probe compensation is clear after briefly studying Equation (4.18). The factor

$$\left(\frac{j\omega R_xC_x+1}{j\omega R_pC_p+1}\right)$$

can be equal to one for all frequencies by adjusting the resistor and capacitor at the probe (and the scope)

$$j\omega R_xC_x+1=j\omega R_pC_p+1 \quad\Rightarrow R_xC_x=R_pC_p \tag{4.20}$$

When (4.20) is satisfied, the gain, $G(\omega)$, is independent of frequency, and the probe response is flat.

Without probe compensation, the gain at low frequencies is independent of the frequency

$$G(\omega)\approx\frac{R_x}{R_x+R_p}\quad\text{if }\omega\ll\frac{1}{R_xC_x}\text{ and }\omega\ll\frac{1}{R_pC_p} \tag{4.21}$$

$C_p := 10\cdot 10^{-12}$ $\quad R_p := 9\cdot 10^{6}$ $\quad R_x := 1\cdot 10^{6}$ $\quad C_x := 35\cdot 10^{-12} + 30\cdot 10^{-12}\cdot 1.83$

$L_c := 0.87\cdot 10^{-6}\cdot 1.83$ $\quad j := \sqrt{-1}$ $\quad R_x\cdot C_x = 8.99\times 10^{-5}$

$x := 10, 10.1 .. 80$ $\quad R_p\cdot C_p = 9\times 10^{-5}$

$$\omega(x) := \left(x + 1 - 10\cdot \mathrm{floor}\left(\frac{x}{10}\right)\right)\cdot 10^{\mathrm{floor}\left(\frac{x}{10}\right)}$$

$$G(\omega) := \frac{\dfrac{R_x\cdot\dfrac{1}{j\cdot\omega\cdot C_x}}{R_x + \dfrac{1}{j\cdot\omega\cdot C_x}}}{\dfrac{R_p\cdot\dfrac{1}{j\cdot\omega\cdot C_p}}{R_p + \dfrac{1}{j\cdot\omega\cdot C_p}} + j\cdot\omega\cdot L_c + \dfrac{R_x\cdot\dfrac{1}{j\cdot\omega\cdot C_x}}{R_x + \dfrac{1}{j\cdot\omega\cdot C_x}}}$$

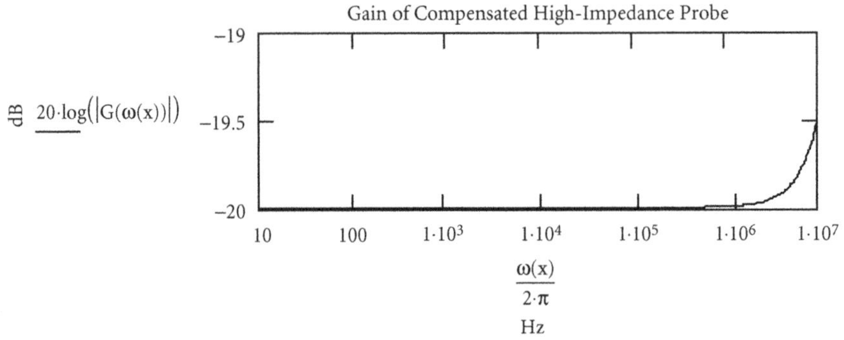

$$20\cdot\log\left(\frac{R_x}{R_x + R_p}\right) = -20 \qquad 20\cdot\log\left(\frac{C_p}{C_p + C_x}\right) = -19.991$$

$$\frac{1}{2\cdot\pi\cdot R_p\cdot C_p} = 1.768\times 10^{3} \qquad \frac{R_x + R_p}{2\cdot\pi\cdot R_x\cdot R_p\cdot(C_x + C_p)} = 1.77\times 10^{3}$$

MATHCAD 4.3 Compensated high-impedance probe.

The frequency must be sufficiently small compared to the cutoff frequency of the probe by itself and the cutoff frequency of the scope network all by itself. As a check, at dc, the capacitors are open circuits and the inductor is a short circuit, and voltage division occurs between the scope and probe resistances. For a 9 MΩ probe and 1 MΩ scope, this gain is 1/10. For measuring higher voltages, a 500 MΩ probe with a gain of about 1/500 can be used. With or without probe compensation, the gain at high frequencies is independent of frequency and is obtained using "capacitor voltage division:"

$$G(\omega) \approx \frac{C_p}{C_p + C_x} \quad \text{if } \omega >> \frac{1}{R_x C_x} \text{ and } \omega >> \frac{1}{R_p C_p} \tag{4.22}$$

For a 10 pF probe and 90 pF scope capacitance, this high-frequency gain is also 1/10. The compensation of a probe should be checked, even for different channels on the same scope: the capacitance between channels can vary (on lower quality instruments).

The analysis in Mathcad 4.3 clearly shows that probe compensation works! The gain is constant over a large frequency range. The capacitance at the scope was increased from 20 to 35 pF to satisfy the equality

$R_x C_x = R_p C_p$. The resonance that is occurring beyond 1 MHz is due to the line inductance that was neglected in (4.19).

The smaller capacitance associated with the 170 Ω line (vs. the standard 50 Ω line) allows the compensation to be controlled entirely at the scope. For example, if the cable capacitance were 97 pF/m, corresponding to the 50 Ω coax previously discussed, the capacitors at both the scope and probe would need to be adjusted for proper probe compensation.

4.8 Testing with a Square Wave

To test the compensation of a passive high-impedance probe, square waves can be sent through the probe, down the line, and examined at the scope. Why is a square wave used instead of a sinusoidal wave? If a 10 MHz square wave is used and the cable is undercompensated, the edges of the wave often overshoot. If a 1 kHz square wave is used instead, the output at the scope will act differently. Undercompensation will round the edges while overcompensation will result in spikes at the edges. Why is this occurring? When a 1 kHz square wave is used as a test signal, the probe is adjusted until the output looks square. With a 10 MHz square wave, however, it is not always possible to obtain a square-shaped output. Why is this? [Smith, '93]

A probe is properly compensated when its gain is constant over frequency. If a sine wave with a particular frequency is used as the test signal, then its amplitude must be accurately known at the input so that it can be compared to the output displayed on the scope. To determine whether the gain of the probe is constant with frequency, the frequency of the sine wave must be varied over the frequency range of interest and the input and output amplitudes compared. This process is very tedious. Note that if a sine wave is used as the test input signal, then the output should also be sinusoidal without any distortion of its shape. Only the amplitude (and phase) can be compared to the input signal.

Square waves have been used as test signals for many years.[5] A review of some elementary signal and systems principles would be helpful before proceeding. A sine wave is a signal with one particular frequency and amplitude. A square wave, however, consists of an infinite number of sinusoidal waves of different frequencies and amplitudes. Thus, a square wave contains more information or intelligence than a sine wave. The Fourier series for a periodic rectangular pulse is given by

$$x(t) = \frac{A\tau}{T} + \sum_{n=1}^{\infty} \frac{A}{n\pi} \sin\left(\frac{2\pi n}{T}\tau\right)\cos\left(\frac{2\pi n}{T}t\right) + \sum_{n=1}^{\infty} \frac{A}{n\pi}\left[1-\cos\left(\frac{2\pi n}{T}\tau\right)\right]\sin\left(\frac{2\pi n}{T}t\right)$$

where A is the amplitude, T is the period, and τ is the width of the pulse. The dc offset or average value of the square wave is the first, frequency independent term, in this expression. The amplitude of this square wave ranges from zero to A. For a square wave to be represented exactly (except at the discontinuities) using sinusoidal waves, the previous equation indicates that an infinite number of terms, (i.e., $n = 1$ to ∞), are required. The frequency of the sinusoid corresponding to $n = 1$ is considered the fundamental frequency (or first harmonic) of the square wave. The frequency corresponding to $n = 2$ is the second harmonic. The $n = 3$ term is the third harmonic. Thus, a square wave consists of a fundamental frequency plus an infinite number of harmonics.

Fortunately, when working with the Fourier series representation of the square wave, an infinite number of terms are not required. As n increases, the amplitudes of these sinusoids decrease. The analysis in Mathcad 4.4 indicates that for this 1 kHz square wave, 100 terms provide a very good representation.

[5]In old TV repair books obtained from the author's grandfather, square waves are recommended as test signals for certain applications.

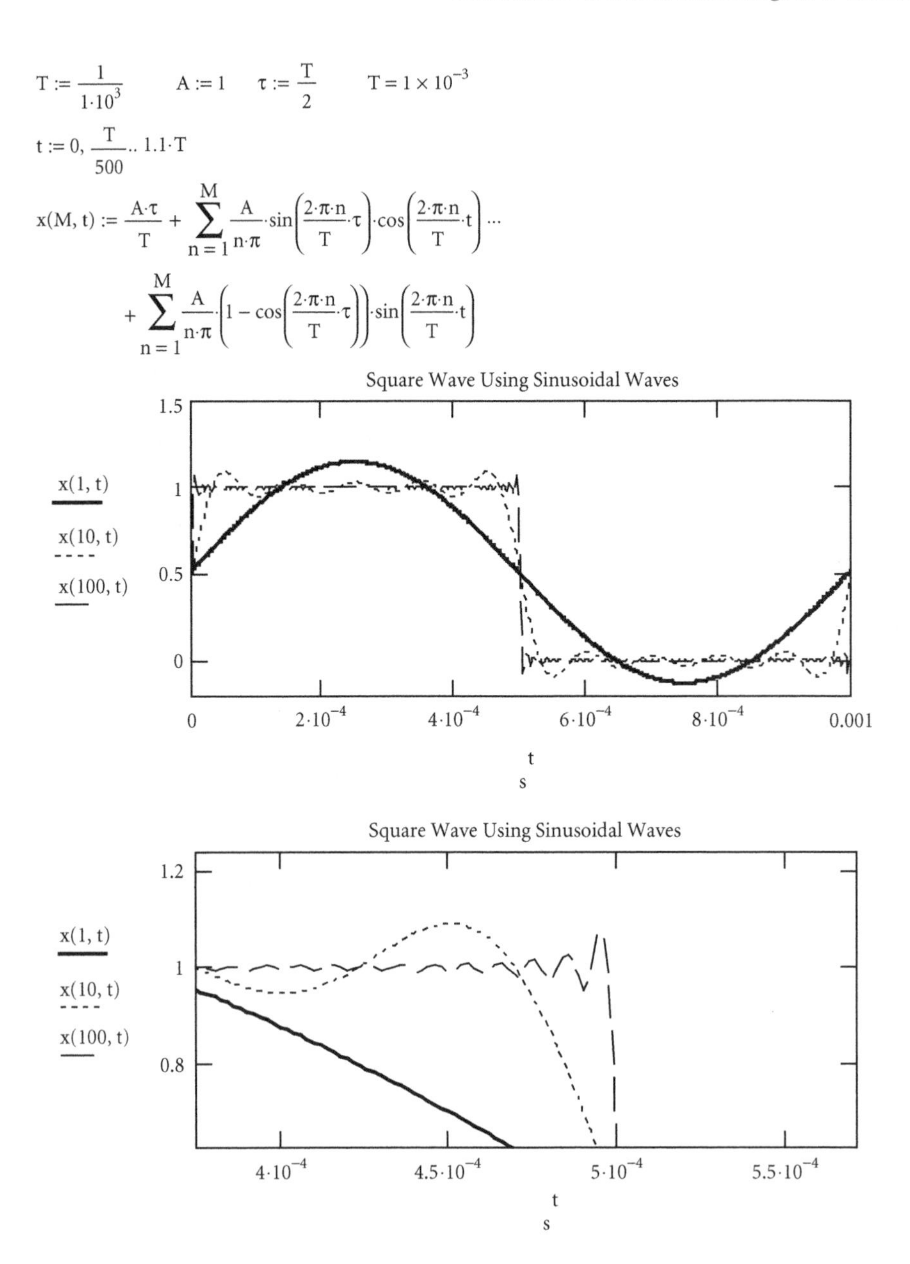

MATHCAD 4.4 Fourier series representation of a square wave using 1, 10, and 100 terms.

The amplitude of each of the frequency components is obtained by taking the square root of the sum of the square of the cosine amplitude and the square of the sine amplitude:

$$c_n = \sqrt{\left[\frac{A}{n\pi}\sin\left(\frac{2\pi n}{T}\tau\right)\right]^2 + \left\{\frac{A}{n\pi}\left[1-\cos\left(\frac{2\pi n}{T}\tau\right)\right]\right\}^2}$$

As is seen in the amplitude spectrum plot given in Mathcad 4.5, the amplitudes after the 20th harmonic, 20 kHz, are small compared to the 1st harmonic, 1 kHz ($n = 1$).

$$T := \frac{1}{1 \cdot 10^3} \qquad A := 1 \qquad \tau := \frac{T}{2}$$

$$n := 0, 1..\, 100$$

$$\mathrm{Amp}(n) := \sqrt{\left(\frac{A}{n \cdot \pi} \cdot \sin\left(\frac{2 \cdot \pi \cdot n}{T} \cdot \tau\right)\right)^2 + \left[\frac{A}{n \cdot \pi} \cdot \left(1 - \cos\left(\frac{2 \cdot \pi \cdot n}{T} \cdot \tau\right)\right)\right]^2}$$

$$c_m(n) := \mathrm{if}\left(n > 0, \mathrm{Amp}(n), \frac{A \cdot \tau}{T}\right)$$

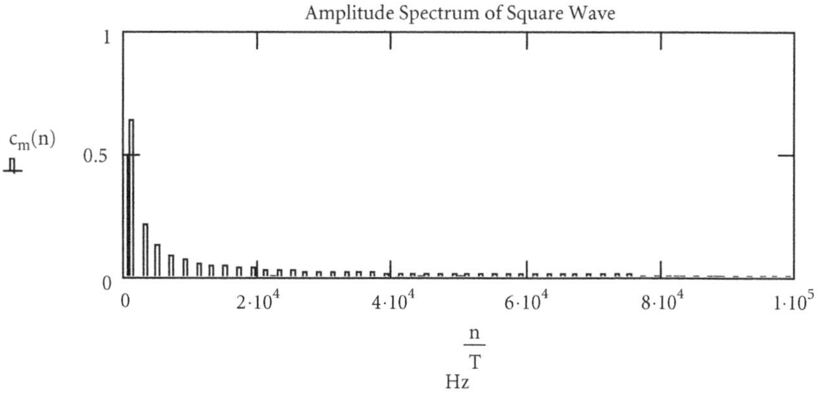

MATHCAD 4.5 Amplitude spectrum of a square wave with a fundamental frequency of 1 kHz.

The response of an improperly compensated probe to a 1 kHz square wave is different from its response to a 10 MHz square wave. The fundamental and major harmonics of a 1 kHz square wave are at different locations in the amplitude spectrum than the fundamental frequency and major harmonics of a 10 MHz square wave. Since an uncompensated high-impedance probe response is a function of frequency, the probe will affect the two square waves differently. Recall that the cutoff frequency of the uncompensated probe previously discussed was about 1 kHz. Since the major harmonics of the 1 kHz square wave exist beyond 1 kHz, distortion is expected.

It is edifying to examine the response of the probe to a 1 kHz square wave. Since the gain expression, $G(\omega)$, was already determined, the output response in the time domain can be written as

$$v(t) = \frac{A\tau}{T} G(0) + \sum_{n=1}^{\infty} 2|F_n| \left| G\left(\frac{2\pi n}{T}\right) \right| \cos\left[\frac{2\pi n}{T} t + \angle F_n + \angle G\left(\frac{2\pi n}{T}\right)\right]$$

An equivalent definition of the Fourier series of a square wave using complex coefficients was used:

$$x(t) = \frac{A\tau}{T} + \sum_{n=1}^{\infty} 2|F_n| \cos\left(\frac{2\pi n}{T} t + \angle F_n\right) \quad \text{where } F_n = \frac{A\tau}{T} e^{-j\frac{\pi n}{T}} \frac{\sin\left(\frac{\pi n}{T}\tau\right)}{\frac{\pi n}{T}\tau}$$

The output response, $v(t)$, was written down by inspection since the individual frequency components and phases of the input signal, $x(t)$, are known, and since the response of the probe to any particular frequency is also known via the gain expression, $G(\omega)$.

The analysis in Mathcad 4.6 for a 1 kHz square wave indicates that for a probe that is undercompensated, (i.e., probe capacitance is too small), the high-frequency components of the square wave are attenuated,

$R_p := 9 \cdot 10^6 \qquad R_x := 1 \cdot 10^6 \qquad C_x := 35 \cdot 10^{-12} + 30 \cdot 10^{-12} \cdot 1.83 \qquad L_c := 0.87 \cdot 10^{-6} \cdot 1.83 \qquad j := \sqrt{-1}$

$T := \dfrac{1}{1 \cdot 10^3} \qquad A := 1 \qquad \tau := \dfrac{T}{2} \qquad M := 100$

$t := 0, \dfrac{T}{500} .. 1.1 \cdot T$

$$RC_{xx}(R_x, C_x, \omega) := \frac{R_x \cdot \dfrac{1}{j \cdot \omega \cdot C_x}}{R_x + \dfrac{1}{j \cdot \omega \cdot C_x}} \qquad RC_{pp}(R_p, C_p, \omega) := \frac{R_p \cdot \dfrac{1}{j \cdot \omega \cdot C_p}}{R_p + \dfrac{1}{j \cdot \omega \cdot C_p}}$$

$$G(\omega, C_p) := \frac{RC_{xx}(R_x, C_x, \omega)}{RC_{pp}(R_p, C_p, \omega) + j \cdot \omega \cdot L_c + RC_{xx}(R_x, C_x, \omega)} \qquad F(k) := \frac{A \cdot \tau}{T} \cdot e^{-j \cdot k} \cdot \frac{\sin(k)}{k}$$

$$v(t, C_p) := \frac{A \cdot \tau}{T} \cdot \frac{R_x}{R_x + R_p} + \sum_{n=1}^{M} 2 \cdot \left| F\left(\frac{\pi \cdot n}{T} \cdot \tau\right) \right| \cdot \left| G\left(\frac{2 \cdot \pi \cdot n}{T}, C_p\right) \right| \cdot \cos\left(\frac{2 \cdot \pi \cdot n}{T} \cdot t + \arg\left(F\left(\frac{\pi \cdot n}{T} \cdot \tau\right)\right) ... + \arg\left(G\left(\frac{2 \cdot \pi \cdot n}{T}, C_p\right)\right) \right)$$

Properly Compensated Probe

$v(t, 10 \cdot 10^{-12})$ — 0.1, 0.05, 0, −0.05; t (s): 0, $2 \cdot 10^{-4}$, $4 \cdot 10^{-4}$, $6 \cdot 10^{-4}$, $8 \cdot 10^{-4}$, 0.001

Under- and overcompensated Probe

$v(t, 5 \cdot 10^{-12})$ —— ; $v(t, 15 \cdot 10^{-12})$ - - - — 0.1, 0.05, 0, −0.05; t (s): 0, $2 \cdot 10^{-4}$, $4 \cdot 10^{-4}$, $6 \cdot 10^{-4}$, $8 \cdot 10^{-4}$, 0.001

MATHCAD 4.6 Properly compensated, undercompensated, and overcompensated high-impedance probe.

resulting in rounded edges. (The high-frequency components are required to generate the sharp transitions at the edges of the wave.) For a probe that is overcompensated (i.e., probe capacitance is too large), the high-frequency components are enhanced, resulting in overshoot in the edges. The fundamental along with the dc component contains most of the energy and amplitude information of the square wave. (The "detailing" of the square wave is accomplished by the harmonics.) Since the fundamental and dc component are not above the cutoff frequency, they are not strongly affected by under- or overcompensation. Thus, when using a 1 kHz square wave as the test signal, the compensator should be adjusted until the shape of the wave is square. If this inspection-style testing were attempted in the frequency domain using a spectrum analyzer, it would be difficult to determine when the wave was square: the spectrum analyzer displays the amplitude of the components vs. frequency.

The fundamental and harmonic frequencies of a 10 MHz square wave are much greater than the fundamental and harmonic frequencies of a 1 kHz square wave. For a 6 ft long cable, the cable is electrically short for frequencies less than about 10 MHz. At the harmonics of the 10 MHz square wave, the transmission line will not appear electrically short. In addition to the gain variation with frequency (the inductance of the cable should be included for these frequencies), there are also reflections due to the mismatch between the line and its load at the scope. Assuming a good 10 MHz square wave is produced by the signal generator, it may not be possible to obtain a "clean" square wave at the scope with sharp edges and without ringing. The compensation of the probe should be an iterative process, adjusting between undercompensation and overcompensation, converging on a reasonable square wave signal at the scope.

4.9 Effect of Inductance on the Probe

The probe tip, ground lead, and circuit board all have inductance. Assume that a passive high-impedance (10X) compensated probe is being used and the total "length" of the three inductances is about 15". The length of the cable connecting the probe to the scope is 6 ft. Plot the gain (in dB) of the probe. What should be done to flatten the response? [Smith, '93]

The 15" of extra wire increase the inductance of the system. If a 20 nH/inch guideline is used, the total additional inductance is

$$L_p = 15''(20\ \text{nH}/\text{inch}) = 0.3\ \mu\text{H}$$

The actual total inductance is a function of factors such as the distance between the leads and the circuit board width. For comparison, 6 ft of the 170 Ω cable (assuming the cable is electrically short) has an inductance of about 1.6 μH. Using (4.16), 6 ft of the cable is electrically short for frequencies less than about 11 MHz. It was seen previously that the inductive effects were negligible below 1 MHz. The inductive effects are not necessarily negligible above 11 MHz. Although the lumped model could probably be used for frequencies slightly greater than 11 MHz, the transmission line model will be used for this example to eliminate any uncertainty and to illustrate another method of attack.

Referring to Figure 4.7, the relationship between the input and output voltage for a lossless transmission line of length l_{th} is

$$\frac{V_s(0)}{V_s(l_{th})} = \frac{1+\rho_L[\cos(2\beta l_{th}) - j\sin(2\beta l_{th})]}{(1+\rho_L)[\cos(\beta l_{th}) - j\sin(\beta l_{th})]} \tag{4.23}$$

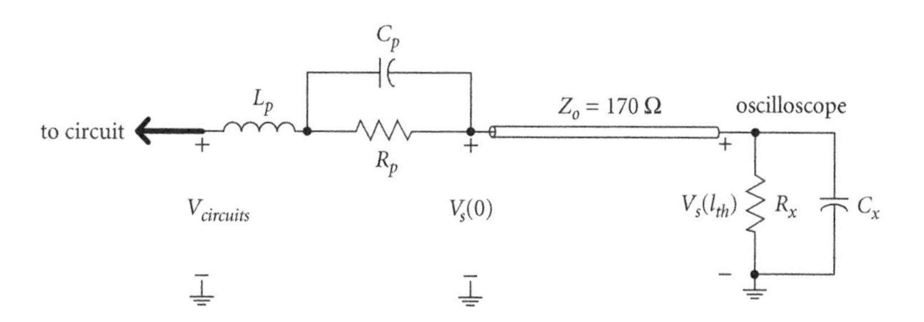

FIGURE 4.7 Modeling the probe, lead, and board inductance.

where the reflection coefficient is

$$\rho_L = \frac{Z_L - Z_o}{Z_L + Z_o} = \frac{\dfrac{R_x}{1 + j\omega R_x C_x} - 170}{\dfrac{R_x}{1 + j\omega R_x C_x} + 170} \tag{4.24}$$

and the phase constant is

$$\beta = \frac{2\pi}{\lambda} = \frac{2\pi}{\dfrac{v}{f}} = \frac{\omega}{\dfrac{1}{\sqrt{LC}}} = \omega C Z_o \tag{4.25}$$

where C is the capacitance per unit length and L is the inductance per unit length of the cable. (The R and G of the cable are not included in this analysis, and Z_o is assumed entirely real. The more realistic lossy cable analysis is left as a learning experience for the reader.) To determine the voltage across the input of the transmission line, $V_s(0)$, of characteristic impedance 170 Ω, it is necessary to determine the impedance looking into the lossless line:

$$Z_{in} = Z_o \frac{Z_L + jZ_o \tan(\beta l_{th})}{Z_o + jZ_L \tan(\beta l_{th})} \tag{4.26}$$

The ratio of the voltage at the input of the cable, $V_s(0)$, to the voltage at the probe, $V_{circuits}$, is

$$\frac{V_s(0)}{V_{circuits}} = \frac{Z_{in}}{Z_{in} + j\omega L_p + \dfrac{R_p}{1 + j\omega R_p C_p}} \tag{4.27}$$

Therefore, the overall gain of the system is

$$G(\omega) = \frac{V_s(l_{th})}{V_{circuits}} = \frac{V_s(l_{th})}{V_s(0)} \frac{V_s(0)}{V_{circuits}}$$
$$= \frac{(1+\rho_L)[\cos(\beta l_{th}) - j\sin(\beta l_{th})]}{1+\rho_L[\cos(2\beta l_{th}) - j\sin(2\beta l_{th})]} \left(\frac{Z_{in}}{Z_{in} + j\omega L_p + \dfrac{R_p}{1 + j\omega R_p C_p}} \right) \tag{4.28}$$

As shown in Mathcad 4.7, the additional inductance has some effect on the frequency response. The inductance of the cable is much greater than this additional inductance. If the extra inductance is increased by a factor of ninety, then the first resonant frequency drops to 10 MHz.

The addition of a resistor in series with the tip of the probe has only a marginal effect on the location of the first resonant frequency of the probe. The resistor does, however, reduce the amplitude of this resonance by decreasing the Q of the circuit. By distributing the resistance along the length of the cable, this lossy cable can more effectively dampen the peaks of the resonance. Resistance in the cable (or probe) will also reduce ringing, which is a time-domain concept.

$C_p := 10 \cdot 10^{-12}$ $\quad R_p := 9 \cdot 10^6$ $\quad R_x := 1 \cdot 10^6$ $\quad C_x := 35 \cdot 10^{-12}$ $\quad L_p := 0.3 \cdot 10^{-6}$ $\quad Z_o := 170$

$x := 60, 60.01 .. 90$ $\quad l_{th} := 6 \cdot 0.305$ $\quad C_c := 30 \cdot 10^{-12}$ $\quad j := \sqrt{-1}$

$$\omega(x) := \left(x + 1 - 10 \cdot \text{floor}\left(\frac{x}{10}\right)\right) \cdot 10^{\text{floor}\left(\frac{x}{10}\right)} \qquad \beta(\omega) := \omega \cdot C_c \cdot Z_o$$

$$Z_L(\omega) := \frac{R_x \cdot \frac{1}{j \cdot \omega \cdot C_x}}{R_x + \frac{1}{j \cdot \omega \cdot C_x}} \qquad \rho_L(\omega) := \frac{Z_L(\omega) - Z_o}{Z_L(\omega) + Z_o} \qquad Z_{in}(\omega) := Z_o \cdot \frac{Z_L(\omega) + j \cdot Z_o \cdot \tan(\beta(\omega) \cdot l_{th})}{Z_o + j \cdot Z_L(\omega) \cdot \tan(\beta(\omega) \cdot l_{th})}$$

$$G(\omega, L_p) := \frac{(1 + \rho_L(\omega)) \cdot (\cos(\beta(\omega) \cdot l_{th}) - j \cdot \sin(\beta(\omega) \cdot l_{th}))}{1 + \rho_L(\omega) \cdot (\cos(2 \cdot \beta(\omega) \cdot l_{th}) - j \cdot \sin(2 \cdot \beta(\omega) \cdot l_{th}))} \cdot \left(\frac{Z_{in}(\omega)}{Z_{in}(\omega) + j \cdot \omega \cdot L_p + \frac{R_p}{1 + j \cdot \omega \cdot C_p \cdot R_p}}\right)$$

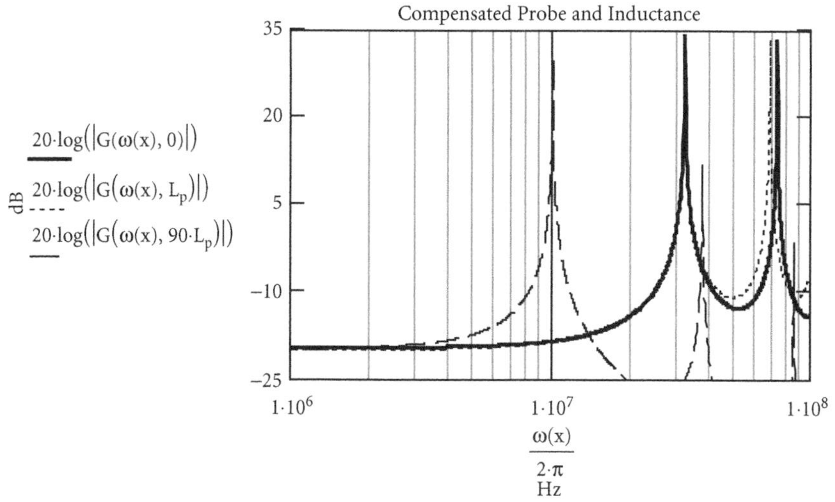

Compensated Probe and Inductance

dB: $20 \cdot \log(|G(\omega(x), 0)|)$; $20 \cdot \log(|G(\omega(x), L_p)|)$; $20 \cdot \log(|G(\omega(x), 90 \cdot L_p)|)$

35, 20, 5, −10, −25

$1 \cdot 10^6$, $1 \cdot 10^7$, $1 \cdot 10^8$

$\frac{\omega(x)}{2 \cdot \pi}$ Hz

MATHCAD 4.7 Compensated probe without and with parasitic inductance.

For this example involving a 170 Ω 6 ft cable, the gain (and input impedance) is not affected much by the lead inductance, L_p. It should, nevertheless, be kept as small as possible. This lead inductance can be a source of ringing. Also, the lead lengths should be kept small to reduce the susceptibility of the probe cable to electrical noise.

5 Cable Shielding and Crosstalk

Near-field interference, or crosstalk, is a major issue in electronic devices and systems. To reduce this crosstalk, as well as far-field interference, transmission lines can be shielded. Because of this topic's complexity, shielding and the grounding of the shield are frequently misunderstood, possibly more than any other concept in interference control. Shielding, however, is not the only method of reducing crosstalk. To help in the understanding and reduction of this crosstalk, in some special electrically-short cases, straightforward coupling models are used.

5.1 Best Cable to Reduce Magnetic Noise

A strong 60 Hz magnetic noise source is present near the following transmission lines: untwisted pair inside a steel tube, untwisted pair, untwisted pair inside a copper tube, twisted pair, twisted pair inside a steel tube, and untwisted pair inside a grounded aluminum tube. Rank these transmission lines in terms of their likely susceptibility to this low-frequency noise.

Classical emission and susceptibility models for conductors clearly show the importance of reducing the length of a transmission line (when the line is electrically short). However, cables can also be shielded and twisted to reduce their emissions and susceptibility.

The most susceptible of the transmission lines listed is the standard untwisted pair (e.g., twin-lead line). The line is obviously electrically short at 60 Hz except for long, power distribution lines. The length of and spacing between the conductors should be as small as possible.

The untwisted pair inside a copper tube has slightly less susceptibility to near-field magnetic noise than the untwisted pair. Although copper is nonmagnetic with a relative permeability of one, a very small induced "bucking" current in the tube will generate a small counter magnetic field.

The susceptibility of the untwisted pair inside a grounded aluminum tube to magnetic fields is almost the same as the untwisted pair inside a copper tube. Actually, since the conductivity of aluminum is 0.61 of the conductivity of copper, and aluminum is also nonmagnetic, the magnitude of the counter magnetic fields would be less. The *grounded* aluminum tube, however, can reduce near-field electric emissions and susceptibility and static charge buildup on the shield. Also, aluminum is currently less expensive than copper. For these reasons, the grounded aluminum tube is marginally favored over the ungrounded copper tube.

The steel tube around the untwisted pair is superior to both the aluminum and copper tube due to steel's magnetic properties. A typical low-frequency value for the relative permeability of steel is 1,000. The high-permeability of steel at these low frequencies has two major consequences: (i) it increases the absorption of the magnetic fields and (ii) it helps redirect the magnetic fields away from the tube's interior.

Interestingly, the twisted pair without any shielding is ranked higher than the untwisted pair in a steel tube. Although twisting the wires may reduce the distance between the wires, the twisting is actually performed for another reason. Recall that a magnetic field will induce a voltage in a loop of wire according

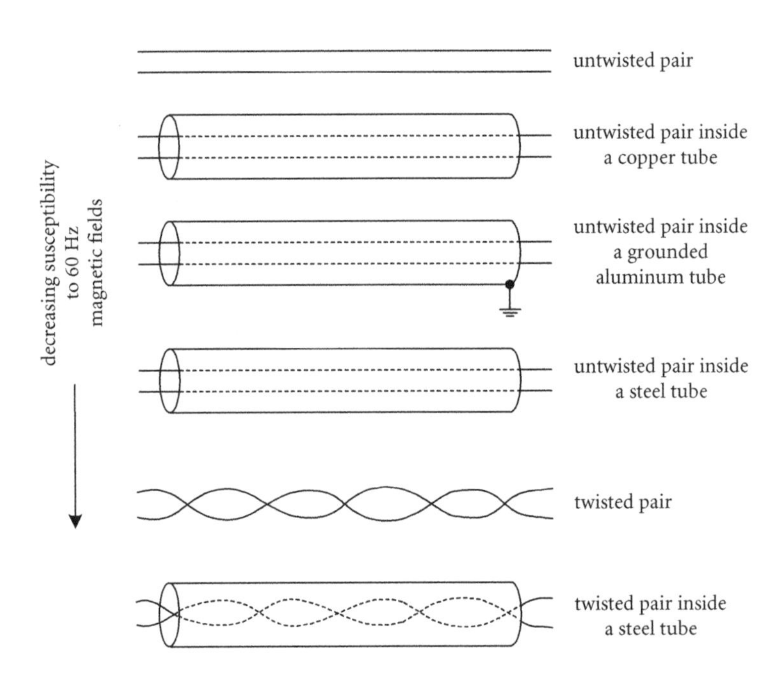

FIGURE 5.1 Six different cables ordered according to their decreasing susceptibility to 60 Hz magnetic noise.

to Faraday's law. The orientation of the loop affects the sign of this voltage. Twisting the two wires forces the induced voltage in neighboring loops to be of opposite polarity. By summing all of the induced voltages from each of the loops generated by the twisting, the net induced noise voltage is significantly less than without the twisting. The sum of the "plus, minus, plus, minus,..." noise-loop voltages is theoretically zero for an even number of loops. Twisting the wire is probably one of the least expensive methods of dramatically decreasing the susceptibility of a cable to magnetic fields.

Of the transmission lines given, the twisted pair inside a steel tube is the least susceptible to magnetic fields. The steel tube absorbs and redirects the magnetic fields. Steel is relatively inexpensive, but it is heavy and has a propensity to rust (unless galvanized). The six transmission lines, ranked from most to least susceptible to 60 Hz noise, are shown in Figure 5.1.

Discuss the reasonableness of the low-frequency noise susceptibilities given in Table 5.1. [Bell, '70]

The analysis of these different cable configurations will illustrate the importance of proper grounding. (In Table 5.1, the shields are nonmagnetic.) The least immune, or most susceptible, cable of those listed to low-frequency noise (both electric and magnetic) is the coaxial structure where both the source and load are grounded and the shield is only connected to the source ground. Although the grounding of the shield will reduce electric field emissions from the center conductor, the outer shield has virtually no influence on the magnetic field susceptibility. Referring to Figure 5.2, the return current for both the signal and noise must be via the ground path between the load and source. This current-path loop can be large. The effective area for either magnetic emissions or magnetic pickup can therefore also be large. This closed current path through the ground is referred to as a ground loop.

It may be a surprise that the shielded twisted pair with both the source and load grounded is a mere 2 dB less susceptibility to low-frequency noise. Although the grounding of the shield will reduce electric field emissions, many of the advantages of using twisted pair are lost since the load and source are not balanced: the load and source are both single-ended grounded. The return current path is divided between the return conductor of the twisted pair and the ground plane as shown in Figure 5.3. At low frequencies, a majority of the current tends to return via the low-impedance ground plane. At higher frequencies, the current tends to return along a path nearest to the forward signal current — the twisted pair conductor.

When the shield or outer conductor of the coaxial cable is grounded at both ends, the return current will divide between the shield and ground plane. If the frequency of the signal is much greater than the

TABLE 5.1 Susceptibilities of Various Cable Configurations [Bell, '70]

Susceptibility (dB)	Cable Type	Ground-Shield Connections
0	Coax	Source and load grounded, shield at source grounded, load and shield not connected 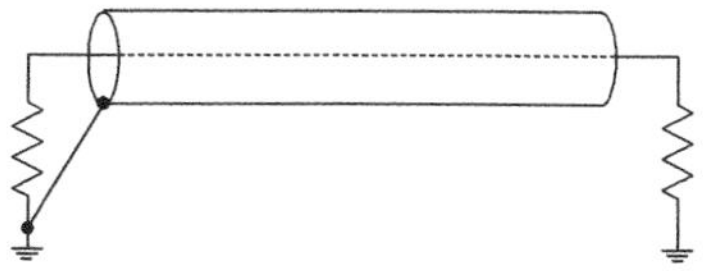
–2	Shielded twisted pair	Source and load grounded, shield at source grounded, load and shield not connected 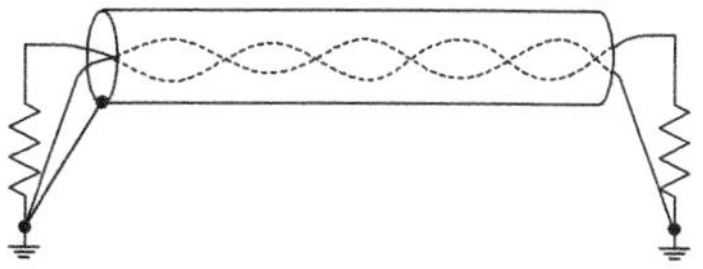
–5	Coax	Source and load grounded, shield at source and load grounded 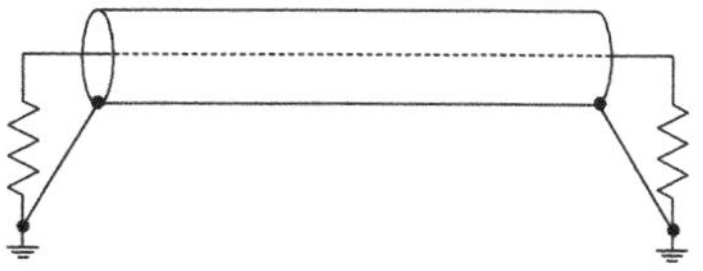
–49	Twisted pair	Source grounded 
–57	Coax	Source grounded, shield at source grounded 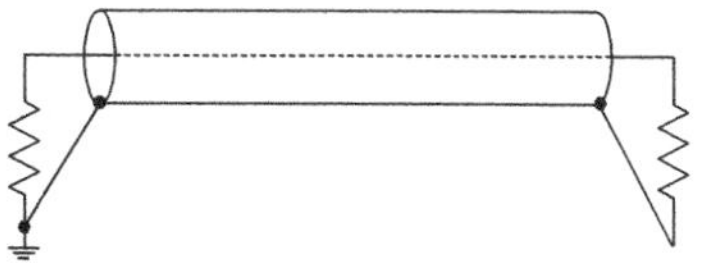
–64	Shielded twisted pair	Source grounded, shield at source grounded, load and shield not connected 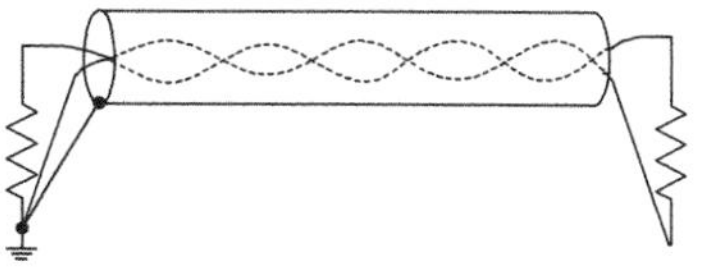
–64	Shielded twisted pair	Source grounded, shield at source and load grounded, load and shield not connected 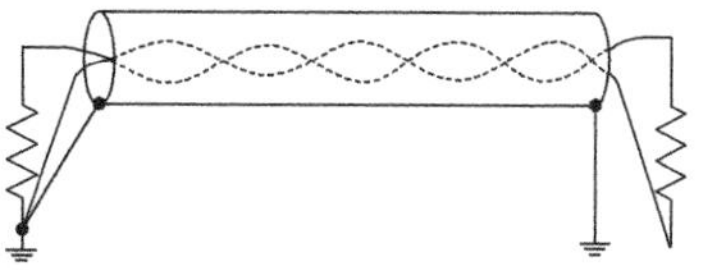

(*Continued*)

TABLE 5.1 Susceptibilities of Various Cable Configurations [Bell, '70] (Continued)

Susceptibility (dB)	Cable Type	Ground-Shield Connections
−71	Shielded twisted pair	Source grounded, shield at source grounded, load and shield connected 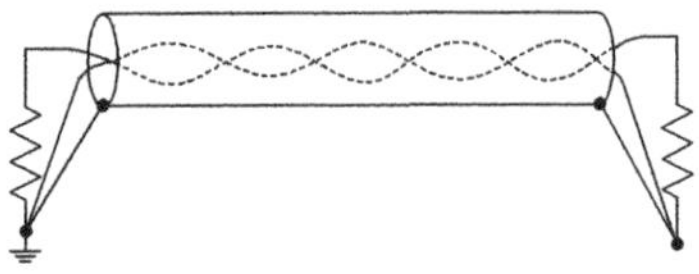 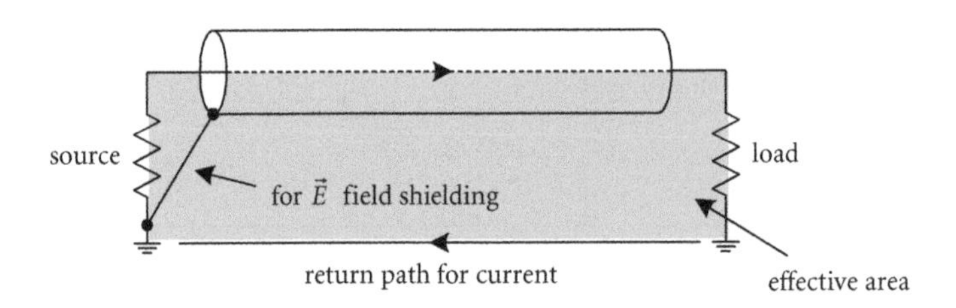

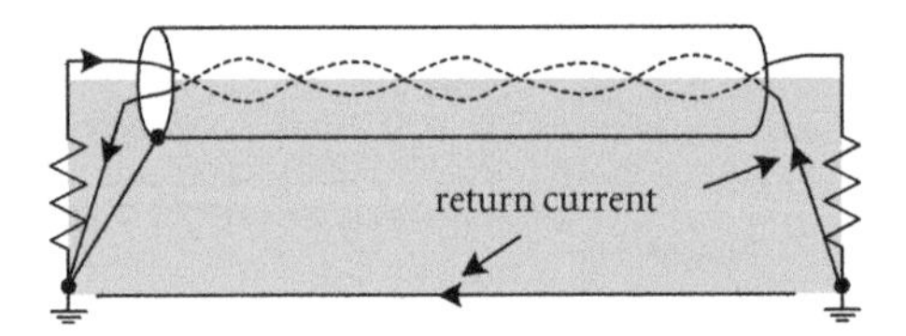

FIGURE 5.2 Effective area for one coaxial cable configuration.

FIGURE 5.3 Return current paths for one shielded twisted pair configuration.

cutoff frequency of the shield, then most of the current will return via the shield. If the frequency of the signal is much less than the cutoff frequency of the shield, then most of the current will return via the ground plane. This situation is slightly better than the shielded twisted pair arrangement because of the lower resistance of the shield relative to the return conductor of the twisted pair.

A dramatic improvement in performance is obtained when a twisted pair is used and the load is balanced.[1] A floating load with neither end of the load connected to ground is shown in Figure 5.4. At low frequencies,

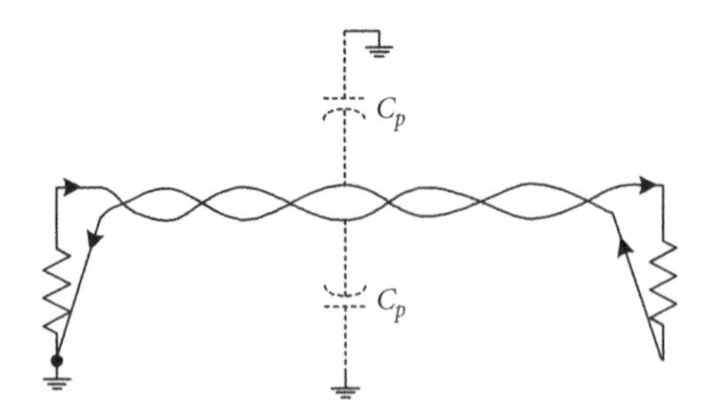

FIGURE 5.4 Stray or parasitic capacitance from each conductor to ground.

[1]However, the entire system is not balanced since the source is single-ended grounded.

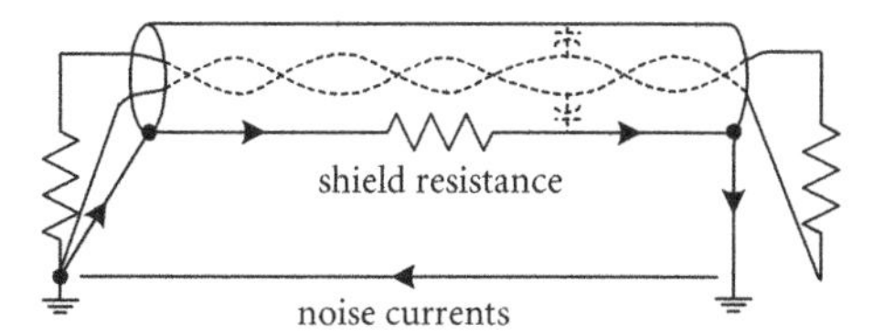

FIGURE 5.5 Noise current passes through the shield and returns through the ground path. Most of the signal current, not shown, returns via the return of the twisted pair conductor.

the parasitic coupling between the load and nearby grounds should be small, and the signal current should mostly return via the return conductor of the twisted pair. The effective pickup area of the complete current path is small, and the twisting produces alternating polarity induced voltages in each of the loops. Since there is no surrounding shield, there is no electric field shielding. However, if the line is balanced, the capacitive coupling (i.e., near-field electric coupling) to each line should be about the same, and the net electric-field induced noise across the load should be negligible.

When the twisted pair in Figure 5.4 is replaced with a coaxial cable, according to the table, there is an improvement in the susceptibility of the cable. As with the twisted pair, there is no ground loop. In addition, the outer conductor, which surrounds the inner conductor, acts as an electric field shield. Because of the symmetry of the coaxial structure (the inner and outer conductors have the same axis), the magnetic fields emanating from a signal on this cable are small. Ideally, outside the cable, the magnetic field from the current-carrying inner conductor is equal but opposite to the field from the current-carrying outer conductor when the two currents are equal but opposite. In addition, the coefficient of magnetic coupling between the inner and outer conductors is close to one, and the induced noise voltage along the outer conductor is close to the induced voltage across the inner conductor if the frequency is not too low. As a result, the net induced voltage across the load, the difference in these voltages, is small.

Shielded twisted pair with the source grounded, load floating, and shield grounded only at one end provides both electric and magnetic field shielding. The signal current returns via the twisted pair since the load is floating. There are no signal currents on the shield, only noise currents. At low frequencies, grounding the shield at both ends provides no real advantage. At higher frequencies, the length of the shield may be electrically long, and multiple grounds could help prevent standing waves from forming on the shield. In Figure 5.5, the shield is grounded at both the source and load. A potential disadvantage of multiple ground points is that noise currents can exist along the shield. In addition, since the shield has a nonzero impedance, the noise voltage can also vary along the shield. Since capacitive coupling exists between the shield and each of the twisted pair conductors, noise will be induced across the load unless the line is perfectly balanced (the capacitive coupling from each conductor to the shield should be nearly the same). This noise will be a function of the distance along the line.

When the load is floating and the shield is connected to the load as shown in Figure 5.6, the symmetry of the concentric shield is utilized. Furthermore, the voltage at the load end of the shield is closer to the voltages along the twisted pair conductors at the load. The currents from the shield to the twisted pair conductors via the parasitic capacitance are less since the voltage across the parasitic capacitance is smaller. The disadvantage of this connection is that any noise current that couples into the system can now exist

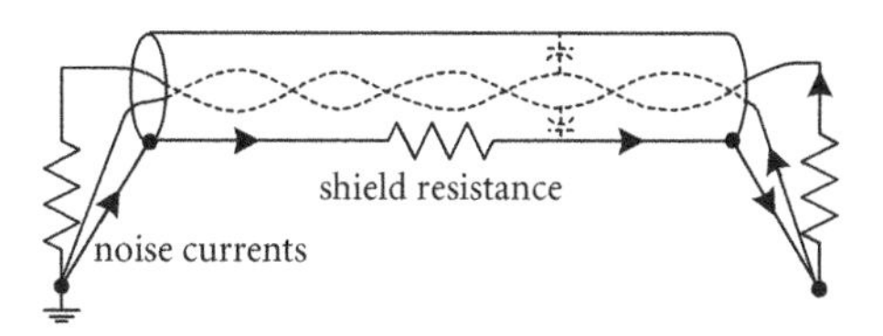

FIGURE 5.6 Interruption of some of the noise current by removing one shield ground.

on the shield and signal return. The conservative approach is to avoid noise currents on the signal and return conductors even when the system is (partially) balanced.

5.2 Connecting Balanced and Unbalanced Systems

Provide recommendations on how to connect balanced and unbalanced systems using shielded two-wire cable. If balanced and unbalanced two-wire cable were used, how would the ground connections change? [Macatee; Morrison, '98; Ott]

In many real-life applications, cables connect two or more devices that may have different grounds. Furthermore, the drivers and receivers for these devices may be balanced, unbalanced, or some combination of balanced and unbalanced. This discussion focuses on different methods of connecting two systems together using a shielded two-wire cable, also referred to as three-conductor cable.[2] The rankings given in Table 5.2 through Table 5.5 are only recommendations (not hard rules) to increase the probability

TABLE 5.2 Connections for a Balanced Driver and Balanced Receiver [Macatee]

Rank	Configuration	Comments
1	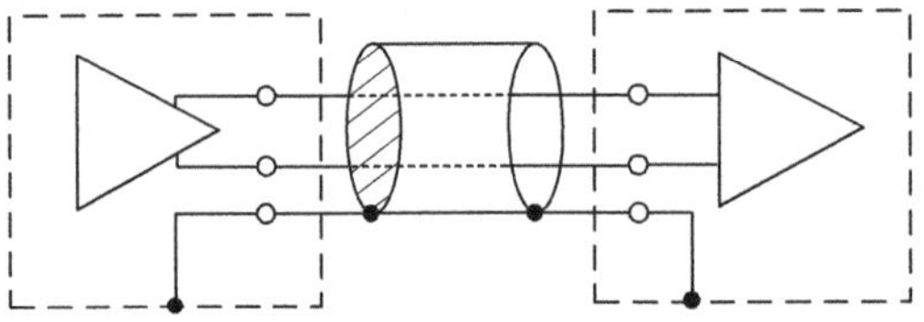	Nearly ideal shielding
2	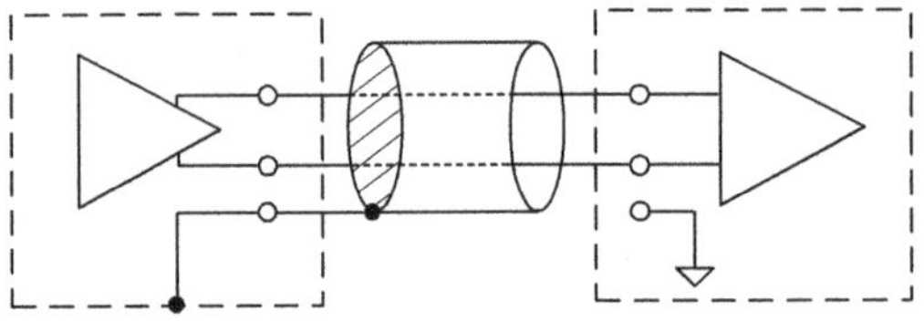	Interrupting the shield current by connecting the shield to only one chassis
2	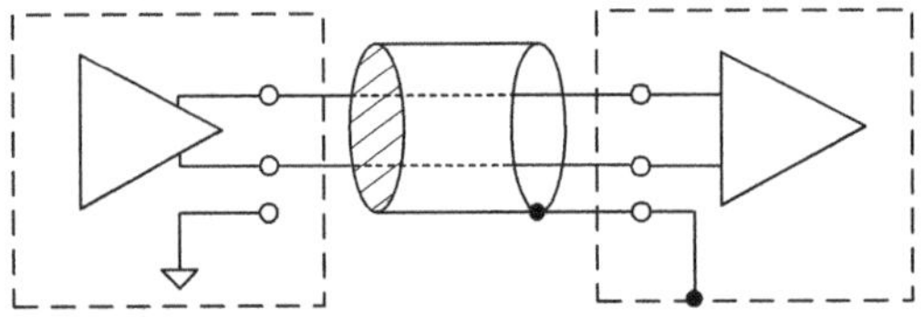	Interrupting the shield current by connecting the shield to only one chassis
3	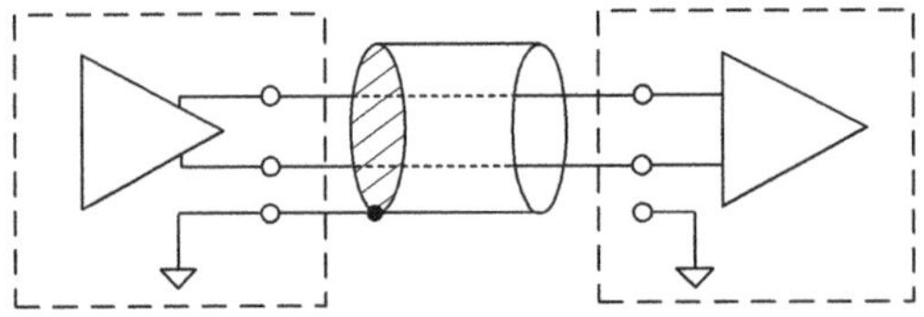	Cable shield connected to signal ground at driver only
3	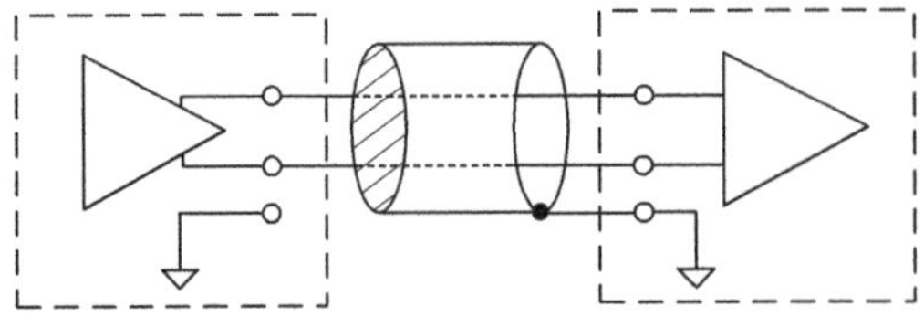	Cable shield connected to signal ground at receiver only

[2]The third conductor is the shield.

of having a low-noise system. They do not guarantee a low-noise system. Although there are other more costly and potentially more effective approaches to connecting systems, such as with isolation transformers, additional shielding, and instrumentation amplifiers, these methods will not be discussed here.

When a shielded two-wire cable, such as shielded twisted pair, connects two systems, the first question typically asked is, "Where should the shield be connected to the ground?" Before recommendations are given, the two different types of ground used in this discussion will be reviewed. The first ground is the signal common, signal reference, or signal return represented by the symbol shown in Figure 5.7. Signal current must return through some path that includes this symbol. The chassis ground or connection to the chassis is represented by the (rake implement) symbol shown in Figure 5.8. The chassis, for safety reasons, is often connected to the earth ground. Frequently, the signal ground is connected to its respective chassis ground at one point. In the tables, the chassis are represented by the dashed line surrounding the driver and receiver. Since the connection to either chassis is clearly shown, the use of the chassis ground symbol would be repetitious. Unless shown, the chassis are not necessarily connected together and their voltages are not necessarily the same. It is very important to note that sometimes the signal return and chassis (and shield) grounds are connected inside the equipment. Users may not be aware of these connections, resulting in disruptive or annoying noise problems. The following tables clearly indicate the signal return and chassis connections.

FIGURE 5.7 Symbol for the common, signal reference, or signal return.

FIGURE 5.8 Symbol for the chassis ground.

Drivers and receivers are generally classified as either balanced or unbalanced. For this particular discussion, a balanced driver and receiver are represented, respectively, by the symbols shown in Figure 5.9. An unbalanced driver and receiver are represented, respectively, by the symbols shown in Figure 5.10.

In these discussions, it is assumed that the cable connecting the driver to the receiver is electrically short. Audio cables are typically (or nearly always) electrically short at audio frequencies. However, the cables may not be electrically short for higher frequency noise signals (e.g., RF) that may exist in the environment. In this case, connecting the cable shield at only one chassis (or even at both chassis) may not be sufficient to prevent the transmission line from acting like an antenna. A hybrid grounding method may be necessary for dealing appropriately with these higher frequency signals.

Only the balanced nature of the last stage of the driving device and first stage of the receiving device is shown in the tables. It is assumed that there are three output lines from the driver and three input lines into the receiver. As is often the case, one of the output lines from the driver and one of the input lines into the receiver (the "ground" conductor) may already be connected to either the chassis ground or signal return. This would limit the number of options available for the "best" set of connections.

FIGURE 5.9 Symbols for a balanced driver and receiver.

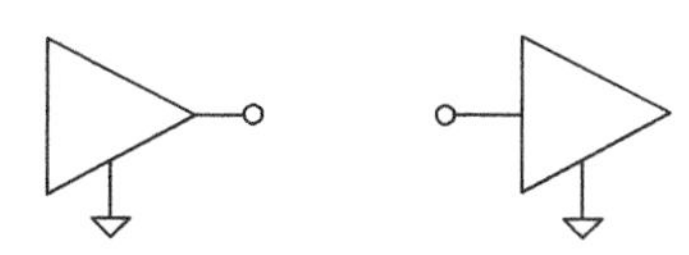

FIGURE 5.10 Symbols for an unbalanced driver and receiver.

Although the same signal return ground symbol is used for both the driver and receiver, these grounds are not necessarily connected unless specifically shown. The signal ground for both the driver and receiver systems is defined as the internal 0 V reference for each system. Each of the four different balancing scenarios will now be discussed.

If the output of one device contains a balanced driver and the input of the other device contains a balanced receiver, then the nearly ideal shielding scheme is the connection of the cable shield to both of the chassis at both ends. If there are no pigtails present, then this shielding configuration is nearly ideal since the entire system is enclosed by one continuous metal shell. The cable can be viewed then as an extension of the chassis. This configuration shown in Table 5.2, which is ranked 1st in performance, is not perfect since the two chassis might be at two different potentials. (Also, the system can still be corrupted and unbalanced by other sources.) If the chassis are connected to the earth ground via the power cord's green/safety conductor, for safety reasons, then the two connections to ground may be at different locations and different potentials.[3] When the chassis are at different potentials, ground current will pass between the chassis and through the cable shield, forming a ground loop. Depending on factors such as the frequency, this current can appear on the inside of the cable shield and chassis. This ground loop current and the field it generates can also interfere with other systems that are not completely shielded.[4] Although the driver and receiver are shown as balanced, they cannot be perfectly balanced, and some of this inner shield current can induce noise into the circuit. Because the length of the connecting cable is often large compared to the largest dimensions of the two chassis, the balancing of the wires inside the cable is probably most critical. However, any current that is present along the inner surface of the two chassis can also induce noise in other nonbalanced (and other nonideal balanced) circuits inside either chassis.

The 2nd ranked configurations connect the cable shield to only one of the chassis. If only one of the chassis are connected to their respective signal grounds, then the other signal ground should not be connected to the cable shield. Otherwise, noise currents induced on the chassis and cable shield would travel inside the device through this signal ground connection, partially defeating the purpose of the shielding. Although the two outputs and two inputs of the balanced driver and receiver are not connected to these signal grounds, the rest of the circuit will probably contain unbalanced circuitry that would use a signal return or ground. When both the driver and receiver chassis are connected to their respective signal grounds, then there is no clear best solution. Usually, the cable shield is *not* left floating. If there is no convenient way of connecting the cable shield to either chassis, the cable shield is connected to one of the signal grounds as shown in the 3rd ranked configurations. Usually, to avoid excessive noise currents on the signal grounds, the cable shield is not connected to both of the signal grounds; that is, one end of the cable shield is left disconnected or not tied to anything. Since the signal ground is likely connected to the chassis ground at a single point, the cable shield is at a low potential. There is a potential benefit of connecting the shield to the signal ground at both ends: the path of the noise current is well known and the crosstalk *might* be better controlled.

In Table 5.3, the driver is unbalanced while the receiver is balanced. Connecting an unbalanced driver to a balanced receiver will decrease the common-mode rejection ratio (*CMRR*) of the entire system. As with completely balanced systems, potentially troublesome noise current can exist along the inner surface of the cable shield and chassis. Since the driver and receiver combination is not balanced in this case, this noise current can couple into the system more easily than in the case with the fully balanced system. For this reason, interrupting the conducting path for these noise currents can result in lower noise levels. Therefore, either of the 2nd ranked configurations may result in lower noise levels than the 1st ranked system. If the option of connecting the cable shield to one chassis is not available, then the cable shield should be connected to one of the signal grounds. Generally, the shield should not be left floating. It is probably a

[3]A current can also be induced in the cable shield and the chassis by external fields through the nonzero pickup loop area generated by the ground path.

[4]For this reason, some experts recommend grounding electrically-short shields on transmission lines at only one end.

TABLE 5.3 Connections for an Unbalanced Driver and a Balanced Receiver [Macatee]

Rank	Configuration	Comments
1	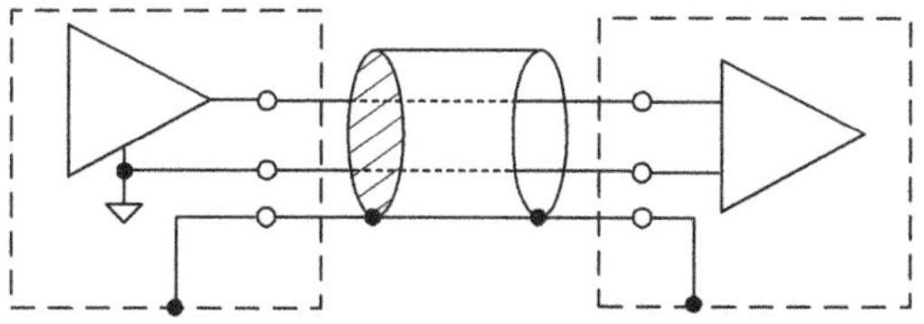	Pseudo-balanced, shield is connected to both chassis
2	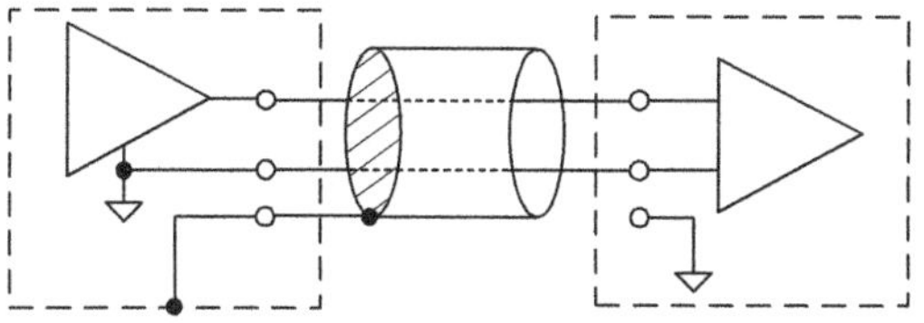	Interrupting the shield current by connecting the shield to only one chassis
2	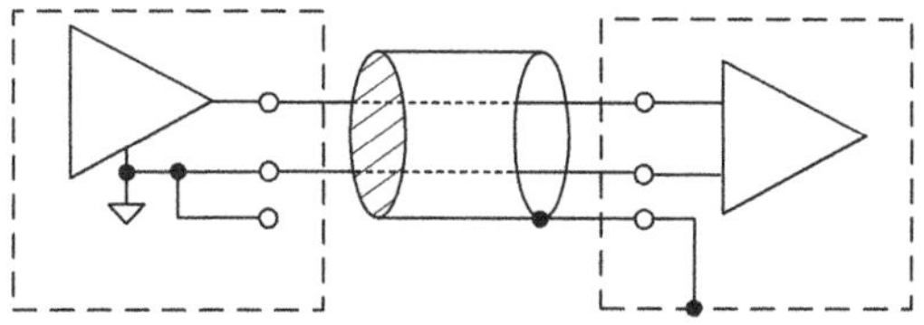	Interrupting the shield current by connecting the shield to only one chassis
3	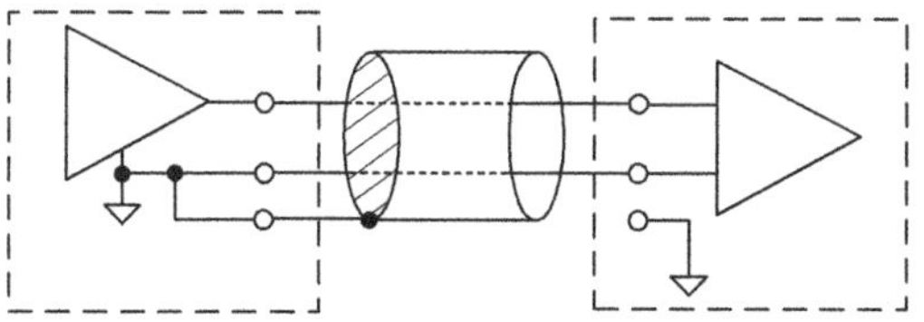	Cable shield connected to signal ground at driver only
3	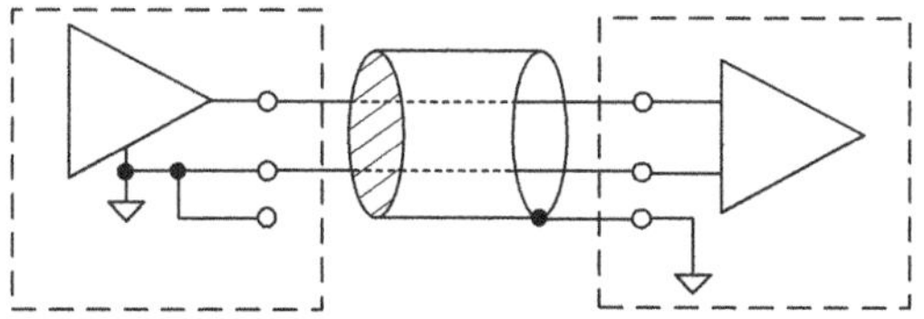	Cable shield connected to signal ground at receiver only

good idea not to connect the cable shield to the signal grounds at both the driver and receiver. Not connecting the shield to both signal grounds will tend to reduce the noise currents on the signal grounds by interrupting the conducting path. The 3rd ranked configuration is commonly used, especially when the actual grounding connections inside the driver and receiver are not known with certainty.

When a balanced driver connects to an unbalanced receiver, the unbalanced receiver easily amplifies noise that is picked up by the system. With balanced receivers, the receiver rejects some of the common-mode noise. For this reason, a balanced driver and unbalanced receiver combination can be very troublesome. As with the previous configurations, the 1st ranked configuration shown in Table 5.4 involves connecting the cable shield at both ends to the chassis. Again, noise currents on the cable shield can induce noise on the two conductors inside the cable. To reduce this noise or crosstalk coupling to the cable's inner conductors, the cable length should be minimized. If this noise level is too great, it may be necessary to disconnect the cable shield from one of the chassis as shown in the 2nd ranked configurations. If connection of the cable shield to a chassis is not possible, then the cable shield can be connected to either signal ground (but not usually both) as shown in the 3rd ranked configurations. Care should be taken when connecting either output of the balanced driver to the

TABLE 5.4 Connections for a Balanced Driver and an Unbalanced Receiver [Macatee]

Rank	Configuration	Comments
1	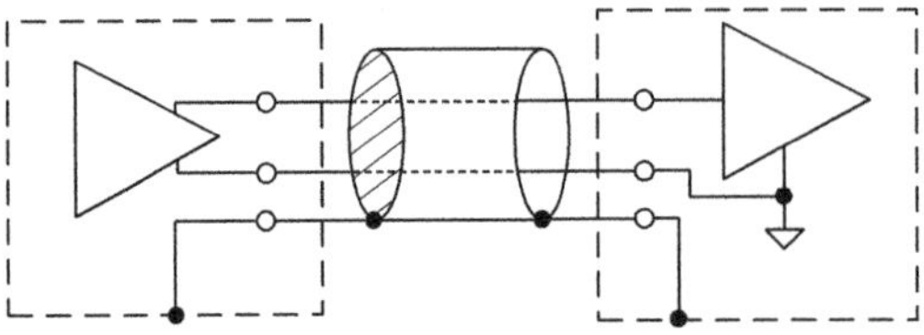	Cable shield extension of the chassis
2	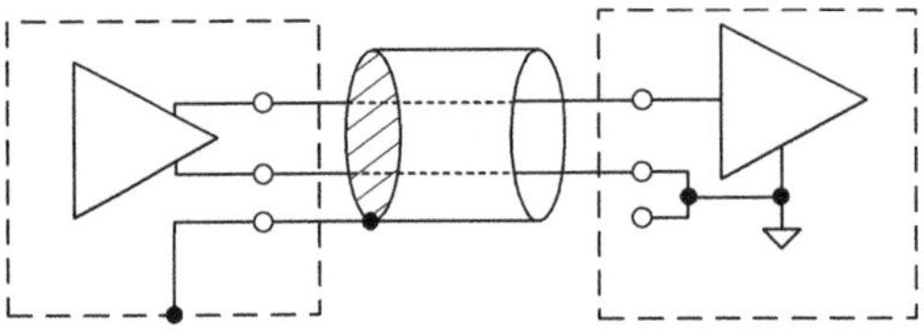	Interrupting the shield current by connecting the shield to only one chassis
2	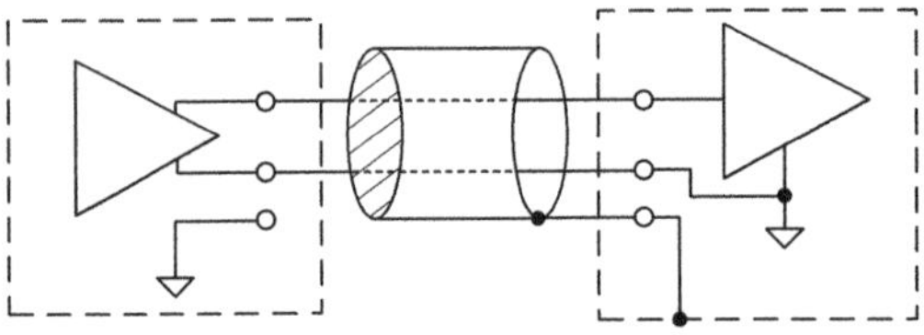	Interrupting the shield current by connecting the shield to only one chassis
3	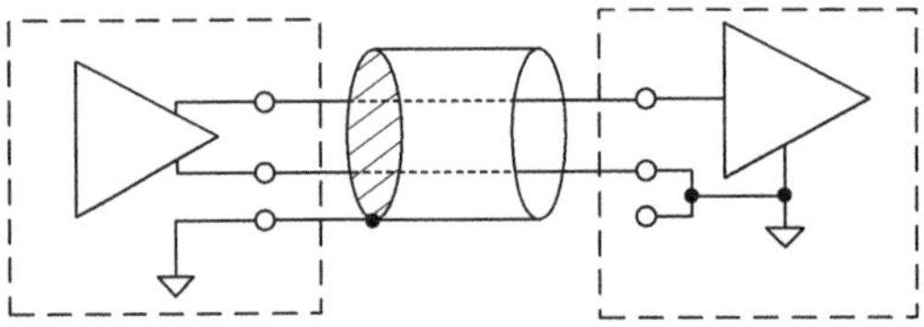	Cable shield connected to signal ground at driver only
3	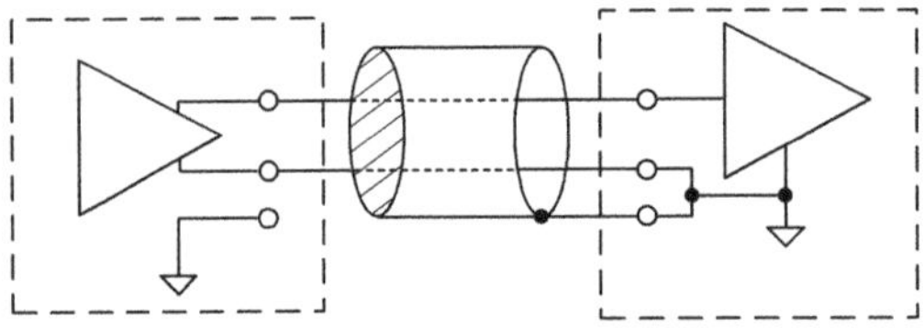	Cable shield connected to signal ground at receiver only

signal ground of the unbalanced receiver. Essentially, the corresponding output is short circuited by this connection, which can damage the output device and cause distortion. Even connecting the "–" output of the balanced driver to the signal ground of the receiver can be troublesome. Balanced floating drivers are available to alleviate some of these "shorting" of the output problems. Although using a balanced driver can help in reducing the emissions from the system, it does not help (much?) for decreasing the susceptibility of the system.

Typically, when an unbalanced driver is connected to an unbalanced receiver, coaxial cable or other two-conductor cable (not three-conductor cable) is used. However, if shielded two-wire cable is used, Table 5.5 can be referred to for grounding recommendations. The previous rationale also applies for these connections.

Generally, shielded two-wire cable when properly connected will be less susceptible to noise and have lower field emissions than unshielded two-conductor cable. The balance nature of the cable is also an important factor. Shielded two-wire cable, such as shielded twisted pair, is often balanced. To help in the balancing of the system, when connecting a balanced driver to a balanced receiver, it is highly recommended

TABLE 5.5 Connections for an Unbalanced Driver and an Unbalanced Receiver [Macatee]

Rank	Configuration	Comments
1	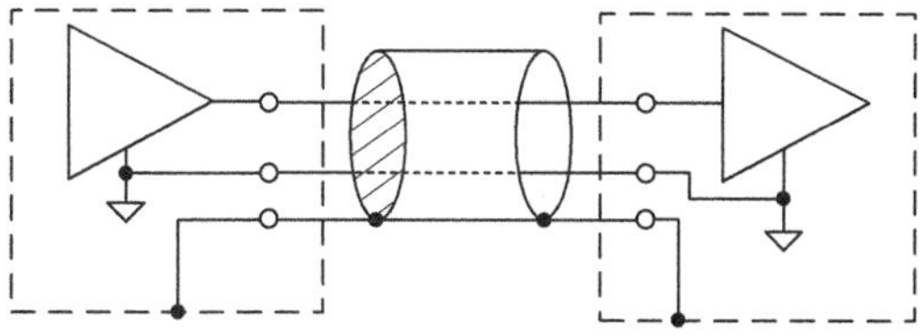	Cable shield extension of the chassis
2	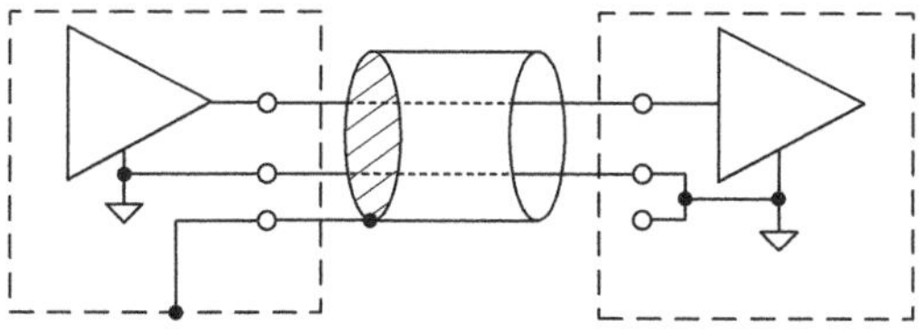	Interrupting the shield current by connecting the shield to only one chassis
2	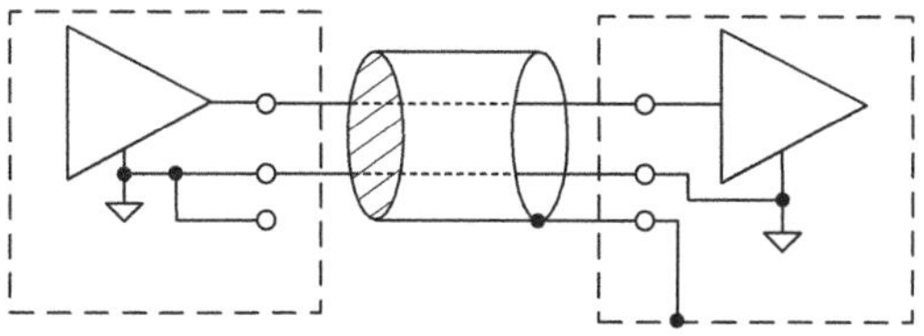	Interrupting the shield current by connecting the shield to only one chassis
3	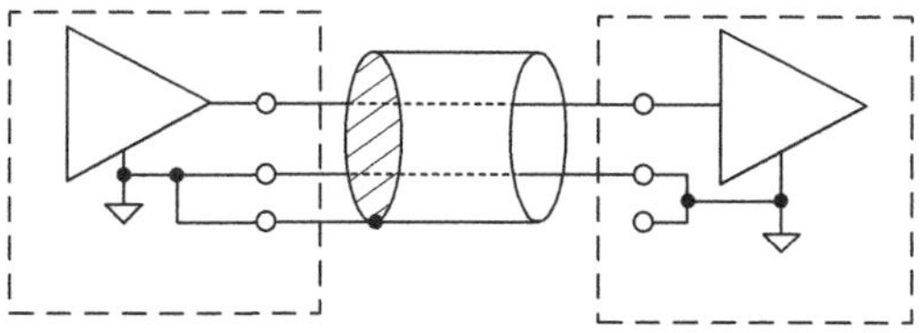	Cable shield connected to signal ground at driver only
3	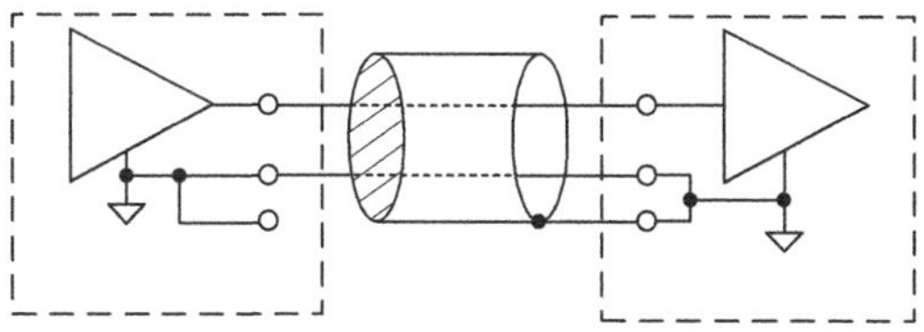	Cable shield connected to signal ground at receiver only

that the connecting cable also be balanced. Although shielded balanced cable is recommended, the following comments are provided in those cases where two-conductor cable is used.

Twisted pair that is not shielded is considered a two-conductor cable. Twisted pair is considered a balanced cable. When balanced twisted pair is used to connect a driver enclosed by a metal chassis to a receiver enclosed by a metal chassis, the cable shield conductor is not present to "continue" the metal enclosure of the chassis. Although not all possible combinations will be discussed, generally when a balanced driver is connected to a balanced receiver, neither side of the twisted pair should be connected, if possible, to either the chassis or signal grounds; otherwise, the balance of the system will be affected. When twisted pair connects an unbalanced driver to a balanced receiver, one conductor of the twisted pair must be connected to the signal ground at the driver to provide a return path for the driver current. Since the signal ground is likely connected to the chassis at one point, this may imply that one conductor of the cable is connected to one chassis (but not both). Neither the signal nor the chassis ground at the balanced receiver should be connected to either of the two twisted-pair conductors. Otherwise, a conductive path is available for any noise currents on the ground system. When twisted pair connects a balanced driver to an unbalanced receiver, one conductor of the twisted pair must be connected to the signal ground at the receiver. Additional connections to ground are normally avoided. Finally, when twisted pair connects an unbalanced driver to an unbalanced receiver, one conductor of the twisted pair must be connected to the signal ground at both the driver and receiver. This implies that the signal (and noise currents) will return on both the cable return conductor and signal ground path.

When systems are connected with coaxial cable, the outer conductor is acting like a shield. Unfortunately, this outer conductor is also the signal return conductor. Coaxial cable has the advantage of providing some shielding, but coaxial cable has the disadvantage of being an unbalanced cable. For this reason, when coax is used to interconnect a balanced driver and receiver, it will tend to decrease the balance of the system more than twisted pair. The outer conductor of coax should be used as the signal return conductor. For this reason, the outer conductor or shield of the coax should be connected to something on both ends and not left floating. The previous recommendations for unshielded twisted pair can be applied to coax. The two conductors for the coax are the inner conductor and outer conductor. The only major difference is that the outer conductor or shield is normally connected to the more negative side of the driver and receiver.

5.3 Bicoaxial Line

In an attempt to reduce the magnetic pickup of two long parallel conductors, each conductor is replaced with a coaxial cable, and the shield of each cable is grounded at the source (the receiver or load is floating). Discuss the effectiveness of this cabling scheme. [Ott]

Often to reduce the emissions and susceptibility of a cable, the system is designed to be balanced.[5] If both the driver (or transmitter) and receiver are balanced, to ensure that the overall system remains balanced, the connecting transmission line should also be balanced. A simple twin-lead or parallel line would not necessarily be balanced because the capacitance from each line to ground (and other nearby objects) would not necessarily be equal. When twisted wire is used between the driver and receiver, the capacitances from each wire to ground (and their positions to ground and other nearby objects) are very similar. Although coaxial cable is considered unbalanced, it can be used in place of each conductor to help keep the system balanced as shown in Figure 5.11. (A similar arrangement of coaxial cables can also be used to construct a balanced probe.) Even though the capacitance of the outer conductor to ground for each of the coaxial cables may be different, the capacitances from the inner conductor of each coax to their respective outer conductor are very similar. Furthermore, by grounding the outer conductor or shield of each coax as shown, the capacitance between the outer shields to ground will be eliminated. The system will appear more balanced with this set of coax than if only twin-lead line were used.

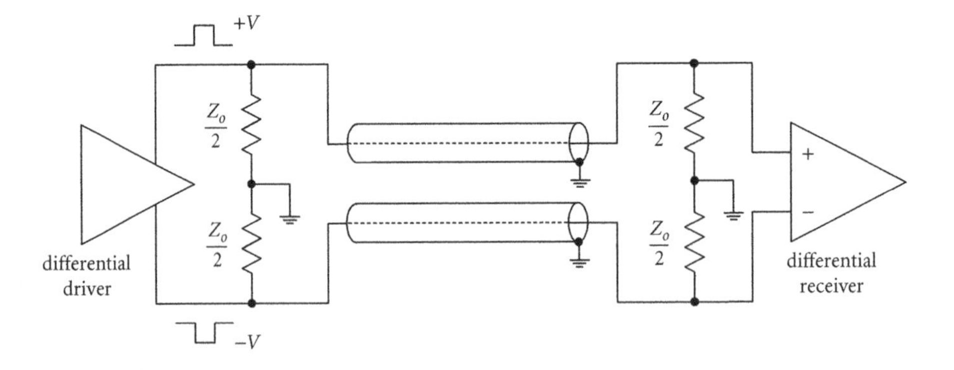

FIGURE 5.11 Balanced line using a pair of coaxial cables.

[5]Briefly, a balanced transmission line is one where the currents in each conductor are equal but opposite in direction and the conductors are symmetrical relative to ground and their environment.

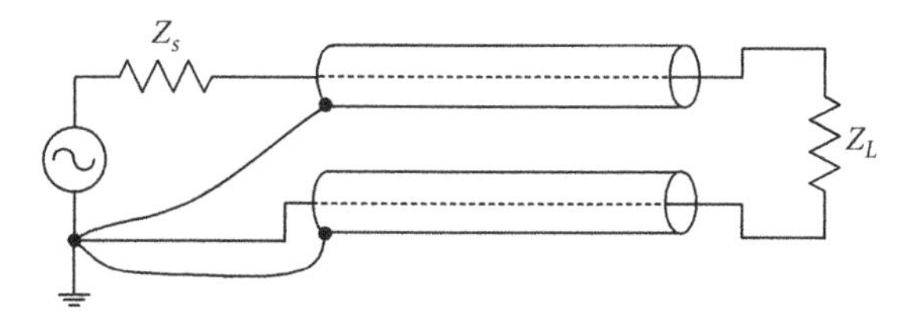

FIGURE 5.12 Pair of coaxial cables connected between an unbalanced source and a floating load.

Now, when the load is floating and the source is single-ended grounded, replacing the parallel wires with a pair of coax should also help keep the capacitance of each center conductor to ground about the same. The outer conductor of each coax should be grounded at the source since grounding at the load might allow ground loop currents to pass on the outer conductors of the coax. If the shields of the coax were only grounded at the load, there would be no direct conductive path between the source and load grounds via the outer shields of the coax; however, a path via the capacitance between the outer shields and the ground at the source would be present. This overall system shown in Figure 5.12 is considered unbalanced since the impedance of each conductor to ground is not the same. (The impedance of the upper conductor to ground includes Z_s while the lower conductor does not.)

Sometimes, for physically long, but still electrically-short, cable runs, two coaxial cables are placed in parallel (the shields are connected together and the inner conductors are connected together). This reduces the ohmic or resistive losses of the conductors (the net resistance of two identical resistors, R, in parallel is $R/2$). However, this also increases the input capacitance (the net capacitance of two identical capacitors, C, in parallel is $2C$). Hence, the RC product, the cutoff frequency of the low-pass filter formed (e.g., $1/RC$), and the bandwidth of the two-cable combination is the same as the single cable.

5.4 Reducing Noise Through Transformers

How can transformers be used to reduce common-mode noise? What other methods are available for reducing noise due to ground loops? [Whitlock, 03; Ott]

In this book, considerable time is devoted to the proper or best way of connecting the signal and chassis grounds to a cable's shield. Connecting a cable shield to the proper ground is often an inexpensive method of reducing noise pickup — when it is effective. Unfortunately, in many situations, especially for longer cables (e.g., 20 ft audio cables), merely grounding the cable's shield may be inadequate. The noise can be due to ground loops, external RF radiation (i.e., plane wave signals), and near-field crosstalk. The common-mode noise that is generated by these sources can often be reduced to an acceptable level by using isolation transformers, baluns, common-mode chokes, optoisolators, differential amplifiers, and instrumentation amplifiers. When selecting between these various methods, various factors must be considered such as the availability of dc power, the frequency, and the bandwidth. Moreover, the field strengths of the undesirable noise signal and desirable signal to be carried on the cable, and the cost, size, and weight of the noise-reduction device must also be considered. Isolation transformers are the focus of this discussion.

Transformers inductively couple time-varying signals between their windings: when the time-varying current in the winding of a transformer is nonzero, a nonzero flux is generated inside the winding core that passes through one or more other windings. Briefly, the reason transformers can provide excellent rejection of common-mode noise signals is that the common-mode signals generate zero net flux in the core of the windings. Referring to Figure 5.13, with equal common-mode currents ($I_{C1} = I_{C2}$) on the primary of a transformer with zero leakage, the total flux within the core is zero since $\Phi_{C1} = \Phi_{C2}$. Therefore, the secondary open-circuit voltage of the transformer is ideally zero. Real transformers can also reduce differential-mode noise signals (and desirable differential-mode signals) if the frequencies of the signals are outside the operating bandwidth of the transformer.

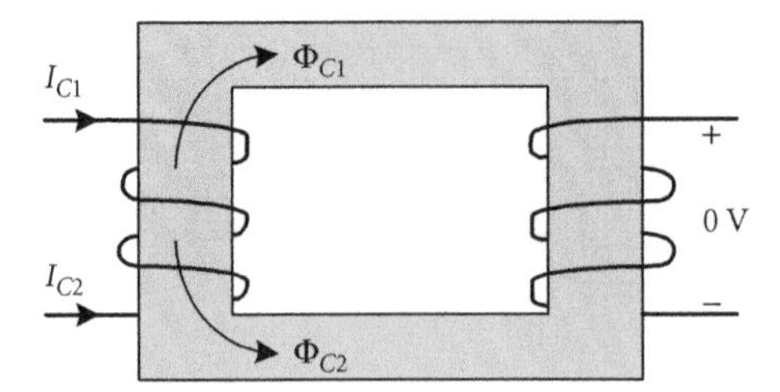

FIGURE 5.13 Common-mode currents and fluxes they generate inside a transformer core.

High-quality isolation transformers can reject a great deal of common-mode noise for unbalanced, balanced, and mixed balanced systems. They are passive reliable devices, which do not require a power source to operate. If properly selected, they can handle large voltage drops across them without arcing or other undesirable forms of breakdown. Finally, trimming or tweaking is not required as with instrumentation amplifiers. They do have their disadvantages, however, including their high cost, large size, and limited frequency response or bandwidth. Furthermore, isolation transformers cannot pass dc signals, which may be important if dc power is to be carried along the cable (as with some devices connected to the external ports of computers).

Real transformers do not reject 100% of common-mode signals. One major reason is that nonzero parasitic capacitance exists between the primary and secondary coils of a transformer. To reduce this capacitance, one or more electrostatic or Faraday shields are used between the windings. These Faraday shields should not be confused with the chassis or frame of a transformer.

Whenever an unbalanced device is connected to an active balanced device, the balance of the balanced device is affected. The imbalance reduces the overall common-mode rejection ratio (*CMRR*). Although the *CMRR* of an amplifier or a receiver is affected by the actual magnitude of the impedance seen by each input, often the *difference* between these two impedances is much more important. If an unbalanced driver is connected to a balanced receiver, the impedance seen by each of the balanced receiver inputs can be very different. By connecting a transformer between the unbalanced driver and balanced receiver, the impedance seen by the balanced driver inputs are nearly identical.[6] This can be easily shown. Referring to Figure 5.14, if the secondary of the transformer has a center-tapped signal reference, then the voltage at both ends of the secondary coil are of equal magnitude but opposite sign relative to this reference.

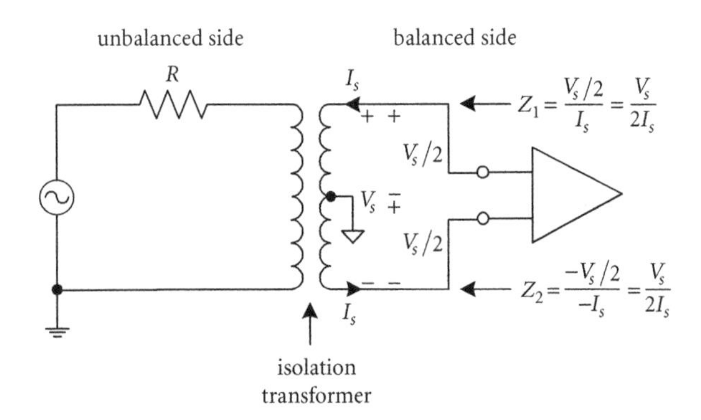

FIGURE 5.14 Impedance seen by each input of the amplifier is identical even though the primary side of the transformer is unbalanced.

[6]Because of the methods used to wind transformers, the parasitic capacitance from one end of the secondary coil of a transformer to the primary coil is not necessarily identical to the capacitance from the other end of the secondary coil to the primary coil.

Since the differential-mode currents from each of these ends of the secondary are of equal sign but opposite direction, the impedance seen by each of the secondary leads are equal since the ratios of the voltage to current are identical:

$$Z_1 = Z_2 = \frac{V_s}{2I_s}$$

If the isolation transformer were not present, the two impedances to ground would be (turning off the source voltage)

$$Z_1 = R, \; Z_2 = 0$$

Isolation transformers can be used at the driver, at the receiver, or at both the driver and receiver. In those rare cases where cost is not an issue, isolation transformers can be used at both the driver and receiver, even if both the driver and receiver are balanced devices. These isolation transformers help balance the transmission line connecting the driver to the receiver and help eliminate any noise picked up by the transmission line from passing to the receiver. They are also used to increase the voltage level (not power level) of weak signals. Sometimes a single resistor, a resistor pad, or even an *RC* network is placed across an isolation transformer. There are various reasons for these networks: (i) to match the load of the transformer to the designed load impedance of the transformer, as sometimes specified through the impedance ratio (e.g., 600 Ω: 600 Ω); (ii) to extend the bandwidth of the transformer; and (iii) to dampen oscillations that might occur at higher frequencies.

Multiple single-ended grounded loads using the same signal or power source must sometimes be connected to different grounds. For example, one load might be 3 ft from the signal source while another load might be 10 ft from the same source. Although the loads might both be floating, frequently they are not and a nearby ground connection is required. Different grounds will likely be at different potentials, especially when the ground connections are distant and the ground impedance is high. In this case, a single transformer may be inadequate for isolation and noise reduction. In the following discussions, the different grounds in the figures are noted by the numbers 1, 2, and 3. In Figure 5.15, a single isolation transformer is used for two different loads, Z_{L1} and Z_{L2}, connected to different grounds. The output of the secondary is shared by both loads and so is a portion of the signal and return conductors. This portion of the shared conductor has a nonzero impedance, Z_{l1}. There are two major EMC-related issues in this circuit. First, because the grounds are at different potentials, a driving voltage will be present between them, which is a source of common-mode coupling. Second, because a portion of the signal and return conductor is shared by the two loads, when the current demand of one load changes, it will produce a voltage drop across the conductors leading to the secondary. This change in voltage will also affect the voltage at the other load—referred to as common-impedance coupling. Thus, even though there may be isolation between the primary and secondary sides of the transformer, there is coupling between the loads at the secondary.

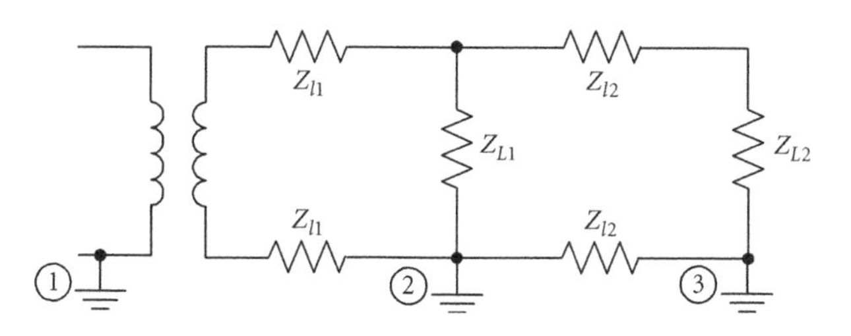

FIGURE 5.15 Model of two loads with different grounds that share both a signal and return conductor to the secondary.

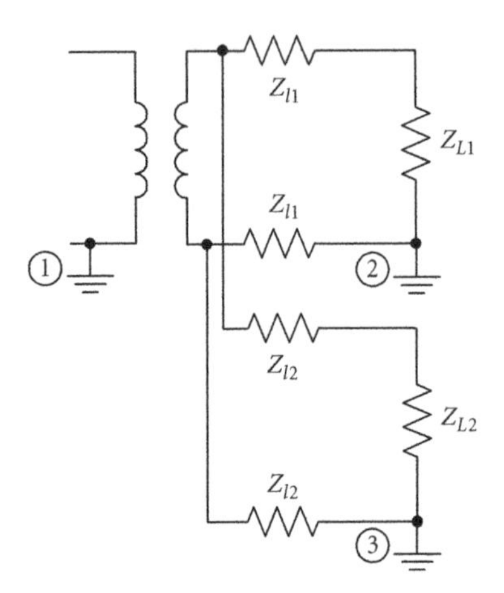

FIGURE 5.16 Model of two loads with different grounds that do not share a signal or return conductor to the secondary.

The circuit connections shown in Figure 5.16 eliminate the direct sharing of a signal and return conductor, but coupling between the two loads can still occur. To understand the origin of this coupling, imagine that a voltage source is connected between the two grounds labeled 2 and 3. This voltage source, which is modeling the potential difference between these two different grounds, will generate a driving current through the secondary-side impedances and, obviously, a ground loop will be formed. In Figure 5.17, a center-tapped secondary is used (appropriately selected to provide the same output voltage to both loads). In this case, only the return conductor, with an impedance of Z_{c1}, is directly shared by the two loads. There might be a small reduction in common-impedance coupling for this circuit compared to that shown in Figure 5.15. If the two loads and their connections to the secondary are balanced, then the current through the shared return would be zero and this portion of the common-impedance coupling would be reduced.

The circuit shown in Figure 5.18 is an improvement over the circuit in Figure 5.17 since no conductor is directly shared by the two loads. The potential difference between the two grounds will still allow for some common-mode coupling between the circuits since they share a common secondary winding. The circuit shown in Figure 5.19 consists of two separate transformers, with their primaries connected in parallel. This arrangement is probably the most expensive and the best of the configurations presented. No conductors are shared between the two different loads, and the individual transformers break the ground loop between the two different grounds 2 and 3.

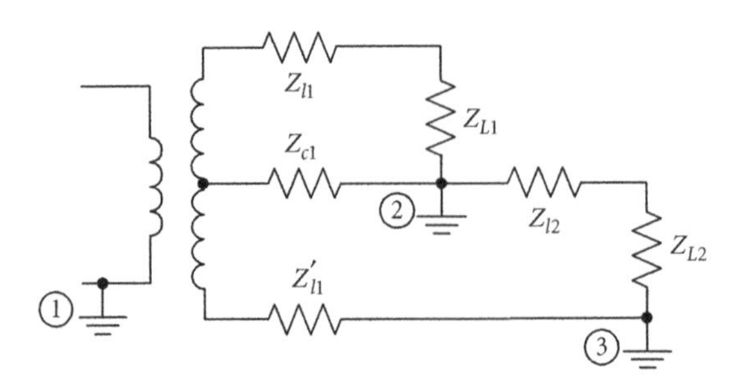

FIGURE 5.17 Model of two loads with different grounds that only share one conductor to the secondary.

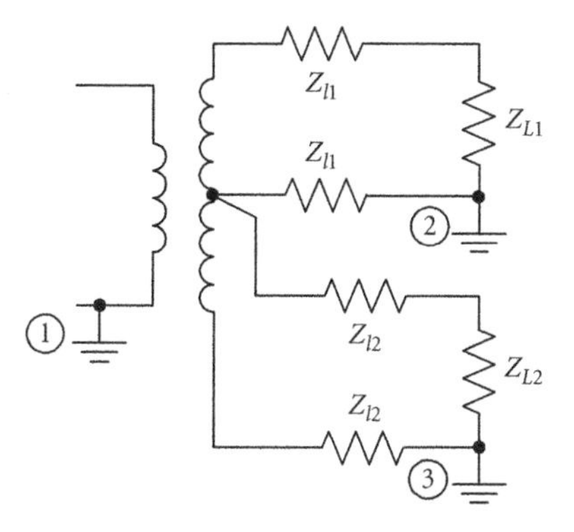

FIGURE 5.18 Model of two loads with different grounds that do not share the same (portion of the) secondary winding.

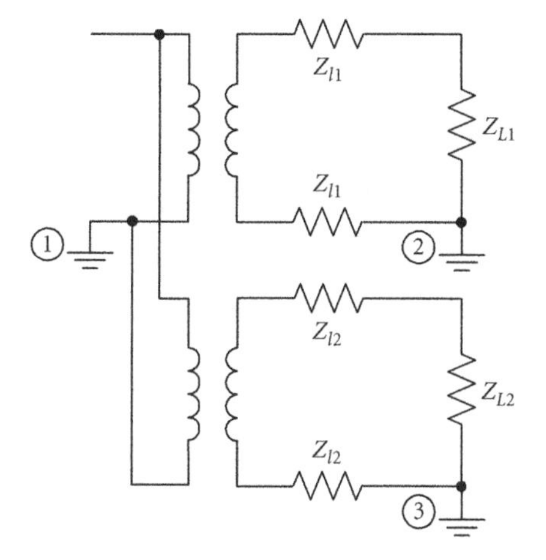

FIGURE 5.19 Model of two loads with different grounds connected to separate transformers.

When possible, it is recommended that separate isolation transformers be used for devices that are very noisy (e.g., refrigerators and air conditioners) or those devices that are very sensitive to noise. Sometimes, special low-noise power outlets are available. These outlets may or may not be using isolation transformers.

5.5 Modeling a Cable as a Transformer

Two parallel wires can be modeled as a 1:1 transformer: each wire is modeled using self partial inductance and the coupling between the wires is modeled using mutual partial inductance. To analyze this 1:1 transformer, separate long wires are connected to each end of the two parallel wires forming two large rectangular loops. A current is injected into one loop, and the induced voltage measured across the other loop (the two loops are in the same plane) as shown in Figure 5.20. Thus, these loops are acting as single-turn windings on a transformer. A coaxial cable can also be analyzed as a transformer. Assume that the shield is the primary and the center conductor is the secondary of the "transformer." Apply a current source to the primary and imagine measuring the induced voltage across the secondary. How does this voltage compare to the voltage across the primary? [Smith, '93; Weston; Grover]

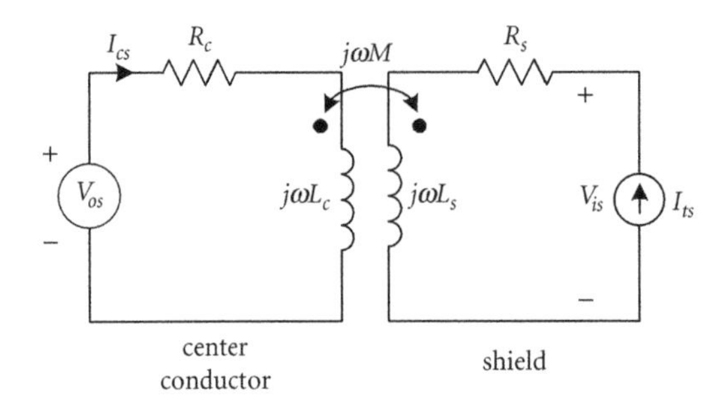

FIGURE 5.20 Experimental setup to determine the magnetic coupling between the center conductor and shield of a coaxial cable.

A cable is rarely classified as a transformer. However, this novel viewpoint helps explain the effects of the shield's nonzero resistance and inductance, and it assists in some "where to ground" problems. Furthermore, the performance of a cable when the currents in the two conductors are opposite but not equal can also be understood using this model. The transformer-equivalent circuit is given in Figure 5.20, assuming the cable is electrically short.

The mutual partial inductance between perpendicular wires is zero. The mutual partial inductances between the two adjacent parallel wires and the parallel lead wires to the current source and the voltmeter are negligible if the distance between these set of wires is great compared to the distance between the center conductor and shield. These conditions greatly simplify the analysis of the circuit.

The voltage across the secondary of this transformer is

$$V_{os} = R_c I_{cs} + j\omega L_c I_{cs} + j\omega M I_{ts}$$

This voltage induced across the center conductor is mostly due to the current through the shield (the outer conductor). If the voltmeter is nearly ideal, which is reasonable for modern instruments, its impedance is very large and the current in the secondary is about zero. The output voltage across the secondary is then

$$V_{os} = j\omega M I_{ts} \tag{5.1}$$

For a coaxial structure, the mutual inductance, M, is about equal to the shield inductance, L_s. The mutual inductance, defined in simple terms, is the flux collected by one loop divided by the current generating the flux from another loop. If one loop is the shield circuit and the other loop is the center-conductor circuit, then most of the flux generated by current in the shield is collected by the center-conductor circuit. (Some flux is not collected around the ends of the cable. Also, leakage flux exists beyond the voltmeter, but this flux should be small for large loops since the magnetic field falls off as $1/r$.) If the inductance of the shield is approximately equal to the inductance of the center conductor, this system is acting as a 1:1 transformer.

The transfer function is easily determined for this circuit. When the secondary is open circuited, the current in the secondary is zero, $I_{cs} = 0$, and the voltage induced across the secondary is only due to the mutual coupling. As before, the voltage across the secondary is

$$V_{os} = R_c I_{cs} + j\omega L_c I_{cs} + j\omega M I_{ts} = j\omega M I_{ts} \approx j\omega L_s I_{ts} \tag{5.2}$$

where R_c is the center conductor's resistance. The total voltage across the primary is

$$V_{is} = R_s I_{ts} + j\omega L_s I_{ts} + j\omega M I_{cs} = R_s I_{ts} + j\omega L_s I_{ts} \tag{5.3}$$

where R_s is the shield's resistance. The gain equation is the ratio of these two expressions:

$$\frac{V_{os}}{V_{is}} = \frac{\frac{j\omega L_s}{R_s}}{\frac{j\omega}{\frac{R_s}{L_s}} + 1} \tag{5.4}$$

Although it may not be immediately apparent, the voltage across the resistance of the shield does not couple into the center-conductor secondary (and vice versa). The ohmic losses of the outer conductor are a function of the voltage across the shielding resistance. The voltage produced across the secondary, the center-conductor circuit, according to Faraday's law is

$$V = -N\frac{d\Phi}{dt}$$

where N is the number of turns and Φ is the magnetic flux through each turn. A time-varying magnetic field from the shield is necessary to produce a voltage across the center conductor. When modeling a circuit, the inductance represents the magnetic flux generating portion of the circuit. The voltage induced across the secondary is a function of the time-varying flux through the secondary loop. The ohmic losses, not the magnetic flux, are modeled through the resistance.

5.6 Break Frequency of Coax

In reference to Figure 5.20, what is the corner frequency of this system? Sketch the Bode magnitude plot of this response. For what frequencies is the induced voltage across the center conductor approximately equal to the voltage across the shield? What is the consequence of a 60 Hz noise signal on the shield? How does this reasoning compare with the common-mode choke model of a coaxial cable? [Ott; Vance]

The corner frequency for the transformer model of the coaxial cable is obtained directly from the transfer function given in (5.4). This corner or cutoff frequency in rad/s is

$$\omega_c = \frac{R_s}{L_s} \tag{5.5}$$

The Bode response shown in Figure 5.21 is sketched by inspection. Well above the corner frequency, the voltage across the center conductor is approximately equal to the voltage across the shield in both magnitude and polarity as to be expected using the 1:1 transformer model. Below the corner frequency,

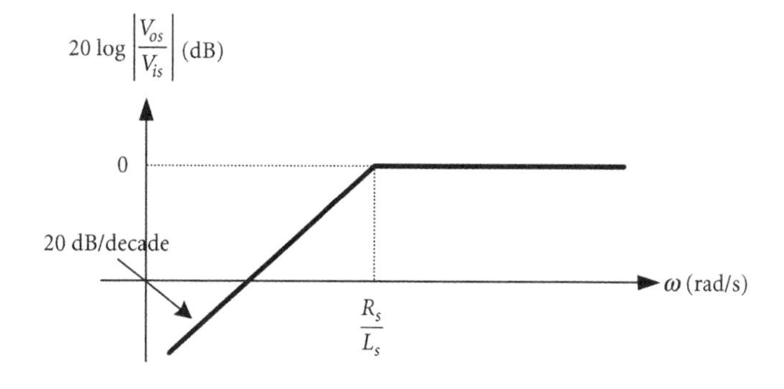

FIGURE 5.21 Bode magnitude plot of (5.4).

the voltages are not the same. Since a 60 Hz noise signal is well below the corner frequency for many cables, a 60 Hz signal on the shield will not induce much of a voltage across the center conductor. The voltage induced across a load for this cable will therefore contain this 60 Hz noise.

A low cutoff frequency is desirable, and the resistance of the shield directly affects this cutoff frequency. For coaxial 50–75 Ω cables, the corner frequency is in the 600 Hz to 2 kHz range. For 125 Ω twisted pair, the corner frequency is higher in the 800 Hz to 7 kHz range. Previously it was shown that the magnetic susceptibility of coaxial cable was slightly less than shielded twisted pair when the source and load were grounded. This was assuming that the shield cutoff frequency of the coax was less than the cutoff frequency of the shielded twisted pair.

A schematic of a common-mode choke is given in Figure 5.22. The 1:1 transformer model of the cable contains this same circuit. Above the corner frequency, the coaxial cable can be viewed as a common-mode choke: it passes normal differential-mode currents but blocks common-mode currents. If a coaxial cable with a grounded source and load is modeled using this common-mode choke or transformer, this choking action and its frequency dependence will be clearly seen. Referring to Figure 5.23, the two mesh equations for this circuit are

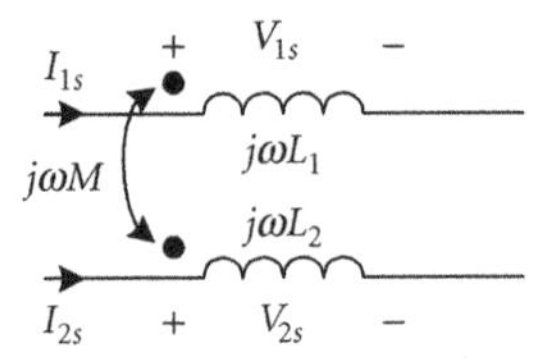

FIGURE 5.22 Common-mode current choke.

$$-V_{Ds} + R_d I_{cs} + R_c I_{cs} + j\omega L_c I_{cs} - j\omega M I_{ss} + R_L I_{cs} = 0$$

$$j\omega L_s I_{ss} - j\omega M I_{cs} + R_s I_{ss} = 0$$

Solving for the shield current in the second equation yields

$$I_{ss} = \frac{j\omega M I_{cs}}{j\omega L_s + R_s} = I_{cs} \frac{\dfrac{j\omega M}{R_s}}{\dfrac{j\omega}{\dfrac{R_s}{L_s}} + 1}$$

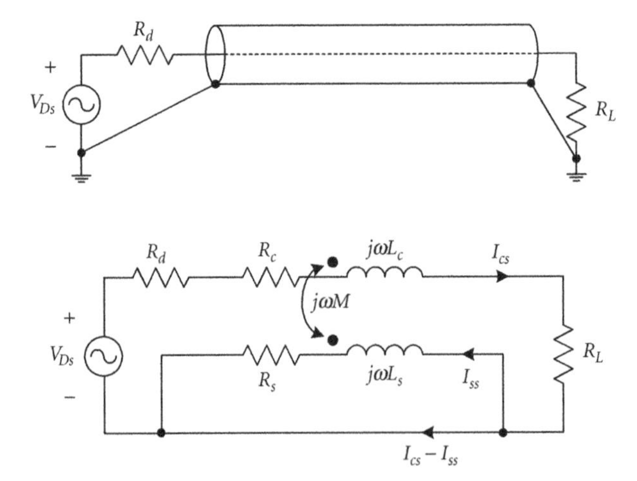

FIGURE 5.23 Transformer model of a coaxial cable connecting a single-ended source to a single-ended load.

and if $M = L_s$,

$$I_{ss} = I_{cs} \frac{\dfrac{j\omega L_s}{R_s}}{\dfrac{j\omega}{\dfrac{R_s}{L_s}} + 1} \tag{5.6}$$

Well above the shield's corner frequency, the shield current is about equal to the current in the center conductor:

$$I_{ss} \approx I_{cs} \quad \text{if } \omega \gg \frac{R_s}{L_s} \tag{5.7}$$

At these higher frequencies, the currents in the two conductors of the coaxial cable are differential. The common-mode currents are zero. The current from the center conductor through the load returns mostly via the shield rather than via the ground plane! The current passing via the ground plane is small if the frequency of the signal is much greater than the corner frequency. Perfect coupling between the shield and center conductor is assumed. Well below the corner frequency,

$$I_{ss} \approx I_{cs} \frac{j\omega L_s}{R_s} \quad \text{if } \omega \ll \frac{R_s}{L_s} \tag{5.8}$$

As the frequency approaches dc, the shield current approaches zero. It is difficult to have a very low susceptibility to low-frequency magnetic fields. If the frequency of the signal is much less than the corner frequency, almost all of the return current is via the ground plane. Unless the cable is lying on the ground, the effective pickup area is much larger for this current path than if the current returned via the shield. A common recommendation for magnetic susceptibility reduction (at lower frequencies) is to ground a cable at only one location. By grounding the cable at only one location, the current path through the ground plane is interrupted or broken.

The actual value of the shield current is obtained by solving for I_{cs} in the first voltage loop equation and substituting it into the second loop equation:

$$I_{ss} = \frac{\dfrac{j\omega L_s}{R_s(R_d + R_c + R_L)} V_{Ds}}{\dfrac{j\omega}{\dfrac{R_s(R_d + R_c + R_L)}{L_s(R_d + R_c + R_s + R_L)}} + 1} \tag{5.9}$$

assuming that $M = L_c = L_s$. This current is equal to the applied voltage divided by the total series resistance of the system if the frequency is much greater than the corner frequency of the *system* not just the cable:

$$I_{ss} \approx \frac{1}{R_d + R_c + R_s + R_L} V_{Ds} \quad \text{if } \omega \gg \frac{R_s(R_d + R_c + R_L)}{L_s(R_d + R_c + R_s + R_L)} \tag{5.10}$$

Since R_c is often of the same magnitude as R_s and the sum of the load and driver resistance is often large compared to these resistances, the corner frequency reduces to

$$\omega_c \approx \frac{R_s}{L_s} \quad \text{if } (R_d + R_L) \gg (R_c + R_s) \tag{5.11}$$

This approximate frequency is identical to the cable's corner frequency.

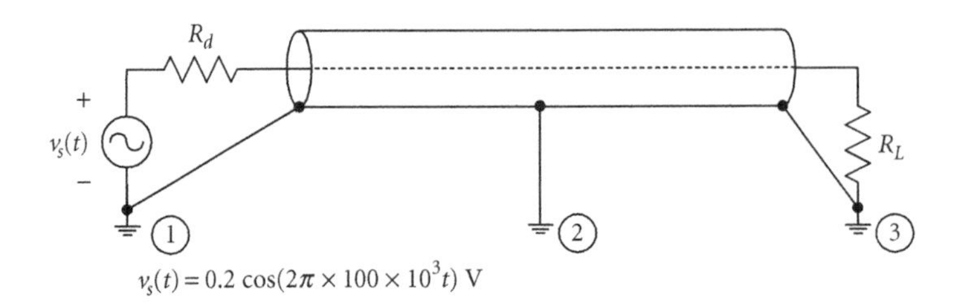

FIGURE 5.24 Multiple ground connections to a coaxial shield.

5.7 Multiple Grounding Points for Coax

As shown in Figure 5.24, the shield of a matched coaxial line with a 100 kHz input signal (0.2 V amplitude) is connected to ground at several locations: at the source, the load, and midway down the line. Because of ground potential differences, however, the voltage between the source ground and the midway ground is 0.5 V, and the voltage between the midway ground and the load ground is 0.7 V. Determine the voltage induced across the center conductor from these ground voltages. Determine the voltage amplitude across the load. [Smith, '93]

Grounding a coaxial cable at multiple locations should normally be avoided at lower frequencies. The grounds have been labeled as 1, 2, and 3 in the figures to indicate that they are not at the same potential even though they are designated with the same ground symbol. Any ground has a nonzero impedance. When current exists in this real ground, a voltage drop exists across the ground.

When the frequency of the noise signal in the ground is much greater than the cutoff frequency of the cable, then the voltage induced across the load should be much less than these ground voltages. The following analysis assumes that the frequency of the noise is much greater than the cutoff frequency (and that the resistances of the shield and center conductor are zero). Also, the load R_L is assumed matched to the transmission line so that there are no reflections. The model of this situation is shown in Figure 5.25, assuming worst-case polarities for the noise voltages. Since the mutual coupling between the center conductor and shield for the coaxial cable is high (i.e., k, the coefficient of coupling between the conductors, is about 1), any voltage that exists across the shield is coupled into the center conductor with the same polarity. This is the transformer-action characteristic of the cable. For example, the voltage across the center conductor segment closest to the 0.2 V source is $j\omega L_s I_{cs} - j\omega L_s I_{2ss}$, and the voltage across the shield segment closest to the source is also equal to $-j\omega L_s I_{2ss} + j\omega L_s I_{cs}$. This voltage must be equal to 0.5 V since 0.5 V is directly in parallel with this shield inductance. Because the self inductances are identical and mutual coupling is perfect, even if the center and shield conductor currents are not the same, the voltages across the coupled inductors are identical.

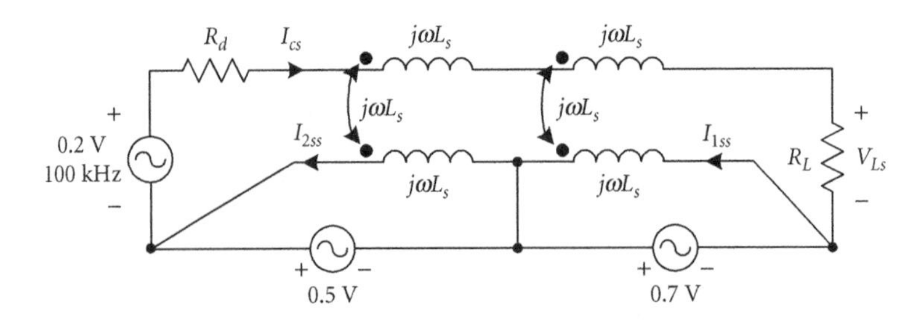

FIGURE 5.25 For the situation in Figure 5.24, the cable is modeled using the perfect-coupling transformer model. The voltages between the ground points are also modeled.

The amplitude of the voltage induced across the first segment of the center conductor is 0.5 V, and the voltage induced across the second segment of the center conductor is 0.7 V. One voltage loop equation around the outer perimeter of the circuit illustrates that the load voltage is independent of the noise voltages across the grounds:

$$-0.2 + R_d I_{cs} + 0.5 + 0.7 + V_{Ls} - 0.7 - 0.5 = 0 \qquad \Rightarrow V_{Ls} = 0.2 - R_d I_{cs}$$

This expression assumed that the noise voltages across the grounds, the 0.5 V and 0.7 V amplitude signals, are of the same frequency as the 0.2 V source signal. This allowed one equation to be written since, in this frequency-domain analysis, the frequency is assumed the same everywhere. If the frequency of the noise voltage is different from the source frequency, then two separate equations would be written and the results summed in the time domain. However, the noise voltages will still cancel even if their frequencies are different from that of the source.

The expressions for the currents are easily obtained by writing three mesh current expressions:

$$-0.2 + R_d I_{cs} + j\omega L_s I_{cs} - j\omega L_s I_{2ss} + j\omega L_s I_{cs} - j\omega L_s I_{1ss} + R_L I_{cs} - 0.7 - 0.5 = 0 \tag{5.12}$$

$$0.7 + j\omega L_s I_{1ss} - j\omega L_s I_{cs} = 0 \tag{5.13}$$

$$0.5 + j\omega L_s I_{2ss} - j\omega L_s I_{cs} = 0 \tag{5.14}$$

Solving for I_{1ss} and I_{2ss} in (5.13) and (5.14), respectively and substituting these results into (5.12), the current in the center conductor is obtained:

$$I_{cs} = \frac{0.2}{R_d + R_L}$$

Substituting this result into (5.13) and (5.14), the two shield currents are

$$I_{1ss} = \frac{0.2}{R_d + R_L} - \frac{0.7}{j\omega L_s}, \quad I_{2ss} = \frac{0.2}{R_d + R_L} - \frac{0.5}{j\omega L_s}$$

When $\omega L_s \gg (R_d + R_L)$, the two shield currents are approximately equal to each other and equal to the current in the center conductor. This was an initial assumption. Also, the resistances of the shield and center conductor were assumed small.

Now, suppose that the shield of a matched coaxial cable is grounded only at its input and output as shown in Figure 5.26. The potential difference between these ground connections is 0.8 V. In addition, the leads between the shield and these ground locations are long. The potential difference between the shield at the source and its ground connection is 0.1 V, and the potential difference between the shield

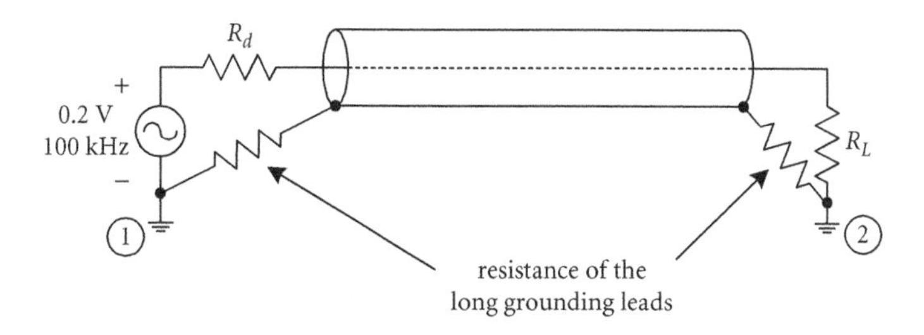

FIGURE 5.26 Nonnegligible resistance of the two ground connections to the shield.

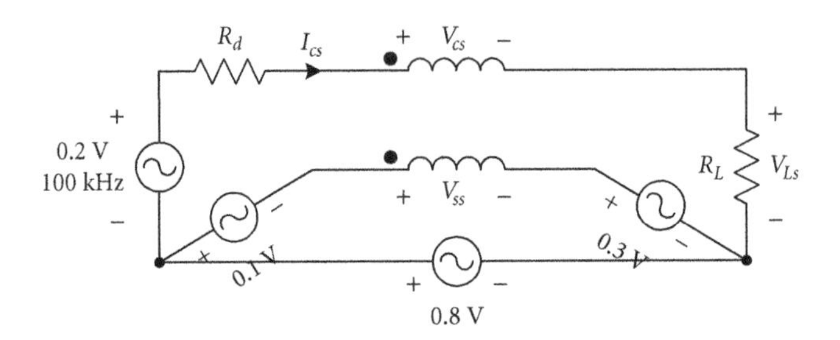

FIGURE 5.27 Voltage drop across the pigtail leads and between the ground connections.

at the load and its ground connection is 0.3 V. Long leads between the shield and ground points can have undesirable consequences. These leads are referred to as "pigtails." At high frequencies, the inductive reactance of these leads is more significant than their ac resistance. The inductance and voltage drop across the leads increase with their length.

Referring to Figure 5.27, the voltage loop equation around the outer perimeter is

$$-0.2 + R_d I_{cs} + V_{cs} + V_{Ls} - 0.8 = 0 \quad \Rightarrow V_{Ls} = 1 - V_{cs} - R_d I_{cs} \tag{5.15}$$

where V_{cs} is the voltage across the center conductor. The voltage across the center conductor is equal to the voltage across the shield assuming perfect coupling. The voltage across the shield is obtained by writing a voltage loop equation that includes the "pigtail" lead voltages:

$$0.1 + V_{ss} + 0.3 - 0.8 = 0 \quad \Rightarrow V_{ss} = V_{cs} = 0.4 \text{ V}$$

Substituting into (5.15), the load voltage is

$$V_{Ls} = 1 - 0.4 - R_d I_{cs} = 0.6 - R_d I_{cs}$$

Again, it was assumed that the frequency of the noise voltages across the shield leads and ground plane is equal to 100 kHz. If the frequencies are not the same, separate equations could be written for each frequency component and the results summed in the time domain (using superposition). Note that the voltage across the load has increased by 0.1 + 0.3 = 0.4 V, the sum of the voltage across both long leads. For this reason, the length of the pigtails should be as small as possible.

5.8 Keeping Noise Off the Shield

Why is the voltage across the output of a coaxial cable proportional to the resistance of the shield but not the inductance of the shield? [Ott]

It will be initially assumed that the load is an open circuit or a high input-impedance measurement device is connected across the output of the cable. It is common to model the difference in voltage across the ground plane or reference as shown in Figure 5.28. This voltage will generate common-mode currents. Other common-mode noise sources can also be modeled in this manner. The two voltage loop equations are

$$j\omega L_s I_{ss} - j\omega M I_{cs} + R_s I_{ss} + V_{CMs} = 0$$

$$R_d I_{cs} + R_c I_{cs} + j\omega L_c I_{cs} - j\omega M I_{ss} + V_{Ls} - V_{CMs} = 0$$

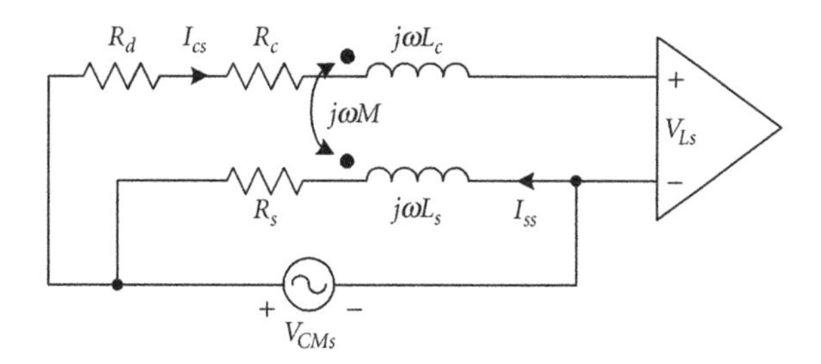

FIGURE 5.28 Common-mode voltage between the input and output of a coaxial cable.

Since $I_{cs} = 0$ for the open-circuited load, the voltage across the output is easily solved for as

$$V_{Ls} = -j\omega L_s I_{ss} - R_s I_{ss} + j\omega M I_{ss}$$

When the coupling is perfect and $L_c = L_s$, then $M = L_s$, and the voltage across the output is only a function of the shield resistance:

$$V_{Ls} = -R_s I_{ss} \tag{5.16}$$

The outer conductor of the coax is designed to carry the signal current. For this reason, noise current on the shield of a coaxial cable should be avoided. The consequence of noise current on the shield is that the output will be a function of the noise. When the noise current cannot be avoided, the shield resistance of the coax should be as small as possible (e.g., using a solid instead of a braided outer conductor). Using only one shield connection to ground will reduce the shield noise current. However, noise current can still couple into the shield via parasitic elements.

Additional insight is obtained by determining the load voltage assuming a finite load resistance. If the output is modeled as a resistive load, R_L, then the two voltage loop equations are

$$j\omega L_s I_{ss} - j\omega M I_{cs} + R_s I_{ss} + V_{CMs} = 0$$

$$R_d I_{cs} + R_c I_{cs} + j\omega L_c I_{cs} - j\omega M I_{ss} + R_L I_{cs} - V_{CMs} = 0$$

If $M = L_c = L_s$ the load voltage is obtained from these two equations after some algebra:

$$V_{Ls} = R_L I_{cs} = \frac{\dfrac{R_L}{R_c + R_d + R_L}}{1 + \dfrac{j\omega}{\dfrac{R_s(R_c + R_d + R_L)}{L_s(R_s + R_c + R_d + R_L)}}} V_{CMs} \tag{5.17}$$

The voltage across the load begins to decrease at a rate of 20 dB/decade after the cutoff frequency

$$\omega_c = \frac{R_s(R_c + R_d + R_L)}{L_s(R_s + R_c + R_d + R_L)} \approx \frac{R_s}{L_s} \quad \text{if } R_L \gg R_s \tag{5.18}$$

The approximation is equal to the shield's corner frequency for high-impedance loads or drivers (compared to the shield and center conductor resistance). The voltage across the load well *below* the cutoff frequency is about

$$V_{Ls} \approx \frac{R_L}{R_c + R_d + R_L} V_{CMs} \quad \text{if } \omega \ll \omega_c \tag{5.19}$$

which for higher resistance loads further reduces to

$$V_{Ls} \approx V_{CMs} \quad \text{if } \omega \ll \omega_c,\ R_L \gg (R_c + R_d) \tag{5.20}$$

At low frequencies, the common-mode voltage appears directly across the high-resistance load. From (5.17), the voltage across the load well *above* the cutoff frequency is about

$$V_{Ls} \approx \frac{R_s R_L}{j\omega L_s (R_s + R_c + R_d + R_L)} V_{CMs} \quad \text{if } \omega \gg \omega_c \tag{5.21}$$

For higher resistance loads this expression is approximately equal to

$$V_{Ls} \approx \frac{R_s}{j\omega L_s} V_{CMs} \quad \text{if } \omega \gg \omega_c,\ R_L \gg (R_s + R_c + R_d) \tag{5.22}$$

The load voltage is clearly a function of both the shield's resistance and inductance. Reducing the shield resistance again reduces the load voltage.

In the previous analysis of the coaxial cable configuration given in Figure 5.24, the common-mode noise voltage coupling to the output was neglected. The solution assumed that the noise frequency was much greater than the cutoff frequency of the coax. As a result, the voltage across the load due to the common-mode voltage was negligible. Assuming a corner frequency of 1 kHz for the coax and a large value for R_L compared to the other resistances in the circuit, from (5.18)

$$\omega_c = 2\pi \times 10^3 = \frac{R_s}{L_s}$$

From (5.22), the load voltage is at least one-hundredth of the common-mode voltage if the frequency of the noise signal is at least 100 kHz:

$$|V_{Ls}| = \frac{2\pi \times 10^3}{2\pi \times 100 \times 10^3} |V_{CMs}| = \frac{|V_{CMs}|}{100}$$

5.9 Switching the Neutral and Hot Wires

One outlet in a laboratory is accidentally miswired: the neutral and ground wire are switched. By connecting one instrument to the miswired outlet and another instrument to a properly wired outlet, show that the potential for noise problems will most likely increase. Assume the equipment is connected via shielded cables.

Commonly, individuals who do not understand the purpose of the safety ground wire will carelessly switch the neutral and ground wires since "they are connected at the same point back at the panel." This misunderstanding has both safety and noise implications. The chassis of both instruments A and B, shown in Figure 5.29, should ideally be at zero volts relative to the earth ground. If an individual touches

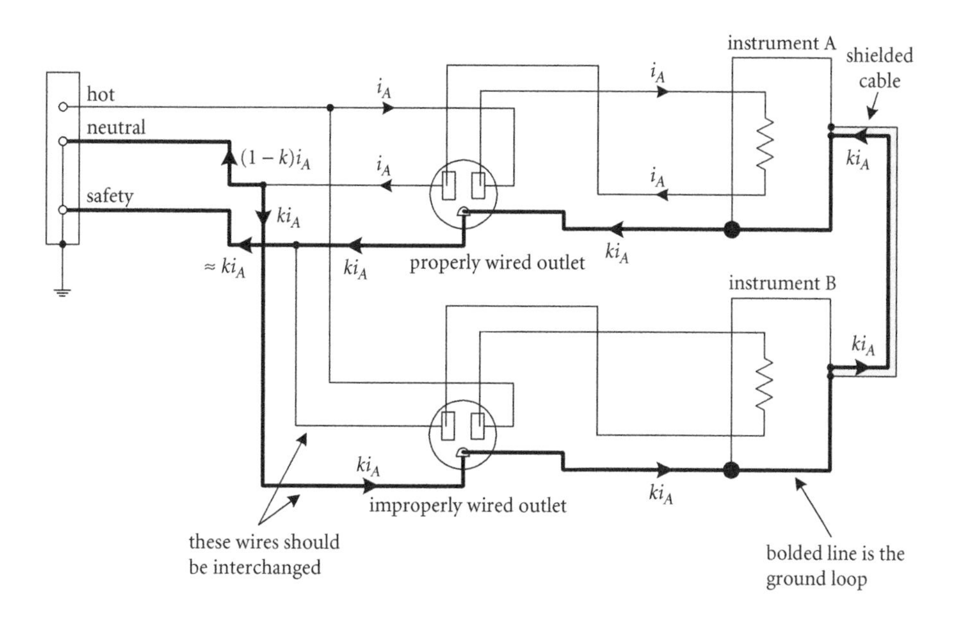

FIGURE 5.29 Two instruments powered through two different outlets. The lower outlet is improperly wired.

both instruments at the same time, the voltage across the individual should be zero (or small). Also, if an individual touches another grounded object and one instrument, the voltage across the individual should also be zero. The safety wires to each chassis are designed to keep the chassis at zero volts during no-fault conditions.[7]

Since the neutral and ground wires are switched for instrument B's outlet, current now exists on the wire designed to keep instrument B's chassis at zero volts. The chassis for B is now at a nonzero potential during normal no-fault conditions. Its potential is a function of the currents along this safety/neutral wire and the impedance of this wire. Notice that the chassis for instrument A is also at a nonzero potential. This situation represents a safety hazard.

There is also a large ground loop present in this circuit. The shielded cable connecting the two instruments provides an alternative path for the return current. Specifically shown in this figure is the current for instrument A. The return current splits between the neutral wire, $(1-k)i_A$, and safety wire for instrument B's outlet, ki_A. The outer conductor of the shielded cable, which is connected to both instrument chassis, provides the return path for the current. This large effective loop area will increase the emissions of the system and will increase the susceptibility of the system to external noise. Miswired outlets can be, and should be, located through "outlet checkers."

5.10 Avoiding Ground Loops and Hum

A stereo tuner, CD player, and cassette[8] player are all connected to a main amplifier. Both the tuner and amplifier have separate ground lugs. All four components have their own three-wire ac power cords. Shielded cables are used between all the components. Discuss how these components should be connected to avoid ground loops and hum pickup.

[7]In those situations where the potential across two objects is not known with certainty to be small, it is recommended that one hand be kept in a pocket (and long pants and shoes be worn).

[8]By the time of this book's publication, the cassette player will probably be as obsolete as the 8-track player!

The term ground loop is used throughout this book. It is frequently given as a "reason" why a system is noisy (especially when the reason for the interference is not understood). Basically, ground loops occur when the electrical grounds in a system are not at the same potential, and these grounds are conductively connected. An ideal ground would have zero impedance, and any current present on an ideal ground would produce zero voltage difference along it. Unfortunately, ground systems are not ideal.

Commonly, safety grounding is blamed as the reason for ground loops. Principally for safety reasons, many appliances, electronic products, and test equipment have their outer chassis connected to an earth ground. Commonly, this earth ground connection is via the power cord to the third prong of the outlet. This earth ground, depending on regulations, can consist of a steel rod driven into the earth or a connection to a cold water supply pipe. Normally, the return current to the source of the power (e.g., electrical distribution panel) for a device does not return on the safety ground wire but on the white or neutral wire. When a dangerous fault occurs to the metal chassis, the earth ground connection provides a low-impedance path for the current back to the distribution system (instead of potentially through any individual contacting the chassis). Many systems must deal with ground loops, especially those involving several pieces of audio and visual components. There is often, unfortunately, a conflict that occurs between safety-related grounding and noise-related grounding. A good grounding scheme adheres to the safety rules while minimizing emissions and susceptibility.

Each outlet[9] contains a separate neutral and ground wire in addition to the hot wire. Depending on how the outlets are wired or whether they have a common breaker at the distribution panel, determines whether these wires are shared by more than one outlet. If the devices do have a common shared neutral or hot wire, then the current used by one device will affect the supply voltages of the other devices since the impedance of these wires is not zero. This variation in the supply voltage can couple into the other devices that are sharing these power lines producing a form of crosstalk technically referred to as common-impedance coupling. Although the common conductor is not necessarily part of a ground loop, common-impedance coupling is a potential source of interference. These shared conductors, however, can allow a ground loop to exist once cables are connected between the devices.

If all of the devices are connected to the same outlet (and the outlet can handle the current required by all of the devices), then the supply voltage at the outlet will be the same for all of the devices connected to that one outlet.[10] This is a single-point star connection. When the devices are connected to different outlets (all sharing the same hot and neutral), it is referred to as a daisy-chain connection. Although daisy-chain connections are nearly everywhere in power systems because of their low cost and ease to make, they are not EMC friendly. If a nearby outlet is not capable of handling the current demands of all of the equipment, then the equipment can be connected to one end of a single, properly rated extension cord connected to a higher rated single outlet.

As previously mentioned, the safety ground or third prong on power cords is provided for safety reasons, but it is also a potential ground loop source. As shown in Figure 5.30, when two or more pieces of equipment are connected with cable that has an outer shield and this outer shield is connected to the chassis of the equipment, then a ground loop can form. When two pieces of equipment are connected by one coaxial cable, for example, the loop formed is through one chassis, through the outer shield of the coax, through the other chassis, through the power safety wire, and then back to the other chassis. If the instruments are powered by the same outlet, then the size of this ground loop will be less than if they were powered by different outlets. The area of this loop should be minimized if the emissions of the signal on the outer shield are to be minimized and the susceptibility of this loop to radio frequencies, for example, is to be minimized. With the ground loop shown, current can circulate between the instruments. Also, noise current on the shield of the cable can induce noise current on the conductor inside the shielded cable. If the cable connecting the instruments is coax, then the shield is also the signal return. Therefore, the

[9]At least for many outlets wired in the U.S. after the advent of color television!

[10]Even if the devices are connected to the same outlet, the chassis voltages can be different because of the different parasitic capacitances between the chassis and the internal circuitry for each device. A small but nonzero leakage current then passes between the chassis and the circuitry.

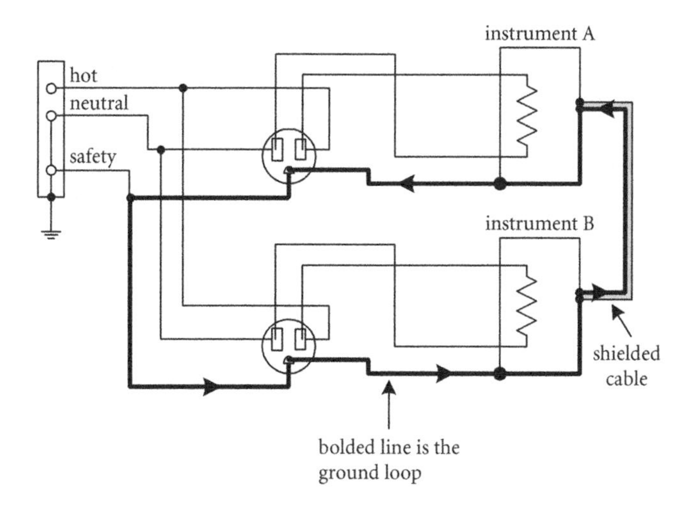

FIGURE 5.30 Ground loop through a shielded cable interconnect.

signal current, including any noise, can also travel on the ground loop. A ground loop can be more tolerated if the system is balanced. If the cable connecting the two instruments were balanced, such as shielded twisted pair, and the I/O were balanced with proper internal grounding, then the effect of this ground loop would be much less. (Hum can also occur in a balanced system if the common-mode rejection of the input stages is insufficient to attenuate the hum.)

Common sources of ground loops in the audio and video field are cable TV connections and external antenna connections with one side grounded (e.g., grounded satellite dish). The outer shield of the coax entering the home is usually not grounded at the power distribution panel but at the cable distribution box or at some other convenient location. Therefore, it is likely that the potential of the outer shield of the coax where it connects to the electrical equipment (e.g., television or recording device) is not at the same potential as the chassis of the electrical equipment that is connected to the safety ground. A ground loop is thus available for 60 Hz cable repeater-power signals that are present on the shield or other nearby noise sources.

The term hum is frequently applied to 60 Hz (or 50 Hz) signals and its harmonics that are coupled into a system. A ground loop is not necessary to pick up these interfering power related signals. These signals can also inductively or capacitively couple into a victim system via crosstalk. This hum can sometimes be heard over the speakers of an audio system as a low-pitch drone or on the screen of a television as slow vertically moving lines or bars. Once this interference is coupled into a system, it can be difficult and frustrating to eliminate. If radio frequency communications are involved, it is best to reduce the hum before any demodulation occurs because of the proximity of the 60 Hz and its harmonics to the baseband signal.

Common sources of hum are ac power sources (including overhead power lines and electrical wiring inside walls), inexpensive plug-in wall transformers, air conditioners, refrigerators (and other devices with large motors), fluorescent lighting, inexpensive light dimmer switches,[11] power supplies with inadequate or defective filtering, and other devices with power transformers. Audible hum can also be generated by the magnetic forces on loose laminations in transformers. This hum can be heard without the aid of any electrical device.

In the design of electrical systems, some issues such as 0.001% vs. 0.005% distortion in an audio receiver are usually not important since the electrical noise is often a much more dominant and critical factor. Because 60 Hz signals are present nearly everywhere, systems are often designed to reduce 60 Hz

[11]Actually, dimmer switches are a source of buzz, which contains higher frequencies than hum.

crosstalk to a tolerable level. Microphones and microphone cables are extra sensitive to hum because of the large amplification that is necessary with the mV-level microphone signal. Similarly, low-level analog instrumentation, such as strain gauges, and audio interface cables are extra sensitive to low-frequency noise. When attempting to reduce hum, shielding, balancing, routing, separating, isolating, orienting, and grounding methods are employed. The following methods, listed in no particular order, are used in multicomponent audio systems, but they are also useful in other systems. With all of the given methods, safety issues must be addressed and given the highest priority. To reduce hum,

1. use balanced cables with balanced I/O (e.g., differential drivers and receivers) to those cables
2. use isolation transformers or isolators (with the appropriate bandwidth response) between balanced and unbalanced systems
3. reduce the length of connecting cables, especially unbalanced cables
4. use passive filters to eliminate the hum before it reaches critical equipment
5. use single-point (star) grounding of the equipment (i.e., connect grounds together at one point such as at a patch panel)
6. avoid unnecessary grounding at multiple points (including accidental contact with other metal objects such as metal columns and coaxial cable plate mountings)
7. reduce the length of ground loops and resultant area formed by the loops
8. reroute the ac power cables
9. avoid parallel runs or routing of sensitive cables with power lines
10. avoid coiling of power cords, since they can act like a transformer or multiturn electrically-small loop (instead bunch the cord in a random disorderly fashion or bunch the cord in the shape of the number 8)
11. orient sensitive and noisy cables (e.g., power lines) so the cables are perpendicular to each other to minimize mutual inductive coupling
12. increase the distance between sensitive cables and power lines and transformers
13. use a separate power feed for noisy equipment (e.g., computers and lighting control)
14. relocate the sensitive equipment
15. remove the noise source
16. add a large ground plane (e.g., conductive mat below all of the equipment)
17. avoid pigtails
18. use shielded cable with a low shield resistance (e.g., thick solid shields)
19. use magnetic shielding such as steel or other high-permeability material (not plastic or aluminum)
20. use battery-operated gear
21. use two-prong equipment (with an internal isolation transformer)
22. connect sensitive equipment to another outlet that has a different route path to the distribution panel
23. ensure that all connections are good (e.g., clean and not corroded)
24. use notch filters to reduce the hum.

With the number of different types of cable in audio and video systems (ac power, speaker, video, microphone, control, data, and interconnect line-level), it is clear why cable grounding, selection, and routing are important.

Sometimes, ground loops are broken by disconnecting the ground connection to the chassis or by using a cheater plug or pigtail adapter. Usually, disconnecting the ground in this manner violates safety regulations because of the very dangerous situations that can arise. One method of attempting to debug a system is to turn off various parts of the system or to disconnect cables, one at a time, to help determine the potential culprit or weakness in the system. Another approach, which is slower, is to remove all cables, including the power cables. Then, each piece of the equipment should be connected to the power source, one piece at a time, and each connecting cable connected, one cable at time, until the hum is heard. For large complex systems, it is a good idea to sketch a block diagram of the system clearly showing all ground and shield connections, connecting cables, and power cords (two and three prong).

The situation described in the initial problem statement will now be briefly discussed. For a stereo tuner, compact disc player, and cassette player that are all connected to a main amplifier, all four components should be powered from the same outlet if the current rating of the outlet is not exceeded. Because three-prong power plugs are used, the chassis are connected to ground and are not floating. It is likely that the cables, if they are unbalanced, already connect the chassis of the amplifier and tuner via the shield of the connecting cable. Connecting the grounding lugs of the amplifier and tuner may provide a smaller ground loop path. If the cable between the amplifier and tuner is balanced (as well as the I/O to this cable), then connecting the lugs is probably still not critical since the equipment is powered by three-prong plugs and the chassis are likely connected together via the safety prong. Usually for simple sound systems involving only a few components, not connected to a cable system, television, video recording receiver, or computer, hum is not a problem.

5.11 Multipoint and Hybrid Grounding

Why would a metal object, including the shield of a transmission line, be earth grounded at multiple locations? How can a shield be grounded at high frequencies but not at low frequencies? [Violette; Paul, '92(b); Weston]

There are several reasons why an object may be earth grounded at one or more locations. The most common reason is probably for safety, including providing a low-impedance path for lightning. A second reason is to provide a path to discharge electrostatic charge buildup. A third reason is to reduce or eliminate the capacitive coupling between conductors (i.e., electric field shielding). A fourth reason is to provide a return path for a signal, such as when providing power to a remote location when only a single power conductor is used (the ground is rarely used as the return today).[12] A fifth reason is to help prevent a conducting object, such as the outer conductor of a coaxial cable, from radiating. This last reason will be the focus of this discussion.

Recall that a standing wave of either the voltage or current along a conductor repeats every one wavelength, λ. If a conducting object of finite conductivity, such as the cylinder shown in Figure 5.31, is connected to an ideal ground using ideal grounding straps, then its potential at these ground connections must be zero. (The potential along the ideal ground and ideal strap must be zero, which implies that they have no inductance or resistance.) Three different standing waves are shown along a conducting cylinder at a specific time. These signals on the cylinder could be induced, for example, from an incident

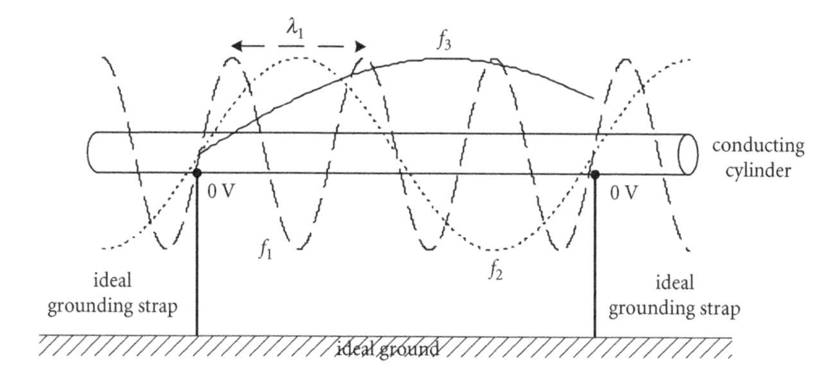

FIGURE 5.31 Voltage waveforms along a conducting cylinder grounded at two locations to help prevent it from radiating. By definition, the voltage must be zero at the grounding locations. At the end of the structure, for the particular waveforms shown, the current must be zero but not the voltage.

[12]Some electric fences use the ground as the return.

plane wave. The wavelength, or distance between two consecutive maximums or two consecutive minimums is $\lambda = v/f$ where v is equal to the speed of light for free-space conditions. The voltage waveform corresponding to f_3 cannot exist along the length of the conducting cylinder since its voltage is not zero at both grounding strap locations. In layman's language, a wave of this size cannot fit between these grounds. If the spacing between the straps is increased, then this waveform could exist on the cylinder, but the zeros of the waveform f_3 must exist at these grounding locations. Notice in this figure that many other higher frequency, smaller wavelength, signals such as f_1 and f_2 can still exist on this conducting cylinder even when it is grounded at two locations. Thus, merely grounding a conducting object does not prevent it from carrying signals and radiating like an antenna. For that matter, even an imperfect ground can act like an antenna or radiate. As the spacing between the two ideal grounding straps connected to the ideal ground is reduced (or the number of nonperiodically spaced straps is increased over a fixed distance), the lowest frequency that can "comfortably" exist on the cylinder increases.

In practice, with a real ground and real grounding straps, multipoint grounding of a metallic structure that is not electrically short does not prevent the structure from radiating. Selecting the zero-volt reference at one of the ground locations, the voltages at the other locations along the real ground are not at zero volts: when current is present along the ground, a voltage drop occurs along the ground's nonzero impedance. Even if all of the straps are connected to one common ground location, which is referred to as single-point grounding, the potential at the other end of the straps along the structure will likely be unequal: the impedances of the straps are nonzero and of different lengths.[13]

There are a few techniques for reducing radiating currents along a metallic structure. One or more common-mode chokes can be placed around and along the structure to help reduce common-mode antenna currents. The metallic structure can also be coated with lossy material to help reduce the amplitude of the current along it. Sometimes, the structure can be physically broken into smaller pieces to prevent the current from passing along the conductor. Guy wires for towers, for example, are broken into smaller segments and connected via insulators. Slots can be introduced into metallic chassis to reduce the current. Laminations are used in the core of transformers to interrupt induced currents.

There are situations where the shield of a cable must be connected to ground at one or more locations. For example, the coaxial connectors at both ends of the cable might contact chassis that are required for safety reasons to be grounded. Sometimes, the safety grounding at one or more locations along a metallic structure can be accomplished without introducing low-frequency ground loops. Instead of connecting every chassis directly to ground through a grounding strap, one or more of the connections can be via a low-impedance inductor as shown in Figure 5.32. For example, if a 1 mH inductor is used, the magnitude of its impedance at 60 Hz (ignoring its ac resistance) is $\omega L \approx 0.4\,\Omega$ while at 1 MHz its impedance magnitude is about 6 kΩ. This ground connection through an inductor is a type of hybrid ground. When using the inductor in this manner, its nonzero impedance may limit fault currents, including transient electrostatic discharges, to unsafe levels.

Imagine that both the source and load are single-ended grounded, and a shield is introduced around the signal conductor to reduce interference; that is, a coaxial cable is used to connect the source and load. Although it is only necessary to ground the electrically-short shield of the coax at one location to

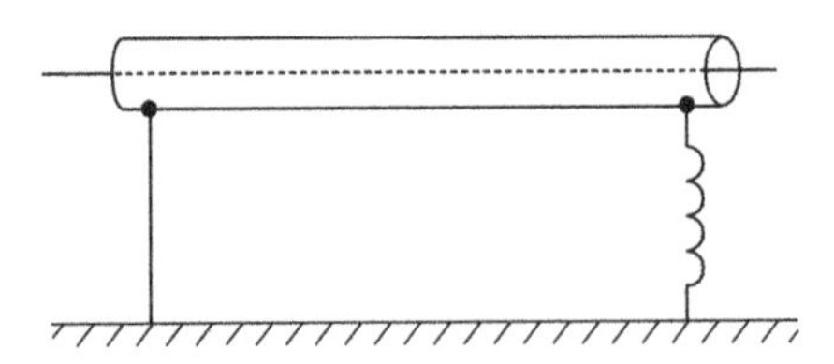

FIGURE 5.32 Inductor-based hybrid ground.

[13]A set of grounding straps of different lengths, cross sections, and electrical properties could, in theory, have the same impedance.

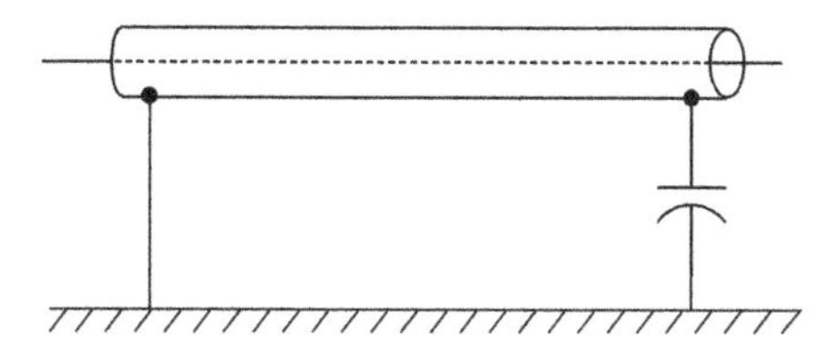

FIGURE 5.33 Capacitor-based hybrid ground.

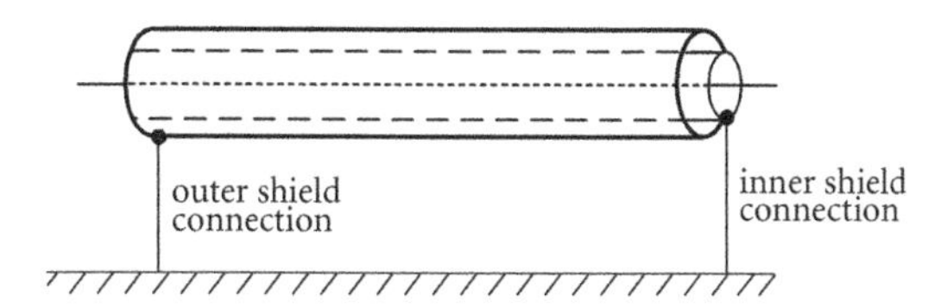

FIGURE 5.34 Triax cable connected to reduce low-frequency ground loops while allowing for a small pickup area for higher frequency signals.

reduce capacitive coupling, both ends of the shield must be connected to ground to reduce inductive coupling. As previously discussed, well above the cutoff frequency of the electrically-short cable, the current will tend to return through the shield rather than through the ground plane. Even when the ground plane is much larger in size than the shield, well above the cutoff frequency the current returns through the shield. However, for the current to return through the shield, the shield must be connected to the ground at both the source and load. In this case, to allow for these higher frequency shield connections to ground while reducing lower frequency ground loops, a capacitor-based hybrid ground can sometimes be used as illustrated in Figure 5.33. A 1 μF capacitor, for example, has an impedance magnitude of $1/\omega C \approx 3\,\mathrm{k}\Omega$ at 60 Hz while an impedance of about 0.2 Ω at 1 MHz. This method of grounding is used in audio circuits and local area networks to reduce low-frequency ground loops. Sometimes, if the capacitance between the inner and outer shields of an electrically-short triax cable is sufficiently large, an additional capacitor is not required. A triax cable contains two concentric cylindrical shields around a center conductor. As shown in Figure 5.34, the inner shield is connected to ground at one end of the cable, and the outer shield is connected to ground at the other end of the cable. There is no direct dc loop through the ground and cable shields. At high frequencies, the path from one ground connection to the other is provided by the parasitic capacitance between the shields. If the shields are sufficiently close (or high-dielectric insulation is used between the shields), the capacitance might be large enough to provide a low-impedance return "path" through the shields.

5.12 Dynamic Range between Systems

Compare the dynamic range for two products that are connected with one cable if the signal ground is connected to the chassis ground for both of the products vs. if the signal ground is not connected to the chassis ground for both of the products. [Whitlock, '98]

Before delving into the noise-grounding issues, the dynamic range (as commonly used in the audio world) should be qualitatively defined. Dynamic range is a measure of the maximum undistorted signal to the noise floor. It is commonly given in terms of dB. The maximum undistorted signal is generally affected, for example, by the supply voltages. For some op-amps, the maximum output voltage is about one volt less than the upper supply voltage and the minimum output voltage is about one volt greater than the lower supply voltage. The dynamic range is also a function of the noise level. This level is a function of natural noise sources (e.g., thermal noise and lightning) and man-made noise sources (e.g., motors). It is interesting

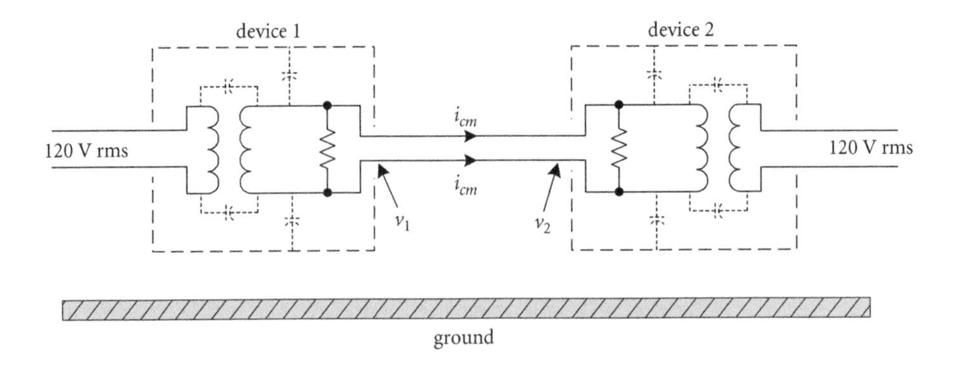

FIGURE 5.35 Common-mode current generated by two devices with floating chassis.

to note that the human ear has a dynamic range of about 120 dB, while high-end audio systems may require 120 dB and video systems may well require 50 dB. Ground loops and common-mode noise can degrade the dynamic range.

If the signal ground is not connected to the chassis ground, which may be connected to the earth ground for safety reasons, then the internal circuitry is considered floating. For a floating conductor, the potential of the conductor relative to the earth is a function of its environment, as well as its shape and size. The proximity of the conductor to the earth and nearby charged objects, such as the hot line of a power cord, will affect the potential of floating conductors. The potential of a floating signal reference or floating signal ground relative to the earth is a function of the parasitic capacitance between the floating reference, chassis, and earth. In Figure 5.35, two devices with floating circuitry are connected with a cable, and the local earth or ground is shown. The internal circuitry beyond the isolation transformer is simply represented by a single resistor in both devices powered by 120 V rms. Because the chassis, internal circuitry, and environments are likely to be different for both devices, the parasitic capacitances shown will also likely be different. Furthermore, the potentials where the cable exits and enters the two products will also likely not be the same. A set of these voltages is shown in the figure as v_1 and v_2. This potential difference between the two products will generate a common-mode current, i_{cm}, between the products. This common-mode current is one source of noise for the entire system. It is an interchassis current.

For safety reasons, the leakage current of each device (for an individual in contact with the device) is limited to *about* 1 mA or less. That is, if a short-circuiting wire is placed between the ground and any normal-consumer accessible point (e.g., on the chassis) on the device, a small but nonzero current will exist on this short-circuiting wire. The open-circuit voltage on the chassis of the device, relative to ground, can be 120 V rms. Also, the dc resistance between the chassis and ground should be very large since the chassis is floating. Because of these properties, the secondary circuitry is sometimes modeled using a current source: the short-circuit current is small, open-circuit voltage is large, and resistance to ground is large. Recall that good current sources have high impedances while good voltage sources have low impedances. Also, the short-circuit current for a current source is finite while the short-circuit current for a voltage source is generally large. This current source model for a floating device clearly shows one source of power frequency noise.

For grounded devices where the chassis ground is also connected to the signal ground in the device, some of the parasitic capacitances between the internal circuitry and chassis and between the internal circuitry and ground will be eliminated. A potential difference is required if the capacitance between two objects is to be relevant. In Figure 5.36, these ground connections are shown. The circuitry, as given, is unbalanced or single-ended grounded. The reason for connecting the signal and chassis grounds is to reduce common-mode noise. Of course, parasitic capacitances still exist between various parts of the internal circuitry and the grounded chassis as shown in this figure. However, at least when $v_{g1} \approx v_{g2}$, then the common-mode currents due to the ground loop formed from the low-side connections are potentially smaller than the floating situation. If the two potentials v_{g1} and v_{g2} are very different, then the common-mode

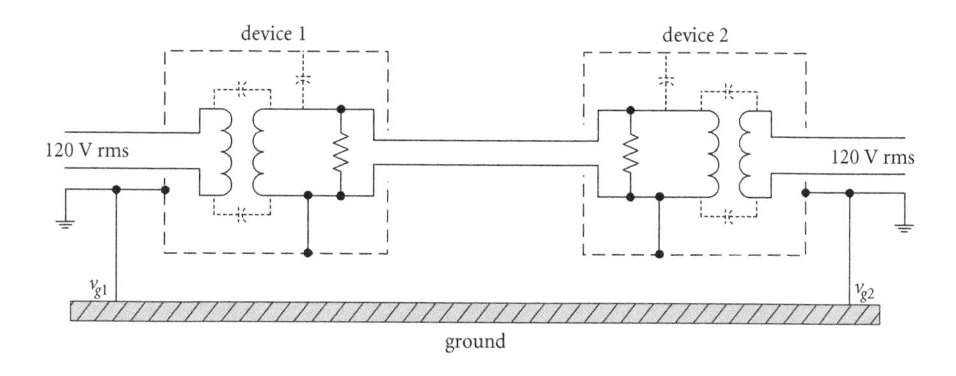

FIGURE 5.36 Grounding the chassis of two interconnected devices can reduce the effects of common-mode currents.

current due to this ground loop could be substantial. In conclusion, connecting the signal and chassis grounds can provide some noise-reduction and dynamic range improvement assuming the potential between the chassis ground connections is small.

By connecting the signal and chassis grounds as shown in Figure 5.36, a ground loop has been introduced. Many times, especially at low frequencies, these ground loops should be avoided. (There are situations, for example, when the connecting cable is shielded twisted pair, where connecting both ends of the cable to the respective chassis can be beneficial. In this case, both the metal chassis and cable shield act as a single shield.) These ground loops are due to potential differences between grounds. To break the ground loop path when the potential difference between the chassis ground connections becomes too large, either the driver or load can be left floating.

Although the internal circuitry shown in these figures involves floating or single-ended grounded devices, similar arguments can be used for balanced drivers and receivers. All balanced devices are not perfect and the parasitic capacitances to the lines are not identical; thus, the potential difference between floating devices can generate noise. This noise will reduce the dynamic range of the system.

5.13 Multiple Returns in Ribbon Cable

Why are multiple returns in a ribbon cable a good design practice? When flat ribbon cables are stacked on top of each other, even multiple returns may not help. Devise a method of reducing the problem associated with stacking ribbon cables. [Ott; Mardiguian, '84; Matisoff; Morrison, '95; Weston; Belden; Paul, '78]

A ribbon cable or flat cable is a multiconductor flat cable used in digital systems. They are used, for example, to connect computer cards designed to perform tasks such as graphics and analog I/O. The number of conductors in a ribbon cable, which are individually insulated and parallel to each other, typically range from 10–60. Two advantages of ribbon cables are the ease with which the insulation can be removed (with the proper tools) from a large number of conductors simultaneously and the ease with which connectors can be placed on the end of the cable. Standard ribbon cables are also very flexible and can often fit in spaces that are considered unusable (e.g., along the side of a cabinet). Furthermore, because of the large surface area of flat cables, they have certain heat dissipation advantages. Compared to a loose collection of wires, ribbon cables are a "controlled" cable since the position and orientation of the conductors are fixed relative to each other.

Technically, ribbon cables are one type of flat cable, although the term ribbon cable is frequently used for many cable types that are flat. As shown in Figure 5.37, a ribbon cable is a group of round wires bonded together in a flat linear array and enclosed by insulation. Ribbon cables are often color coded for quick identification of the individual conductors. Other types of flat cables include parallel rectangular conductors surrounded by insulation, twisted-pair flat cables, shielded flat cables, and sets of coaxial lines.

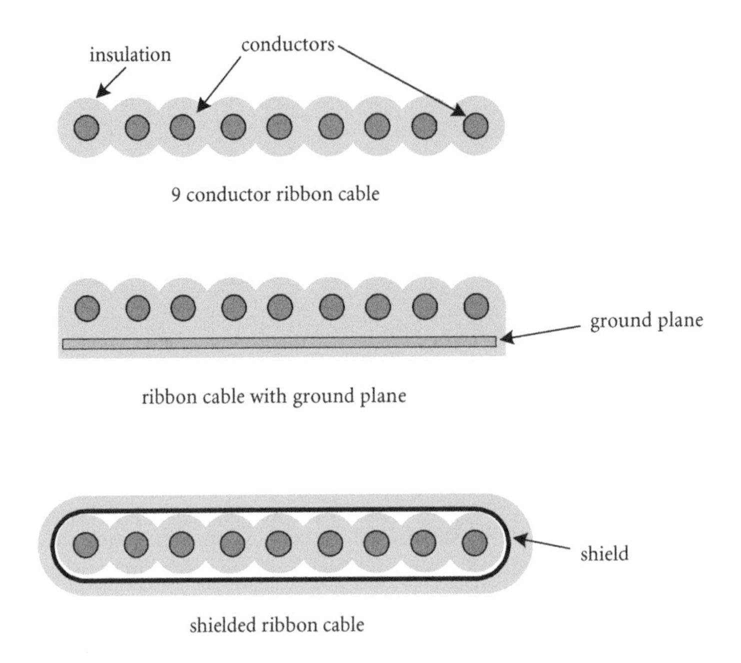

FIGURE 5.37 Ribbon cable without a ground plane, with a ground plane, and with a surrounding shield.

There are disadvantages associated with the use of ribbon cables that are mostly EMC related. Probably the major disadvantages are the crosstalk that occurs between the various conductors and the radiation from and susceptibility of the cable. This crosstalk occurs between not only adjacent conductors but between all of the conductors to various degrees. To understand why this occurs, first briefly study Figure 5.38 where only one of the conductors in the cable is used as the return or ground line for all of the other

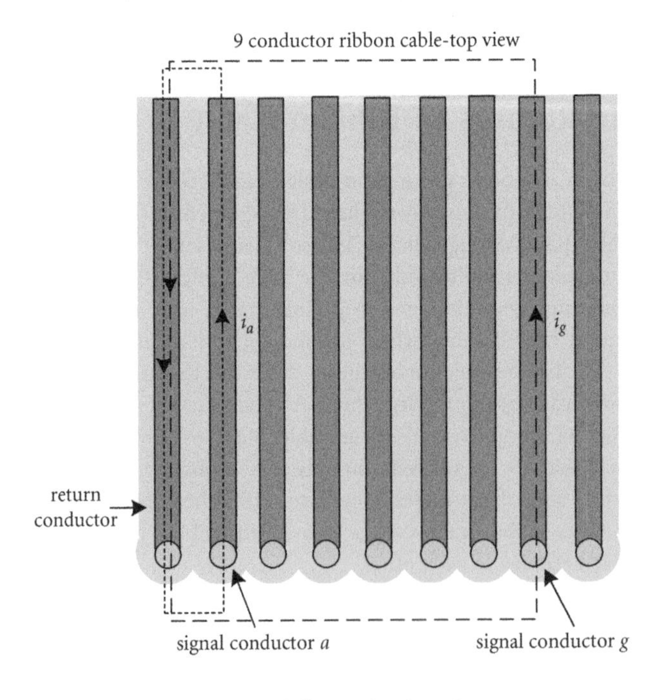

FIGURE 5.38 Common return conductor for two different signal conductors.

signal conductors. The major current loops for two different signals on conductors *a* and *g* are shown. Note that the loop generated by conductor *g*'s current surrounds the loop generated by conductor *a*'s current. Therefore, the time-varying field generated by current i_g will easily induce a current in conductor *a*. Similarly, the field generated by a time-varying current i_a will induce a current in conductor *g*. The actual induced currents (and voltages) are a complicated function of the mutual capacitance and inductance between each of the conductors and the source and load impedances between each conductor and the return conductor. It has been shown that to predict accurately the crosstalk between each of the wires, the interactions between all of the wires on the cable must be considered, including those inactive and floating wires. Because of the large number of conductors in a ribbon cable, accurate crosstalk analysis should normally be based on a reliable numerical program.

At lower frequencies, common-impedance coupling is an issue. The impedance of the return conductor (i.e., its resistance and inductive reactance) becomes important. The voltage drop along the return conductor is a function of all of the currents returning along it. Therefore, the voltage of the return conductor, which is the reference conductor for all of the signal lines, will vary with the signal currents. This variation of the voltage across or current through a common conductor is referred to as common-impedance coupling.

Two other important issues are the emissions or radiation from ribbon cables and the susceptibility of ribbon cables to external noise. For electrically-short cables, both the emissions and susceptibility increase with the length of the cable and with the loop area generated by each conductor and its return. When only a single return conductor is used, the loop area generated by the conductors that are not adjacent or near to the return conductor can generate significant emissions and be quite susceptible to external noise.

There are various methods that are used to improve the performance of ribbon cable. One method is to use every other conductor or more than one conductor as a return or ground. Two possible schemes are *GSGSGS*... or *GSSGSSG*... where *S* and *G* represent signal and ground conductors, respectively. These schemes essentially reduce the loop area for the signal and its return, reducing emissions, susceptibility, and crosstalk. Although a percentage of the current for a signal can pass through returns that are not nearby, most of the current will return through the path of least impedance. In this case, the path of the least impedance is generally where the loop area and, hence, inductance is the smallest. Common-impedance coupling is also reduced when multiple returns are used. It is not always necessary to use a separate or nearby return for every signal conductor. A nearby return should be used for critical lines such as enabling or strobing signals. The obvious cost of having multiple return conductors is a loss in conductors that could be used for carrying more data.

Another method of extending the range of ribbon cables (i.e., their usable length and upper frequency of operation) is to use balanced differential sources and receivers. Most students are familiar with unbalanced single-ended sources; that is, sources where one side of the supply is grounded and, therefore, the return for the signal is also grounded. When balanced sources (and receivers) are used, neither side is connected directly to the ground or signal reference. To maintain the balance of the sources, two conductors are required for each source.

As shown in Figure 5.37, flat cables are also available with flat conducting returns. If a return has to be shared among several signals, then the impedance of the return should be as small as possible. The addition of this large return conductor can substantially reduce the mutual capacitance and inductance between the signal conductors. It also reduces the loop area generated by the signal and its return current. The return current for each signal will be concentrated directly under the wire in the ground plane. It is necessary, however, to terminate properly the ground plane at both ends of this type of ribbon cable with a full-width connection to the system ground. Shielded ribbon cables are also available but require a full 360° connection for effectiveness; otherwise, pigtail-related problems can arise. It is important to restate that the link between the shield and return plane of the cable and the equipment should be as complete and continuous as possible. A single pigtail or drain wire is normally inadequate, especially at higher frequencies.

Another type of flat cable uses multiple sets of twisted pair in a flat package (referred to as "Twist-'n-flat" or "Vari-Twist"). By twisting the pairs, the differential radiation (radiation from the currents in the signal and return that are in opposite directions) from the wires is substantially reduced. The radiation from the common-mode signal (radiation from the currents in the signal and return that are in the same direction) is not affected (much) by the twisting of the wires. Unfortunately, these twisted-pair multi-conductor cables can have flat termination areas spaced along the cable for termination or mounting. In these untwisted areas of the cable, the EMC advantages of twisted lines are lost.

Flat cables are also available that consist of multiple, miniature, parallel coaxial cables, each with its own inner conductor, concentric outer conductor, and drain wire(s).[14] This ribbon coaxial cable was designed for high-speed computer applications. Finally, ribbon cable is also available that offers the ease of installation at the terminations but the possible flexibility of a round cable. The ribbon cable is essentially squeezed or spiraled into a round jacket. Outer shields are also available. All of these non-standard ribbon cables add to the cost of the cable and its installation.

The disadvantages of a ribbon cable's "large" capacitance are occasionally debated. When the capacitance of ribbon cables is measured, it is seen that it is somewhat larger than many other cables (and larger than an unbundled set of wires). Although this larger capacitance can limit the useful length of ribbon cables because of the distortion and signal source loading that can occur with excessive capacitance, it is more likely that crosstalk will limit the useful length of the ribbon cable. A rule-of-thumb seen for the maximum length for ribbon cables is around 10 ft. However, various techniques, including increasing the rise and fall time of the signals on the conductors, can extend this length.

When flat cables are stacked, coupling does occur between the different cable layers. Increasing the distance between the layers, using individually shielded flat cables, or inserting a shield between the layers can reduce the crosstalk between cables.

5.14 Loose Wires as a Cable

Two loose wires are used to connect a video circuit to a scope. Why is this not a good idea?

Not only is the characteristic impedance not very predictable or even well defined for the two loose wires, which is most relevant when the wires are not electrically short, but the wires are likely to be susceptible to inductive coupling. This coupling is strongly a function of the loop area of the victim circuit. Loose wires, unlike cables such as coax and twisted pair, are likely to have a much greater pickup loop area. Also, the video circuit and its connection to the scope are more likely to be a source of emissions or interference to other circuits.

5.15 Transfer Impedance

What is the transfer impedance, how does it typically vary with frequency, and what are typical values? [Tsaliovich; Vance; DeGauque; Goedbloed; Chatterton; Schelkunoff; Guofu; Young; Hoeft; Haus]

There are many types of shielded transmission lines, and sometimes it is difficult to compare their shielding and crosstalk-reduction effectiveness. The outer shield, for example, can be solid, braided, or taped. The shield material can be nonmagnetic, such as copper, or magnetic, such as steel.[15] There can be one or multiple shields. In addition, there can be more than one conductor within the shield, and pairs of these conductors, such as twisted pair, can also be individually shielded. With the exception of a few cases, it is extremely difficult to determine the effectiveness of these shields starting from Maxwell's equations. Thus, experimental methods were developed to determine the effectiveness of these shielded cables in screening external electric and magnetic fields. One measure of the effectiveness of a shield is its transfer impedance.

[14] A drain wire is a conductor that runs parallel to and in contact with the outer conductor of a coaxial cable. Connecting to the drain wire at the ends of a coaxial cable is much easier than connecting to the typical braided or delicate-foil outer shield of the cable. However, the resultant pigtails produced have certain EMC disadvantages.

[15] As throughout this book, a magnetic material is one with a relative permeability usually much greater than one.

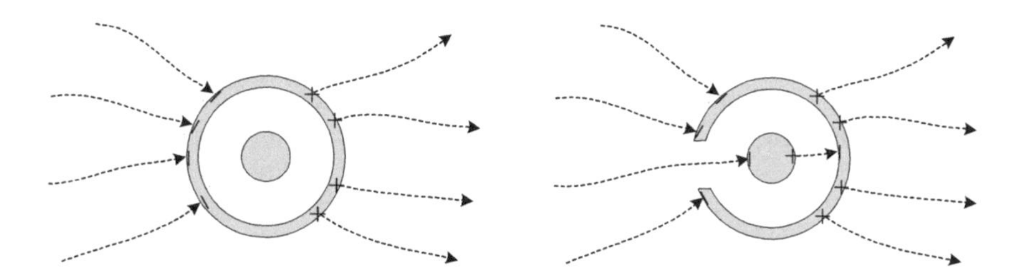

FIGURE 5.39 Cross-sectional view of a coaxial cable, without and with an aperture, immersed in an electric field.

A brief qualitative discussion will be provided of the effect of conductors on electric and magnetic fields. Imagine that a long coaxial cable is immersed in a near-field electric field as shown in Figure 5.39. The original field is not necessarily uniform. Since the cable is conducting (and probably also constructed using insulators with dielectric constants greater than one), the cable will distort the field. If the outer shield is a decent solid conductor, then the electric field inside the shield due to this applied external field is extremely small even for thin shields. (It is zero for electrostatic fields.) This is sometimes referred to as Faraday shielding. For electric fields, many conductors are considered good or even nearly ideal, and the electric field is mostly normal to the surface of the conductor. If the shield is not solid but contains apertures or openings, then the electric field can induce charges on the outer surface of the inner conductor (and the inner surface of the shield). Assuming that the inner conductor was initially charge neutral, for every negative charge induced on the inner conductor, an equal number of positive charges must appear elsewhere on the inner conductor. Notice in Figure 5.39 that an electric field exists between the inner and outer conductor. If the external electric field is time varying, the sign and magnitude of this induced charge will vary. This induced field will be modeled later using a capacitor and current source.

In many ways, it is much easier to understand electric field coupling than magnetic field coupling. A conducting nonmagnetic shield, for example, usually has little effect on low-frequency magnetic fields. A conducting magnetic shield, on the other hand, can produce some field "ducting," routing a portion of the external low-frequency magnetic fields around it. Unlike electric fields, magnetic fields can be both normal and tangential to conductors. Therefore, the effect of both tangential and transverse (or perpendicular) magnetic fields to the axis of the coaxial structure will be examined. In Figure 5.40, the external magnetic field, H_{tex}, is entirely tangential to a long coaxial structure. The internal magnetic field, H_{tin}, is also tangential. The strength of the internal magnetic field is a function of the thickness, radius, conductivity, and permeability of the outer shield and the frequency of the magnetic field. When the magnetic field is time varying, it induces a circulating current on the outer shield (and inner conductor). The magnetic field and induced current are not parallel. In real conductors, these currents generate ohmic power loss. The major consequence of these longitudinal magnetic fields between the inner and outer

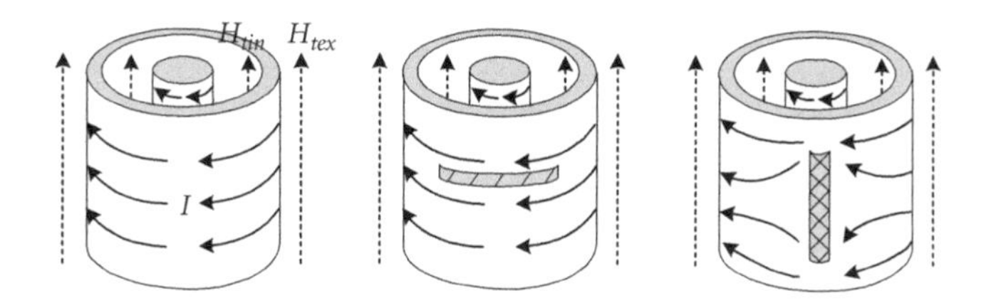

FIGURE 5.40 Coaxial cable, without an aperture and with two different apertures, immersed in a tangential magnetic field.

conductors is the resultant induced EMF or voltage. According to Faraday's law, the induced voltage produced by a time-varying magnetic field passing through a closed loop is proportional to the area of the loop and strength and frequency of the applied field. The circulating current around the outer conductor is essentially a one-turn loop. The nonzero inductance of this loop has a nonzero impedance and, hence, a voltage drop across it when a current is passing through it. Thus, this field "leakage" through the shield will be later modeled as an inductance. When a slot or an aperture is present, its effect on both the internal and external fields is a function of the size, shape, and orientation of the aperture. For example, for the circumferential slot shown in Figure 5.40, the effect on the magnetic fields is less than with the longitudinal slot shown. If the circumferential slot is thin, then the current on the outer shield will still be mainly circumferential. However, when the slot is longitudinal, the slot disrupts the current along its entire length: conduction current cannot "jump" across the gap of the slot. The current distribution on the conductor and its magnetic field will be disturbed or distorted. The apertures will produce other field components besides the tangential components shown. Again, the aperture increases, at least locally near the opening, the field between the conductors.

A magnetic field transverse to the axis of the coaxial shield also passes easily through the shield unless the frequency is very high or magnetic material is used. Actually, a field typically has both a radial, or normal, component to the shield and a circumferential, or tangential, component to the shield. A sketch of the magnetic field distribution of a thin, conducting cylindrical shell immersed in a transverse uniform magnetic field is shown in Figure 5.41. The current in the shell, at a particular instant of time, is out of the page along the bottom half of the cylinder and into the page along the top half of the cylinder. If the length of the cylinder is electrically short, then significant current cannot exist along the length of a cylinder with open ends (i.e., the current must go to zero at each end).[16] However, if both ends have short-circuiting caps, or the currents deviate significantly from the z direction at the ends (i.e., end effects through parasitic capacitance), then the current can pass in one direction down the length of the cylinder and return down the other side. Although not shown, if a conductor exists inside this cylindrical shield, longitudinal current can also be induced along it assuming the current has some return path, such as the outer shield or some other external circuit. One consequence of longitudinal current is power loss. This power loss is a function of frequency since the distribution of current from the outside to the inside of the cylinder will be a function of the skin depth (or similar parameters). The strength of the magnetic field inside the shell is also a function of frequency, decreasing with increasing frequency. Unlike the fields shown in Figure 5.40, this one-turn inductor does not pick up any external flux since the normal to its cross-sectional area is entirely perpendicular to the magnetic field. Of course, apertures in the shield can distort the field, generating a longitudinal component. Often with an electrically-short shield, the

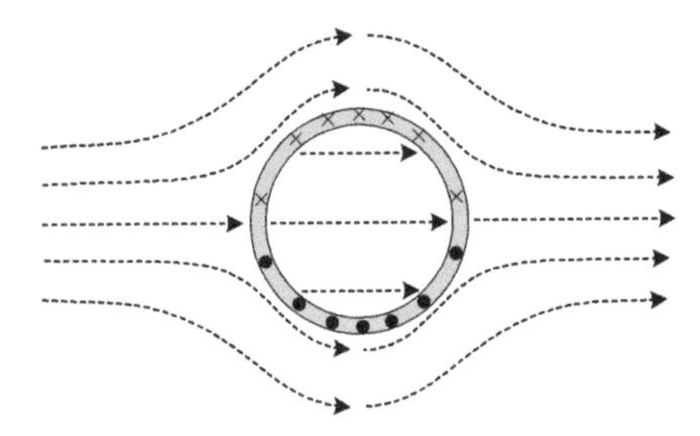

FIGURE 5.41 Cross-sectional view of cylindrical shell immersed in a uniform magnetic field. A sketch of the direction and distribution of the current induced in the shell is also shown.

[16]In a circuit, which is electrically small, a closed path is necessary for current to be nonzero. A closed path is not necessary for electrically-long structures where the current can be zero at the ends and yet nonzero at other locations in between.

current is mainly in one direction and the return path is not the shield itself but some other external circuit.

Although qualitatively determining the field distributions for a shielded cable is insightful, it does not provide a quantitative measure of the cable's shielding effectiveness. Electrical engineers generally prefer to work with currents, voltages, capacitances, and inductances rather than magnetic and electric fields. Although other figure of merits for a shield will and have been discussed, one popular measure of a shielded cable (including connectors, wire meshes, and gaskets) is its transfer impedance. Although it will be discussed more fully later, the transfer impedance of a cable oriented in the z direction is defined as[17]

$$Z_t = \frac{1}{I_e}\frac{dV_t}{dz} \ \Omega/\text{m} \tag{5.23}$$

where I_e is the current injected into the outer shield by an external circuit and V_t is the voltage generated across the inner surface of the outer shield. The transfer impedance is an impedance per unit length. Because of the differential nature of this expression, it is normally used for shield lengths that are electrically small. If the length of the cable under test is l_{th}, then the voltage across the inner surface of the shield is approximately

$$Z_t \approx \frac{1}{I_e}\frac{V_t}{l_{th}} \quad \Rightarrow V_t = I_e l_{th} Z_t \tag{5.24}$$

It is not convenient to measure directly the voltage along the inner surface of the shield. One common method of indirectly measuring the voltage induced across the inner side of the shield is shown in Figure 5.42. An external source injects current into the shield and provides the necessary closed path. The voltage is measured between the center conductor and shield at one end of the cable while the two conductors are short circuited at the other end. It might initially appear that the voltage measured is not actually V_t but the voltage across the shield and center conductor. However, the current in the center conductor is zero (or, at least, small for an electrically-short sample and a high-impedance voltage measurement instrument). Therefore, the voltage measured is indeed V_t. If the total injected current, I_e, and length, l_{th}, are known, the transfer impedance of the cable can be determined from (5.24).

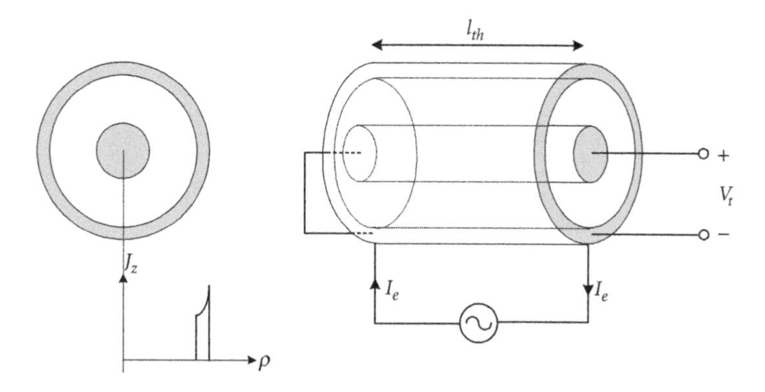

FIGURE 5.42 Sketch of the distribution of current in the conductors of a coaxial cable when current is intentionally injected into the shield.

[17]Throughout this section, the *s* subscript typically used in this book for phasor or frequency-domain quantities is omitted.

It is not always possible to determine the actual current along the surface of a shield due to an external electromagnetic field or to predict the common-mode current in a shield for every cable application. For that matter, the actual current in the shield or the resultant noise voltage that it generates in the source or load of a cable will vary from one situation to the next. The injected current, I_e, is simulating a disturbance. By determining the transfer impedance of a cable by intentionally injecting current into the shield, one measure of the "goodness" of a cable can be determined.

It might surprise some readers that the voltage measured across the outer surface of the shield can be different from the voltage measured across the inner surface of the shield. At low frequencies where the current is essentially uniformly distributed through the shield, these voltages are the same and about equal to $I_e R_{DC}$ where R_{DC} is the dc resistance of the shield. For higher frequencies, the current distribution is not uniform over the thickness of the outer shield. When the current is injected into the shield as shown in Figure 5.42, the current density, J_z, is greatest along the outer surface and decreases with increasing distance into the shield. For very high frequencies, where the skin depth, $\delta = 1/\sqrt{\pi f \mu \sigma}$, is small compared to the thickness of the shield, the current density along the inner surface is small. For electrically-short lengths, the voltage over a length l_{th} of the shield's surface is

$$V = -\int \vec{E} \cdot d\vec{L} \approx -E_z l_{th}$$

where E_z is the electric field tangential to the surface. (For a perfect conductor, this tangential electric field is zero.) Since $J_z = \sigma E_z$ where σ is the conductivity of the shield, E_z and V are both smaller across the inner surface.

For a solid shield, the transfer impedance decreases with frequency. A small Z_t is desirable from a shielding standpoint since it implies that little energy from the induced current on the shield passes through to the inner side of the shield. The exact expression for the transfer impedance for a cylindrical shield of inner radius r_i, external radius r_e, conductivity σ, and permeability μ is

$$Z_t = \frac{1}{2\pi\sigma r_i r_e [I_1(\gamma r_e) K_1(\gamma r_i) - K_1(\gamma r_e) I_1(\gamma r_i)]} \quad \text{where } \gamma = \sqrt{j\omega\mu\sigma} = \frac{1}{\delta} + j\frac{1}{\delta} \tag{5.25}$$

where δ is the skin depth, I_1 is the modified Bessel function of the first kind of order one, and K_1 is the modified Bessel function of the second kind of order one. Because of the complexity of this expression, the following approximation is usually used instead:

$$Z_t \approx R_{DC} \frac{(1+j)\frac{\Delta}{\delta}}{\sinh\left[(1+j)\frac{\Delta}{\delta}\right]} \quad \text{where } R_{DC} = \frac{1}{2\pi a \sigma \Delta}, \ \delta = \frac{1}{\sqrt{\pi f \mu \sigma}} \tag{5.26}$$

where $a = \sqrt{r_i r_e}$ and $\Delta = r_e - r_i$. It is easily obtained by using the large argument approximations $I_1(x) \approx e^x/\sqrt{2\pi x}$ and $K_1(x) \approx e^{-x}/\sqrt{2x/\pi}$. This approximate expression assumes that the cylindrical shell is a good conductor and the shell thickness is small or thin relative to the radius: $a \gg \Delta$. In addition, it assumes that $a \gg \delta$ so that the cylindrical shell appears like a flat surface. Even (5.26) is a complex quantity with a magnitude and phase, which are a function of the frequency. At low frequencies, (5.26) reduces to

$$Z_t \approx R_{DC} \frac{(1+j)\frac{\Delta}{\delta}}{(1+j)\frac{\Delta}{\delta}} = R_{DC} \quad \text{if } \delta \gg \Delta \tag{5.27}$$

since $\sinh x \approx x$ when x is small. This transfer impedance is just the dc resistance of the shield at low frequencies as previously stated. At high frequencies, (5.26) reduces to

$$Z_t \approx \frac{1}{2\pi a\sigma\Delta}\frac{2(1+j)\frac{\Delta}{\delta}}{e^{(1+j)\frac{\Delta}{\delta}}} = \frac{1}{\pi a\sigma\delta}(1+j)e^{-(1+j)\frac{\Delta}{\delta}} \quad \text{if } \delta \ll \Delta \tag{5.28}$$

since $\sinh x \approx e^x/2$ for large positive x. The magnitude of (5.28) is

$$|Z_t| = \frac{1}{\pi a\sigma\delta}\sqrt{2}\left|e^{-\frac{\Delta}{\delta}}e^{-j\frac{\Delta}{\delta}}\right| = \frac{\sqrt{2}e^{-\frac{\Delta}{\delta}}}{\pi a\sigma\delta} \quad \text{if } \delta \ll \Delta \tag{5.29}$$

since $|e^{jx}| = 1$. One source stated that the transfer impedance decreases at a rate of 20 dB/decade. However, the transfer impedance does not necessarily decrease at this rate:

$$\begin{aligned} 20\log|Z_t| &= 20\log\left(\frac{\sqrt{2}\sqrt{\pi f\mu\sigma}e^{-\Delta\sqrt{\pi f\mu\sigma}}}{\pi a\sigma}\right) = 20\log\left(\frac{\sqrt{2\mu}\sqrt{f}}{\sqrt{\pi\sigma}ae^{\Delta\sqrt{\pi f\mu\sigma}}}\right) \\ &= 20\log\left(\frac{1}{a}\sqrt{\frac{2\mu}{\pi\sigma}}\right) + 10\log f - 20\log\left(e^{\Delta\sqrt{\pi f\mu\sigma}}\right) \\ &= 20\log\left(\frac{1}{a}\sqrt{\frac{2\mu}{\pi\sigma}}\right) + 10\log f - 20\Delta\sqrt{\pi f\mu\sigma}\log e \\ &\approx 20\log\left(\frac{1}{a}\sqrt{\frac{2\mu}{\pi\sigma}}\right) - 15\Delta\sqrt{\mu\sigma}\sqrt{f} \quad \text{if } 20\Delta\sqrt{\pi f\mu\sigma}\log e \gg 10\log f \end{aligned} \tag{5.30}$$

For a factor of ten change in the frequency, the change in the transfer impedance in dB is a function of the frequency and the thickness and electrical properties of the shield:

$$\begin{aligned} \Delta_{dB} &= 20\log|Z_t(10f)| - 20\log|Z_t(f)| \\ &\approx 20\log\left(\frac{1}{a}\sqrt{\frac{2\mu}{\pi\sigma}}\right) - 15\Delta\sqrt{\mu\sigma}\sqrt{10f} - \left[20\log\left(\frac{1}{a}\sqrt{\frac{2\mu}{\pi\sigma}}\right) - 15\Delta\sqrt{\mu\sigma}\sqrt{f}\right] \\ &= 15\Delta\sqrt{\mu\sigma}\left(\sqrt{f} - \sqrt{10f}\right) \end{aligned} \tag{5.31}$$

If f = 1 MHz, Δ = 1 mm, $\mu = 4\pi \times 10^{-7}$, and $\sigma = 5.8 \times 10^7$, then $\Delta_{dB} \approx -280$ dB while if f = 10 MHz, $\Delta_{dB} \approx -880$ dB![18] Although, obviously, the transfer impedance does not suddenly pass from the low-frequency to the high-frequency region, a reasonable transition frequency is where $\delta = \Delta$ or

$$\frac{1}{\sqrt{\pi f_c \mu\sigma}} = \Delta \quad \Rightarrow f_c = \frac{1}{\Delta^2 \pi\mu\sigma}$$

[18] Yes, these values are only theoretical "paper" results. As with all real shields, seams and other openings limit the effectiveness to values well below these numbers.

It is interesting to note that near this transition frequency, the transfer impedance does decrease at a rate, from (5.31), of about 20 dB/decade:

$$\Delta_{dB} = 15\Delta\sqrt{\mu\sigma}\left(\sqrt{\frac{1}{\Delta^2\pi\mu\sigma}} - \sqrt{\frac{10}{\Delta^2\pi\mu\sigma}}\right) = \frac{15}{\sqrt{\pi}}(\sqrt{1} - \sqrt{10}) \approx -18\text{ dB} \tag{5.32}$$

The solid-shield transfer impedance expression is important, not since most shields are solid, since many are not, but since it provides a well-defined limit for a single, solid ideal shield. For this reason, the solid shield cable can be used as a standard, reference, or calibration cable. Other expressions will now be provided for shields with openings. All of these transfer impedances are greater in magnitude than (5.26) since they allow additional magnetic fields through the shield. The magnetic field coupling is often modeled as a mutual inductance, M. Since the impedance of this mutual inductance is $j\omega M$, the shield transfer impedance will eventually increase with frequency rather than decrease. In general, the transfer impedance is the sum of a diffusion term and this aperture mutual-inductive coupling term:

$$Z_t = Z_d + j\omega M \tag{5.33}$$

Real shields, as will be seen, show this behavior. Again, for an ideal solid shield, $M = 0$ and Z_d is given by (5.25). For a thin perforated pipe of radius a and thickness Δ,

$$Z_t = R_{DC}\frac{(1+j)\dfrac{\Delta}{\delta}}{\sinh\left[(1+j)\dfrac{\Delta}{\delta}\right]} + j\omega\mu_o v\frac{d^3}{24\pi^2 a^2} \tag{5.34}$$

where v is the number density of circular apertures or holes per unit length, each of diameter d, in the pipe. (This expression was derived by replacing each circular hole with a magnetic dipole and using the concept of magnetic polarizability.) The value of the dc resistance in (5.34) is sometimes modified (i.e., increased) to account for holes and resultant loss of material in the shield.

A braided shield consists of several belts (referred to as carriers) of parallel wires or strands. These carriers are weaved together to form the braided shield. For a braided shield, when the apertures between braids are modeled as elliptical in shape even though they are more like a rhombus or diamond, the transfer impedance is

$$Z_t = \frac{4}{\pi d_w^2 N \wp \sigma\cos\psi}\frac{(1+j)\dfrac{d_w}{\delta}}{\sinh\left[(1+j)\dfrac{d_w}{\delta}\right]} + j\omega\frac{v\alpha_m\mu_o}{4\pi^2 a^2} \tag{5.35}$$

where a is the average radius of the shield, v is the number of apertures per unit length, $\wp$ is the number of carriers, N is the number of wires per carrier, α_m is the magnetic polarizability of each aperture, d_w is the diameter of each wire, σ is the conductivity of each wire, and δ is the skin depth in each wire. Referring to Figure 5.43, for each aperture, L_g is the major axis and L_p the minor axis of the elliptical approximation of the diamond-shaped opening. The magnetic polarizability, in terms of $K(e)$ and $E(e)$, the complete elliptical functions of the first and second kind, is

$$\alpha_m = \begin{cases} \dfrac{\pi L_g^3}{24}\dfrac{(1-e^2)e^2}{E(e)-(1-e^2)K(e)} & \text{if } \psi < 45° \\ \dfrac{\pi L_g^3}{24}\dfrac{e^2}{K(e)-E(e)} & \text{if } \psi > 45° \end{cases} \tag{5.36}$$

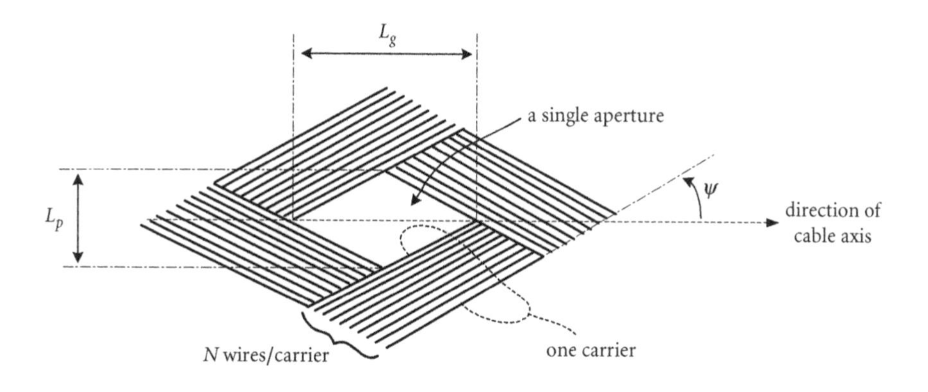

FIGURE 5.43 Few of the parameters used to describe the braid of a cable screen.

where $e=\sqrt{1-(L_p/L_g)^2}$ is the eccentricity of the opening. The relationship between e and ψ, the pitch angle of the weave measured from the cable's axis, is

$$e=\begin{cases}\sqrt{1-\tan^2\psi} & \text{if } \psi<45^\circ\\ \sqrt{1-\cot^2\psi} & \text{if } \psi>45^\circ\end{cases} \tag{5.37}$$

The first term in (5.35) is the diffusion term. This result assumes that the dc resistance of the braided shield is equal to the dc resistance of a no-aperture cylindrical shield. It also assumes that the high-frequency characteristics are the same as a one-wire diameter thick no-aperture shield. The second term in (5.35) is the leakage term due to the apertures.

One final theoretical transfer impedance expression will be provided—for spiral and tape shields. These shields are used when flexibility is important. However, they tend to act like a current-carrying solenoid wound about the center conductor, which negatively impacts their ability to shield. They can be used to complement other shield types such as the braided shield. Like any inductor, when the voltage drop per "turn" is sufficiently large, arcing can occur between turns, which is a source of electrical noise. Referring to Figure 5.44, the transfer impedance of the helical-wound nonoverlapping-tape shield is

$$\begin{aligned}Z_t=&\frac{\tan\psi}{\sigma\Delta w}\frac{(1+j)\dfrac{\Delta}{\delta}}{\sinh\left[(1+j)\dfrac{\Delta}{\delta}\right]}\left\{1+\tan^2(\psi)\cosh\left[(1+j)\frac{\Delta}{\delta}\right]\right\}\\&+j\omega\frac{\mu_o}{4\pi}\left[\left(1-\frac{c^2}{a^2}\right)\tan^2\psi+\frac{1}{N}\left(\frac{s}{4a}\right)^2\frac{1}{\cos\psi}\right]\end{aligned} \tag{5.38}$$

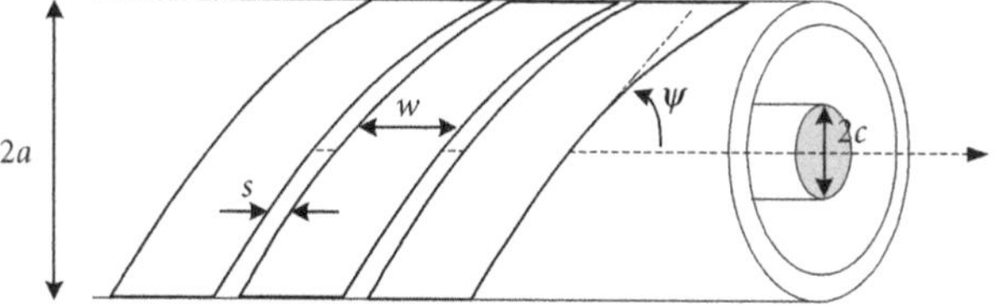

FIGURE 5.44 Parameters of a tape-wound shield.

$$a := 0.005 \quad \Delta := 10^{-3} \quad d := 10^{-3} \quad c := \frac{a}{4} \quad w := 0.005 \quad s := 0.001 \quad \psi := \frac{\pi}{4} \quad N := 1$$

$$\mu := 4\cdot\pi\cdot10^{-7} \quad \mu_o := 4\cdot\pi\cdot10^{-7} \quad \sigma := 5.8\cdot10^{7} \quad j := \sqrt{-1}$$

$$x := 20, 20.1 .. 70$$

$$f(x) := \left(x + 1 - 10\cdot\text{floor}\left(\frac{x}{10}\right)\right)\cdot10^{\text{floor}\left(\frac{x}{10}\right)} \qquad \delta(f) := \frac{1}{\sqrt{\pi\cdot f\cdot\mu\cdot\sigma}}$$

$$Z_{tsolid}(f, \Delta) := \frac{1}{2\cdot\pi\cdot a\cdot\sigma\cdot\Delta}\cdot\frac{(1+j)\cdot\frac{\Delta}{\delta(f)}}{\sinh\left[(1+j)\cdot\frac{\Delta}{\delta(f)}\right]}$$

$$Z_{tholes}(f, \Delta, v) := \frac{1}{2\cdot\pi\cdot a\cdot\sigma\cdot\Delta}\cdot\frac{(1+j)\cdot\frac{\Delta}{\delta(f)}}{\sinh\left[(1+j)\cdot\frac{\Delta}{\delta(f)}\right]} + j\cdot2\cdot\pi\cdot f\cdot\mu_o\cdot v\cdot\frac{d^3}{24\cdot\pi^2\cdot a^2}$$

$$Z_{ttape}(f, \Delta, w, s, \psi) := \left[\frac{\tan(\psi)}{\sigma\cdot\Delta\cdot w}\cdot\frac{(1+j)\cdot\frac{\Delta}{\delta(f)}}{\sinh\left[(1+j)\cdot\frac{\Delta}{\delta(f)}\right]}\cdot\left[1+\tan(\psi)^2\cdot\cosh\left[(1+j)\cdot\frac{\Delta}{\delta(f)}\right]\right]\right] \ldots$$

$$+ j\cdot2\cdot\pi\cdot f\cdot\frac{\mu_o}{4\cdot\pi}\cdot\left[\left[\left(1-\frac{c^2}{a^2}\right)\cdot\tan(\psi)^2\right] + \frac{1}{N}\cdot\left(\frac{s}{4\cdot a}\right)^2\cdot\frac{1}{\cos(\psi)}\right]$$

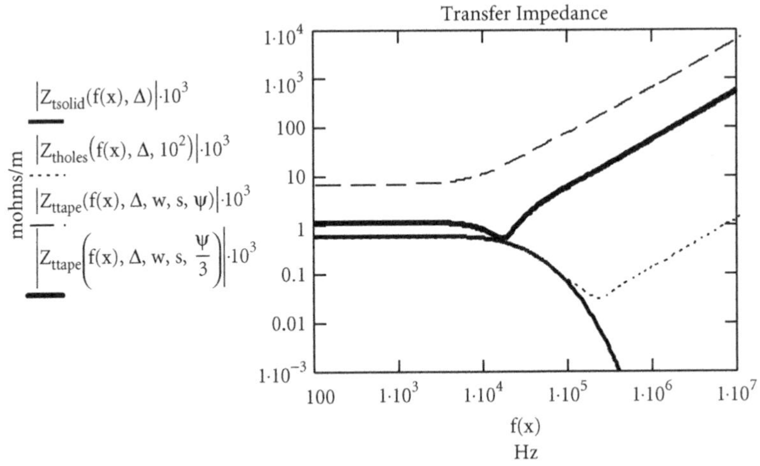

Transfer Impedance

mohms/m: $|Z_{tsolid}(f(x), \Delta)|\cdot10^3$, $|Z_{tholes}(f(x), \Delta, 10^2)|\cdot10^3$, $|Z_{ttape}(f(x), \Delta, w, s, \psi)|\cdot10^3$, $\left|Z_{ttape}\left(f(x), \Delta, w, s, \frac{\psi}{3}\right)\right|\cdot10^3$

Vertical axis: $1\cdot10^{4}$, $1\cdot10^{3}$, 100, 10, 1, 0.1, 0.01, $1\cdot10^{-3}$

Horizontal axis: 100, $1\cdot10^{3}$, $1\cdot10^{4}$, $1\cdot10^{5}$, $1\cdot10^{6}$, $1\cdot10^{7}$

f(x)
Hz

MATHCAD 5.1 Transfer impedance vs. frequency for a solid, perforated, and tape-wound shield.

where a is the radius of the shield, c is the radius of the inner conductor, N is the number of helical tapes wound in the same direction, ψ is the spiral angle of the tapes relative to the axis of the cable, w is the width of, s is the spacing between, Δ is the thickness of, σ is the conductivity of, and δ is the skin depth in each tape.

In Mathcad 5.1, the magnitude of the transfer impedance is plotted vs. frequency for a solid, perforated, and tape-wound shield. At lower frequencies, the impedance is dominated by diffusion while at higher frequencies, with the exception of the solid shield, magnetic coupling dominates. The radius and thickness of all of the shields is 0.5 cm and 1 mm, respectively. Although not shown, it is also common to plot the ratio Z_t/R_{DC} so that shields with the same low-frequency or dc resistance can be readily compared.

The measured transfer impedance of the solid shield is in excellent agreement with the theoretical expression. However, when a shield has holes, the theoretical expressions should be used with caution. There are many nonideal or additional factors with real cables that are not (easily) accounted for in the theoretical expressions. For example, a real single-layer braided cable has magnetic leakage between the strands in addition to between the carriers. The contact resistance between strands or any other joints can increase the transfer impedance. Also, with aging and flexing due to, for example, temperature changes, wind, and corrosion, the transfer impedance of a cable will change. Shields constructed of aluminum foil can be very fragile. A seam along a "solid" shield will also increase the transfer impedance. There are many other shield types that are difficult to model, including shields with multiple layers of braids and tapes.

When magnetic materials are introduced, the modeling is further complicated because of their nonlinear characteristics. Measurements have confirmed that a solid shield constructed of nickel, iron, Hipernom alloy, or even steel can provide a lower transfer impedance than a solid copper shield at higher frequencies (e.g., from 10 kHz to 1 MHz depending on the material). At lower frequencies, because of its higher conductivity, copper has a lower transfer impedance than these magnetic materials. Some cable manufacturers provide plots of the magnitude of the transfer impedance vs. frequency. However, there are several methods of performing these tests, and the results can be a function of the method used.[19] The length of the sample used in the test should be known since it will determine the upper frequency limit of the test before the electrically-short assumption is violated. For example, a 1 m long cable (assuming a dielectric constant of one) is electrically short up to about 30 MHz while a 1 cm cable is electrically short up to about 3 GHz. Generally, when resonances are seen in the plot, the cable is not electrically short at these higher frequencies.

Although the actual value of the dc resistance, R_{DC}, and leakage mutual inductance, M, will vary with the cable type, for coaxial cables, R_{DC} is in the 1–20 mΩ/m range and M is in the 0.1–1 nH/m range. A sketch of typical measured transfer impedances is shown in Figure 5.45. A solid copper shield is shown for reference with a low-to-high frequency transition frequency of about 1 MHz. The actual low-frequency impedance and cutoff frequency will vary with the thickness and material type. As is apparent for those cable types shown, the aluminum and mylar foil layer shield has the greatest low-frequency impedance.[20]

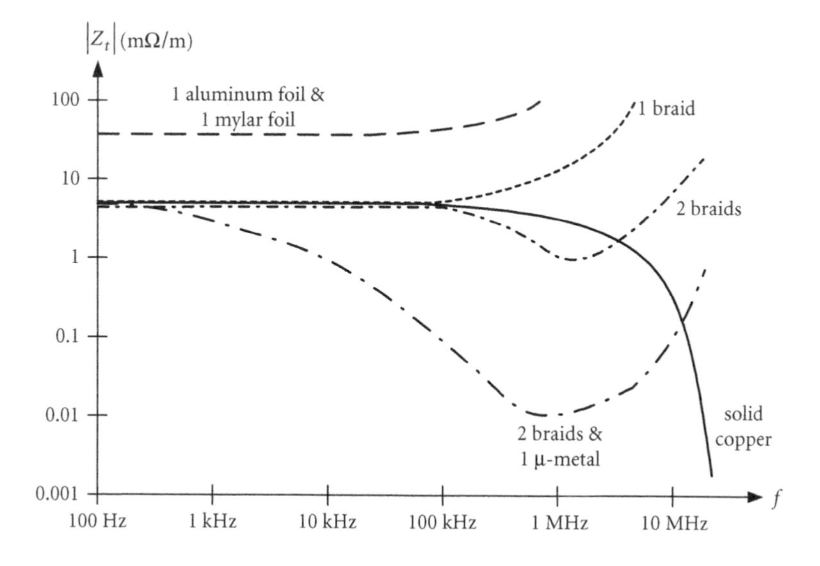

FIGURE 5.45 Typical measured transfer impedances. [Raychem]

[19]Every test method has a finite bandwidth associated with it.

[20]Actually, unlike the other cables shown, this two-layer foil shield does not correspond to a coaxial cable. There are two inner conductors, which were connected together during the measurement, inside the two-layer shield.

Although there are various methods to join foil along the length of the cable, even a longitudinal-fold of the foil will be a source of leakage. The single-braid shield has similar characteristics to the solid shield until around 100 kHz, where the leakage from the holes in the braid become important. A two-braid shield can actually outperform the solid copper shield over a limited frequency range. The initial dip seen in the response is due to the skin effect, resulting in a smaller current density on the inner surface of the shield. Eventually, though, the leakage of the braids dominate. Also shown is one "superscreen" or low-transfer impedance (LTI) shield composed a layer of μ-metal sandwiched between two braids. Mu-metal is a magnetic material with a permeability that is much greater than one over a certain frequency range. Double "superscreen" cables are also available. Magnetic materials can be very effective in reducing magnetic fields; however, their permeability is nonlinear and frequency-dependent. Generally, it is not difficult to obtain cables with transfer impedances similar to a solid copper shield from dc to 1–10 MHz. In more demanding applications involving, for example, electromagnetic pulses and secure government communications, there are methods of further reducing the transfer impedance of a cable. Not surprisingly, these methods often increase the complexity, cost, size, weight, and stiffness of the cable (and the connections to the cable). These methods, some of which are related, include

1. increasing the thickness and conductivity of the shield
2. increasing the optical coverage of the shield (i.e., reducing the number of holes by using a braid with a tighter weave)
3. using multiple layers of different shield types (e.g., braid, foil, and tape)
4. using a combination of nonmagnetic and magnetic layers (e.g., "superscreen" cables)
5. isolating the shield layers (e.g., triaxial cable)
6. reducing (or, ideally, eliminating) the gap of any seams.

Although the focus of this section has been coaxial cables, the transfer impedance of other cable types, wire meshes, gaskets, connectors, and the skin of an aircraft can also be measured.

When selecting a cable for a specific application, there are many factors to consider in addition to its shielding effectiveness. In some cases, it can be strongly argued that a cable's characteristic impedance is the most important electrical property of the cable. In other applications, the cable's power handling capability, capacitance, and bandwidth can be as important.

5.16 Loss Impedances and Transfer Admittance

Define the internal and external loss impedances for a cable. What precaution should be taken when measuring these impedances? Define the transfer admittance and determine its relationship to the transfer capacitance. [Tsaliovich; Vance; DeGauque]

In some cases, the transfer impedance can be used to estimate the coupling of an external source into a cable. However, although it is an excellent indicator of the effectiveness of a cable shield, it is not a complete measure of a cable's shielding performance for all circumstances.

To understand fully the origin of the various terms in the coupling model soon to be presented, the transfer impedance definition given in (5.23) will be redefined. It is still assumed that the cable under test is electrically small. Referring to Figure 5.42, recall that the transfer impedance was determined by injecting a current into the outer shield and measuring the open-circuit voltage between the inner conductor and shield. This transfer impedance will now be redefined as the immunity transfer impedance:[21]

$$Z_{it} = \frac{1}{I_e}\frac{dV_{it}}{dz}\ \Omega/\text{m} \tag{5.39}$$

[21]Throughout most of this section, the s subscript typically used in this book for phasor or frequency-domain quantities is omitted.

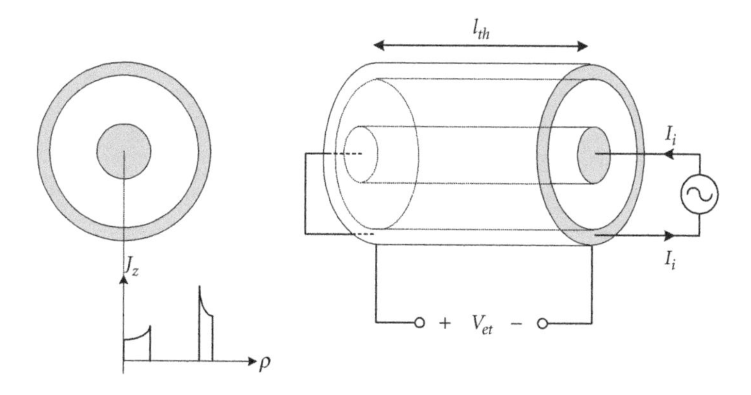

FIGURE 5.46 Measurement of the emissions transfer impedance.

where V_{it} is the voltage measured between the inner and outer conductors. The test voltage source, injecting the current into the shield, can also be applied between the inner and outer conductors of the cable, with the other end short circuited. By then measuring the voltage across the outer surface of the cable, as shown in Figure 5.46, the emissions transfer impedance can be determined:

$$Z_{et} = \frac{1}{I_i}\frac{dV_{et}}{dz}\ \Omega/\mathrm{m} \tag{5.40}$$

In this test, equal but oppositely directed currents exist in the inner and outer conductors. (For the immunity transfer impedance test, no current exists in the inner conductor.) Assuming the two conductors are concentric, the magnetic field outside the cable is zero. If the field were not zero, a time-varying magnetic field would pass through the loop formed by the outer shield and test leads, which would modify the voltage measurement via Faraday's law. If the voltmeter used to measure V_{et} has a high impedance, the current through the voltmeter leads should be small.

For certain situations, such as with nonmagnetic, thin solid tubes of homogeneous material, the impedances obtained from (5.39) and (5.40) are identical. If a shield consists of a nonmagnetic outer layer and magnetic inner layer, then these two transfer impedances would not necessarily be the same. When determining Z_{et} using a given current injection amplitude, the inner magnetic layer might saturate. However, when using the same injection current amplitude in determining Z_{it}, this magnetic layer may not saturate since the current density through it would be smaller. Reciprocity requires linearity.

For test purposes, when determining the immunity and emissions transfer impedances, the current through the voltmeter is assumed small. In any real application, current also exists along the inner and outer conductors due to other internal or external sources. For example, a desirable signal return current probably exists along the shield in addition to an external noise current. The total voltage measured across the inner surface of the shield would then consist of two voltage components. To account for both currents, two additional impedances will be defined. The internal loss impedance is

$$Z_{is} = \frac{1}{I_i}\frac{dV_{is}}{dz}\ \Omega/\mathrm{m} \tag{5.41}$$

and the external loss impedance is

$$Z_{es} = \frac{1}{I_e}\frac{dV_{es}}{dz}\ \Omega/\mathrm{m} \tag{5.42}$$

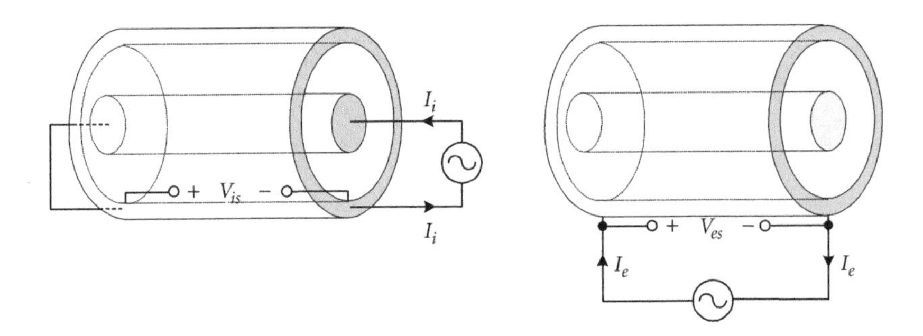

FIGURE 5.47 Voltages and currents used to define the internal and external loss impedances.

These variables are shown in Figure 5.47. The measurements of these voltages must be performed with greater care since, in both cases, there is a time-varying magnetic flux passing between the loop generated by the shield and voltmeter leads. This flux contribution can be reduced by decreasing the pickup area of the loop (e.g., keeping the leads close to the shield). This magnetic field contribution can be understood using one of Maxwell's equations. Faraday's law in the frequency domain states that

$$\oint_C \vec{E}_s \cdot d\vec{L} = -j\omega \iint_S \vec{B}_s \cdot d\vec{s}$$

The clockwise-directed, closed-path line integration is shown in Figure 5.48. The cross-hatched region represents the surface corresponding to this closed path, which is in the –*z* direction. The magnetic field is shown into the page for a particular instant of time. Therefore,

$$I_e Z_{es} - V_{es} = -j\omega \iint_S B_{zs} dx\, dy$$

The measured voltage is thus the sum of the desired voltage across the shield and the induced voltage due to the time-varying magnetic field passing through the closed path:

$$V_{es} = I_e Z_{es} + j\omega \iint_S B_{zs} dx\, dy \tag{5.43}$$

To repeat, the actual voltage measured by the voltmeter is the sum of the actual voltage across the shield and an induced time-varying magnetic field contribution. Even though V_{es} (i.e., the voltmeter) and Z_{es} are in parallel, at least in a circuits sense, the voltage across them is not the same. As the area of the

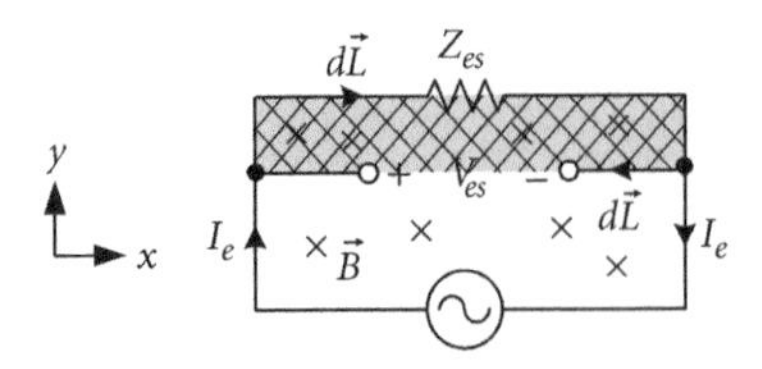

FIGURE 5.48 Closed line integration used to determine the magnetic flux contribution to the measured voltage.

closed path containing the surface of the cable and voltmeter leads increases, the strength of this flux contribution increases (assuming the magnetic field is in the same direction everywhere through the surface). Since Z_{es} can be very small, this magnetic flux contribution to the overall voltage can be important, especially for longer leads and at higher frequencies. Again, this magnetic field pickup applies whenever the leads are in a time-varying magnetic field (and the area they generate is not entirely tangential to the magnetic field). This magnetic flux can provide unexpected "violations" of KVL when not properly modeled. To reduce its effect, the pickup area of the test leads should be small (or oriented so the field does not pass through the area generated by the leads). Twisting of the leads is helpful. Usually, instead of applying Faraday's or Lenz's law to describe the effect of the time-varying magnetic field, the much more familiar concepts of self and mutual inductance are used.

Now that the various impedances have been defined, the total differential voltage with respect to length across the inner surface of the shield and outer surface of the shield can be given when current exists on the shield due to both internal and external sources:

$$\frac{dV_i}{dz} = I_i Z_{is} + I_e Z_{it}, \quad \frac{dV_e}{dz} = I_i Z_{et} + I_e Z_{es} \tag{5.44}$$

In matrix form,

$$\begin{bmatrix} \dfrac{dV_i}{dz} \\ \dfrac{dV_e}{dz} \end{bmatrix} = \begin{bmatrix} Z_{is} & Z_{it} \\ Z_{et} & Z_{es} \end{bmatrix} \begin{bmatrix} I_i \\ I_e \end{bmatrix} \tag{5.45}$$

If the shield is linear and reciprocity can be applied, then $Z_{it} = Z_{et} = Z_t$. Furthermore, if the return path for the noise current is entirely external so that I_i is zero, then

$$\frac{dV_i}{dz} = I_e Z_t \tag{5.46}$$

or if the noise current is entirely internal so that I_e is zero, then

$$\frac{dV_e}{dz} = I_i Z_t \tag{5.47}$$

Although the electric field coupling was previously discussed and illustrated in Figure 5.39, the modeling of this component of the coupling has not yet been presented. When a shield contains holes or other nonconducting apertures, the external electric fields can induce charges on the inner conductor and induce a voltage across the inner conductor and shield. The current induced on the inner conductor from this electric field coupling can be determined by applying a voltage between an external reference and the shield. The current passing through the inner conductor when the inner conductor and shield are connected at one side of the cable, or short circuited, is a measure of this electric field coupling.[22] The transfer admittance is formally defined as

$$Y_t = \frac{1}{V_e}\frac{dI_t}{dz} \ 1/\Omega\text{-m} \tag{5.48}$$

(Sometimes, the transfer admittance is defined as the negative of this expression.) It is probably best to introduce the concept of capacitance when explaining this parameter. For a solid shield with no openings,

[22]The accurate measurement of this current at high frequencies is not a trivial task.

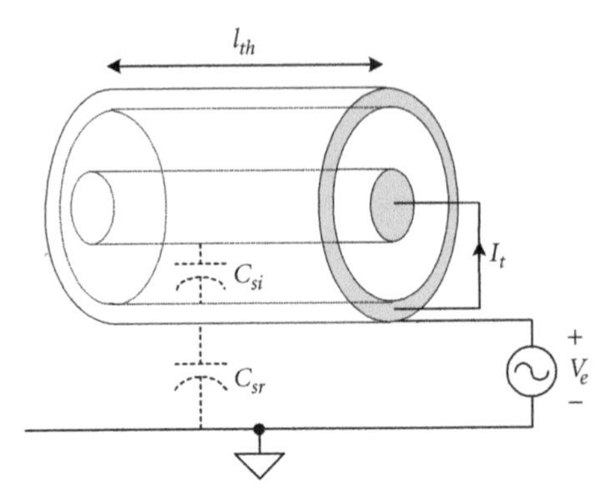

FIGURE 5.49 Transfer admittance measurement when there are no openings in the shield.

there is a capacitance between the shield and inner conductor, C_{si}, and between the shield and outside reference, C_{sr}. The capacitance between the inner conductor and outside reference, C_{ir}, is zero when there are no openings in the shield. For the test setup shown in Figure 5.49, the voltage difference between the inner conductor and shield is zero, I_t is zero, and Y_t is zero. The return path for the current from V_e is entirely via C_{sr}. The line length, l_{th}, is assumed electrically short.

When there are one or more openings in the shield, a capacitance exists between the center inner conductor and reference as shown in Figure 5.50.[23] Although the voltage difference between the inner conductor and shield is still zero, current can now pass from the shield to the reference via C_{sr} and from the inner conductor to the reference via C_{ir}. When C_{ir} is small compared to C_{sr}, most of the return current is through C_{sr}, and I_t is small. (The sinusoidal steady-state current through C_{ir} is $V_e/(1/j\omega C_{ir}) = j\omega C_{ir} V_e$.) The capacitance between the inner conductor and reference and between the shield and reference are both a function of the size, shape, and position of the reference. Generally, the closer the reference, the greater these capacitances. (Other nearby objects will also affect these capacitances.) Sometimes, to provide a known reference, the reference conductor is taken as a cylindrical shell concentric to the shield. Because of the insightful nature of capacitance, the transfer capacitance, C_t, is introduced:

$$Y_t = j\omega C_t = j2\pi f C_t \tag{5.49}$$

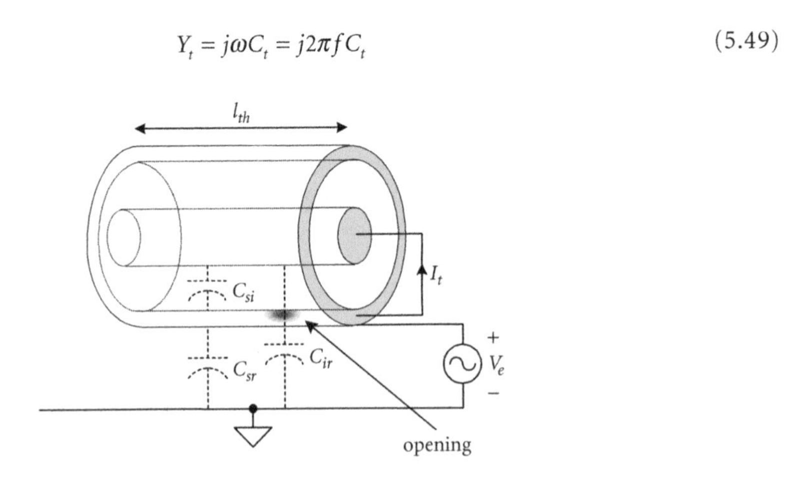

FIGURE 5.50 Transfer admittance measurement when there is an opening in the shield. A capacitance exists between the inner conductor and reference.

[23]The capacitances C_{si} and C_{sr} in this figure are not necessarily equal to the corresponding capacitances without the opening in the shield.

The transfer capacitance is the capacitance between the inner conductor and external reference conductor, given as C_{ir} in Figure 5.50. For a well-designed cable, it is often very small relative to the cable's capacitance, C_{si}. Again, both the transfer admittance and capacitance are a function of the environment. The transfer impedance is not (at least not usually) a function of the external environment. (The transfer impedance is not the reciprocal of the transfer admittance.) Actually, the transfer admittance defined by (5.48) can contain both a real and an imaginary part. The imaginary part corresponds to the reactance of the transfer capacitance. The real part corresponds to the conductance between the inner conductor and external reference electrode. In most cases, the resistance of the dielectric of the cable and the resistance from the cable to the reference are very large and can be neglected. (In other words, the conductance is very small.) The transfer admittance is usually not as important as the transfer impedance, especially if the optical coverage of the shield is high.[24] Measured results for the transfer capacitance are not as common as for the transfer impedance, but a few simple analytical expressions exist. For a braided shield of average radius a with ellipse-shaped openings,

$$C_t = g\frac{v\alpha_e C_{si} C_{sr}}{4\varepsilon_c \pi^2 a^2} \quad \text{where } \alpha_e = \frac{\pi L_g^3}{24}\frac{1-e^2}{E(e)},\ g = \frac{2\varepsilon_c}{\varepsilon_c + \varepsilon_e} \tag{5.50}$$

where α_e is referred to as the electric polarizability of the ellipse-shaped holes, ε_c is the permittivity of the insulation between the inner conductor and shield, and ε_e is the permittivity of the insulation between the shield and external reference. The capacitances were previously defined and are per unit length. Refer to Figure 5.43 and the related discussion for a description of the remaining variables. $E(e)$ is equal to $\pi/2$ when e is zero (corresponding to a circle). Thus, for circular holes of diameter d, (5.50) reduces to

$$C_t = \frac{vC_{si}C_{sr}d^3}{24\varepsilon_c \pi^2 a^2} \tag{5.51}$$

where it is further assumed that $\varepsilon_c \gg \varepsilon_e$. For the helical tape shown in Figure 5.44,

$$C_t = \frac{g}{\cos\psi}\left(\frac{s}{4a}\right)^2 \frac{C_{si}C_{sr}}{4\pi\varepsilon_c} \quad \text{where } g = \frac{2\varepsilon_c}{\varepsilon_c + \varepsilon_e} \tag{5.52}$$

Because of the dependence of the transfer capacitance on the capacitance (and permittivity) between the shield of the cable and external reference, sometimes the charge transfer elastance is suggested as an alternative measure of the electric field coupling. This parameter is defined as

$$S_s = \frac{C_{ir}}{C_{si}C_{sr}} \tag{5.53}$$

(The variable K_t is also seen for this parameter.) What is interesting about this parameter is that it is not a (strong) function of the distance between the cable and outside reference conductor. To determine the reasonableness of this statement, assume that the distance from the shield to the reference is large (or $\varepsilon_c \gg \varepsilon_e$). In this case, the leakage capacitance, C_{ir}, is mainly determined by the dimensions of the shield and reference,[25] the same as C_{sr}. Therefore, S_s is somewhat independent of the position of the outside reference. It is not, however, independent of the permittivity ε_e. To relate S_s with the current disturbance on the inner conductor, note that for short lines of length l_{th}, (5.48) can be written as

$$\frac{I_t}{l_{th}} = j\omega C_{ir} V_e$$

[24]Simply defined, the optical coverage is the ratio of the metal portion of the shield to the total area of the shield.
[25]The net capacitance of a large and small capacitor in series is approximately equal to the smaller capacitance.

The charge injected onto the shield from V_e is equal to $Q_e = C_{sr}V_e$. Therefore,

$$\frac{I_t}{l_{th}} = j\omega C_{ir}\frac{Q_e}{C_{sr}} = j\omega C_{ir}\frac{Q_e}{\dfrac{C_{ir}}{S_s C_{si}}} = j\omega S_s C_{si} Q_e \tag{5.54}$$

This current is a function of the cable capacitance (and cable dimensions) but not a (strong) function of the position and shape of the reference conductor. The tradeoff is that this current is now a function of the injected charge on the shield instead of the easily measured applied voltage. Although the actual value of S_s will vary with the cable type, for coaxial cables, it is in the 5–100 m/μF range.

5.17 The Coupling Model

How can the transfer impedance and admittance be used to model simply the coupling to a source or load on a shielded cable? [Vance]

If the transfer impedance and admittance are available and the strength of the interfering current, I_{es}, and voltage, V_{es}, on the shield are known, then the simple coupling model shown in Figure 5.51 can sometimes provide reasonable results. This model assumes that (i) the length of the cable, l_{th}, is electrically short, (ii) the coupling is weak so that the strength of the external injected signal is large compared to the *reaction* of the victim circuit on the source of the injected signal, (iii) the shield is reasonably good so that it attenuates much of the direct coupling to the inner conductor, and (iv) the conductor and dielectric losses of the cable are negligible. The variables L and C are the inductance and capacitance per unit length, respectively, of the cable. The source and load impedances connected to the cable are Z_s and Z_L, respectively, in this model.

A few comments concerning the reasonableness of this model might be helpful to some readers. The Thévenin equivalent of a (linear electrically-small) circuit at a port of interest is obtained by measuring the open-circuit voltage and short-circuit current at this port. The port of interest in this coupling model is between the inner conductor and shield of the cable. Interestingly, the voltage source in the model is a function of Z_t, which was obtained by measuring the open-circuit voltage between the inner conductor and shield of the cable. The current source in this model is a function of Y_t, which was obtained (in theory) by measuring the short-circuit current between the inner conductor and shield. However, the source of excitation for these two measurements was not the same. Although the strict requirements for obtaining the Thévenin equivalent are not met, this does not imply that the model is inadequate or that there are not multiple methods of obtaining a reasonable Thévenin equivalent. The open-circuit induced voltage is determined through the use of Z_t while the short-circuit current is determined through the use of Y_t. Using superposition, current division, and voltage division, the expressions for the voltages across the source and load impedances in the frequency

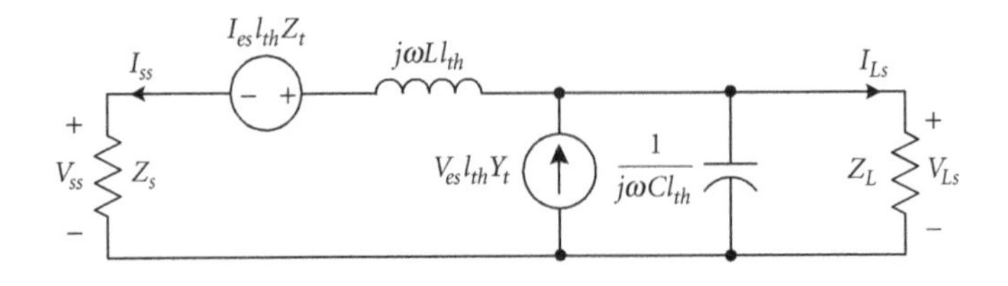

FIGURE 5.51 Simple shielded-cable coupling model.

domain are quickly obtained:

$$V_{ss} = -I_{es} l_{th} Z_t \frac{Z_s}{Z_s + j\omega L l_{th} + \left(\frac{1}{j\omega C l_{th}} \middle\| Z_L \right)} + V_{es} l_{th} Y_t Z_s \frac{\frac{1}{j\omega C l_{th}} \| Z_L}{Z_s + j\omega L l_{th} + \left(\frac{1}{j\omega C l_{th}} \middle\| Z_L \right)} \tag{5.55}$$

$$V_{Ls} = I_{es} l_{th} Z_t \frac{\frac{1}{j\omega C l_{th}} \| Z_L}{Z_s + j\omega L l_{th} + \left(\frac{1}{j\omega C l_{th}} \middle\| Z_L \right)} + V_{es} l_{th} Y_t \left(\frac{1}{j\omega C l_{th}} \middle\| Z_L \right) \frac{Z_s + j\omega L l_{th}}{Z_s + j\omega L l_{th} + \left(\frac{1}{j\omega C l_{th}} \middle\| Z_L \right)} \tag{5.56}$$

A number of special cases of these equations are seen in the literature. In many applications, the transfer impedance's effect (induced voltage) dominates over the transfer admittance's effect (induced current); that is, the magnetic coupling is stronger than the electric coupling. In this case, (5.55) and (5.56) reduce to

$$V_{ss} \approx -I_{es} l_{th} Z_t \frac{Z_s}{Z_s + j\omega L l_{th} + \left(\frac{1}{j\omega C l_{th}} \middle\| Z_L \right)} \tag{5.57}$$

$$V_{Ls} \approx I_{es} l_{th} Z_t \frac{\frac{1}{j\omega C l_{th}} \| Z_L}{\left(\frac{1}{j\omega C l_{th}} \middle\| Z_L \right) + Z_s + j\omega L l_{th}} \tag{5.58}$$

If the source and load impedances are not too "extreme" compared to the cable's series and parallel impedances, then

$$\frac{1}{j\omega C l_{th}} \| Z_L \approx Z_L, \quad Z_s + j\omega L l_{th} \approx Z_s \tag{5.59}$$

and (5.57) and (5.58) reduce to

$$V_{ss} \approx -I_{es} l_{th} Z_t \frac{Z_s}{Z_s + Z_L}, \quad V_{Ls} \approx I_{es} l_{th} Z_t \frac{Z_L}{Z_s + Z_L} \tag{5.60}$$

Other expressions seen for the source and load voltage are obtained by setting $Z_s = Z_L$ in (5.60):

$$V_{ss} \approx -\frac{I_{es} l_{th} Z_t}{2}, \quad V_{Ls} \approx \frac{I_{es} l_{th} Z_t}{2} \tag{5.61}$$

From (5.60), the currents through the source and load impedances are

$$I_{ss} = \frac{V_{ss}}{Z_s} = -I_{es} l_{th} \frac{Z_t}{Z_s + Z_L}, \quad I_{Ls} = \frac{V_{Ls}}{Z_L} = I_{es} l_{th} \frac{Z_t}{Z_s + Z_L} \tag{5.62}$$

and the ratio of the injected current, I_{es}, to the induced current, $I_{is} = -I_{ss} = I_{Ls}$ is

$$\left|\frac{I_{es}}{I_{is}}\right| = \left|\frac{Z_s + Z_L}{l_{th} Z_t}\right| \tag{5.63}$$

For electrically-short cables, this ratio is one measure of shielding effectiveness. In dB,

$$SE_{dB} = 20\log\left|\frac{I_{es}}{I_{is}}\right| = 20\log\left|\frac{Z_s + Z_L}{l_{th} Z_t}\right| \tag{5.64}$$

At high frequencies, the inductive coupling dominates over diffusion through the shield. Using (5.33),

$$SE_{dB} \approx 20\log\left|\frac{Z_s + Z_L}{j\omega M l_{th}}\right| \tag{5.65}$$

Referring to Figure 5.52, assume that the end termination between the outside of the shield and reference is equal to Z_{oe}, where Z_{oe} is equal to the characteristic impedance of the shield and *reference*. Therefore, for this matched condition, $V_{es} = Z_{oe} I_{es}$. (This also assumes low-loss conditions.) Using equations (5.55) and (5.56), the currents in the source and load are

$$I_{ss} = \frac{V_{ss}}{Z_s} = \frac{I_{es} l_{th}}{Z_s + j\omega L l_{th} + \left(\frac{1}{j\omega C l_{th}} \Big\| Z_L\right)} \left[-Z_t + Z_{oe} Y_t \left(\frac{1}{j\omega C l_{th}} \Big\| Z_L\right)\right] \tag{5.66}$$

$$I_{Ls} = \frac{V_{Ls}}{Z_L} = \frac{I_{es} l_{th}}{Z_L} \frac{\frac{1}{j\omega C l_{th}} \Big\| Z_L}{Z_s + j\omega L l_{th} + \left(\frac{1}{j\omega C l_{th}} \Big\| Z_L\right)} [Z_t + Z_{oe} Y_t (Z_s + j\omega L l_{th})] \tag{5.67}$$

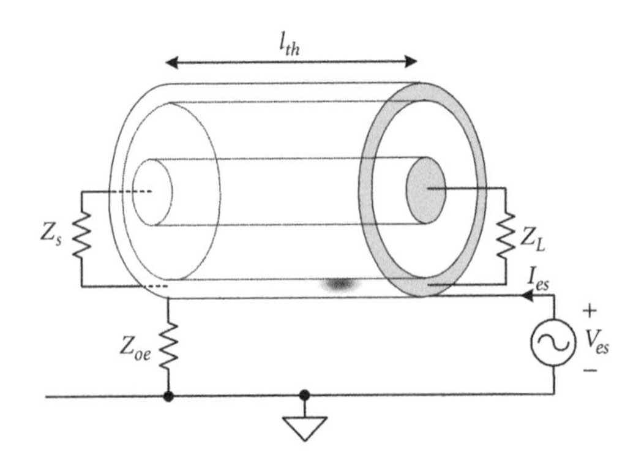

FIGURE 5.52 Matched termination for the transmission line between the shield and reference.

Again, assuming that the conditions in (5.59) are satisfied,

$$I_{ss} \approx \frac{I_{es}l_{th}}{Z_s + Z_L}(-Z_t + Z_{oe}Y_t Z_L),\ I_{Ls} \approx I_{es}l_{th}\frac{1}{Z_s + Z_L}(Z_t + Z_{oe}Y_t Z_s)$$

and

$$\frac{I_{es}}{I_{ss}} = \frac{Z_s + Z_L}{l_{th}(Z_{oe}Z_L Y_t - Z_t)},\ \frac{I_{es}}{I_{Ls}} = \frac{Z_s + Z_L}{l_{th}(Z_{oe}Z_s Y_t + Z_t)} \tag{5.68}$$

These equations do not neglect the transfer admittance. As seen, the currents at the source and load are not necessarily the same. Thus, measurement of the shielding effectiveness at only the source or load can generate misleading results.

5.18 Pigtails and Connectors—Weak Links in a System

It is stated that the weak links in a shielded cable system are the pigtails and connectors. Explain. [Violette; Chatterton; Perez; Paul, '92(b); Tsaliovich]

A shield around the source conductors, the victim conductors, or, even better, around both the source and victim conductors can be quite effective in reducing crosstalk. However, a cable shield must be terminated at its ends via some type of connector. A careless connection can increase the emission level and susceptibility of the system. There are various methods of performing this termination to help preserve the integrity of the shield. Probably, the worst method of terminating is the use of a single, small diameter, long pigtail wire.

One major weakness of a real shield is openings in or cables (or conductors) entering the shield. If a conductor passes through an opening of a shielded enclosure, then any undesirable current induced on the conductor can pass directly into the shielded enclosure dramatically reducing the shield's effectiveness. As shown in Figure 5.53, the violating conductor can consist of a single wire or even the shield (or inner conductor) of a coaxial cable. For that matter, any conductor, whether shielded or not, can be a sneaky route for current through the shield. This includes grounding wires.

Little can be done for a single unshielded conductor passing through a shielded enclosure (besides using a conductive filter such as a feed-through capacitor). However, when the conductor or conductors have a single shield, the classical grounding approach is to connect the shield to the *outside* surface of the shielded enclosure. A pigtail wire is a single conductor connecting the outer shield of the cable to the shielded enclosure as shown in Figure 5.54. To reduce the impedance of the pigtail connection so as to encourage the current to pass through it to the enclosure, multiple pigtails (in parallel) can be used. The length of the pigtails should be as small as possible to minimize their impedance. The best connection, but sometimes the most costly, is the use of a complete, 360° bond between the cable shield and enclosure.

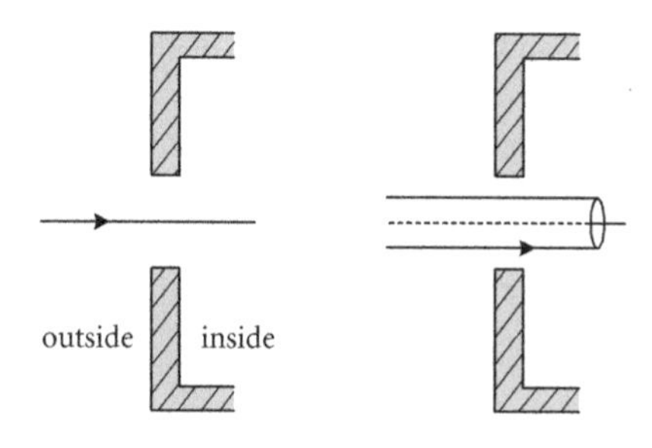

FIGURE 5.53 Undesirable current passing directly into a region to be shielded.

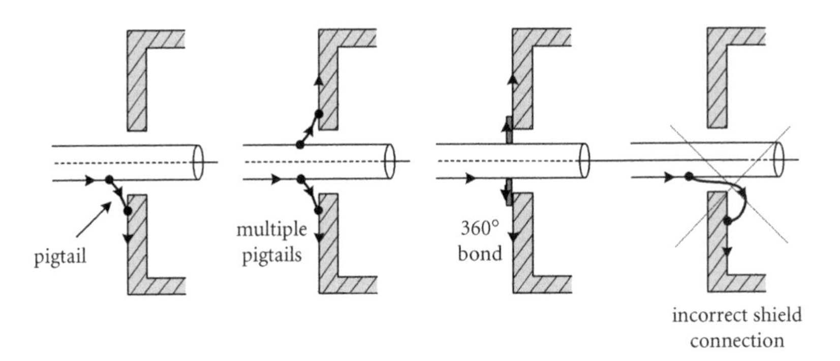

FIGURE 5.54 Single pigtails, multiple pigtails, complete circumferential bond, and improper pigtail connection.

This connection essentially extends or continues the shield barrier from the cable to the enclosure.[26] It is important to realize that even with a complete circumferential bond, a fraction of the current on the outside will diffuse through the metal to the inner surface of the enclosure. The transfer impedance is a measure of this diffusion. In some situations, it may not be possible to connect the shield to the outer surface of the enclosure. Depending on the actual system, it might be advantageous to connect the shield to the inner surface of the enclosure. Of course, in this case, the noise current is also on both the shield and inner surface of the enclosure. By connecting the pigtail to a reference inside the enclosure, such as a high-frequency analog reference, the pigtail might be an effective radiating antenna!

Although the cable shields in Figure 5.54 are shown passing through the aperture of the enclosure, often the cable is terminated at the enclosure or some type of junction box or patch panel. Some cables, especially those with fragile outer shields constructed of foils, contain a drain wire that is in contact with the shield. This drain wire is sometimes used as the pigtail even though it is often of small radius. Because coaxial cables are so commonly used and relatively easy to work with, inexpensive connectors are available that are effective if properly connected to the outer shield (whether braided or solid). For ribbon and other multiwire cables, this connection is not always as simple and a pigtail may be unavoidable. By dedicating several conductors of a ribbon cable as the ground, multiple conductors are available to reduce the impedance of the pigtail to the enclosure. Cables are available containing individually shielded wires, such as twisted pair, that are enclosed in a common shield. The outer common shield is connected to the outer surface of the enclosures. The shields of the individually shielded wires would not necessarily be connected to the outer surface of the enclosure since this would allow current to pass within the cable. These internal shields are frequently connected together inside the cable. At lower frequencies, connection of these shields to a ground reference may actually be desirable only at one end to avoid ground loops.

The pigtail wire and signal conductor can be modeled at lower frequencies using the model shown in Figure 5.56, assuming the length of the pigtail is electrically short and coupling to it is weak. When the cable connected to the pigtail is not electrically short, the impedance at the other end of the cable must be properly transformed to the location of the pigtail when using this model. As expected, experiments have shown that the emissions level (and the susceptibility of the system) increases with the frequency or electrical length of the pigtail.

As is apparent to anyone who has connected multiple antennas to a single television through a switching box, connectors will introduce signal loss in a system. Furthermore, especially at high frequencies, connectors will introduce an impedance mismatch or discontinuity, which will generate reflections. Also, depending on the inductance and capacitance of the connection, additional resonant frequencies can be introduced.

[26]Sometimes, to improve a pigtail connection, metal tape is used around the termination.

5.19 Capacitive or Inductive Crosstalk?

Under what conditions is the crosstalk between two conductors mainly capacitive? Under what conditions is the crosstalk mainly inductive? [Paul, '94; Ott]

Crosstalk requires at least three conductors: a generator (or culprit), a victim, and a reference conductor. The model in Figure 5.55 includes both inductive and capacitive coupling between two conductors with a common reference. The source and load are resistive for both the generator circuit and victim circuit, and the line resistances have been set to zero.[27] Of course, to model conductors using lumped parameters, their length, l_{th}, should be electrically small.

The general solutions for the voltages across the victim resistors, R_{sv} and R_{Lv}, are quite complex. However, the sinusoidal steady-state solutions when the line is electrically short and coupling between the lines is weak are much simpler. Weak coupling implies that

$$L_m \ll \sqrt{L_s L_v} \quad \text{and} \quad C_m \ll \sqrt{(C_s + C_m)(C_v + C_m)} \tag{5.69}$$

Under these conditions, the voltages across the source and load resistance of the victim circuit are

$$V_{svs} = j\omega V_{ss}(M_{inds} + M_{cap}) \tag{5.70}$$

$$V_{Lvs} = j\omega V_{ss}(M_{indL} + M_{cap}) \tag{5.71}$$

where

$$M_{cap} = C_m \frac{R_{sv} R_{Lv}}{R_{sv} + R_{Lv}} \frac{R_L}{R_s + R_L} \tag{5.72}$$

$$M_{inds} = L_m \frac{R_{sv}}{R_{sv} + R_{Lv}} \frac{1}{R_s + R_L} \tag{5.73}$$

$$M_{indL} = -L_m \frac{R_{Lv}}{R_{sv} + R_{Lv}} \frac{1}{R_s + R_L} \tag{5.74}$$

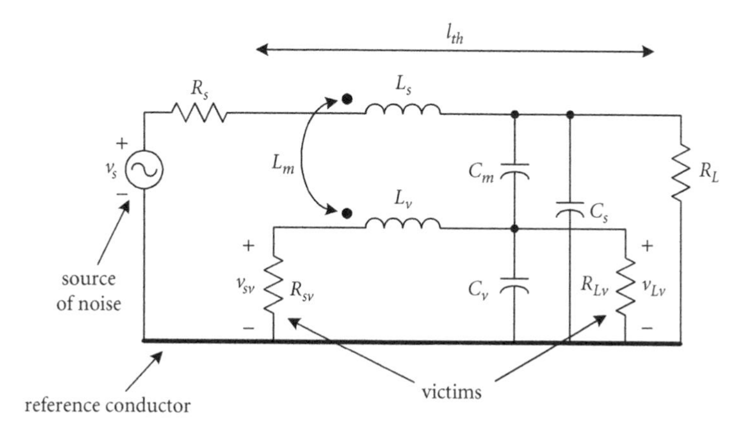

FIGURE 5.55 Self and mutual inductances and capacitances for a three-conductor system.

[27]At lower frequencies, below the cutoff frequency, the impedance of the return conductor can be an important source of common-impedance coupling. Common-impedance coupling is not modeled in this circuit.

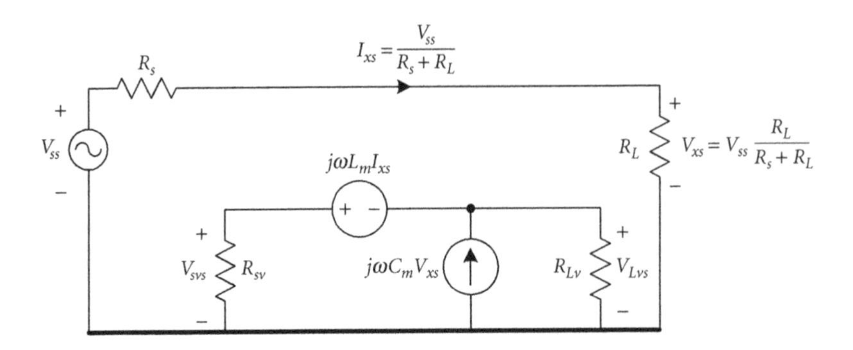

FIGURE 5.56 Circuit in Figure 5.55 when the coupling is weak.

The capacitance and inductance variables in these equations are the total capacitance and inductance, respectively, not the capacitance and inductance per unit length. Note that the voltages across the victim resistances are directly proportional to the frequency of the signal. These equations can be remembered by examining the crosstalk model shown in Figure 5.56, where the sources are a function of the coupling capacitor, C_m, and coupling inductance, L_m. The voltages across the victim resistors in this figure are identical to the results obtained from (5.70)–(5.74). The dependent sources in the victim circuit are a function of both the noise current and noise voltage via the inductive coupling and capacitive coupling, respectively.

If the crosstalk is mainly capacitive, then for the source resistance of the victim circuit

$$|M_{inds}| \ll |M_{cap}|$$

$$L_m \frac{R_{sv}}{R_{sv} + R_{Lv}} \frac{1}{R_s + R_L} \ll C_m \frac{R_{sv} R_{Lv}}{R_{sv} + R_{Lv}} \frac{R_L}{R_s + R_L} \Rightarrow \frac{L_m}{C_m} \ll R_{Lv} R_L \tag{5.75}$$

and for the load resistance of the victim circuit

$$|M_{indL}| \ll |M_{cap}|$$

$$L_m \frac{R_{Lv}}{R_{sv} + R_{Lv}} \frac{1}{R_s + R_L} \ll C_m \frac{R_{sv} R_{Lv}}{R_{sv} + R_{Lv}} \frac{R_L}{R_s + R_L} \Rightarrow \frac{L_m}{C_m} \ll R_{sv} R_L \tag{5.76}$$

When the capacitive coupling dominates, the dependent voltage source in Figure 5.56 is negligible and may be turned off. The noise voltages across the source and load resistances of the victim are a function of the current source. Both inequalities given in (5.75) and (5.76) are a function of the load resistance of the generator circuit. The product of the generator load resistance and the corresponding load or source resistance of the victim should be large compared to the ratio of the mutual inductance to the mutual capacitance, L_m/C_m. If the medium surrounding the conductors is homogeneous (i.e., uniform electrical properties), this ratio of mutual coupling terms is equivalent to

$$\frac{L_m}{C_m} = \sqrt{\frac{L_v}{C_v + C_m} \frac{L_s}{C_s + C_m}} = Z_{ov} Z_{os} \tag{5.77}$$

The variables Z_{ov} and Z_{os} are referred to as the characteristic impedance of the victim and source circuits, respectively, in the presence of the other circuit. For weak coupling, the mutual capacitance is small compared to the self capacitances of the two circuits. The standard characteristic impedances of the source and victim line may be used in this case.

The crosstalk is also given in dB. It can be defined in terms of a source voltage or source current to a victim voltage or victim current. The crosstalk, in dB, is normally negative when defined as 20log of the ratio of the victim voltage to the source voltage. When crosstalk in dB is positive, it is probably defined as 20log of the ratio of the source voltage to the victim voltage. Then, it is sometimes called crosstalk isolation. To complicate matters further, the dB reference used for crosstalk is sometimes seen as dBx (or DBX). A 90 dB power loss from the culprit to the victim circuit is at times assumed as a reference, and the dBx value is the crosstalk loss above this 90 dB power value:

$$dBx = 90 - crosstalk_{dB} \tag{5.78}$$

For example, if the crosstalk between two circuits is 40 dB, then the crosstalk in dBx is 90 – 40 = 50 dBx.

For capacitive coupling to dominate for the source resistance of the victim circuit

$$Z_{ov} Z_{os} \ll R_{Lv} R_L \tag{5.79}$$

and for the load resistance of the victim circuit

$$Z_{ov} Z_{os} \ll R_{sv} R_L \tag{5.80}$$

For high-resistance terminations, the capacitive coupling dominates. For inductive coupling to dominate for the source resistance of the victim circuit

$$Z_{ov} Z_{os} \gg R_{Lv} R_L \tag{5.81}$$

and for the load resistance of the victim circuit

$$Z_{ov} Z_{os} \gg R_{sv} R_L \tag{5.82}$$

For low-resistance terminations, the inductive coupling dominates. Again, these statements assume electrically-short lines and weak coupling.

The total voltage across the load resistance of the victim circuit can be made small even when both the mutual capacitance and mutual inductance are large. This can occur because the sign of the mutual inductance term is negative:

$$V_{Lvs} = j\omega V_{ss}(M_{indL} + M_{cap}) = j\omega V_{ss}\left(-L_m \frac{R_{Lv}}{R_{sv} + R_{Lv}} \frac{1}{R_s + R_L} + C_m \frac{R_{sv} R_{Lv}}{R_{sv} + R_{Lv}} \frac{R_L}{R_s + R_L}\right)$$

The load voltage across the victim is zero when

$$L_m \frac{R_{Lv}}{R_{sv} + R_{Lv}} \frac{1}{R_s + R_L} = C_m \frac{R_{sv} R_{Lv}}{R_{sv} + R_{Lv}} \frac{R_L}{R_s + R_L} \quad \Rightarrow \quad \frac{L_m}{C_m} = R_{sv} R_L$$

In those (rare) cases when this relationship is satisfied, the inductive and capacitive contributions exactly cancel at the load of the victim circuit.

Frequently, it is stated that for low-impedance loads, the inductive coupling dominates, and for high-impedance loads, the capacitive coupling dominates. When the load impedance of the generator or noise source circuit is low, it seems reasonable that the current tends to be high for a fixed source voltage. In addition, high currents generate high magnetic fields. Also, when the load impedance of the generator or noise circuit is high, it seems reasonable that the current tends to be low for a fixed source voltage. Low currents might imply larger voltages and, hence, high electric fields. However, the expressions in this section clearly show that the generator's load impedance is not the only factor in determining whether inductive or capacitive coupling dominates.

5.20 Measurement Tools

The noise voltage across the source and load resistance is about the same. Is the noise coupling capacitive or inductive? The noise voltage across the source and load resistance is about proportional to their respective resistance. Is this noise coupling capacitive or inductive?

The coupling between the noise source circuit and victim circuit can be modeled as shown in Figure 5.57. Note that if the coupling is mainly capacitive, the circuit reduces to that shown in Figure 5.58. Since the source and load resistances are in parallel, the voltage across them is the same. Therefore, if the coupling is mainly capacitive, the voltage across the source and load of the victim circuit is about the same:

$$\frac{V_{Lvs}}{V_{svs}} \approx 1 \tag{5.83}$$

If the coupling is mainly inductive, the circuit reduces to that shown in Figure 5.59. In this case, the voltage across the source and load is proportional to their resistances:

$$\frac{V_{Lvs}}{V_{svs}} \approx \frac{-j\omega L_m I_{xs} \dfrac{R_{Lv}}{R_{Lv}+R_{sv}}}{j\omega L_m I_{xs} \dfrac{R_{sv}}{R_{Lv}+R_{sv}}} = -\frac{R_{Lv}}{R_{sv}} \tag{5.84}$$

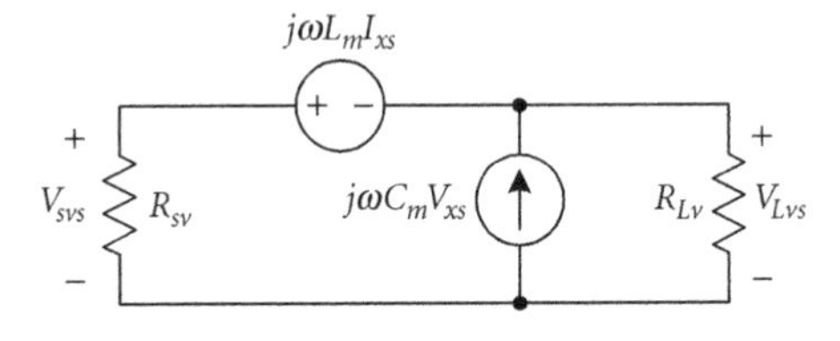

FIGURE 5.57 Model for weak coupling and electrically-short conductors.

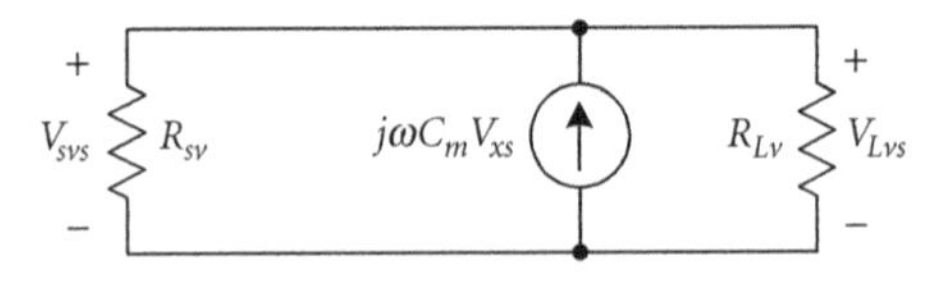

FIGURE 5.58 Model when the coupling is mainly capacitive.

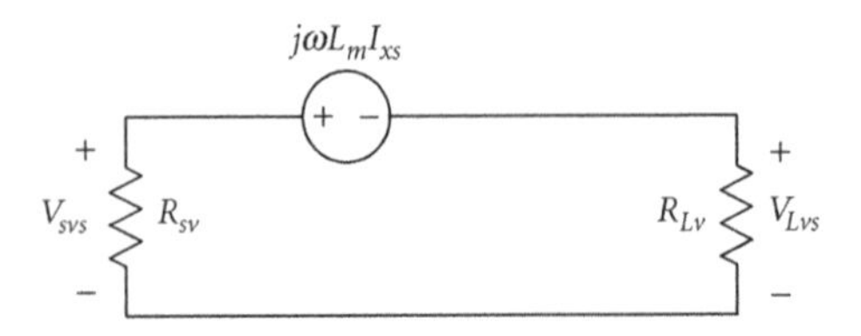

FIGURE 5.59 Model when the coupling is mainly inductive.

This relationship can be employed to determine whether the crosstalk is mainly capacitive or inductive. Once it is determined that the crosstalk is mainly electric or magnetic, the appropriate method of attack can be determined to reduce the coupling.

According to (5.83) and (5.84), when the source and load resistances of the victim are about the same size, the magnitude of the voltages across them will be the same whether the coupling is mainly capacitive or inductive. Another test would be necessary to determine whether the coupling is mainly electric or magnetic. For example, assuming the coupling is either inductive or capacitive, the source or load resistance can be removed. If noise still exits across the remaining resistor, the model in Figure 5.58 implies that the coupling is capacitive. If the noise disappears across the remaining resistor, the model in Figure 5.59 implies that the coupling is inductive.

5.21 Susceptibility of High and Low Resistances

Determine the susceptibility of a circuit to electric and magnetic fields for low-resistance and high-resistance sources and loads.

When the source and load resistances of a victim circuit have extreme values corresponding to a short or an open circuit, the susceptibility of the circuit to capacitive or inductive coupling can be approximately determined. Referring to Figure 5.57, when the source and load resistances of the victim circuit are small, the voltages are (unless the induced current source is very large)

$$V_{svs} \approx j\omega V_{ss} L_m \frac{R_{sv}}{R_{sv}+R_{Lv}} \frac{1}{R_s+R_L} \tag{5.85}$$

$$V_{Lvs} \approx -j\omega V_{ss} L_m \frac{R_{Lv}}{R_{sv}+R_{Lv}} \frac{1}{R_s+R_L} \tag{5.86}$$

The induced voltage is split between the victim's source and load resistances. When the source resistance is small and load resistance is large (e.g., a decent voltage source connected to a decent amplifier),

$$V_{svs} \approx j\omega V_{ss}\left(L_m \frac{R_{sv}}{R_{Lv}} \frac{1}{R_s+R_L} + C_m R_{sv} \frac{R_L}{R_s+R_L}\right) \quad \text{if } R_{Lv} \gg R_{sv} \tag{5.87}$$

$$\begin{aligned} V_{Lvs} &\approx j\omega V_{ss}\left(-L_m \frac{1}{R_s+R_L} + C_m R_{sv} \frac{R_L}{R_s+R_L}\right) \quad \text{if } R_{Lv} \gg R_{sv} \\ &\approx j\omega V_{ss}\left(-L_m \frac{1}{R_s+R_L}\right) \quad \text{if } R_{Lv} \gg R_{sv},\ L_m \gg C_m R_{sv} R_L \end{aligned} \tag{5.88}$$

If the capacitive contribution in (5.88) is negligible, then it is also negligible in (5.87) (but not necessarily small relative to the inductive portion of (5.87)). Because $R_{sv}/R_{Lv} \ll 1$, $|V_{Lsv}| \gg |V_{svs}|$. Under these conditions, the load is more susceptible to magnetic interference or coupling (the L_m term) than electric interference or coupling (the C_m term). When the source resistance is large and load resistance is small,

$$\begin{aligned} V_{svs} &\approx j\omega V_{ss}\left(L_m \frac{1}{R_s+R_L} + C_m R_{Lv} \frac{R_L}{R_s+R_L}\right) \quad \text{if } R_{sv} \gg R_{Lv} \\ &\approx j\omega V_{ss}\left(L_m \frac{1}{R_s+R_L}\right) \quad \text{if } R_{sv} \gg R_{Lv},\ L_m \gg C_m R_{Lv} R_L \end{aligned} \tag{5.89}$$

$$V_{Lvs} \approx j\omega V_{ss}\left(-L_m \frac{R_{Lv}}{R_{sv}} \frac{1}{R_s+R_L} + C_m R_{Lv} \frac{R_L}{R_s+R_L}\right) \quad \text{if } R_{sv} \gg R_{Lv} \tag{5.90}$$

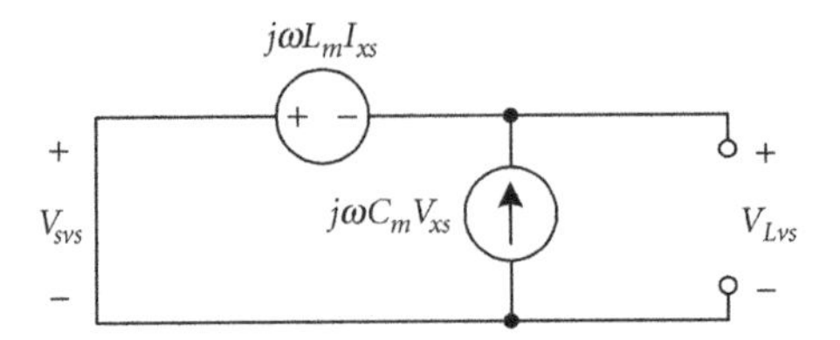

FIGURE 5.60 Coupling model for a victim with a low source impedance and high load impedance.

If the capacitive contribution in (5.89) is negligible, then it is also negligible in (5.90) (but not necessarily small relative to the inductive portion of (5.90)). Because $R_{sv}/R_{Lv} \gg 1$, $|V_{svs}| \gg |V_{Lsv}|$. Under these conditions, the source is more susceptible to magnetic interference than electric interference. An example of a low-impedance load is a current-to-voltage converter used to convert the current output of a photodiode to a voltage. The weld location for an arc welder is another example of a low-impedance load. Of course, arc welders are not very susceptible to interference, but they are powerful interference sources.

When the source and load resistances of the victim circuit are both high,

$$V_{svs} \approx j\omega V_{ss}\left(L_m \frac{R_{sv}}{R_{sv}+R_{Lv}} \frac{1}{R_s+R_L}\right) \quad \text{if } L_m \gg C_m R_{Lv} R_L \tag{5.91}$$

$$V_{Lvs} \approx j\omega V_{ss}\left(-L_m \frac{R_{Lv}}{R_{sv}+R_{Lv}} \frac{1}{R_s+R_L}\right) \quad \text{if } L_m \gg C_m R_{sv} R_L \tag{5.92}$$

both the source and load are generally susceptible to magnetic field (L_m term) coupling. Of course to satisfy the inequalities given in (5.91) and (5.92), the load resistance of the source circuit must be very small. An electrically-short monopole antenna is an example of a high-impedance source (with a capacitive input impedance).

These relationships can be remembered by merely replacing the appropriate resistance with a short circuit or an open circuit in the crosstalk model. For example, when the source is a low resistance and the load is a high resistance, the model reduces to that shown in Figure 5.60. The voltage across the source is small while the voltage across the load is only a function of the inductive term (the voltage source) since all the current from the capacitive coupling source passes through the short circuit.

5.22 Susceptibility of Scopes

The 1 MΩ input of an oscilloscope is connected to a 22 kΩ load. The level of the 60 Hz hum generated from the 120 V rms power line (capacitive coupling only) is 2 mV peak to peak as seen on the scope. Estimate the capacitance from the leads of the scope to the power line. Are high-impedance or low-impedance loads more likely to be affected by capacitive coupling?

Using the model shown in Figure 5.57, it is easy to see that when the mutual inductive term, L_m, is set to zero, the total voltage across the two parallel resistors, representing the 22 kΩ load and 1 MΩ scope input resistance, is

$$V_{Lvs} = V_{svs} = j\omega C_m V_{xs}(R_{sv} \| R_{Lv}) \tag{5.93}$$

Solving for the mutual capacitance,

$$C_m = \frac{V_{Lvs}}{j\omega V_{xs}(R_{sv} \| R_{Lv})}$$

Since the phase of the voltage across the scope, relative to the power line, is not required to determine the capacitance,

$$C_m = \frac{|V_{Lvs}|}{\omega |V_{xs}| (R_{sv} \| R_{Lv})} = \frac{\dfrac{2\times 10^{-3}}{2\sqrt{2}}}{2\pi \times 60(120)\left[\dfrac{(22\times 10^3)10^6}{22\times 10^3 + 10^6}\right]} \approx 0.73 \text{ pF}$$

To simplify this analysis, it was assumed that the power line and scope have one conductor in common. This allows the three-conductor crosstalk model to be used. Otherwise, the actual conductor configuration (e.g., size, spacing, and orientation of all four conductors) for both the power line and scope leads would have to be known.

Assuming that the inductive coupling can be neglected, (5.93) clearly shows that the induced voltage across the victim circuit is directly proportional to the parallel combination of the source and load resistances of the victim. Therefore, "larger" loads in the victim circuit are more susceptible to capacitive interference. (This assumes that the inductive coupling is negligible.)

5.23 Foam Encapsulation

Circuits in satellites are occasionally encapsulated in plastic foam with a dielectric constant of 1.1 to 1.2. Why? Will the foam increase or decrease the inductive coupling and the capacitive coupling?

The foam helps protect the circuit from its environment, including physical shocks and other disturbances. The mutual inductance or inductive coupling is not a function of the dielectric constant or relative permittivity. However, the mutual capacitance or capacitive coupling is a strong function of the dielectric constant. When the foam completely surrounds the circuit and it is thick compared to the dimensions of the circuit, the increase in mutual capacitance between elements in the circuit will be directly proportional to the dielectric constant of the foam. Hence, it is important that the dielectric constant of the foam be nearly equal to one, corresponding to free space. A dielectric constant of 1.1 to 1.2 is quite good considering that many of the foams used in cabling have dielectric constants of around 3.

5.24 Inductive Crosstalk and the 3-W Guideline

Determine the crosstalk between two circuit traces if only inductive coupling is considered. What is the 3-W guideline, and what is its origin? [Johnson, '93; Montrose, '99; Walker; Haus]

Crosstalk between two circuits can be due to both capacitive and inductive coupling. For this discussion, the coupling is assumed to be entirely inductive. For crosstalk to occur, at least three conductors must be involved. For this two-trace situation, the third conductor will be assumed to be a common, large, flat ground plane or return of width s as shown in Figure 5.61. The victim circuit is assumed to be trace b (with the large flat plane as its return conductor). The mutual

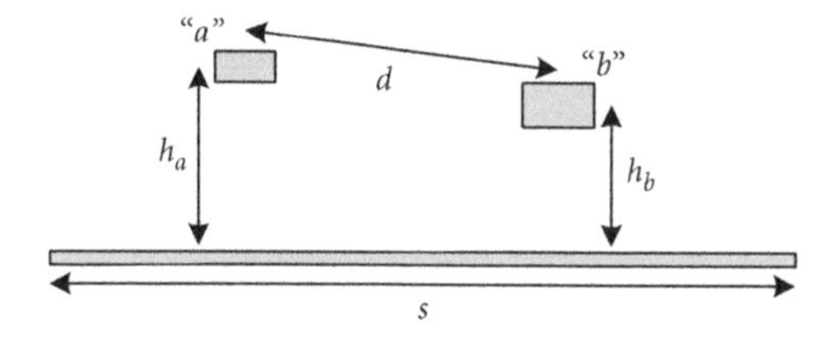

FIGURE 5.61 Three-conductor system used in the crosstalk analysis.

inductance between two rectangular conductors, a and b, that are not too close to the plane conductor or to each other is

$$L_m = 10^{-7} l_{th} \ln\left(1 + 4\frac{h_a h_b}{d^2}\right) \quad \text{if } l_{th} \gg \max(h_a, h_b, d), \quad s \gg \max(h_a, h_b, d) \tag{5.94}$$

where l_{th} is the length of the conductors. (Sometimes the variable M is used for mutual inductance.) The open-circuit voltage across the victim trace is

$$V_{ocs} = j\omega L_m I_{xs} \tag{5.95}$$

where I_{xs} is the current in the interfering circuit, which in this example is the current in trace a. Of course, this expression assumes the interfering signal on trace a is in sinusoidal steady state. Again, this voltage expression neglects the capacitive coupling between traces a and b. Equation (5.95) can be obtained directly from the circuits definition for the mutual inductance between two loops. The open-circuit voltage induced in one loop due to the time-varying current in the other loop is

$$v_2 = L_m \frac{di_1}{dt} \Rightarrow V_{2s} = L_m j\omega I_{1s}$$

Crosstalk between two traces can be defined in several different ways. If it is defined as the ratio of the open-circuit voltage induced in the victim circuit, trace b, to the current in trace a, then

$$crosstalk = \frac{V_{ocs}}{I_{xs}} = j\omega L_m \tag{5.96}$$

(The crosstalk can also be defined as the victim voltage divided by the source voltage.) Therefore, the crosstalk due to this mutual inductance is, neglecting the capacitive coupling,

$$crosstalk = j\omega L_m = j\omega 10^{-7} l_{th} \ln\left(1 + 4\frac{h_a h_b}{d^2}\right) \tag{5.97}$$

If the trace heights are the same,

$$crosstalk = j\omega L_m = j\omega 10^{-7} l_{th} \ln\left[1 + \left(\frac{2h}{d}\right)^2\right] \quad \text{if } h = h_a = h_b \tag{5.98}$$

Sometimes, another expression is seen for this crosstalk that does not contain a logarithmic function. This expression is an approximation based on the return current density along a large return plane in the presence of a current-carrying circular wire. The current density tangential to the return plane is (assuming the frequency is sufficiently high so that the magnetic flux density normal to the conductors is negligible)

$$|J| = \frac{I}{\pi h}\left[\frac{1}{1 + \left(\frac{d}{h}\right)^2}\right] \tag{5.99}$$

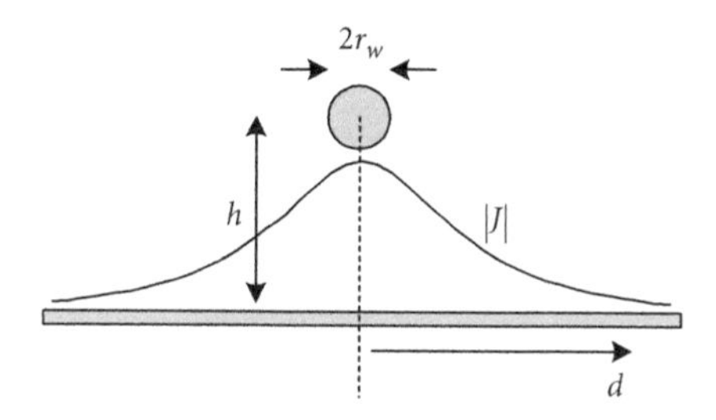

FIGURE 5.62 Sketch of the current density along the large return conductor.

where d is the position along the return plane as shown in Figure 5.62. As expected, the current density is the greatest directly under the wire, and it decreases with distance from the wire. The coupling between this wire and another wire at the same height h located at a distance d from this wire will be a function of the strength of this current density. The magnetic field tangential to the return conductor is a function of the current density, and the mutual inductance is a function of the magnetic field. The actual mutual inductance between two wires above a ground plane is determined by integrating the actual magnetic field present between the victim wire and return plane. However, the mutual inductance and, hence, crosstalk can be roughly approximated by assuming that they are proportional to (5.99):

$$crosstalk \propto \frac{1}{\pi h}\left[\frac{1}{1+\left(\frac{d}{h}\right)^2}\right] = \frac{1}{\pi}\left[\frac{\frac{h}{d^2}}{\left(\frac{h}{d}\right)^2+1}\right] \tag{5.100}$$

Obviously, the sinusoidal steady-state crosstalk is also a function of the frequency, length of the wires, and free-space permeability, μ_o (since the total flux is a function of $\vec{B}=\mu_o\vec{H}$ and the area). Although (5.100) appears much different from (5.98), they are similar in form when $d \gg h$. The current-density version of the crosstalk can be approximated as

$$crosstalk \propto \frac{1}{\pi}\frac{h}{d^2}\left[1+\left(\frac{h}{d}\right)^2\right]^{-1} \approx \frac{1}{\pi}\frac{h}{d^2}\left[1-\left(\frac{h}{d}\right)^2\right] \quad \text{if } d \gg h \tag{5.101}$$

using the binomial expansion, $(1+x)^n \approx 1+nx$ for small values of x. The original crosstalk expression based on the mutual inductance expression can be approximated as

$$\begin{aligned} crosstalk &= j\omega 10^{-7} l_{th} \ln\left[1+\left(\frac{2h}{d}\right)^2\right] \approx j\omega 10^{-7} l_{th}\frac{1}{2}\left(\frac{2h}{d}\right)^2\left[2-\left(\frac{2h}{d}\right)^2\right] \\ &\approx j\omega 10^{-7} l_{th} 4\left(\frac{h}{d}\right)^2\left[1-2\left(\frac{h}{d}\right)^2\right] \quad \text{if } d \gg h \end{aligned} \tag{5.102}$$

where the Taylor series approximation was used for the natural logarithm, $\ln(1+x) \approx x - x^2/2$, for small values of x. Adding the frequency, length, height, and μ_o correction factors, (5.101) and (5.102) are very similar. Note that (5.94) is actually the expression for the mutual inductance between two circular wires above a flat plane. Assuming the rectangular traces are not too close and their aspect ratio not too extreme, their mutual inductance is approximately that of two circular wires.

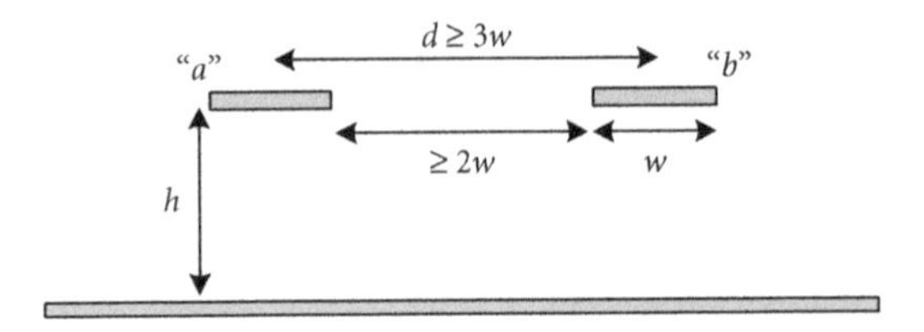

FIGURE 5.63 3-W guideline.

To minimize the crosstalk between the two traces, the distance, d, between the traces should be large and their height above the ground should be small. The mutual inductance decreases with increasing d (since the flux density decreases with distance from a current-carrying wire), and the mutual inductance also decreases with decreasing h (since the loop area formed by the victim wire and return plane decreases with decreasing h).

As a simple example of the use of the crosstalk equation, the open-circuit voltage induced on a victim trace will be determined. Assume that the width of each trace is $w = 15$ mils, their height above the return plane is $h = 62$ mils, the center-to-center distance between them is 30 mils, and their lengths are 8 cm. The amplitude of the sinusoidal 1 MHz current on one of the traces is 10 mA. The induced open-circuit voltage, obviously of the same frequency, across the other trace circuit is

$$|V_{oc}| = |j\omega L_m I_{xs}| = (2\pi \times 10^6)(10^{-7})(0.08)\ln\left[1+\left(\frac{2\times 62}{30}\right)^2\right]\times 0.01 \approx 1.5\text{ mV}$$

This voltage would be measured, neglecting the capacitive coupling, between the victim trace and return plane, assuming the victim load (or source) is open circuited.

For the inductive coupling between two "high-threat" traces to be tolerable, the 3-W guideline states that the center-to-center distance between the two traces should be at least three times the width of the single trace. This guideline is shown in Figure 5.63. "High-threat" traces are typically those traces where emissions from them (or to them) might be important or critical. Clock traces, reset lines, and differential pairs are examples of typical traces that use the 3-W guideline. (In some high-density boards, this guideline cannot be met and instead a 2-W guideline is used.) When using the 3-W guideline, other conductors should not be placed between these two traces (separated by the 3-W spacing).

This guideline is likely based on the ratio of the magnetic flux collected by the victim trace and return plane to the total magnetic flux generated by the source trace and return plane. In terms of inductances, this ratio is equal to the mutual inductance between the two traces, with a return plane present, divided by the self inductance of a single trace above the return plane:

$$\frac{M}{L} = \frac{10^{-7} l_{th} \ln\left[1+\left(\frac{2h}{d}\right)^2\right]}{10^{-7} l_{th} \ln\left\{1+\frac{32h^2}{w^2}\left[1+\sqrt{1+\left(\frac{\pi w^2}{8h^2}\right)^2}\right]\right\}} \tag{5.103}$$

Using this ratio, the percent flux *not* coupled into the nearby trace circuit is plotted vs. the ratio of the center-to-center spacing to trace width for various normalized values of trace height. If $w = 15$ mils and $h = 62$ mils, then $h \approx 4w$, and the percent flux not coupled to the nearby victim trace is, referring to Mathcad 5.2, about 70% when the 3-W guideline is used (i.e., $d/w = 3$). If 90% of the flux is not to be coupled to the nearby trace, then d/w should be about 10. In some cases, coupling between two traces may be appropriate, such as with a differential pair.

w := 1

d := 2, 2.1.. 10

$$\text{flux\%}(h, d) := 100 - \frac{\ln\left[1 + \left(\frac{2 \cdot h}{d}\right)^2\right]}{\ln\left[1 + \frac{32 \cdot h^2}{w^2} \cdot \left[1 + \sqrt{1 + \left(\frac{\pi \cdot w^2}{8 \cdot h^2}\right)^2}\right]\right]} \cdot 100$$

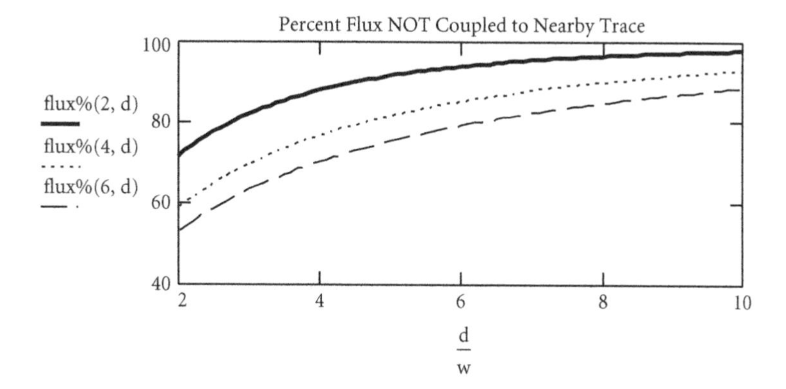

MATHCAD 5.2 Percent magnetic flux not coupled to a nearby trace as a function of the separation distance to trace width for three different heights.

The obvious cost of using this guideline is the loss of physical real estate or board space. Also, the actual spacing between traces may have to be adjusted based on the desired characteristic impedance of the trace, expected current level on the trace, and breakdown voltage between traces. However, the 3-W guideline does provide a designer a rule-of-thumb when laying out printed circuit boards. This rule-of-thumb is also used for other crosstalk and coupling situations. For example, when twin-lead transmission line is passed along a metallic tower, the distance between the twin-lead line and tower should be large (e.g., about three times the width of the cable). This distance should result in reasonably low electrical interaction between the twin-lead line and metallic tower. Although the analysis in this example was for two traces above a flat ground return, the same approach can be used for other configurations such as two circular wires surrounded by a circular ground conductor.

5.25 Capacitive Crosstalk and the 3-W Guideline

Determine the crosstalk between two circuit traces if only capacitive coupling is considered. What is the 3-W guideline, and what is its origin? [Ott; Mardiguian, '88(a); Montrose, '00]

The total crosstalk voltage in a victim circuit is a function of both the mutual inductance and mutual capacitance. For this discussion, the coupling is assumed to be entirely capacitive. The victim circuit is assumed to be trace *b* (with the large flat plane as the return conductor) for the configuration shown in Figure 5.64. If a sinusoidal signal source, of amplitude *A*, is applied between trace *a* and the return plane, conductor *c*, the voltage induced across the victim trace, *b*, and conductor *c*, is easily obtained through voltage division:

$$V_{bcs} = A\frac{\frac{1}{j\omega C_{bc}}}{\frac{1}{j\omega C_{ab}} + \frac{1}{j\omega C_{bc}}} = A\frac{C_{ab}}{C_{ab} + C_{bc}} \tag{5.104}$$

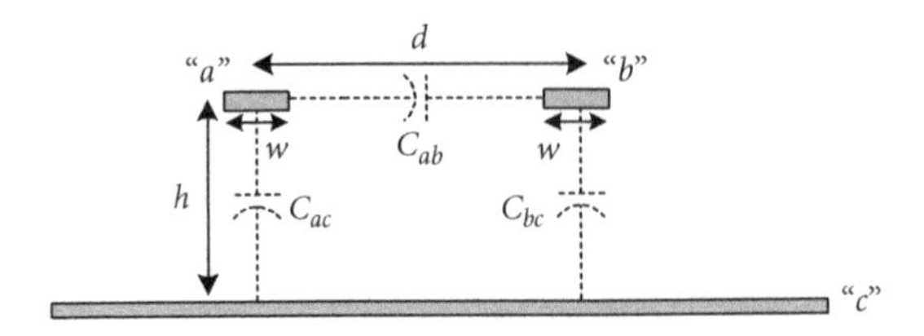

FIGURE 5.64 Capacitive coupling between all three conductors.

It is more realistic to include a resistance in parallel with C_{bc}, possibly representing the parallel combination of the source and load resistances for trace b. Labeling this resistance R_{bc}, the voltage is now

$$V_{bcs} = A\frac{\dfrac{1}{j\omega C_{bc}}\Big\|R_{bc}}{\dfrac{1}{j\omega C_{ab}}+\left(\dfrac{1}{j\omega C_{bc}}\Big\|R_{bc}\right)} = \frac{j\omega R_{bc}C_{ab}A}{1+j\omega R_{bc}(C_{ab}+C_{bc})} = \frac{j\omega R_{bc}C_{ab}A}{1+\dfrac{j\omega}{\dfrac{1}{R_{bc}(C_{ab}+C_{bc})}}} \tag{5.105}$$

The last form of this equation, which is in standard Bode magnitude form, clearly shows that the noise voltage increases with frequency but levels off to the voltage of (5.104) much above the cutoff frequency given by

$$\omega_c = \frac{1}{R_{bc}(C_{ab}+C_{bc})} \tag{5.106}$$

The magnitude of the voltage (5.105) is

$$|V_{bcs}| = \frac{\omega R_{bc}C_{ab}A}{\sqrt{1+[\omega R_{bc}(C_{ab}+C_{bc})]^2}} \tag{5.107}$$

For frequencies much less than the cutoff frequency, the magnitude of the interfering voltage is

$$|V_{bcs}| \approx \omega R_{bc}C_{ab}A \quad \text{if } \omega \ll \omega_c \tag{5.108}$$

For frequencies much less than the cutoff frequency, the noise voltage increases with frequency, load resistance, mutual capacitance, and signal amplitude. For frequencies much greater than the cutoff frequency, the noise voltage is frequency-independent:

$$|V_{bcs}| \approx A\frac{C_{ab}}{C_{ab}+C_{bc}} \quad \text{if } \omega \gg \omega_c \tag{5.109}$$

Again, this result is neglecting the inductive coupling (and any resonances that may occur).

The "circuit capacitances," assuming that a and b are not too close to c or to each other are

$$C_{ab} = 4\pi\varepsilon_o l_{th}\frac{\ln\left[1+\left(\dfrac{2h}{d}\right)^2\right]}{\left\{\ln\left\{1+\dfrac{32h^2}{w^2}\left[1+\sqrt{1+\left(\dfrac{\pi w^2}{8h^2}\right)^2}\right]\right\}\right\}^2 - \left\{\ln\left[1+\left(\dfrac{2h}{d}\right)^2\right]\right\}^2} \tag{5.110}$$

$$C_{ac} = 4\pi\varepsilon_o l_{th} \frac{\ln\left\{1+\frac{32h^2}{w^2}\left[1+\sqrt{1+\left(\frac{\pi w^2}{8h^2}\right)^2}\right]\right\}}{\ln\left\{1+\frac{32h^2}{w^2}\left[1+\sqrt{1+\left(\frac{\pi w^2}{8h^2}\right)^2}\right]\right\}^2 - \left\{\ln\left[1+\left(\frac{2h}{d}\right)^2\right]\right\}^2} - C_{ab} \tag{5.111}$$

$$C_{bc} = C_{ac} \tag{5.112}$$

where the thickness of conductors a and b is small compared to h, and l_{th} is the length of the conductors. The width of the ground plane is large compared to h, w, and d. The crosstalk in dB vs. frequency is given in Mathcad 5.3 for 1 cm long traces with $w = 15$ mils and $h = 60$ mils, and for a victim load resistance of 100 Ω. The center-to-center distance is varied. The dielectric constant of the substrate is assumed

$w := 15 \cdot 2.54 \cdot 10^{-5}$ $\quad l_{th} := 0.01$ $\quad \varepsilon_o := 8.854 \cdot 10^{-12}$ $\quad A := 1$ $\quad R_{bc} := 100$

$x := 60, 62..120$

$$\omega(x) := \left(x + 1 - 10 \cdot \text{floor}\left(\frac{x}{10}\right)\right) \cdot 10^{\text{floor}\left(\frac{x}{10}\right)}$$

$$C_{ab}(w,h,d) := 4\cdot\pi\cdot\varepsilon_o\cdot l_{th}\cdot \frac{\ln\left[1+\left(\frac{2\cdot h}{d}\right)^2\right]}{\ln\left[1+\frac{32\cdot h^2}{w^2}\cdot\left[1+\sqrt{1+\left(\frac{\pi\cdot w^2}{8\cdot h^2}\right)^2}\right]\right]^2 - \ln\left[1+\left(\frac{2\cdot h}{d}\right)^2\right]^2}$$

$$C_{bc}(w,h,d) := 4\cdot\pi\cdot\varepsilon_o\cdot l_{th}\cdot \frac{\ln\left[1+\frac{32\cdot h^2}{w^2}\cdot\left[1+\sqrt{1+\left(\frac{\pi\cdot w^2}{8\cdot h^2}\right)^2}\right]\right]}{\ln\left[1+\frac{32\cdot h^2}{w^2}\cdot\left[1+\sqrt{1+\left(\frac{\pi\cdot w^2}{8\cdot h^2}\right)^2}\right]\right]^2 - \ln\left[1+\left(\frac{2\cdot h}{d}\right)^2\right]^2} - C_{ab}(w,h,d)$$

$$\text{crsstlk}(w,h,d,\omega) := 20\cdot\log\left[\frac{\omega\cdot R_{bc}\cdot C_{ab}(w,h,d)\cdot A}{\sqrt{1+\left[\omega\cdot R_{bc}\cdot\left(C_{ab}(w,h,d)+C_{bc}(w,h,d)\right)\right]^2}}\right]$$

MATHCAD 5.3 Capacitive crosstalk between two traces vs. frequency for three different separation distances.

$w := 1 \quad \varepsilon_o := 8.854 \cdot 10^{-12} \quad l_{th} := 1$

$d := 2, 2.1 .. 10$

$$C_{ab}(w,h,d) := 4 \cdot \pi \cdot \varepsilon_o \cdot l_{th} \cdot \frac{\ln\left[1 + \left(\frac{2 \cdot h}{d}\right)^2\right]}{\ln\left[1 + \frac{32 \cdot h^2}{w^2} \cdot \left[1 + \sqrt{1 + \left(\frac{\pi \cdot w^2}{8 \cdot h^2}\right)^2}\right]\right]^2 - \ln\left[1 + \left(\frac{2 \cdot h}{d}\right)^2\right]^2}$$

$$C_{bc}(w,h,d) := 4 \cdot \pi \cdot \varepsilon_o \cdot l_{th} \cdot \frac{\ln\left[1 + \frac{32 \cdot h^2}{w^2} \cdot \left[1 + \sqrt{1 + \left(\frac{\pi \cdot w^2}{8 \cdot h^2}\right)^2}\right]\right]}{\ln\left[1 + \frac{32 \cdot h^2}{w^2} \cdot \left[1 + \sqrt{1 + \left(\frac{\pi \cdot w^2}{8 \cdot h^2}\right)^2}\right]\right]^2 - \ln\left[1 + \left(\frac{2 \cdot h}{d}\right)^2\right]^2} - C_{ab}(w,h,d)$$

$$\text{flux\%}(w,h,d) := 100 - \frac{C_{ab}(w,h,d)}{C_{bc}(w,h,d) + C_{ab}(w,h,d)} \cdot 100$$

MATHCAD 5.4 Percent electric flux not coupled to a nearby trace as a function of separation distance to trace width for three different heights.

to be unity. More complex capacitance expressions, taking into account a nonunity dielectric constant for the substrate material between the traces and return plane, are available. It should not be assumed that these results can be used at the higher frequencies since the traces may not be electrically short and the inductive coupling may be substantial. The equations are plotted at these high frequencies to show the theoretical location of (5.106).

For the capacitive coupling between two high-threat traces to be tolerable, the 3-W guideline states that the center-to-center distance between the two traces should be at least three times the width of the single trace. The previous inductive coupling discussion should be consulted for additional information concerning the usage of this guideline. Since (5.111) is available, the capacitance between the trace and return plane in the presence of the other trace will be used instead of that of an isolated trace above a return plane. The percent electric flux *not* coupled to the victim trace is plotted vs. d/w in Mathcad 5.4. As with inductive coupling, the degree of coupling between the two traces decreases with increasing center-to-center spacing. Furthermore, as the distance between the trace and return plane decreases, the capacitive coupling between the two traces also decreases. Because of the large size and close proximity of the return plane, many of the electric flux lines terminate on or are "grabbed" by the return plane.

Sometimes, it is recommended that guard traces be used instead of the 3-W guideline for critical traces. Referring to Figure 5.65, guard traces are grounded traces surrounding (and in close proximity to) both

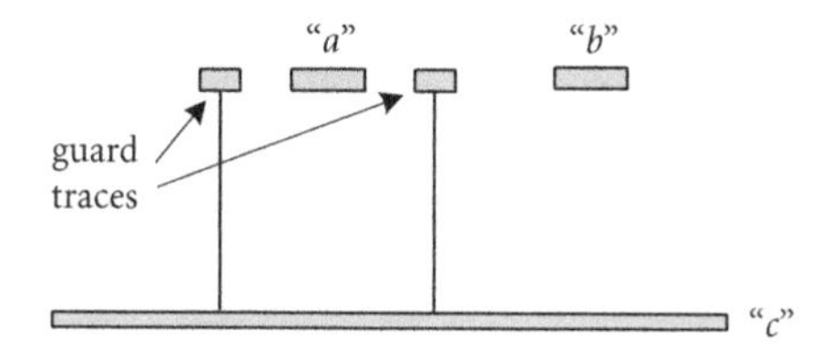

FIGURE 5.65 Guard traces around trace *a*.

sides of the signal trace. The guard traces are connected to the ground plane at one or more locations using vias. These guard traces will generally reduce electric field coupling to other traces (and have some effect on magnetic field coupling). They will also affect the characteristic impedance of the line. Before these additional grounded traces are implemented, their effectiveness should be carefully compared to the crosstalk reduction obtained using the 3-W guideline.

5.26 Long Lines vs. Close Lines

Determine which of the following situations is worse from the standpoint of crosstalk: i) 1 cm long parallel run traces separated by 1 mm (≈39 mils) or ii) 10 cm long parallel run traces separated by 1 cm (≈390 mils). Assume the distance to the return plane is variable. [Mardiguian, '84]

Since the source and load resistances of both the source and victim circuits are not given, the general expressions for the coupling between weakly coupled electrically-short conductors cannot be used. However, the mutual inductance and capacitance between the source and victim circuit can be determined for both of the given situations as a function of the width of the traces and their height above the ground plane. The voltages induced across the source and load resistances of the victim circuit are a function of the mutual inductance and capacitance.

First, the mutual inductance will be examined. Using (5.94), the mutual inductance for both situations is plotted in Mathcad 5.5. For trace-to-ground heights above about 4 mm, the mutual coupling of the

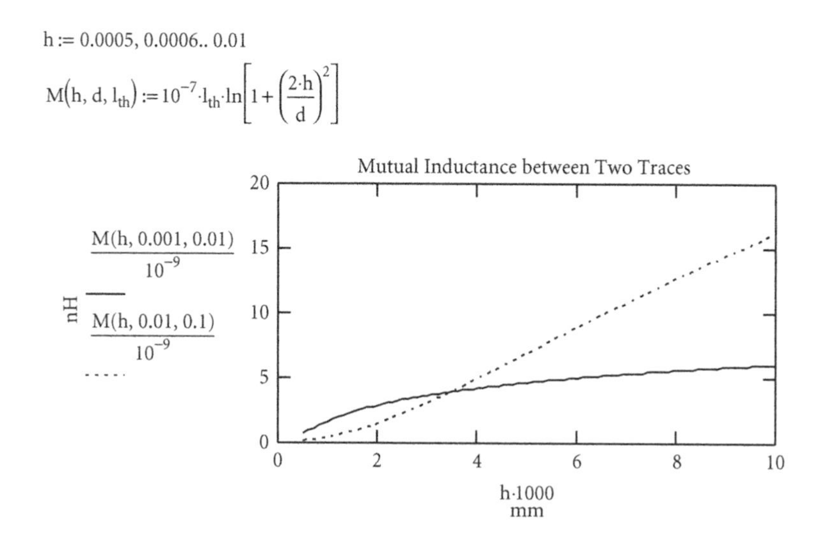

MATHCAD 5.5 Mutual inductance between two traces as a function of trace-to-ground height for two different trace separation/length combinations.

10 cm long traces with a 1 cm separation is greater than the 1 cm long traces with a 1 mm separation. Notice that the height, h, and trace separation, d, variables are inside the natural logarithm while the length variable, l_{th}, is outside the logarithm. Doubling the length will double the mutual inductance, but doubling the height will not double the mutual inductance. The length of the traces should be as short as possible (unless the transformer action between the traces is desirable). The height is varied from about 20 to 390 mil in Mathcad 5.5.

As seen in Mathcad 5.6, the mutual capacitance between the longer traces is also greater than the mutual capacitance between the shorter but closer traces for heights greater than about 4 mm. The width of the traces is 0.4 mm ($\approx$ 16 mils) in the first plot and 0.8 mm ($\approx$ 31 mils) in the second plot.

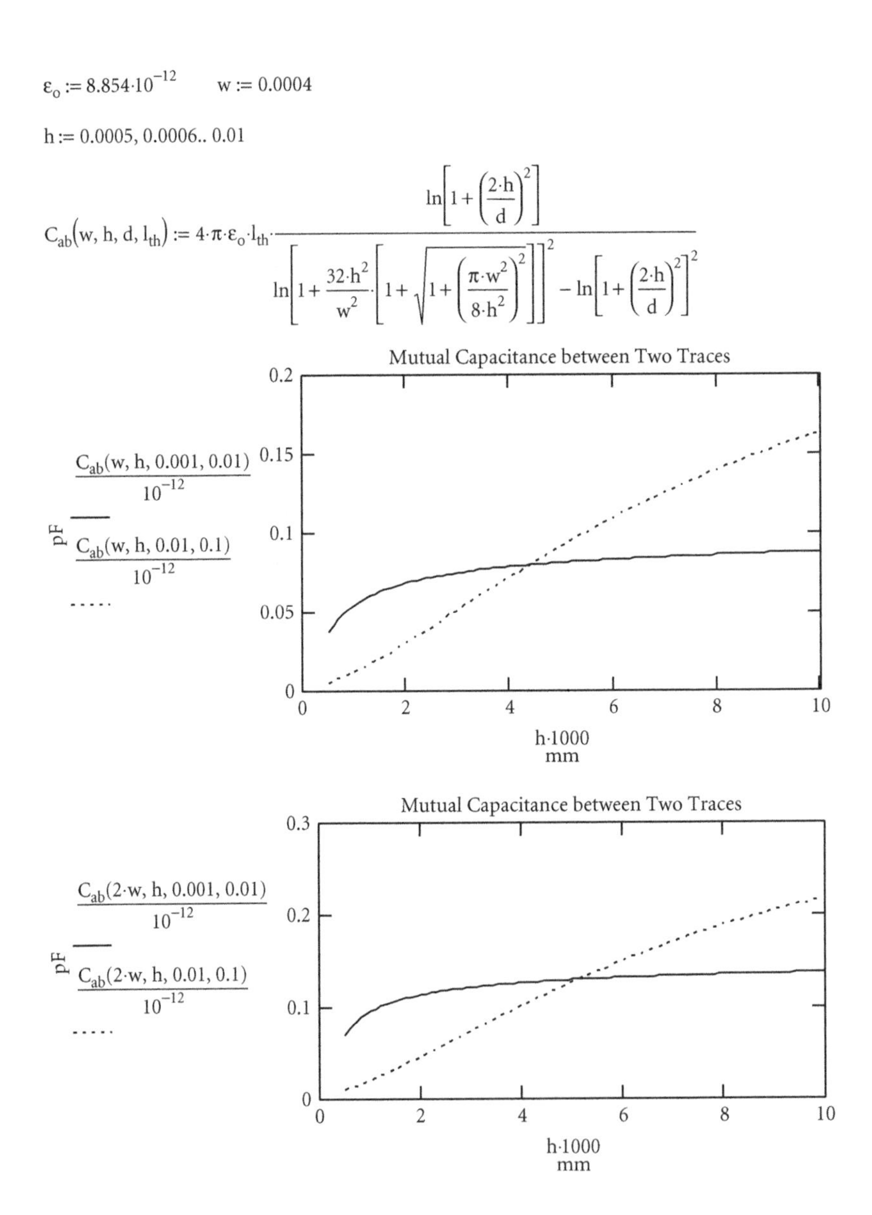

MATHCAD 5.6 Mutual capacitance between two traces as a function of trace-to-ground height for two different trace separation/length combinations. In the second plot, the trace width is doubled.

5.27 6" Guideline for Telephone Lines

In a handbook it is recommended that telephone lines be at least 6" from standard #12 AWG nonmetallic-sheathed power lines in homes. Determine the reasonableness of this guideline. [Freeman, '94]

The resistances of the source and load of the power circuit and the telephone circuit, which is also the victim, are not provided in the statement. To simplify the analysis, only the mutual inductance between the power and telephone circuits will be examined in this discussion. It is further assumed that approximately three #12 AWG wires can be inserted between the bare hot and bare neutral wires in the power cable, and that the telephone lines are #22 AWG wires.

Referring to Figure 5.66, the mutual inductance is between the loop formed by conductors 1 and 2, which will be considered the power line circuit, and the loop formed by conductors 3 and 4, which will be considered the telephone line circuit. The telephone wires are assumed untwisted, and the parameter s, the distance between the wires, will be varied. The center-to-center distance, d, between the #22 telephone wires is approximated as 76 mils ($\approx$1.9 mm), which implies that about two #22 wires can be inserted between the two wires. The center-to-center distance, h, between the #12 hot and neutral wires of the power line is approximated as 320 mils ($\approx$8.1 mm), which implies that about three #12 wires can be inserted between the hot and neutral wires in the power cable. Both the telephone and power cables are in the same direction, and the dimensions h, s, and d are assumed small compared to the length of the lines.

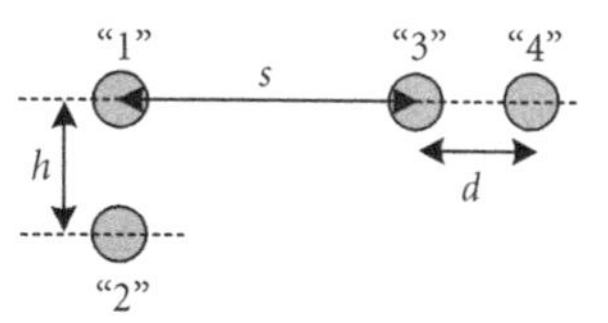

FIGURE 5.66 Conductors 1–2 form the power line and conductors 3–4 the telephone line.

The mutual inductance between the two pair of conductors shown in Figure 5.66 is

$$M = 2\times10^{-7} l_{th} \ln\left[\frac{(d+s)\sqrt{s^2+h^2}}{s\sqrt{(s+d)^2+h^2}}\right] \tag{5.113}$$

where l_{th} is the length of the conductors. Both the mutual inductance and open-circuit voltage are plotted in Mathcad 5.7 as the distance s is varied from 2 to 12 inches. The length of the lines is 1 m, and the current on the power lines is 20 A rms. The expression

$$v_{oc} = M\frac{di}{dt} \quad \Rightarrow \quad V_{ocs} = Mj\omega I_s$$

is used to obtain the open-circuit voltage. Obviously, the actual voltage induced across the telephone line is also a function of the source and load resistance in the telephone circuit and of the mutual capacitance. The mutual inductance for a 6" spacing is less than one-tenth of the 2" spacing. Typical, average, voice voltage levels on a telephone line are in the 40 mV range. As seen, the open-circuit inductive coupling contribution is small compared to 40 mV for the given parameters and range in s. Based only on this simple analysis, it appears that the 6" guideline is overly conservative. However, this assumes that the coupling is entirely due to the 60 Hz signal. Although the 60 Hz is (hopefully) the dominant frequency on the power line, harmonics and many other higher frequency signals are also present on the power line. These other signals also include transients (e.g., voltage surges) that can contain significant high-frequency energy. Since the open-circuit voltage due to inductive coupling is directly proportional to the frequency, these transients can produce significant noise on the telephone line. The 6" guideline is probably based more on real-life considerations, such as typical distances to nearby power lines and typical convenient locations for telephone and power outlets, than theoretical expressions.

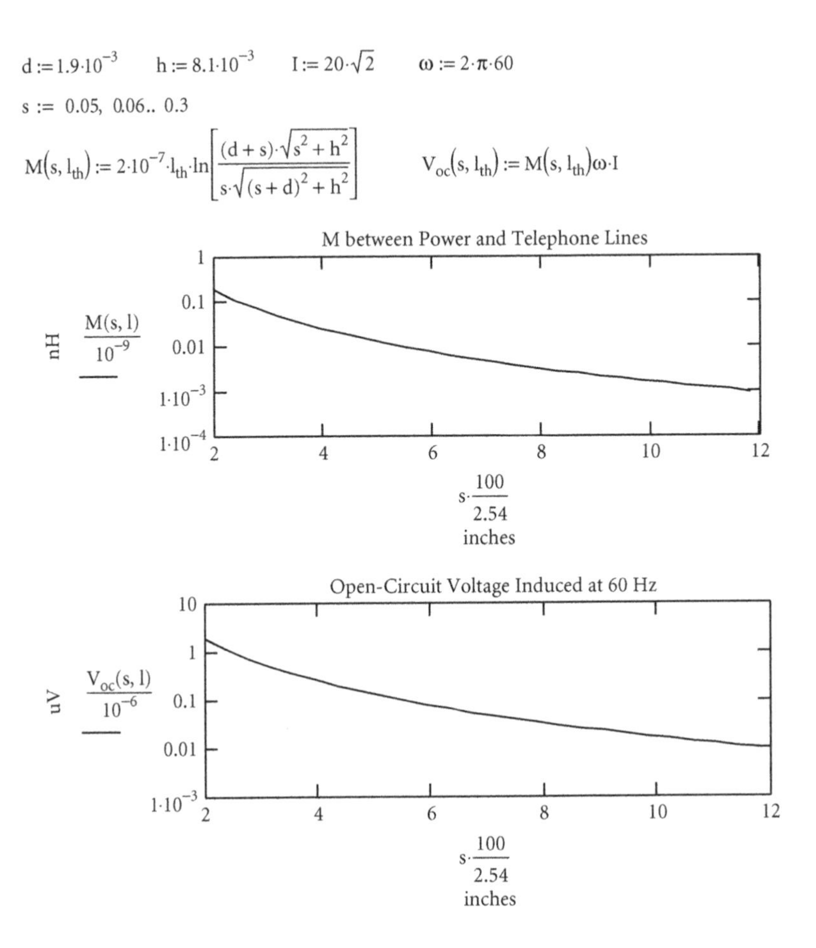

MATHCAD 5.7 Mutual inductance and induced open-circuit voltage between two circuits as a function of the spacing between them.

5.28 Four-Conductor Trace Layout

Compare the mutual inductance between two sets of parallel traces on one or both sides of a printed circuit board. [Rogers; Walker]

To assist in crosstalk reduction when laying out traces on circuit boards, it is helpful to examine the mutual inductance between two sets of traces on a single or double-sided printed circuit board. Although the mutual inductance is not the only factor affecting the crosstalk between conductors, it is often an important factor. Also, typical substrate materials are currently nonmagnetic, which implies the substrate material does not affect the mutual inductance between or the self inductance of the conductors.

The mutual inductance expressions for circular wires are generally used with good results for rectangular traces. (Both M and L_m are commonly used variables for mutual inductance.) Therefore, in the given figures, rectangular conductors can replace the circular conductors. The variable h is the substrate or board thickness, and it has a value of 45 mils in all of the plots. When $h = 0$ or is not present, the traces are located on the same side of the board. Conductors 1 and 2 compose one circuit and conductors 3 and 4 compose the other circuit. Therefore, the coupling or inductive crosstalk is between the pair of conductors 1 and 2 and the pair of conductors 3 and 4. A time-varying current passing through conductors 1 and 2 will induce an open-circuit voltage across conductors 3 and 4 and vice versa. The mutual inductance can be positive, negative, or zero, but only the magnitude is plotted. A negative inductance implies a sign reversal in the induced voltage. These equations are valid as along as the conductors are not

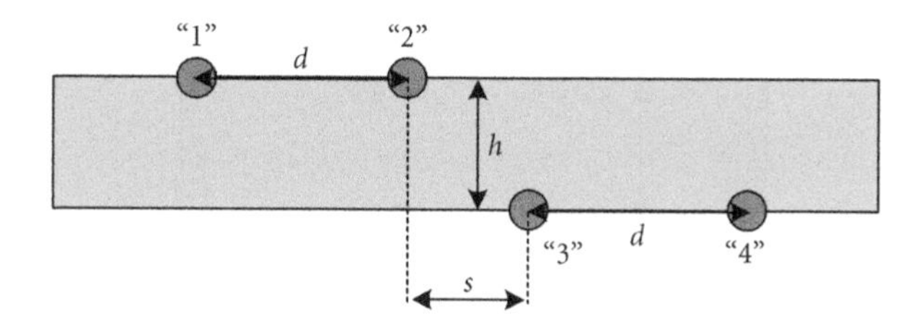

FIGURE 5.67 Trace configuration 1.

too close to each other (≈ 3–5 radii away) and the length of the conductors is long compared to the separation distances. The length of the traces were selected as 1 cm so that the per-unit-length inductance would have units of nH/cm. These mutual inductance expressions neglect end effects (and the wires connecting the traces together). When the actual conductor length is not large compared to the largest conductor separation distance, the end effects should be considered.

For configuration 1 shown in Figure 5.67, the generator and victim circuits are on opposite sides of the substrate or printed circuit board. The mutual inductance is given by

$$M_1 = 2\times10^{-7} l_{th} \ln\left[\frac{\sqrt{(2d+s)^2+h^2}\sqrt{s^2+h^2}}{(d+s)^2+h^2}\right] \tag{5.114}$$

and is plotted in Mathcad 5.8 for d = 30 mils. Beyond a critical separation distance, s, the mutual inductance decreases with increasing s: the magnetic field from a circuit drops off with distance. Also, the mutual inductance is greatest when the set of traces are vertically aligned with each other ($s = -30$ mils). Interestingly, the mutual inductance can be made quite small even when the distance s is relatively small by carefully selecting the separation distance, s. This occurs when the total flux through the loop formed by conductors 3 and 4 is small. The total flux through the circuit formed by conductors 3 and 4 is the

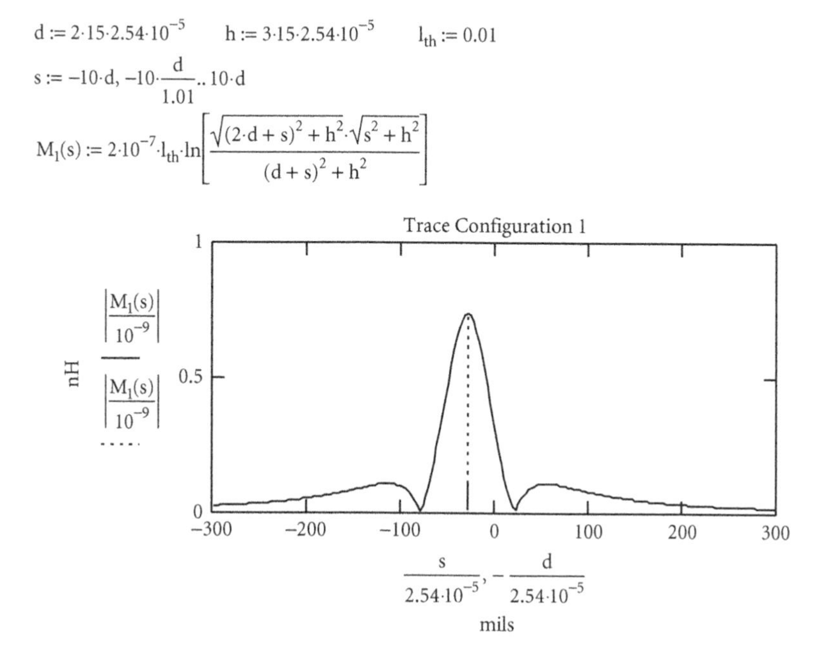

MATHCAD 5.8 Mutual inductance for trace configuration 1 as s is varied.

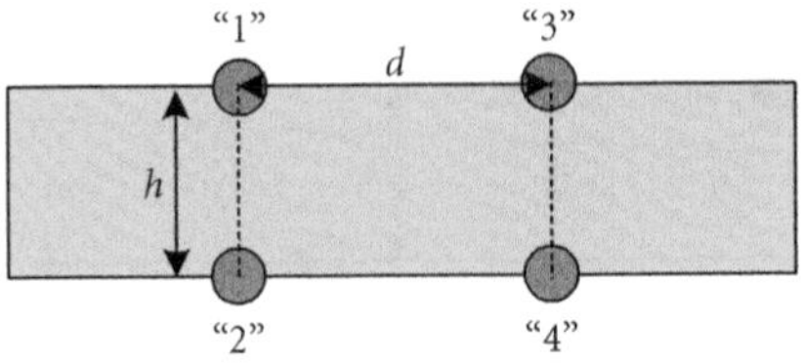

FIGURE 5.68 Trace configuration 2.

sum of the magnetic fluxes from current-carrying conductors 1 and 2. Because the currents in 1 and 2 are of the same magnitude but opposite sign, the fluxes generated by these two conductors can completely, or nearly, cancel for a certain range of separation distances, *s*.

For configuration 2 shown in Figure 5.68, the generator or transmitting circuit is located on the left and the victim or receiving circuit is located on the right. The return conductors 2 and 4 are now located on the same side of the board. The mutual inductance, given by,

$$M_2 = 2\times10^{-7} l_{th} \ln\left(\frac{d^2+h^2}{d^2}\right) \tag{5.115}$$

is plotted in Mathcad 5.9. As expected, the mutual inductance decreases with increasing separation distance. Compared to configuration 1, where the return conductors 2 and 4 are on opposite sides of the board, configuration 2 can have substantially lower mutual inductance (assuming this is a useful comparison). This lower inductance is clearly seen in Mathcad 5.10 where $s = -d$ in (5.114), for configuration 1. The logarithmic scale is used for the *y* axis for this plot so the difference can be clearly seen.

$d_x := 15\cdot2.54\cdot10^{-5}$ $\quad h := 3\cdot15\cdot2.54\cdot10^{-5}$ $\quad l_{th} := 0.01$

$d := 2\cdot d_x, 2.1\cdot d_x .. 10\cdot d_x$

$$M_2(d) := 2\cdot10^{-7}\cdot l_{th}\cdot\ln\left(\frac{d^2+h^2}{d^2}\right)$$

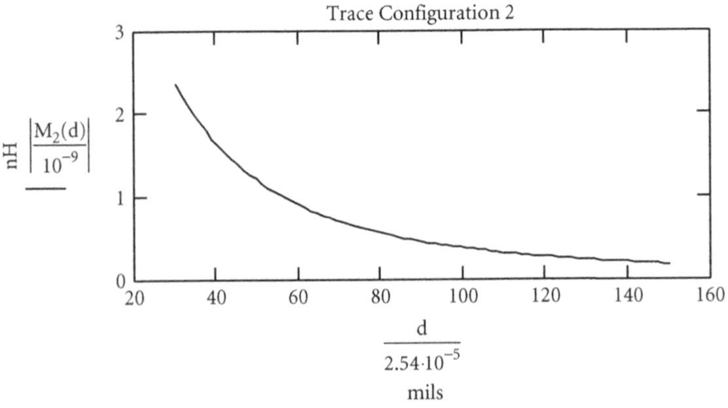

MATHCAD 5.9 Mutual inductance for trace configuration 2 as *d* is varied.

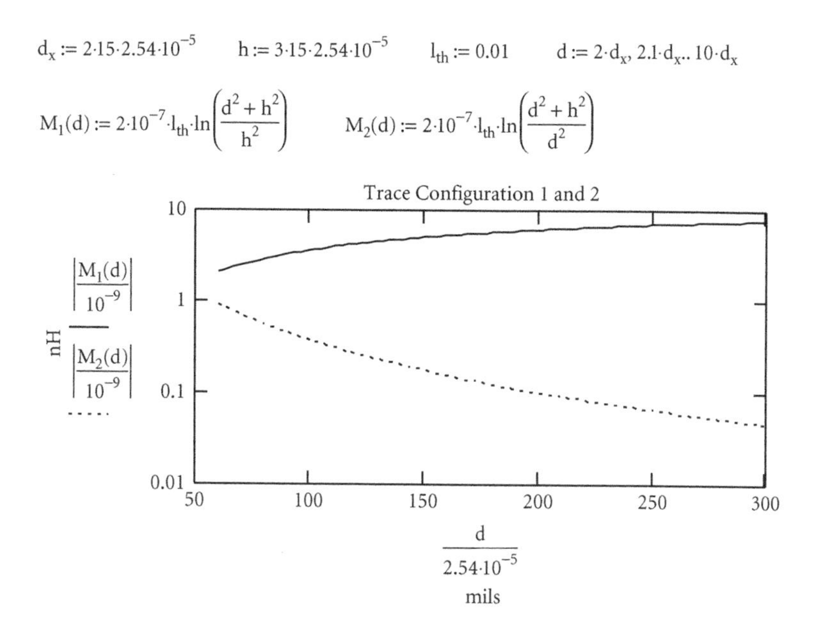

MATHCAD 5.10 Mutual inductance for trace configurations 1 and 2 as d is varied.

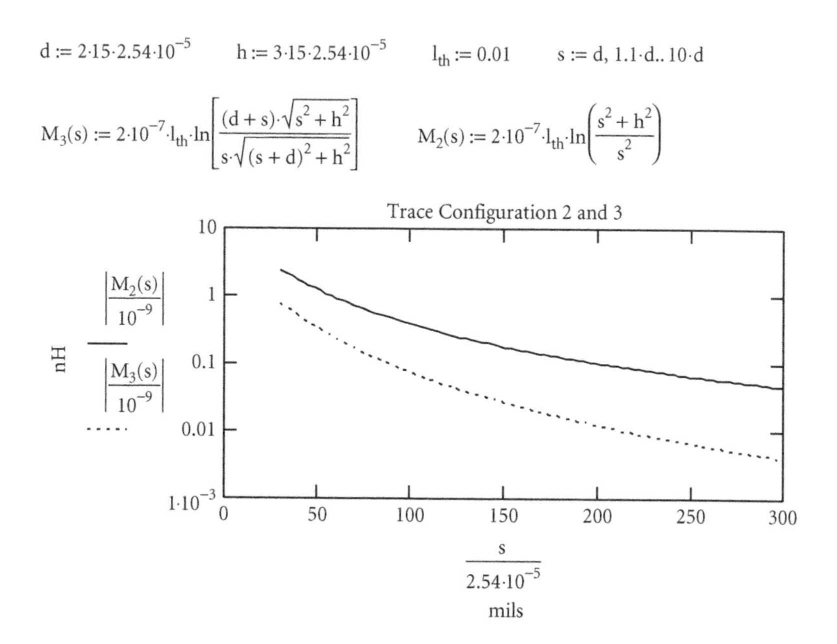

MATHCAD 5.11 Mutual inductance for trace configuration 2 and 3.

As seen in Mathcad 5.11, there is a slight improvement in configuration 3, shown in Figure 5.69, over configuration 2. This improvement is due to the greater distance of conductor 4 from the source of the flux from conductors 1 and 2. The cost of configuration 3 is more board space, at least on one side of the board. The mutual inductance is

$$M_3 = 2\times10^{-7} l_{th} \ln\left[\frac{(d+s)\sqrt{s^2+h^2}}{s\sqrt{(s+d)^2+h^2}}\right] \tag{5.116}$$

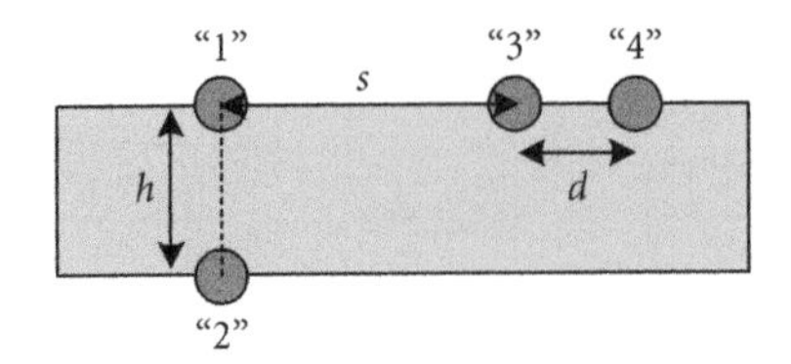

FIGURE 5.69 Trace configuration 3.

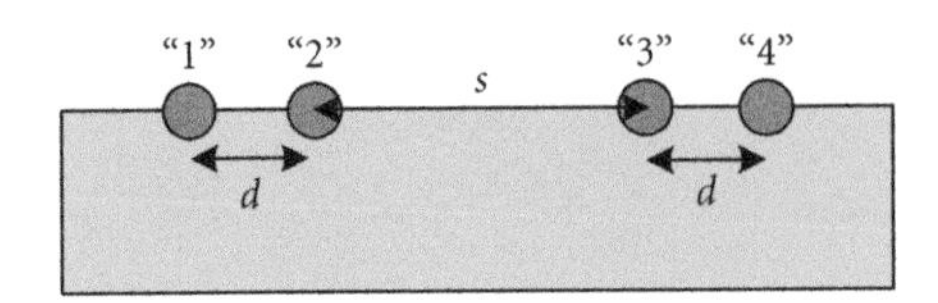

FIGURE 5.70 Trace configuration 4.

Trace configuration 4, shown in Figure 5.70, is a special case of configuration 1 with $h = 0$. The mutual inductance

$$M_4 = 2\times 10^{-7} l_{th} \ln\left[\frac{s(2d+s)}{(d+s)^2}\right] \tag{5.117}$$

is plotted in Mathcad 5.12. For smaller values of s, configuration 1 has somewhat less mutual inductance than configuration 4. This lower inductance is mainly due to the greater distance between the generating circuit and victim circuit in configuration 1 (for nonzero values of h).

$d := 2\cdot 15\cdot 2.54\cdot 10^{-5}$ $\quad h := 3\cdot 15\cdot 2.54\cdot 10^{-5}$ $\quad l_{th} := 0.01$

$s := d, 1.1\cdot d .. 10\cdot d$

$M_4(s) := 2\cdot 10^{-7}\cdot l_{th}\cdot \ln\left[\frac{s\cdot(2\cdot d + s)}{(d+s)^2}\right]$ $\quad M_1(s) := 2\cdot 10^{-7}\cdot l_{th}\cdot \ln\left[\frac{\sqrt{(2\cdot d+s)^2+h^2}\cdot\sqrt{s^2+h^2}}{(d+s)^2+h^2}\right]$

Trace Configuration 1 and 4

nH: $\left|\frac{M_1(s)}{10^{-9}}\right|$ (solid), $\left|\frac{M_4(s)}{10^{-9}}\right|$ (dotted)

0.6, 0.4, 0.2, 0

0, 50, 100, 150, 200, 250, 300

$\frac{s}{2.54\cdot 10^{-5}}$

mils

MATHCAD 5.12 Mutual inductance for trace configuration 1 and 4 as s is varied.

$d := 3\cdot 15\cdot 2.54\cdot 10^{-5} \quad l_{th} := 0.01$

$s := d, 1.1\cdot d .. 10\cdot d$

$M_5(s, l_{th}) := 4\cdot 10^{-7}\cdot l_{th}\cdot \ln\left(1 + \frac{s}{d}\right)$

Trace Configuration 5

1000

500

0

$\frac{|M_5(s, 1)|}{10^{-9}}$

nH

0 50 100 150 200 250 300

$\frac{s}{2.54\cdot 10^{-5}}$

mils

MATHCAD 5.13 Mutual inductance for trace configuration 5.

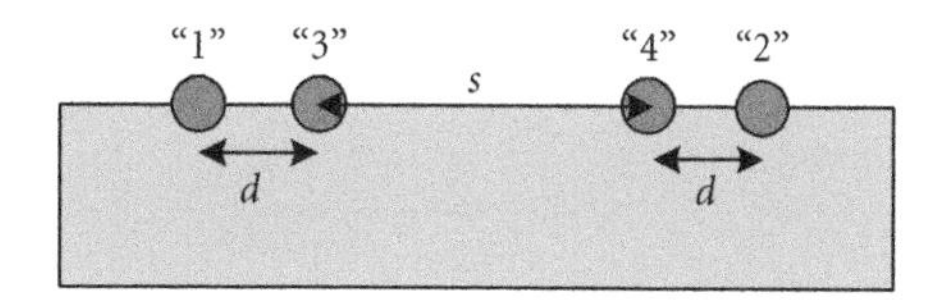

FIGURE 5.71 Trace configuration 5.

As seen by the large values of mutual inductance shown in Mathcad 5.13, configuration 5 is to be avoided (unless a high degree of inductive coupling is desired such as with a transformer). The mutual inductance for the configuration shown in Figure 5.71 is

$$M_5 = 4\times 10^{-7} l_{th} \ln\left(1+\frac{s}{d}\right) \tag{5.118}$$

Since one circuit is inside the other circuit, a large percentage of the magnetic flux generated by one circuit loop is picked up by the other circuit loop.

5.29 377 Ω Guideline

Sometimes the ratio

$$\frac{dv/dt}{di/dt} = \frac{dv}{di} \tag{5.119}$$

of the signal on the source or generator circuit is compared to 377 Ω. When this ratio is much less than 377 Ω, the coupling is said to be most likely magnetic, while if this ratio is much greater than 377 Ω, the coupling is said to be most likely electric. Determine the validity of this guideline. [Upton; Paul, '94]

Expression (5.119) at first seems very reasonable, especially if the reader is somewhat familiar with the concept of wave impedance. The ratio of the transverse components of the electric to the magnetic fields (*E*/*H*) for a current-carrying electrically-small loop and charged electrically-small straight wire is less than 377 Ω near the loop while greater than 377 Ω near the straight wire. The loop is inductive or magnetic in nature while the wire is capacitive or electric in nature. A real circuit, however, is better modeled by a resistively loaded loop (i.e., a resistor in series with the loop). The wave impedance, not surprisingly, can be varied by adjusting this resistance. For small values of resistance, much less than 377 Ω, the wave impedance is inductive. For large values of resistance, much greater than 377 Ω, the wave impedance is capacitive. Given these relationships, it would initially seem reasonable that the coupling to another circuit may also be related to 377 Ω. Unfortunately, it is not quite this simple.

Previously in (5.70)–(5.74), the expressions for the voltage across the victim circuit for a three-conductor system were given for sinusoidal steady-state conditions (assuming the wires are weakly coupled and electrically short). Because of the correspondence between $j\omega$ multiplication in the frequency domain and differentiation in the time domain, it follows that $j\omega \Leftrightarrow d/dt$. Thus, referring to (5.70)–(5.74), the time-domain crosstalk expressions for the voltage across the source and load of the victim circuit are

$$v_{sv}(t) = (M_{inds} + M_{cap})\frac{dv_s(t)}{dt} \tag{5.120}$$

$$v_{Lv}(t) = (M_{indL} + M_{cap})\frac{dv_s(t)}{dt} \tag{5.121}$$

where

$$M_{cap} = C_m \frac{R_{sv}R_{Lv}}{R_{sv}+R_{Lv}}\frac{R_L}{R_s+R_L}, \quad M_{inds} = L_m \frac{R_{sv}}{R_{sv}+R_{Lv}}\frac{1}{R_s+R_L}$$

$$M_{indL} = -L_m \frac{R_{Lv}}{R_{sv}+R_{Lv}}\frac{1}{R_s+R_L}$$

The corresponding coupling model for the victim or receptor circuit is shown in Figure 5.72. As in the frequency domain, for the capacitive coupling to dominate at the victim's source, $v_{sv}(t)$, $L_m/C_m \ll R_{Lv}R_L$, and for the capacitive coupling to dominate at the victim's load, $v_{Lv}(t)$, $L_m/C_m \ll R_{sv}R_L$. For the inductive coupling to dominate, the inequalities are reversed. These expressions involve the load resistance of the source or generator circuit, the source and load resistances of the victim circuit, and the mutual inductance and capacitance between the source and victim circuit. The 377 Ω guideline involves just the change of the source circuit voltage with respect to the source circuit current. If the equivalent circuit seen by the source voltage is determined, dv/di for this equivalent circuit is the dynamic resistance (or impedance)

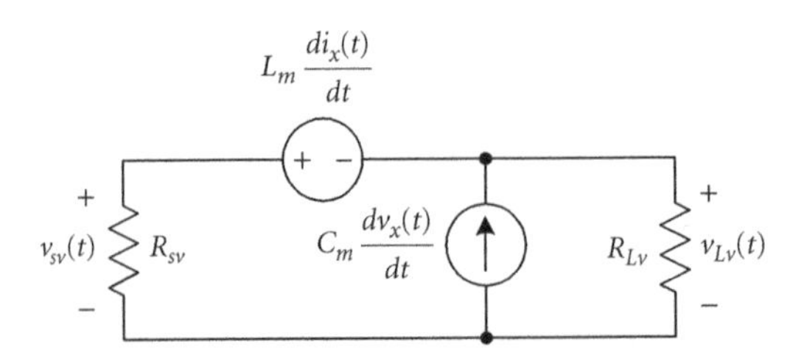

FIGURE 5.72 Time-domain victim coupling model for weak coupling and electrically-short conductors.

$w := 1 \quad \varepsilon_o := 8.854 \cdot 10^{-12} \quad l_{th} := 1 \quad h := 4$

$d := 2, 2.1 .. 10$

$$C_M(w, h, d) := 4 \cdot \pi \cdot \varepsilon_o \cdot l_{th} \cdot \frac{\ln\left[1 + \left(\frac{2 \cdot h}{d}\right)^2\right]}{\ln\left[1 + \frac{32 \cdot h^2}{w^2} \cdot \left[1 + \sqrt{1 + \left(\frac{\pi \cdot w^2}{8 \cdot h^2}\right)^2}\right]\right]^2 - \ln\left[1 + \left(\frac{2 \cdot h}{d}\right)^2\right]^2}$$

$$L_M(w, h, d) := 10^{-7} \cdot l_{th} \cdot \ln\left[1 + \left(\frac{2 \cdot h}{d}\right)^2\right]$$

MATHCAD 5.14 Mutual impedance between two identical traces above a common ground plane.

seen by the source. Even if the ratio L_m/C_m happens to be about 377 Ω, the coupling is still a function of the impedances of the victim circuit.

For systems involving more than two conductors, there are several different characteristic impedances. The mutual (or coupling) characteristic impedance between the source and victim circuit for a three-conductor system will be defined in this example as

$$Z_m = \sqrt{\frac{L_m}{C_m}} \tag{5.122}$$

In Mathcad 5.14, this mutual characteristic impedance is plotted vs. d/w for two identical traces of width w, a distance of h above a return plane. The variable d is the center-to-center spacing between the traces. In the given plot, $h/w = 4$. For the coupling or crosstalk to be capacitive, the geometric mean of the generator's load resistance and victim's source or load resistance should be much greater than the mutual characteristic impedance: $Z_m \ll \sqrt{R_v R_L}$. In this plot, for the given range of d/w, the mutual impedance is never equal to 377 Ω. In conclusion, the given 377 Ω guideline should not be used unless its theoretical basis is known (and is reasonable).

5.30 Why Twisting Often Helps

Why does twisting of two wires reduce their emissions and susceptibility? Compare the magnetic fields from parallel and twisted wires. [Paul, '79; Paul, '92(b); Moser, '68(a); Weston; Shenfeld, '69]

With twisted pair, the twisting reduces the magnetic field or inductive coupling. (The twisting can also reduce the effects of capacitive coupling if the system is balanced.) To understand qualitatively how

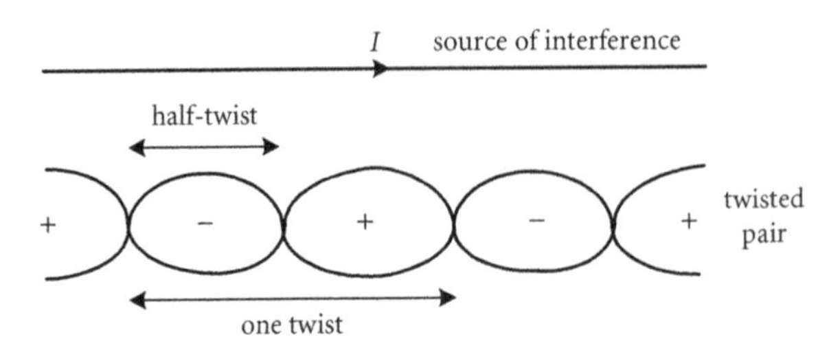

FIGURE 5.73 Oppositely signed voltages induced in neighboring loops of twisted pair.

twisting reduces inductive coupling, Faraday's law should be resurrected. This law states that the open-circuit voltage induced in a loop of N turns with a time-varying magnetic flux $\Phi(t)$ passing through it is

$$v(t) = -N\frac{d\Phi(t)}{dt}$$

The time-varying flux is due to the current from the generator or source circuit. The flux passing through each loop is a function of the orientation of the loop. If the flux is parallel to the plane of the loop, no flux passes through the loop and the induced voltage is zero. The maximum induced voltage is obtained when the flux passes perpendicular to the plane of the loop. The total time-varying flux collected by a single loop is a function of the orientation of the loop since

$$\Phi = \iint \vec{B} \cdot d\vec{s}$$

and $d\vec{s}$, being a vector, has direction. By twisting two wires, neighboring loops are physically 180° out of phase with each other, and the induced voltages in neighboring loops are also of opposite polarity. The "+" and "–" symbols in Figure 5.73 are indicating the sign of the induced voltage in each loop at one instant of time. As will be seen, when these voltages are summed, they can nearly cancel each other. In theory, an even number of half-twists would result in a total induced voltage of zero, while an odd number of half-twists would result in a total induced voltage corresponding to one loop. As seen in the figure, one twist is defined as two loops while one-half of a twist is defined as one loop.

In this discussion, it is assumed that the line lengths are electrically short. When the generator or interfering circuit and the victim circuit are electrically short, the magnitude (and phase) of the interfering current is approximately the same along the length of the generator, and the magnitude of the induced voltage for each twist of the victim line is approximately the same. Also, in the following analysis, the coupling between the generator wire and twisted pair is assumed weak.

When modeling crosstalk involving one generator wire, two victim twisted wires, and a common reference plane, it is common to use a model similar to the weakly coupled, electrically-short three-conductor model shown in Figure 5.75. Unfortunately, because twisted pair is a nonuniform transmission line, the mutual capacitance and inductance between the generator circuit and each of the conductors of the twisted pair will vary along the length of each loop. To simplify matters, each loop is assumed to be mostly rectangular in shape as shown in Figure 5.74 and the transition between loops is assumed abrupt. With this simplification, the common expressions for mutual inductance and capacitance between two wires above a return plane can be used. The mutual terms at the nonuniform sections near each of the conductor crossover points can be ignored.

In Figure 5.74, the generator line and twisted pair are parallel and above a large flat plane. There are four different conductors in this model. Although the given figure shows the generator line directly above the victim conductors, the position of the generator line relative to the victim conductors can be nearly anything as long as they are parallel and the distance between the generator line and victim conductors

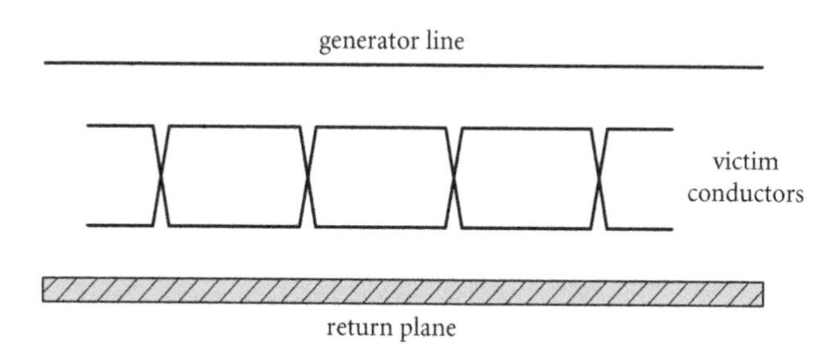

FIGURE 5.74 Twisted pair modeled as abruptly changing rectangular loops.

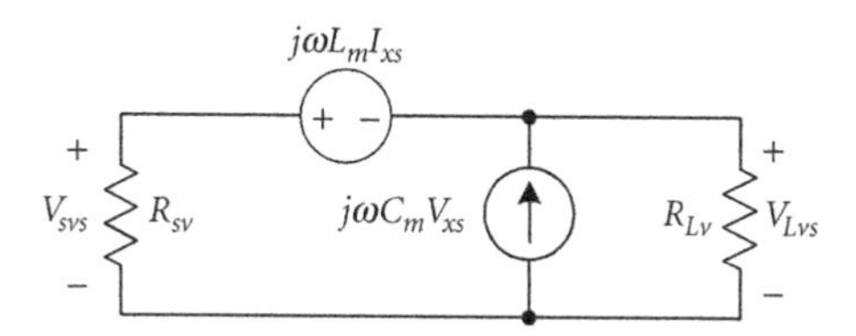

FIGURE 5.75 Weakly coupled model used for each conductor in the twisted pair.

are not too close (so they are weakly coupled). For example, the distance of the generator line to the return plane can be equal to the distance of both victim conductors to the return plane, and all of the conductors can be in the same horizontal plane.

To model the coupling, the victim crosstalk model for weakly coupled electrically-short lines shown in Figure 5.75 can be used. The variable I_{xs} is the current in the generator line, and V_{xs} is the driving voltage across the generator line and return plane. This model can be used for *each* segment of *each* loop of the twisted pair. (As stated previously, the coupling to the segments at the abrupt transitions will not be modeled.) The termination resistances at the near and far ends of the overall twisted pair are not shown in Figure 5.76. The variables L_m and C_m shown in this figure are the mutual inductance and mutual capacitance, respectively, over the entire length of *one* segment of the half-twist. Note that the mutual inductance and capacitance for *each wire* segment of a loop are different. The distance between the generator and each of the victim wires, as well as the distances of the generator line and victim wires to the return plane, are not

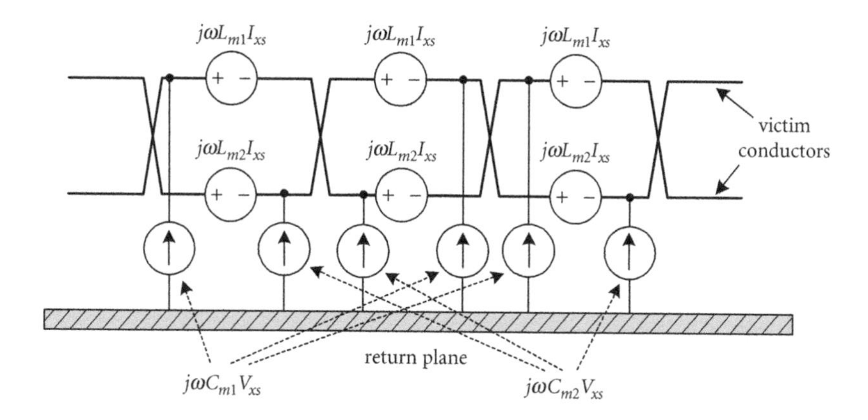

FIGURE 5.76 Coupling to each segment of each victim conductor is modeled using both an induced voltage and current source.

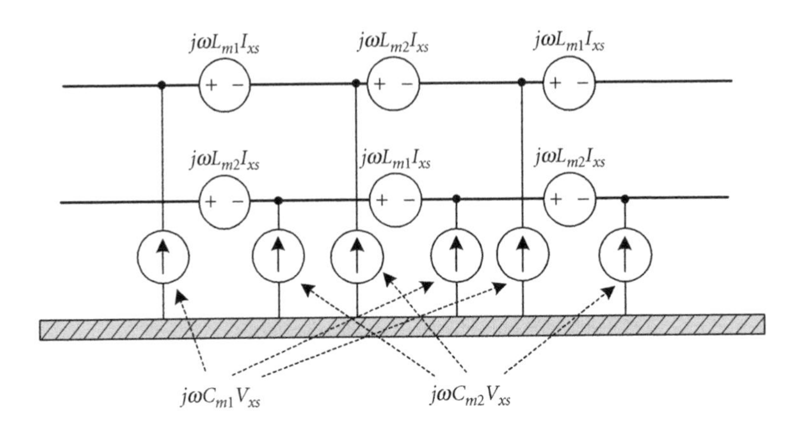

FIGURE 5.77 Coupling model of untwisted wires.

necessarily the same. Since the coupling to the victim conductors has been modeled, the wires can be untwisted as shown in Figure 5.77 to simplify the model visually.

The voltage sources are due to the magnetic field coupling (mutual inductance) while the current sources are due to the electric field coupling (mutual capacitance). The previous qualitative discussion involving Faraday's law included only the mutual inductance coupling. If the current sources are turned off (i.e., replaced by open circuits), then the total driving force due to only inductive crosstalk can be determined. Imagining that an impedance is present at both the near and far ends (i.e., the source and load sides of the victim circuit), the total driving voltage due to the magnetic coupling for three half-twists is

$$\begin{aligned} V_{Ts} &= j\omega L_{m1}I_{xs} + j\omega L_{m2}I_{xs} + j\omega L_{m1}I_{xs} - j\omega L_{m2}I_{xs} - j\omega L_{m1}I_{xs} - j\omega L_{m2}I_{xs} \\ &= j\omega(L_{m1} - L_{m2})I_{xs} \end{aligned} \tag{5.123}$$

while for two half-twists this voltage is

$$V_{Ts} = j\omega L_{m1}I_{xs} + j\omega L_{m2}I_{xs} - j\omega L_{m1}I_{xs} - j\omega L_{m2}I_{xs} = 0$$

It is easy to conclude that for an odd number of half-twists, the total voltage is equal to (5.123) whereas for an even number of half-twists the total voltage is zero. The mutual inductances L_{m1} and L_{m2}, which are the mutual inductances between the generator wire and each of the twisted wires, can be quite similar for twisted pair. Hence, for an odd numbers of half-twists, the induced voltage due to magnetic coupling can be small. For comparison purposes, the total voltage induced across a parallel set of wires, with no twist, is

$$V_{Ts} = j\omega N(L_{m1} - L_{m2})I_{xs} \tag{5.124}$$

where N is the number of segments each with a length equal to the length of the half-twist segment. In (5.124), L_{m1} and L_{m2} are the mutual inductances between the generator wire and each of the untwisted parallel wire segments. Even if the mutual inductances are similar in value, this expression is multiplied by N while the twisted-pair version is not.

The previous results clearly illustrate the reduction in the inductive portion of the crosstalk when twisting a set of wires. However, the total crosstalk for a weakly coupled electrically-short line also includes capacitive coupling, which is modeled through the current supplies. Whether the capacitive or inductive coupling dominates is a complicated function of the mutual inductances, mutual capacitances, and termination impedances. In some cases, for low-impedance terminations, the inductive coupling dominates,

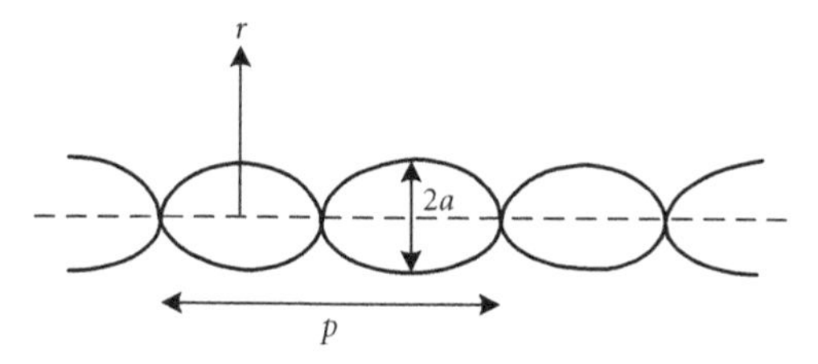

FIGURE 5.78 Pitch, loop radius, and radial distance in (5.126).

and for high-impedance terminations, the capacitive coupling dominates. Sometimes, an engineer with some cable or EMC experience will change a parallel line to a twisted pair in an attempt to reduce the emissions and susceptibility of the line. The engineer may be surprised that, in some cases, the noise pickup of the line will change very little: the crosstalk may be predominantly capacitive. By twisting the lines, mainly the inductive crosstalk is reduced, but the inductive crosstalk may only be a small fraction of the overall crosstalk. For high-impedance terminations (with unbalanced systems) where the coupling is mostly capacitive, twisting the lines will have little effect on the crosstalk. For low-impedance terminations (with unbalanced systems) where the coupling is mostly inductive, twisting the lines will have an effect on the crosstalk.

Twisting a pair of conductors can improve the balance of a system if the driver and receiver for the victim circuit are also balanced (or differential). In other words, twisting can affect the capacitive coupling of a nearly balanced system. Referring to Figure 5.77, the capacitive coupling contribution is zero when $C_{m1} = C_{m2}$.

The maximum magnetic field from twisted pair is obtained by modeling the wires as a long bifilar helix. Each of the helices has the same pitch, p, and radius, a, and they are 180 spatial degrees from each other. The pitch is the distance of one full cycle as shown in Figure 5.78. The wires are assumed infinitely thin with a constant current along each wire. The actual expressions for the magnetic flux density components from twisted pair are complex. An approximation for the maximum magnetic flux density of any of the components for $a/p \le 1/(3\pi)$ in T[28] is

$$B_{max} = \frac{10^{\left[\frac{-54.5\frac{r}{p}+20\log_{10}a-30\log_{10}p-10\log_{10}r-22.10+20\log I}{20}\right]}}{10^4} \tag{5.125}$$

or in dBT

$$B_{max,dB} = -54.5\frac{r}{p} + 20\log_{10}a - 30\log_{10}p - 10\log_{10}r - 102.1 + 20\log I \tag{5.126}$$

where I is the magnitude of the current in the wires. Referring to Figure 5.78, the variable r is the radial distance from the axis of the twisted pair. Decreasing the loop radius, a, or pitch, p, reduces the magnetic field emissions from twisted pair. A tighter twist results in lower emissions. Obviously, the magnetic flux density increases with the amplitude of the current. For comparison purposes, two values of the flux density for an infinitely long two-wire parallel line *at the two specific positions shown* in Figure 5.79 are

$$B_r = \frac{\mu_o Ia}{\pi(r^2+a^2)}, \quad B_\theta = \frac{\mu_o Ia}{\pi(r^2-a^2)} \tag{5.127}$$

where $2a$ is the distance between the two wires and r is the radial distance from their center. Again, the current along each of the wires is assumed constant. Just the radial components of the two fields

[28]T = tesla = Wb/m^2 is an SI unit.

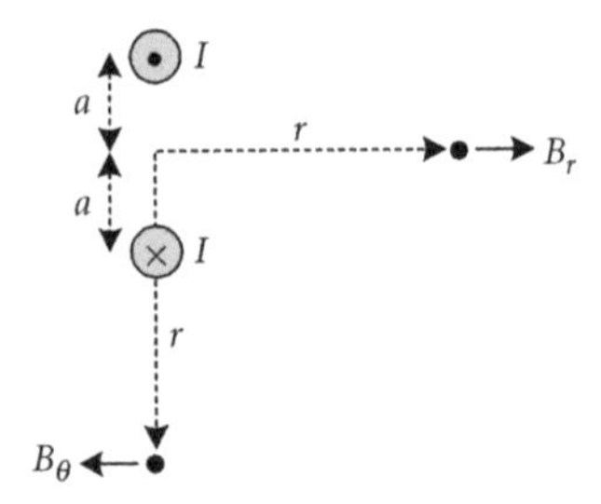

FIGURE 5.79 Magnetic flux density direction (at the two specified locations) from two current-carrying conductors.

can also be compared. The ratio of the radial twisted-pair flux density to the radial parallel-line flux density is

$$\frac{B_{r,twisted}}{B_{r,parallel}} \leq 2\pi^2\left(\frac{r}{p}\right)^{\frac{3}{2}} e^{-\left(\frac{2\pi}{p}\right)r} \tag{5.128}$$

which has 10% or less error when $r > 3a$ and $2\pi a/p < 2/3$. This ratio is plotted in Mathcad 5.15. The magnetic field emissions from twisted pair are obviously less than from untwisted parallel wires. The magnetic field decreases rapidly with distance from the twisted pair.

The previous analysis assumed that the current in the two wires was of the same amplitude but oppositely directed. These currents are referred to as differential-mode currents. Frequently, most of the radiation or emissions from twisted pair are due to the common-mode currents on the line. The common-mode currents are in the same direction along both wires. Of course, these currents must have a return path to their source, and this return path can be through parasitic capacitances to ground. The various mutual capacitances were previously discussed. It is important to state that often these undesirable common-mode currents are the dominant emission source from a twisted pair, and twisting the wires has little effect on these common-mode emissions (besides an improvement in the balance of the system and possible reduction in the spacing between the conductors). Balancing the system, generally reduces

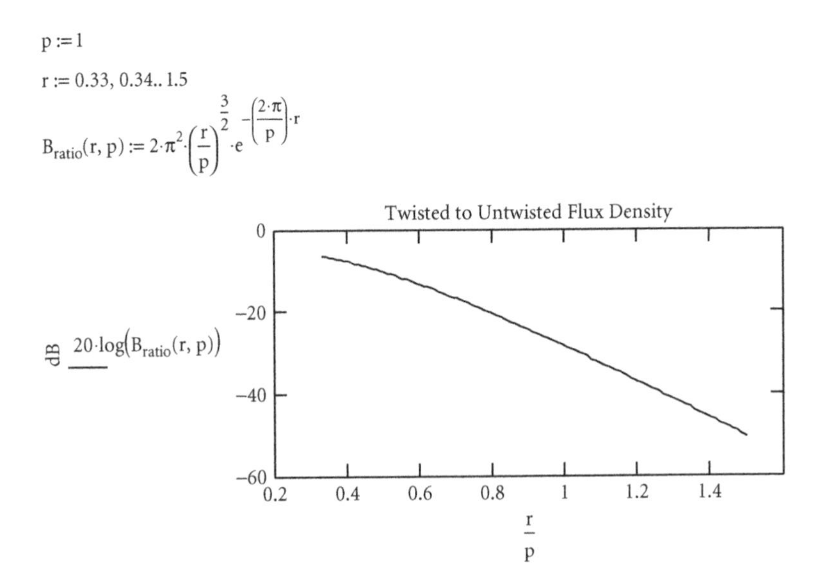

MATHCAD 5.15 Ratio of the radial twisted-pair flux density to the radial parallel-line flux density vs. the ratio of radial distance to pitch.

the common-mode currents and capacitive crosstalk to the twisted pair. Two potential disadvantages of twisting are that it places strain on the conductors, and unless the twisting pitch is carefully controlled, it might produce undesirable "bumps" in the characteristic impedance of the twisted pair. Nonuniformity in the characteristic impedance can be important at higher frequencies.

When the lengths of the conductors are not electrically short, the magnitude and phase of the currents along the wires are not constant. For electrically-long lines, the phase of the current changes along the length of the conductors. Therefore, the magnetic fields from the wires do not cancel as effectively as (is expected) with an electrically-short line. Furthermore, for very high frequencies, such as in the GHz range, the pitch of the wires, p, can be equal to one wavelength. In this situation, the magnetic fields from neighboring loops can *add*, resulting in magnetic emissions greater than with an untwisted pair![29]

5.31 *RC* Circuit and Crosstalk

Crosstalk is usually directly related to a change with time of the current or voltage in another circuit. If a gate with an output voltage $v(t)$ is driving a parallel RC circuit, show that the maximum change of current into this circuit with respect to time is

$$\frac{di(t)}{dt} = \frac{1}{R}\frac{dv(t)}{dt} + C\frac{d^2v(t)}{dt^2} \approx \frac{1}{R}\frac{\Delta V}{\tau_r} + 1.5C\frac{\Delta V}{\tau_r^2} \tag{5.129}$$

where τ_r is the 10–90% rise time of the gate voltage that changes from 0 to ΔV. [Johnson, '93]

The change of the current (and voltage) with respect to time for transient signals is important since the highest frequency of interest and, hence, energy content is a function of the rise and fall time of the signal. Generally, the faster the transient signal (i.e., the smaller the rise and fall time), the greater the highest frequency of interest. Recall that the coupling to a victim circuit is directly proportional to the frequency, for sinusoidal steady-state signals. For a three-conductor system, the voltages across the victim's source and load resistances in the time domain are

$$v_{sv}(t) = (M_{inds} + M_{cap})\frac{dv_s(t)}{dt}$$

$$v_{Lv}(t) = (M_{indL} + M_{cap})\frac{dv_s(t)}{dt}$$

Both are proportional to the change of the source voltage, $v_s(t)$, with respect to time. If the source voltage is changing quickly with time, this derivative is large.

The differential equation relating the current and voltage for the parallel RC circuit shown in Figure 5.80 is easily obtained by using KCL:

$$i(t) = \frac{v(t)}{R} + C\frac{dv(t)}{dt}$$

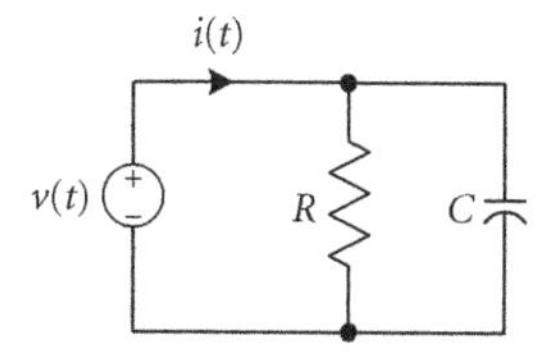

FIGURE 5.80 Approximation for the change of $i(t)$ with respect to time is developed.

[29]The current along a conductor with a length of one wavelength, λ, will be positive over one-half of its length and negative over the other half of its length.

Differentiating both sides with respect to t,

$$\frac{di(t)}{dt} = \frac{1}{R}\frac{dv(t)}{dt} + C\frac{d^2v(t)}{dt^2}$$

The change of the current with respect to time from the supply is a function of the change in the supply voltage with respect to time (i.e., its first derivative) and the change of the change in the supply voltage with respect to time (i.e., its second derivative). Expressions for the first and second derivatives of the supply voltage signal with respect to time are thus needed.

The following expression, involving the error function, will be used to model the low-to-high transition of the gate output voltage for $t \geq 0$:

$$v(t) = \frac{\Delta V}{2} + \frac{\Delta V}{2} erf\left(\frac{t-\tau_r}{k\tau_r}\right) = \frac{\Delta V}{2} + \frac{\Delta V}{2}\frac{2}{\sqrt{\pi}}\int_0^{\frac{t-\tau_r}{k\tau_r}} e^{-\lambda^2} d\lambda \tag{5.130}$$

This function gradually rises from *about* 0 to ΔV. The coefficient k is set equal to 0.56. This function, along with its first and second derivatives (with respect to t), are plotted in Mathcad 5.16. The rise time is arbitrarily set to 0.5 s and the maximum output voltage is set to 5 V. Notice the large amplitudes of

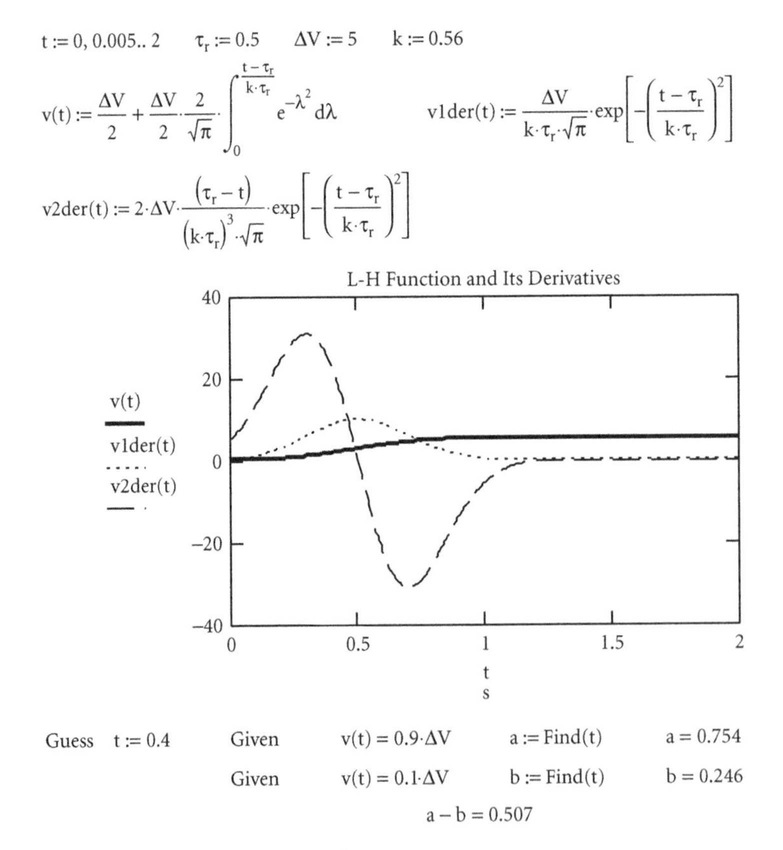

MATHCAD 5.16 Model used for the voltage source and its first and second derivatives.

the first and second derivatives. Because the variable t is in the limit of the integral, Leibniz's formula should be used to determine the first derivative of this function:

$$\frac{d}{dx}\int_{\alpha(x)}^{\beta(x)} f(x,y)dy = \frac{d\beta(x)}{dx} f[x,\beta(x)] - \frac{d\alpha(x)}{dx} f[x,\alpha(x)] + \int_{\alpha(x)}^{\beta(x)} \frac{df(x,y)}{dx} dy$$

Substituting the given function,

$$\frac{dv}{dt} = 0 + \frac{d}{dt}\left(\frac{\Delta V}{2}\frac{2}{\sqrt{\pi}}\int_0^{\frac{t-\tau_r}{k\tau_r}} e^{-\lambda^2} d\lambda\right)$$

$$= \frac{\Delta V}{2}\frac{d}{dt}\left(\frac{t-\tau_r}{k\tau_r}\right)\frac{2}{\sqrt{\pi}} e^{-\left(\frac{t-\tau_r}{k\tau_r}\right)^2} - \frac{\Delta V}{2}\frac{d}{dt}(0)\frac{2}{\sqrt{\pi}} e^{-(0)^2} + \frac{\Delta V}{2}\frac{2}{\sqrt{\pi}}\int_0^{\frac{t-\tau_r}{k\tau_r}} \frac{d}{dt}(e^{-\lambda^2})\, d\lambda \tag{5.131}$$

$$= \frac{\Delta V}{2}\frac{1}{k\tau_r}\frac{2}{\sqrt{\pi}} e^{-\left(\frac{t-\tau_r}{k\tau_r}\right)^2} - 0 + \frac{\Delta V}{2}\frac{2}{\sqrt{\pi}}\int_0^{\frac{t-\tau_r}{k\tau_r}} 0 d\lambda$$

$$= \frac{\Delta V}{k\tau_r\sqrt{\pi}} e^{-\left(\frac{t-\tau_r}{k\tau_r}\right)^2}$$

The maximum value of this first derivative occurs at $t = \tau_r$:

$$\left.\frac{dv}{dt}\right|_{t=\tau_r} = \frac{\Delta V}{k\tau_r\sqrt{\pi}} \tag{5.132}$$

The second derivative of the function is obtained in the standard manner:

$$\frac{d^2v}{dt^2} = \frac{\Delta V}{k\tau_r\sqrt{\pi}} e^{-\left(\frac{t-\tau_r}{k\tau_r}\right)^2}\left[-2\left(\frac{t-\tau_r}{k\tau_r}\right)\frac{1}{k\tau_r}\right] = \frac{2\Delta V(\tau_r - t)}{(k\tau_r)^3\sqrt{\pi}} e^{-\left(\frac{t-\tau_r}{k\tau_r}\right)^2} \tag{5.133}$$

The maximum value of the second derivative does not occur at the same time as the maximum of the first derivative. The maximum and minimum values of the second derivative are obtained by first setting the third derivative equal to zero and solving for the time. These times are

$$t = \tau_r \pm \frac{k\tau_r}{\sqrt{2}} = \tau_r\left(1 \pm \frac{k}{\sqrt{2}}\right) \tag{5.134}$$

The minimum and maximum values of the second derivative are, respectively,

$$\left.\frac{d^2v}{dt^2}\right|_{t=\tau_r\left(1+\frac{k}{\sqrt{2}}\right)} = \frac{-2\Delta V}{(k\tau_r)^2\sqrt{2\pi}} e^{-\frac{1}{2}}, \quad \left.\frac{d^2v}{dt^2}\right|_{t=\tau_r\left(1-\frac{k}{\sqrt{2}}\right)} = \frac{2\Delta V}{(k\tau_r)^2\sqrt{2\pi}} e^{-\frac{1}{2}} \tag{5.135}$$

Although the maximum values of the two derivatives do not occur at the same time, an overestimate of the maximum change of the current from the supply would be

$$\left.\frac{di(t)}{dt}\right|_{\max} < \frac{1}{R}\frac{\Delta V}{k\tau_r\sqrt{\pi}} + C\frac{2\Delta V}{(k\tau_r)^2\sqrt{2\pi}}e^{-\frac{1}{2}} \approx \frac{1}{R}\frac{\Delta V}{\tau_r} + 1.5C\frac{\Delta V}{\tau_r^2} \tag{5.136}$$

This expression clearly shows the importance of the rise time of the input signal. As the rise time decreases, the change of the current with respect to time increases. If the rise time is halved, then the resistive contribution to di/dt is doubled and the capacitive contribution to di/dt is quadrupled.

As an example, if $R = 100\ \Omega$, $C = 50$ pF, $\Delta V = 4$ V, and $\tau_r = 4$ ns, then

$$\left.\frac{di(t)}{dt}\right|_{\max} < \frac{1}{R}\frac{\Delta V}{\tau_r} + 1.5C\frac{\Delta V}{\tau_r^2} \approx 10\times 10^6 + 19\times 10^6 = 29\times 10^6 \text{ A/s}$$

Since the inductive coupling between two circuits (i.e., the open-circuit voltage induced in a nearby circuit) is proportional to this change with respect to time, small rise times can lead to very large crosstalk.

It is important and interesting to note that the capacitive coupling between two circuits is proportional to the change in the voltage with respect to time:

$$\left.\frac{dv}{dt}\right|_{t=\tau_r} \propto \frac{\Delta V}{\tau_r} \tag{5.137}$$

Although this derivative does increase with decreasing rise time, it does not contain the τ_r^2 term in (5.136). Possibly, this is the reason some texts state that the mutual inductive coupling is often more troublesome than mutual capacitive coupling.

5.32 Summary of Methods to Reduce Crosstalk

Summarize the major methods that can be used to reduce crosstalk. [Montrose, '99; Kimmel, '95; Goedbloed]

Crosstalk, which results from near-field coupling of the electric and magnetic fields between transmission lines, is sometimes referred to as the "hidden" transmission line. A signal on one transmission line, through parasitic inductance and capacitance, can couple into another transmission line. Although the total crosstalk between two lines is a function of the impedances on the generator and victim lines, it is common to separate, sometimes incorrectly, the crosstalk into capacitive and inductive coupling terms. Because of the complex nature of crosstalk, it is helpful to have a listing of recommendations for reducing it. Notice that the layout of the circuitry is probably the most important factor for controlling crosstalk.

To reduce the capacitive or electric field coupling between two lines (referred to as the generator and victim),

1. **decrease the length of the generator and victim circuits** (i.e., keep the lines short)
2. reduce the generator signal level
3. increase the distance between the generator and victim
4. avoid parallel runs of the generator and victim
5. route signal traces orthogonally for dual strip lines
6. decrease the distance between the generator and its return

7. decrease the distance between the victim and its return
8. reduce the highest frequency of interest of the signal on the generator (or increase the rise and fall times of the generator signal)
9. decrease the impedances of the generator and victim (which may, unfortunately, increase the inductive coupling)
10. balance the victim circuit
11. terminate, properly, to reduce reflections
12. use a shield or guard trace on the generator or victim circuit, with the appropriate grounding (i.e., use electric field shielding)
13. avoid pigtails
14. avoid parallel runs of high-speed logic signals with low-level analog signals
15. separate fast repetitive signals from slower aperiodic signals
16. use the 3-W rule.

Nearly all of these methods suggested for reducing the capacitive coupling will also reduce the inductive coupling. One difference is that the impedances of the generator and victim should be increased to reduce inductive coupling (which may, unfortunately, increase the capacitive coupling). In addition, to reduce inductive coupling,

1. use twisted pair for the generator and/or victim
2. use some form of shield that can return the current
3. use a high-permeability magnetic shield between the generator and victim (i.e., use magnetic field shielding).

Most of these recommendations for reducing crosstalk will also reduce the radiated emissions and susceptibility of a transmission line. Of course, there are other ways of reducing crosstalk. For example, the proper use of a double-sided or multilayer board will often significantly reduce crosstalk.

Finally, common-impedance coupling, which is usually important at lower frequencies, also affects the crosstalk. Common-impedance coupling is due to current that passes on the common return for the generator and victim. The following suggestions should reduce this form of crosstalk:

1. avoid a common or shared return for the generator and victim
2. if a common return cannot be avoided, reduce its impedance (e.g., increase its width and thickness)
3. decrease the current on the common return
4. use single-point grounding at lower frequencies.

5.33 Fiber's Weakness

Although optical fiber cables are immune to EMI, what practical factors typically weaken this immunity? [Goedbloed]

Optical fiber *cables* are extremely immune to external fields with extremely low emission levels. Furthermore, low-frequency ground loops can be virtually eliminated. Assuming the budget allows for the optical fiber cable and the connecting drivers and receivers (and environmental factors permit), they can be very effective from an EMC standpoint (at least for longer links). A optical fiber cable is most susceptible at its input and output where the driver and receiver are located. Even though the cable itself is immune to external noise, any connecting circuitry at the cable's ports is still potentially susceptible. The optical emissions from an optical cable increase with cable deformation (e.g., bending) and cracking.

6 Radiated Emissions and Susceptibility

Interference that is received via the surrounding nonconducting medium is referred to as radiated interference. If the victim and source are about a wavelength or more apart, the interference is referred to as far-field or just radiated interference. If the victim and source are electrically close, the interference is referred to as near-field interference or crosstalk.

6.1 Radiated or Conducted Vehicle Interference?

A noise source in an automobile is interfering with the car's own radio. What simple tests can be performed to determine whether the interference is conducted via the power cable or radiated via the speaker wires and antenna?

Interference that is received via the surrounding nonconducting medium is referred to as radiated interference. Interference that is received via direct contact with a conducting object is referred to as conducted interference. Sometimes, though, it may not be clear whether the interference is radiated, conducted, or both. For example, the product and a cable entering the chassis of the product might both be exposed to an undesirable radiated signal. Some of this signal might pass directly through the chassis and be a source of radiated interference to the product. Some of the signal might be picked up by the cable and pass into the chassis as a source of conducted interference. Then, the cable once inside the chassis could reradiate the signal and be another cause of radiated interference.

As illustrated in Figure 6.1, interference to the radio may enter through one or several of the following routes:

1. through the "front door," via the antenna or antenna's transmission line
2. through the "back door," via the power leads
3. through the speaker leads (and other inputs or outputs)
4. though the radio's chassis.

Several classical tests can be performed and the interference level observed to determine possibly the pathway for the interference. Again, the interference may be entering the product from multiple pathways complicating the results of the tests. Although the focus of this brief discussion is locating the pathway of interference to an automobile radio, these methods can also be employed for a variety of other products.

To determine whether the interference is entering via the antenna leads, the antenna leads can be removed and replaced at the radio with a dummy load (so that the sensitivity of the system does not change). To determine whether the interference is entering via the power leads, the power leads can be replaced with much shorter leads and then connected directly to a battery of the appropriate rating. Another method of determining whether the interference is entering via the power leads is with a portable battery-powered radio in the vicinity of the car radio. If the interference is picked up on the portable radio, it is

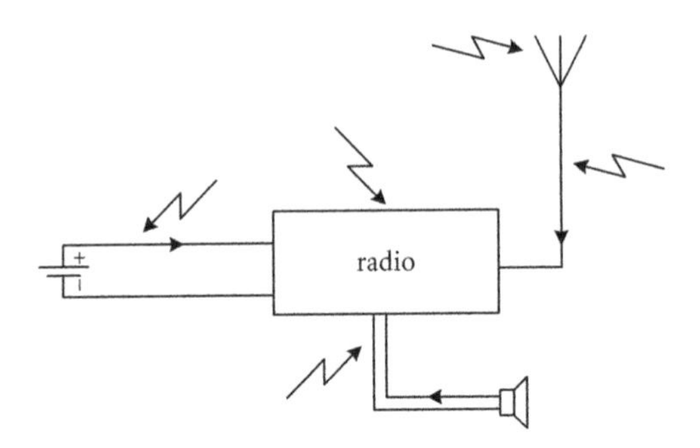

FIGURE 6.1 Many systems, such as a simple radio, are exposed to a variety of undesirable radiated and conducted signals.

likely to have a radiated component. To determine whether the interference is entering the speaker leads, these leads can also be reduced in length or rerouted. If the radio has a headphone jack, the speaker leads can be temporarily removed and the radio's output monitored through (short-lead) headphones. Common-mode chokes and filters may also be used on any of the incoming leads to reduce the incoming conducted interference. To determine whether the interference is radiated, directly passing to the radio's components, a low-cost shield (e.g., copper or even aluminum foil) can be placed around the chassis. A nonmagnetic shield would not be effective, however, in reducing near-field magnetic interference.

The source of the interference can be internal or external to the car. External sources included other automobiles, cell phones, and radar guns. Considering that some automobiles have over 50 motors[1] and motors are a strong source of magnetic fields, the automobile itself could easily be the source of interference. There are many other potential sources of internal interference such as a turn signal actuator, hazard flasher, light dimmer, and clock of an embedded computer.

6.2 The Automobile Noise Mystery

For each of the following automobile radio interference clues, determine one or more potential noise sources. Explain your reasoning and provide possible interference cures. [Nelson; Bosch; Carr, '73]

Interference Clues

High pitch whining that increases with speed
Crackling that occurs when driving over rough roads but disappears temporarily when the brake is slightly depressed
Humming noise that appears on rainy days even when the engine is off
Erratic noise that occurs when the automobile is rocked, the engine is stopped, and the ignition is on
Crackling and other noises that appear when the engine and all other devices are turned off
Crackling that is relatively constant with engine speed but decreases at very fast speeds
Popping sound that increases tempo with speed and stops instantly when the ignition key is turned off at fast idle
Noise that is very apparent on rough but not smooth roads
Hissing and crackling that is worse when driving on rough and dry roads but does not disappear after the engine is stopped

As illustrated in Figure 6.2, in an automobile there are many sources of electrical noise (and interference) and potential victims of this noise. Interference can be either radiated, conducted, or both.

[1]Wiper, blower, starter, power window, and power seat motors are just a few examples.

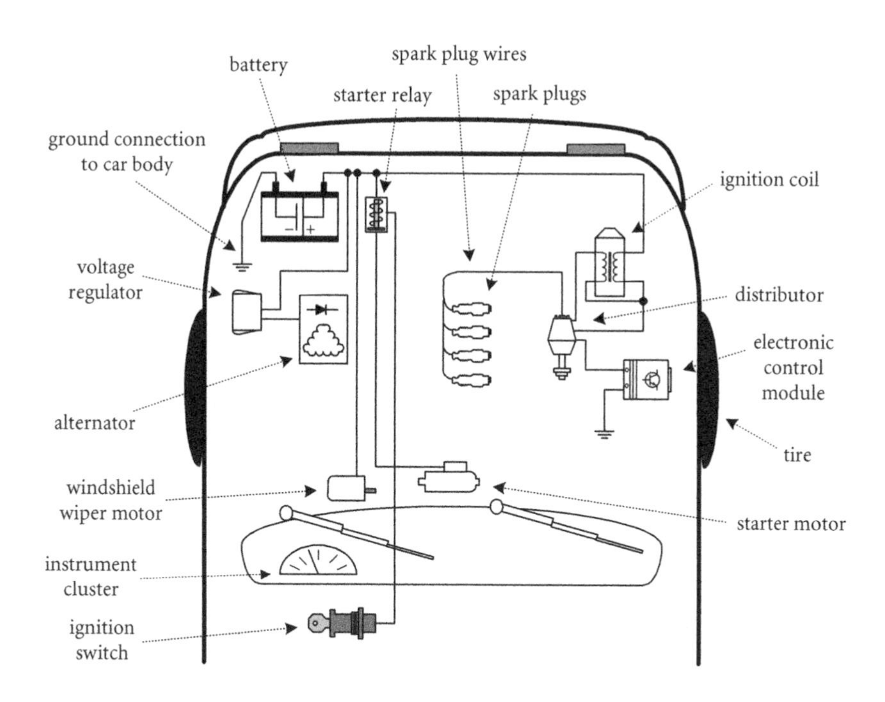

FIGURE 6.2 Numerous electrical noise sources in an automobile.

This discussion will focus on the potential sources of the interference (and possible quick fixes) rather than whether the coupling is radiated or conducted. For those individuals desiring more mathematical rigor, the adage, "good things come to those who wait," applies.

The major sources of interference in a vehicle are the i) ignition system, ii) generator system, iii) small motors and switches, iv) static discharges, v) electronic controls, and vi) instruments. The first step in eliminating interference is the identification or determination of the noise source. To assist in this process, a brief review of relevant automobile electrical basics will be provided. Although there are many methods of generating and distributing power in a vehicle, only one of these methods will be reviewed.

The ignition system portion of an automobile ignites the air-fuel mixture in the cylinders in the proper time sequence. It is typically broken into two circuits: the primary low-voltage side and the secondary high-voltage side. The ignition coil is a step-up transformer (e.g., 200 turns on the primary and 20,000 turns on the secondary) with a soft iron core that transforms a low voltage on the primary to a high voltage on the secondary (the voltage across the secondary can reach 20–40 kV). Sufficient energy is then available on the secondary side to ionize the air in the spark plug gap and ignite the air-fuel mixture. The ignition switch, starter relay, and battery are contained on one side of the ignition coil, and the electronic control module and distributor are contained on the other side of the ignition coil. The starter relay connects the starter to the battery when the ignition is turned "ON." The starter or cranking motor is used to turn the crankshaft, which starts the engine. (The starter is the major acoustic noise source when attempting to start a car that is "flooded.") The electronic control module contains a switching transistor that interrupts the current in the primary side of the coil. In order for voltage to be produced across the secondary of the ignition coil *transformer*, the current in the primary must be time varying. The distributor distributes the high voltage produced across the secondary to the spark plugs in the proper sequence.

The secondary side of the ignition coil contains the spark plug wires, spark plugs, and rotor. The rotor is on top of the distributor. It rotates to different positions to connect the high voltage of the ignition coil to the appropriate spark plug. Some electrical systems do not contain a distributor. Sometimes, multiple ignition coils are used: one coil is dedicated to each spark plug or to a pair of spark plugs. The spark plugs produce the necessary spark across their air gap to ignite the air-fuel mixture. It produces this spark several thousand times per minute. Spark plug wires are heavily insulated wires designed to transfer the high voltage from the ignition coil to the spark plug.

The generator system converts some of the mechanical energy supplied from the engine into electrical energy where it is used to keep the battery charged. The battery is used, of course, for the ignition system, for motors, and for accessories (e.g., stereo system). There are various names used to describe generators including alternator, ac generator, and charger. There are two basic methods of generating a time-varying voltage from mechanical rotation: a coil can be moved through a stationary magnetic field or a magnetic field can be moved through a stationary coil. In old dc generators, the coil assembly of the armature is rotated in a stationary magnetic field. In the modern ac generators, a rotor turns a magnet around a stationary coil, referred to as the stator. The alternating current produced by an alternator is converted to direct current to replenish the battery and supply the various dc accessories. Generators, brushes, and commutators are used to convert the ac to dc. With modern alternators, diodes are contained inside the alternator to rectify the ac to dc. The output voltage of an alternator changes with the speed of the engine. So as not to damage or adversely affect the battery or other systems using the battery's voltage, a voltage regulator is used.

"high pitch whining that increases with speed"

This type of noise is usually from the generator or alternator. The whine produced by an older dc generator is less pure than an alternator. Both generator and alternator whine pitch will tend to be a function of the engine speed since they are connected to the engine's crankshaft via a drive belt.

Shunt capacitors are often used to reduce noise from the alternator. A 0.05 μF suppression capacitor across the generator output is often the first approach tried to reduce the effects of the noise. A 0.5 μF capacitor can also be tried at the accessory terminal in the fuse box. If the noise is still present, a power line filter can be installed at the radio.

Temporarily removing the drive belt should eliminate the noise. Diodes for ac-to-dc rectification are contained inside the alternator. Open-circuited, short-circuited, or otherwise defective diodes could be the source of the noise. A short-circuited, an open-circuited, or a grounded stator could also be the source of this noise inside the alternator.

"crackling that occurs when driving fast over rough roads but disappears temporarily when the brake is slightly depressed"

The source of this noise is likely electrostatic discharge. Tribocharging between the tires and road is a function of the contact area, speed, and humidity level. The charge buildup on the tires can discharge to nearby metal objects such as the brakes. A voltage difference can also exist between the wheel trim and car body. Depressing the brakes can discharge some of this static buildup. Even after the automobile has stopped, the crackling can still be present since the time constant associated with the discharge is not zero. The key to reducing this noise is preventing the charge buildup in the first place. Probably the only reasonable method of reducing this noise (besides driving more slowly) is through proper grounding, especially of floating metal objects.

"humming noise that appears on rainy days even when the engine is off"

The windshield wiper motor is probably responsible for this swish noise. The simplest of all possible tests is to turn off the wipers and note whether the noise disappears. This approach can also be used to determine the source of switch pop noise from brake lights, turn signals, cigarette lighters, and other accessories. Due to the frictional contact between the rubber blades and glass windshield, the noise could also be from ESD. Antistatic blades are available that have a lower resistivity than standard rubber blades. Often wiper motors already have suppression capacitors. Additional filters may be installed at the motor, or the motor and leads can be shielded.

"erratic noise that occurs when the automobile is rocked, the engine is stopped, and the ignition is on"

Rocking the automobile will slosh the gasoline in the tank in addition to jarring any loose metallic connections. The source of this erratic popping or "sloshing sound" could be the fuel gauge. In some fuel gauges, the free end of a bimetallic strip is used, surrounded by a heating coil, to control the gauge pointer. The level of current through the heating coil, which is a function of the fuel level in the tank, controls the bending of the bimetallic strip. A voltage regulator is also used with this gauge to provide a constant 5 V to the gauge. Loose or corroded mechanical contacts can be the source of this noise since current passes from the fuel sensor to the gauge. Once this path has been examined and weak connections strengthened, a 0.5 μF capacitor across the gasoline sensor wire and ground can be used to reduce this noise.

"crackling and other noises that appear when the engine and all other devices are turned off"

External noise sources such as thunderstorms, sun spots, and high-voltage lines are possible sources of this noise. It is a function of the automobile's environment. For example, when driving under high-voltage distribution lines, the noise heard on an AM radio increases dramatically due to the corona around the wires. Possibly, multiple antennas coupled with signal processing can reduce the effects of this noise.

"crackling that is relatively constant with engine speed but decreases at very fast speeds"

This ragged, rasping, crackling noise is sometimes described as sounding like the frying of eggs. The purpose of the voltage regulator is to regulate the voltage level of the output of the alternator since its output does change with speed. Ideally, the regulator's output should not vary with speed of the engine. Actually, its output does change with the speed, number of loads, and power consumption of the loads. This nonideality is most noticeable at high speeds. Heavily loading down the regulator (e.g., turning on the bright lights and other accessories), should affect the voltage level. Generally, electronic voltage regulators are reliable and are replaced when they malfunction.

"popping sound that increases tempo with speed and stops instantly when the ignition key is turned off at fast idle"

The ignition system could be the source of this noise. Although the noise from the alternator is also speed dependent, after the ignition key is turned off, alternator noise may continue for a short period. The alternator is connected via a belt to the engine's crankshaft, and the crankshaft will continue to move due to momentum for a short period after turning off the ignition. Ignition noise, however, typically will stop instantly since it is a function of the electrical current through the spark plugs, spark plug wires, ignition coil, and electronic control module. When the ignition is turned off, the current to these devices will quickly stop.

Ignition noise is the main source of interference in an automobile. Sparks are a source of high-frequency noise, and sparks are present in the spark plug gaps and in the distributor contacts. At the spark plugs, the peak voltage is in the 10–30 kV range. (The current in the primary of the coil can reach 100 A.) Spark plug wires, carrying these transients, act as antennas.

There are several classical methods of reducing ignition noise. The most common is the use of resistor spark plugs. Resistor plugs are quite effective in reducing interference. A resistor is placed inside the spark plug, at the source of the interference, to reduce the rate of the current. This reduces the radiation from the cables (and, of course, increases the electrical losses in the system). An interference suppression resistor may also be used at the distributor rotor to suppress the high-frequency noise. Filters between the distributor contacts and ground may also be used. Two other methods used to reduce this interference are shielded spark plug wires and screened spark plug connectors. These braided conducting shields can reduce electric field emissions and high-frequency radiation.[2]

The entire ignition system should be examined when attempting to locate a defective device. The distributor should be examined for carbon-coated, cracked, broken, worn, or loose contacts or wiper. The spark plug wires should not be broken, twisted, or crimped. (A simple resistance measurement may

[2]Shielded spark plug wires are standard equipment on aircraft engines.

be used to ensure that the wires are not open circuited. The resistance of good wires is not necessarily small, however, especially if resistive wires are used.) Spark plugs should not be oil or carbon fouled, spotted, blistered, or cracked. (The voltage required to produce a spark will vary with the type of fouling. The high voltage at the spark plug can partially travel or flashover to other objects.) Finally, the distributor coil casing should be grounded to the car's ground system.

"noise that is very apparent on rough but not smooth roads"

The rough roads increase the mechanical vibration of the automobile. If the noise only appears during these conditions, it is often an indication of loose or corroded contacts. Corroded contacts may be acting as a nonlinear junction. A nonlinear junction can be formed between two dissimilar metals or two similar metals with an oxidized layer between them. Loose junctions may be a source of sparking. A loose connection between the antenna and transmission line is one example of a poor connection.

As a practical side note, the engine block and chassis are used as the ground reference for the system. When connecting one side of a filter or other add-on devices to the body, it is important to make a good electrical connection to the chassis or engine block. Paint should be scratched off (and grease and dirt removed) at the bonding point.

"hissing and crackling that is worse when driving on rough and dry roads but does not disappear after the engine is stopped"

This interference is likely due to electrostatic discharges. Although tribocharging of the tires is one source of ESD, charge can also be generated on belts, joints, plastic components, seats, and instrument cluster faces. If the charge builds up to a sufficiently large level, it can arc across to nearby objects. It is therefore important that devices be properly connected to the automobile's ground, through short, wide grounding straps. (Objects should not be floating.) Parts that have been covered with antirattle putty compound should be carefully examined for proper grounding.

6.3 Copper Plane Addition

A product only occasionally "acts up." It is believed to be due to an EMC problem. Would the addition of a large copper ground plane below the product increase the number of occurrences of this malfunction?

The addition of a ground plane *can* increase the number of occurrences. For example, if the EMC problem is believed to be ESD related, an ESD gun or "zapper" can be used to locate weaknesses in the product. By including a ground plane below the product, as shown in Figure 6.3, some of the radiation from the gun will reflect off the ground plane and thus impact the product from various angles and with various polarizations, phases, and amplitudes. This may increase the exposure of the product to this radiation and, thus, increase the likelihood of repeating the intermittent problem. It is important to realize, however, that reflected radiation from the ground plane may be constructive or destructive in

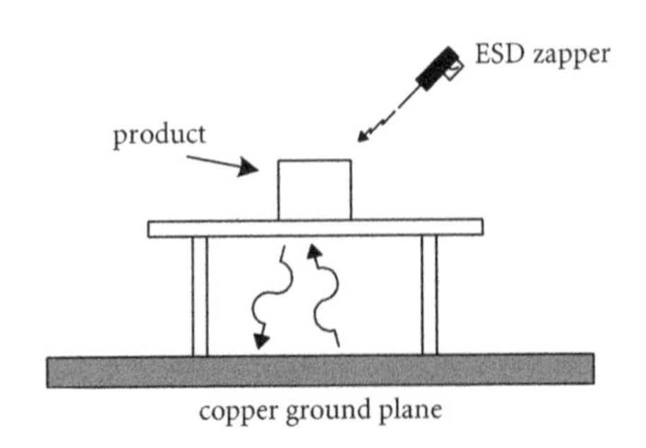

FIGURE 6.3 Large copper ground plane might increase the propensity of a product to malfunction during an ESD test.

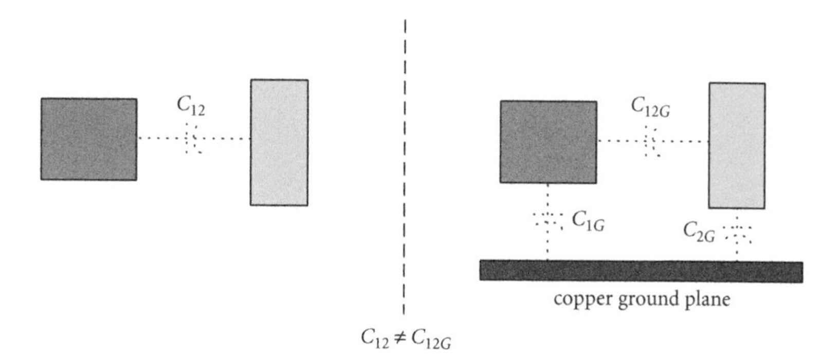

FIGURE 6.4 Addition of a ground plane will change the capacitive coupling between nearby objects.

nature: the sum of the incident and reflected waves may increase or decrease the radiation at the product, respectively, based on the total path length of the waves.

The copper ground plane will increase the electric field coupling between the product and the ground if the ground plane is connected to the product's safety ground. If the interference is due to current on the ground line, this may increase the exposure of the product to the noise. The level of exposure is a function of the electrical and physical properties of the product's chassis. Capacitive coupling between two components in the product may actually decrease with the addition of the ground plane. The change in capacitance between two conducting objects with the addition of a local ground plane is shown in Figure 6.4. The addition of a conducting ground plane may reduce the crosstalk between two components or subsystems.

If the chassis of the product is nonmetallic with one or more cables leading to it, then the addition of a nearby ground plane can reduce the susceptibility of the product to an external electrostatic discharge event. The parasitic capacitance provided by the ground plane can provide a lower impedance path for the discharge current. In conclusion, unless the product is well understood, it is not clear whether the addition of a ground plane will increase, decrease, or have no effect on the product's susceptibility to radiated and conducted noise.

6.4 Emissions from Twin-Lead Line

In the far field from an omnidirectional antenna, oriented parallel to the z axis, the electric field is of the form

$$E_{\theta s} = M_s I_s \frac{e^{-j\beta_o r}}{r} F(\theta) \tag{6.1}$$

where I_s is the current at the center of the antenna, r the distance from the center of the antenna, and β_o is the free-space phase constant. The variables M_s and $F(\theta)$ are a function of the antenna type (e.g., half-wave dipole). From this expression, determine the total far-field electric field from the two parallel wires of length l_{th} shown in Figure 6.5, separated by a distance d and carrying currents I_{1s} and I_{2s}. [Paul, '92(b); Kraus, '88]

The far-field radiation emitted from two parallel wires is often used to model many unintentional antennas such as a pair of traces on a printed circuit board or twin-lead transmission line. Surprisingly, there are situations where the transmission line leading to an antenna can actually radiate more than the antenna itself. The total field, as a function of the spacing between the two wires and the current in each wire, can be obtained by summing the fields from each wire:

$$E_{\theta totals} = M_{1s} I_{1s} \frac{e^{-j\beta_o r_1}}{r_1} F_1(\theta) + M_{2s} I_{2s} \frac{e^{-j\beta_o r_2}}{r_2} F_2(\theta) \tag{6.2}$$

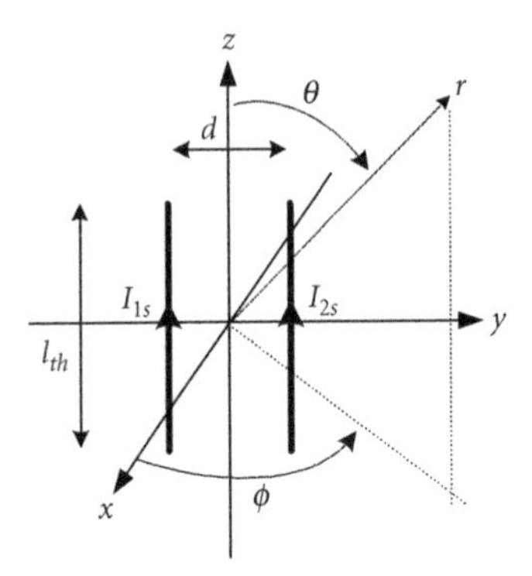

FIGURE 6.5 Two parallel current-carrying straight wires in the spherical coordinate system.

(Note that the fields vary as $1/r$, which is expected in the far field.) The radiation pattern, $F(\theta)$, for each wire and M_s are the same for two identical wires. Furthermore, in the far field from these two wires, the distance from the point of interest in the far field to each antenna is about

$$r_1 \approx r + \Delta, \; r_2 \approx r - \Delta \tag{6.3}$$

where Δ is a function of the spacing of the wires, d, and position of the far-field point of interest (as a function of θ and ϕ in spherical coordinates):

$$\Delta = \frac{1}{2} d \sin\theta \sin\phi \tag{6.4}$$

Notice that if $\theta = 0°$ then $\Delta = 0$, which is expected since the point of interest in the far field is midway between the two wires. The difference between r_1 and r_2 is the greatest broadside, $\phi = 90°$, and in the plane, $\theta = 90°$, of the wires.

Substituting these variables into the total field expression, (6.2), yields

$$\begin{aligned} E_{\theta totals} &\approx M_s \left[I_{1s} \frac{e^{-j\beta_o(r+\Delta)}}{r+\Delta} + I_{2s} \frac{e^{-j\beta_o(r-\Delta)}}{r-\Delta} \right] F(\theta) \\ &= M_s e^{-j\beta_o r} \left(I_{1s} \frac{e^{-j\beta_o \Delta}}{r+\Delta} + I_{2s} \frac{e^{j\beta_o \Delta}}{r-\Delta} \right) F(\theta) \end{aligned} \tag{6.5}$$

One further approximation is typically performed on (6.5). However, it must be carefully applied. In the far field from the two wires, the distance to either wire is about the same, $r_1 \approx r_2 \approx r$. The largest difference between these distances is $|r_1 - r_2| \approx d$. Now, if the Δ term is *incorrectly* assumed zero or negligible *everywhere* in this last expression, the total field would be

$$M_s \frac{e^{-j\beta_o r}}{r} (I_{1s} + I_{2s}) F(\theta) \tag{6.6}$$

If the currents in each wire are equal but opposite (i.e., entirely differential), $I_{1s} = -I_{2s}$, then the total field everywhere in the far field, according to (6.6), would be zero. If the spacing between the wires were zero, which it cannot be, then this would be a reasonable result. However, the spacing between any two real wires is not zero and, hence, the two fields should not exactly cancel in the far field at all far-field positions.

The denominators in (6.5) involve distance, which is always a positive quantity. The arguments of the *complex* exponential terms involve the phase. Although the magnitude of the complex exponential is one, $|e^{\pm j\beta_o\Delta}| = 1$, it can have negative components as seen by Euler's formula:

$$e^{\pm j\beta_o\Delta} = \cos(\beta_o\Delta) \pm j\sin(\beta_o\Delta)$$

If the product $\beta_o\Delta$ is between 90° and 270°, then the real part of this complex exponential is negative. If the complex exponentials are not removed from the equation, the fields from each wire can add or subtract in the far field depending on the magnitude *and* phase of the current in each wire. Thus, it is important not to eliminate the complex exponential terms from the approximation. The appropriate approximation is therefore

$$E_{\theta totals} \approx M_s \frac{e^{-j\beta_o r}}{r}\left(I_{1s}e^{-j\beta_o\Delta} + I_{2s}e^{j\beta_o\Delta}\right)F(\theta) \tag{6.7}$$

6.5 Differential-Mode Current Emissions from Twin-Lead Line

Determine the electric field in the far field from two electrically-short wires if the currents in each wire are equal but in opposite directions. [Paul, '92(b)]

The expression for the total electric field from two parallel wires given in (6.7) can be used to determine the field strength in the far field due to differential-mode currents. A simple expression for the field strength is obtained if the wires are assumed electrically short. In this case, the Hertzian dipole characteristics may be used (where the current distribution along the wires is approximately uniform):

$$M_s = j\frac{\eta_o\beta_o}{4\pi}l_{th} \quad \text{and} \quad F(\theta) = \sin\theta \tag{6.8}$$

where l_{th} is the wire length. Since the currents are differential,

$$I_{1s} = -I_{2s} = I_{Ds}$$

The electric field in the far field for two electrically-short parallel wires, carrying equal but opposite currents, is therefore

$$\begin{aligned} E_{\theta totals} &= j\frac{\eta_o\beta_o}{4\pi}l_{th}\frac{e^{-j\beta_o r}}{r}\left(I_{Ds}e^{-j\beta_o\Delta} - I_{Ds}e^{j\beta_o\Delta}\right)\sin\theta \\ &= j\frac{\eta_o\beta_o}{4\pi}l_{th}\frac{e^{-j\beta_o r}}{r}I_{Ds}\left[\cos(\beta_o\Delta) - j\sin(\beta_o\Delta) - \cos(\beta_o\Delta) - j\sin(\beta_o\Delta)\right]\sin\theta \\ &= \frac{\eta_o\beta_o}{2\pi}l_{th}\frac{e^{-j\beta_o r}}{r}I_{Ds}\sin(\beta_o\Delta)\sin\theta \end{aligned} \tag{6.9}$$

This expression can be further simplified by realizing that since the spacing between the wires is also electrically small, $d/\lambda_o \ll 1$, the argument of the sinusoidal function is much less than one, and $\sin x \approx x$:

$$\sin\left(\beta_o\frac{1}{2}d\sin\theta\sin\phi\right) = \sin\left(\frac{2\pi}{\lambda_o}\frac{1}{2}d\sin\theta\sin\phi\right) = \sin\left(\pi\frac{d}{\lambda_o}\sin\theta\sin\phi\right) \approx \pi\frac{d}{\lambda_o}\sin\theta\sin\phi$$

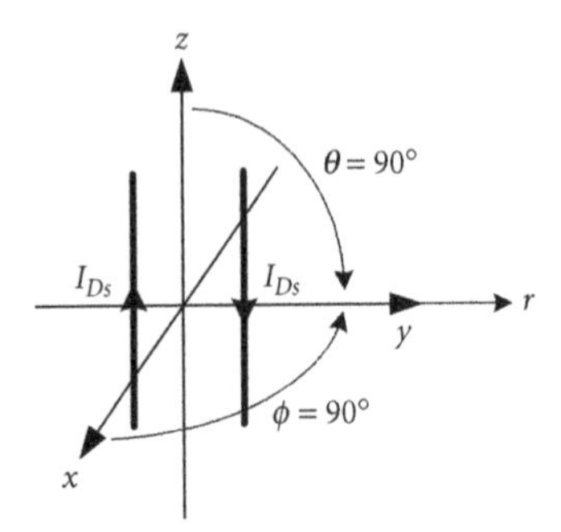

FIGURE 6.6 Maximum far-field value of the electric field from these differential-mode currents is along the y axis.

The far field is then

$$E_{\theta totals} \approx \frac{\eta_o \beta_o}{2\pi} l_{th} \frac{e^{-j\beta_o r}}{r} I_{Ds} \pi \frac{d}{\lambda_o} \sin^2\theta \sin\phi = \frac{d\eta_o \pi I_{Ds} l_{th}}{\lambda_o^2} \frac{e^{-j\beta_o r}}{r} \sin^2\theta \sin\phi \quad (6.10)$$

The maximum value of this field occurs when $\theta = 90°$ and $\phi = 90°$ (i.e., broadside to the wires and in their plane as shown in Figure 6.6). Its magnitude is

$$|E_{\theta totals}|_{\max} \approx \frac{d(377) f^2 \pi |I_{Ds}| l_{th}}{(3\times 10^8)^2} \frac{|1}{r} \approx 1.3\times 10^{-14} \frac{|I_{Ds}| f^2 l_{th} d}{r} \quad (6.11)$$

where r is the distance from the origin (midway between the wires) to the far-field point of interest. At this particular far-field location, the electric field is in the z direction, parallel to the wires.

6.6 Common-Mode Current Emissions from Twin-Lead Line

Determine the electric field in the far field from two electrically-short wires if the currents in each wire are equal and in the same direction. [Paul, '92(b)]

Again, (6.7) can be used to determine the field strength in the far field due to common-mode currents. As with differential-mode currents, a simple expression for the field strength is obtained if the lines are assumed electrically short. Again, the Hertzian dipole characteristics given in (6.8) may be used (where the current distribution along the wires is approximately uniform). Since the currents are common-mode, they are equal and in the same direction:

$$I_{1s} = I_{2s} = I_{Cs}$$

The electric field in the far field for two electrically-short parallel wires, carrying equal but opposite currents is

$$\begin{aligned} E_{\theta totals} &= j\frac{\eta_o \beta_o}{4\pi} l_{th} \frac{e^{-j\beta_o r}}{r} \left(I_{Cs} e^{-j\beta_o \Delta} + I_{Cs} e^{j\beta_o \Delta}\right) \sin\theta \\ &= j\frac{\eta_o \beta_o}{4\pi} l_{th} \frac{e^{-j\beta_o r}}{r} I_{Cs} [\cos(\beta_o \Delta) - j\sin(\beta_o \Delta) + \cos(\beta_o \Delta) + j\sin(\beta_o \Delta)] \sin\theta \quad (6.12) \\ &= j\frac{\eta_o \beta_o}{2\pi} l_{th} \frac{e^{-j\beta_o r}}{r} I_{Cs} \cos(\beta_o \Delta) \sin\theta \end{aligned}$$

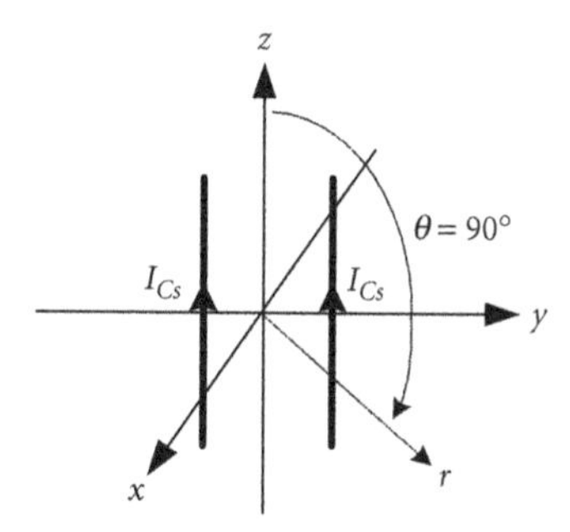

FIGURE 6.7 Maximum far-field value of the electric field from these common-mode currents is in the xy plane.

This expression can be further simplified by realizing that since the spacing between the wires is also electrically small, $d/\lambda_o \ll 1$, the argument of the sinusoidal function is much less than one, and $\cos x \approx 1$:

$$\cos\left(\beta_o \frac{1}{2} d \sin\theta \sin\phi\right) = \cos\left(\frac{2\pi}{\lambda_o}\frac{1}{2} d \sin\theta \sin\phi\right) = \cos\left(\pi \frac{d}{\lambda_o} \sin\theta \sin\phi\right) \approx 1$$

The far field is then

$$E_{\theta\,totals} \approx j\frac{\eta_o \beta_o}{2\pi} l_{th} \frac{e^{-j\beta_o r}}{r} I_{Cs} \sin\theta \tag{6.13}$$

The maximum value of this field occurs when $\theta = 90°$ (in the xy plane):

$$|E_{\theta\,totals}|_{\max} = \frac{377}{\lambda_o} l_{th} \frac{1}{r} |I_{Cs}| = \frac{377}{3\times 10^8} f l_{th} \frac{|I_{Cs}|}{r} \approx 1.3\times 10^{-6} \frac{|I_{Cs}| f l_{th}}{r} \tag{6.14}$$

The variable r is the distance from the origin (midway between the wires) to the far-field point of interest as shown in Figure 6.7. In the xy plane and in the far field, the electric field is again in the z direction, parallel to the wires.

6.7 Reducing Emission Levels

Discuss the differences between far-field emissions from common-mode and differential-mode currents on a twin-lead electrically-short cable. How can the emission levels from each of these current types be reduced? [Paul, '92(b)]

As was previously derived, at any far-field location, the emissions for common-mode currents are

$$|E_{\theta\,totals}| \approx 1.3\times 10^{-6} \frac{|I_{Cs}| f l_{th}}{r} \sin\theta$$

and the emissions from differential-mode currents are

$$|E_{\theta\,totals}| \approx 1.3\times 10^{-14} \frac{|I_{Ds}| f^2 l_{th} d}{r} \sin^2\theta \sin\phi$$

One difference between these two fields is the differential-mode expression's dependence on the angular position, ϕ. In the far field, two straight wires carrying equal currents in the same direction essentially appear like one current-carrying wire. Two straight wires carrying equal currents in opposite directions, however, do not appear like one current-carrying wire since the fields can destructively interfere with each other.

A simple diagnostic tool utilizes the dependency of the differential-mode emissions on ϕ. To determine whether an emission from a twin-lead cable is mainly due to a common-mode or differential-mode signal, the cable can be rotated about its axis and the level of noise monitored. If the noise level changes significantly, assuming the noise is measured in the far field of the cable, the noise is likely to be from differential-mode currents. Differential-mode emissions are sensitive to the rotation of the cable (ϕ) while common-mode emissions are not. On the other hand, if the load connected to the twin-lead line is removed and the noise level is mostly unaffected, this would imply that the noise is mostly from common-mode currents on the line. Common-mode currents can return via parasitic capacitance to their source.

Another difference between the two emissions is the dependence on the frequency. If the frequency is doubled, the common-mode emissions will double, but the differential-mode emissions will quadruple. Also, the ratio of the common-mode to differential-mode coefficient multipliers for the two emissions is large:

$$\frac{1.3\times10^{-6}}{1.3\times10^{-14}}=10^{8}$$

If the magnitude of the common-mode and differential-mode currents are the same,

$$|I_{Cs}|=|I_{Ds}|$$

the field emissions from the common-mode currents are usually much greater than the emissions from the differential-mode currents. Common-mode currents are also referred to as antenna currents since currents of this type are common on antennas, which are designed to radiate (and receive) fields. Assuming the magnitudes of the common-mode and differential-mode currents are identical, when the product of the frequency, f, and distance between the wires, d, is less than 10^8,

$$1.3\times10^{-6}\frac{|I_{Ds}|\,fl_{th}}{r}>1.3\times10^{-14}\frac{|I_{Ds}|\,f^{2}l_{th}d}{r}\quad\text{or}\quad 10^{8}>fd \tag{6.15}$$

the common-mode emissions will be greater than the differential-mode emissions. For a properly designed transmission line, however, the differential-mode currents are much greater than the common-mode currents.

If the distance, r, to the line is fixed, the deadly field emissions from common-mode currents can be reduced by

1. decreasing the common-mode current level
2. decreasing the frequency of the common-mode currents
3. decreasing the length of the line.

The emission level from differential-mode currents can be reduced by

1. decreasing the differential-mode current level
2. decreasing the frequency of the differential-mode currents
3. decreasing the length of the line
4. decreasing the spacing between the conductors of the line[3]
5. changing the orientation of the line.

[3]Decreasing the spacing between the conductors will also change the characteristic impedance of the twin-lead transmission line.

A numerical example is nearly always helpful. Imagine that a flat cable consists of #28 AWG copper wires with a center-to-center distance between neighboring wires of 39 mils (≈ 0.991 mm). Two adjacent wires are carrying a differential-mode current of 1 mA. The emissions from this differential-mode current will be compared to the emission levels for the same cable carrying a common-mode current of 1 mA. It will be assumed that the flat cable is electrically short so the previously derived equations for the emission levels can be used. The emissions due to differential-mode currents in the plane of the cable and broadside to the cable are

$$|E_{\theta\,totals}|_{\max} \approx 1.3\times10^{-14}\frac{(1\times10^{-3})f^2 l_{th}(0.991\times10^{-3})}{r} \approx 13\times10^{-21}\frac{f^2 l_{th}}{r}$$

The emissions due to the common-mode currents are

$$|E_{\theta\,totals}|_{\max} \approx 1.3\times10^{-6}\frac{(1\times10^{-3})f l_{th}}{r} = 1.3\times10^{-9}\frac{f l_{th}}{r}$$

In dB, the ratio of the differential-mode to common-mode emissions is

$$20\log\left(13\times10^{-21}\frac{f^2 l_{th}}{r}\right) - 20\log\left(1.3\times10^{-9}\frac{f l_{th}}{r}\right) = -220 + 20\log f \text{ dB} \tag{6.16}$$

The emissions are identical at a frequency of 100 GHz:

$$0 = -220 + 20\log f \quad \Rightarrow f = 100\,\text{GHz}$$

Of course, for the cable to be electrically short at this frequency, its length should be less than about 0.2 mm:

$$\frac{\lambda}{10} = \frac{v}{10f} = \frac{\dfrac{3\times10^8}{\sqrt{\varepsilon_{reff}}}}{10(100\times10^9)} = \frac{\dfrac{3\times10^8}{\sqrt{2.5}}}{10(100\times10^9)} \approx 0.2\text{ mm}$$

assuming an effective relative permittivity of 2.5 for the cable. A length of 0.2 mm or less is unrealistic for a flat cable. The maximum common-mode emission level is (nearly) always greater than the maximum differential-mode emission level when the current magnitudes are identical and the cable is electrically short.

6.8 Susceptibility of Twin-Lead Line

Beginning with Maxwell's equations, determine the susceptibility model for an electrically-short twin-lead line.[4] [Paul, '76]

When a plane wave is incident onto a twin-lead line, it produces a voltage across and current in the line, which is an induced noise. The magnitude and sign of these voltages and currents are obtained through Maxwell's equations. As shown in Figure 6.8, the distance between the wires is d and the length of the electrically-short line is Δx. Because the final model, and various derivatives of this model are used in this book, all of the steps in its derivation will be given.

[4]This section closely follows the referenced article authored by Professor Paul.

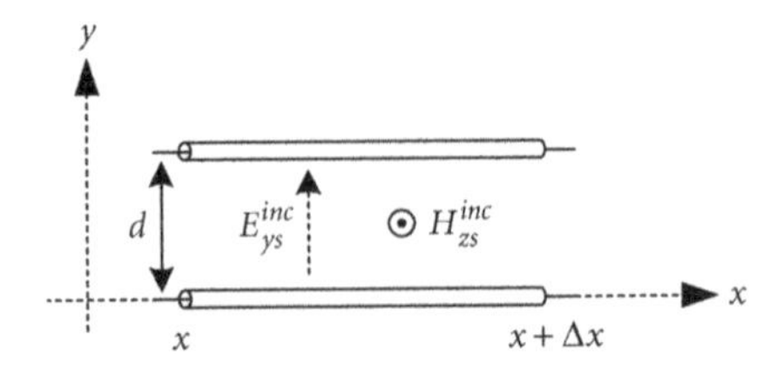

FIGURE 6.8 Parallel wires of length Δx illuminated by an electromagnetic field.

When a plane wave is incident upon a twin-lead line, the electric field component that is across the wires and magnetic field component that is normal to the plane containing the wires will produce noise on the line. Maxwell's equations can be used to determine this noise level. Beginning with one of Maxwell's equations (referred to as Faraday's law), written in the frequency domain for free-space conditions,

$$\nabla \times \vec{E}_s = -j\omega \vec{B}_s$$

and rewriting it using Stokes' theorem, results in the expression

$$\oint \vec{E}_s \cdot d\vec{L} = -j\omega \iint \vec{B}_s \cdot d\vec{s} \tag{6.17}$$

The expression on the left represents the total induced emf (voltage) produced by the total magnetic flux density, $\vec{B}_s$, normal to the plane (of the paper). (The vector $d\vec{s}$ is normal to the plane, and the dot product ensures that only the normal component of the magnetic flux density contributes to the collected magnetic flux.) If the integral on the left of (6.17) is evaluated along a counter clockwise closed path around the transmission line of length Δx, the following equation is obtained:

$$\int_0^d [E_{ys}(x+\Delta x, y) - E_{ys}(x,y)]dy = -j\omega\mu_o \int_0^d \int_x^{x+\Delta x} H_{zs}(x,y)\,dx\,dy \tag{6.18}$$

This assumes that both the upper and lower conductors are perfectly conducting so that the tangential electric fields along them are zero. Therefore, the only remaining nonzero electric fields contributing to this integral are across the output and input of the Δx segment of the line. The voltage anywhere across the conductors is defined as (assuming d is electrically small)

$$V_s(x) = -\int_0^d E_{ys}(x,y)\,dy \tag{6.19}$$

and the derivative of the function $V_s(x)$ with respect to x is

$$\frac{dV_s(x)}{dx} = \lim_{\Delta x \to 0}\left[\frac{V_s(x+\Delta x) - V_s(x)}{\Delta x}\right]$$

Therefore,

$$\frac{dV_s(x)}{dx} = -\lim_{\Delta x \to 0}\left\{\frac{1}{\Delta x}\int_0^d [E_{ys}(x+\Delta x, y) - E_{ys}(x,y)]dy\right\}$$

Substituting this result into (6.18),

$$\frac{dV_s(x)}{dx} = j\omega\mu_o \int_0^d \lim_{\Delta x \to 0} \left[\frac{1}{\Delta x} \int_x^{x+\Delta x} H_{zs}(x,y)dx \right] dy = j\omega\mu_o \int_0^d H_{zs}(x,y)dy \tag{6.20}$$

(If a function is integrated then differentiated, the result is the original function since differentiation and integration are "inverse" operations.)

The expression on the right side of (6.20) is the total magnetic flux (per Δx) normal to the plane of the transmission line. The magnetic field under this integral, $H_{zs}(x, y)$, is not just the incident magnetic field, H^{inc}. It is the sum of the incident term and scattered term, H^{scat}. The scattered magnetic field, which is usually not mentioned in elementary electromagnetic courses, is due to the current in the wires induced from the applied or incident plane wave. (If an electrically-short transmission line is open circuited or the source or load impedance is large, then this induced current is small.) Equation (6.20) can be written as a function of the incident and scattered fields:

$$\frac{dV_s(x)}{dx} = j\omega\mu_o \int_0^d \left[H_{zs}^{inc}(x,y) + H_{zs}^{scat}(x,y) \right] dy \tag{6.21}$$

The scattered magnetic field is related to the standard differential-mode currents via the inductance of the line. The partial flux collected as a function of x is

$$\Phi_s^{scat}(x) = LI_s(x) = -\int_0^d \mu_o H_{zs}^{scat}(x,y)dy \tag{6.22}$$

Substituting this expression into (6.21) and rearranging yields

$$\frac{dV_s(x)}{dx} + j\omega LI_s(x) = j\omega\mu_o \int_0^d H_{zs}^{inc}(x,y)dy \tag{6.23}$$

This differential equation will be used to determine one-half of the susceptibility model—that due to the applied magnetic field.

In addition to the magnetic field component of the incident plane wave, the electric field component of the plane wave also contributes to the current and voltage generated along the line. Another of Maxwell's equations (referred to as Ampère's law), written in the frequency domain, is helpful in the second half of this analysis:

$$\nabla \times \vec{H}_s = j\omega\varepsilon_o \vec{E}_s \tag{6.24}$$

It is assumed that the space between the wires is lossless or conduction free (i.e., the conduction current, J, is zero between the wires). Also, it is assumed that the electric field is entirely y directed across the wires and is independent of y, which is a reasonable assumption when $d \ll \lambda$. From (6.24), the y-component of the curl of $\vec{H}_s$ is related to the y-component of $\vec{E}_s$ via

$$[\nabla \times \vec{H}_s(x,y)]_y = \frac{\partial H_{xs}(x,y)}{\partial z} - \frac{\partial H_{zs}(x,y)}{\partial x} = j\omega\varepsilon_o E_{ys}(x,y)$$

$$\Rightarrow E_{ys}(x,y) = \frac{1}{j\omega\varepsilon_o} \left[\frac{\partial H_{xs}(x,y)}{\partial z} - \frac{\partial H_{zs}(x,y)}{\partial x} \right] \tag{6.25}$$

Again, the definition for the voltage across the line is used along with (6.22):

$$
\begin{aligned}
V_s(x) &= -\int_0^d E_{ys}(x,y)dy = -\frac{1}{j\omega\varepsilon_o}\int_0^d \left[\frac{\partial H_{xs}(x,y)}{\partial z} - \frac{\partial H_{zs}(x,y)}{\partial x}\right]dy \\
&= \frac{1}{j\omega\varepsilon_o}\int_0^d \frac{\partial H_{zs}^{scat}(x,y)}{\partial x}dy - \frac{1}{j\omega\varepsilon_o}\int_0^d \frac{\partial H_{xs}^{scat}(x,y)}{\partial z}dy \\
&\quad - \frac{1}{j\omega\varepsilon_o}\int_0^d \left[\frac{\partial H_{xs}^{inc}(x,y)}{\partial z} - \frac{\partial H_{zs}^{inc}(x,y)}{\partial x}\right]dy \\
&= -\frac{1}{j\omega\varepsilon_o}\frac{1}{\mu_o}\frac{d}{dx}[LI_s(x)] - \frac{1}{j\omega\varepsilon_o}\int_0^d \frac{\partial H_{xs}^{scat}(x,y)}{\partial z}dy - \int_0^d E_{ys}^{inc}(x,y)dy
\end{aligned}
\tag{6.26}
$$

Again, the total magnetic field is the superposition of the scattered and incident fields. The integral involving H_{xs}^{scat} is zero since the scattered magnetic field is zero along the wire. The current along the wires is entirely in the x direction:

$$
\vec{J}_s^{scat}(x,y) = \sigma E_{xs}^{scat}(x,y)\hat{a}_x = \nabla \times \vec{H}_s^{scat}(x,y) = \sigma\left[\frac{\partial H_{zs}^{scat}(x,y)}{\partial y} - \frac{\partial H_{ys}^{scat}(x,y)}{\partial z}\right]\hat{a}_x \tag{6.27}
$$

The current that results from the incident field generates a magnetic field only in the y and z directions; in other words, current in the x direction does not generate a magnetic field in the x direction. Using this result in (6.26) yields the integral-differential equation

$$
\frac{dI_s(x)}{dx} + \frac{j\omega\mu_o\varepsilon_o}{L}V_s(x) = -\frac{j\omega\mu_o\varepsilon_o}{L}\int_0^d E_{ys}^{inc}(x,y)\,dy \tag{6.28}
$$

Finally, if the medium surrounding the wires is homogeneous, then the inductance per unit length, L, and capacitance per unit length, C, of the line are related to the properties of the surrounding medium via the expression valid for TEM signals

$$
LC = \mu\varepsilon
$$

For free space around the wires, $LC = \mu_o\varepsilon_o$. Thus, the two integral-differential equations relating the incident electric and magnetic fields to the induced current and voltage on the parallel lines are

$$
\frac{dI_s(x)}{dx} + j\omega CV_s(x) = -j\omega C\int_0^d E_{ys}^{inc}(x,y)\,dy = I_{ds}(x) \tag{6.29}
$$

$$
\frac{dV_s(x)}{dx} + j\omega LI_s(x) = j\omega\mu_o\int_0^d H_{zs}^{inc}(x,y)dy = V_{ds}(x) \tag{6.30}
$$

These two equations are the standard transmission line equations with driving functions (or source functions) $I_{ds}(x)$ and $V_{ds}(x)$. These sources are a function of the incident electric field and incident magnetic field to the line. The susceptibility model for an electrically-short twin-lead line is obtained directly from

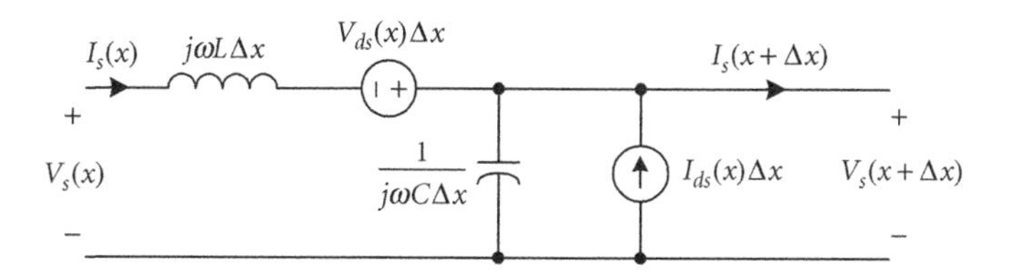

FIGURE 6.9 Susceptibility model (in the frequency domain) for electrically-short parallel conductors of length Δx.

these two equations. It is given in Figure 6.9 and derived later. Notice the similarity of this model to the one used to model crosstalk between weakly coupled lines. The solution for (6.29) and (6.30) when the line is not electrically short is discussed later.

6.9 Small-Loop and Hertzian Dipole Models

Referring to the susceptibility model given in Figure 6.9, why is the current source frequently omitted for electrically-small loops? Why is the voltage source frequently omitted for electrically-small dipoles? [Adams; Koucherng]

It is reasonable to assume that the response of an electrically-small circular loop is about the same as a rectangular-shaped loop assuming they have equal areas. Under this assumption, if the electrically-small loop is modeled as a twin-lead line with a short-circuited load, as shown in Figure 6.10, then it is clear that the current source is irrelevant: all of its current passes through the short. From (6.29), the current source is equal to

$$I_{ds}(x) = -j\omega C \int_0^d E_{ys}^{inc}(x,y)\,dy \tag{6.31}$$

It is a function of both the capacitance of the line and electric field across the line. The capacitance is nearly always neglected for an electrically-small loop.[5] The difference in voltage between one location and another on this loop is very small since it is essentially a single piece of wire. The open-circuit voltage across the input of the short-circuited line is entirely a function of the voltage source that is equal to

$$V_{ds}(x) = j\omega\mu_o \int_0^d H_{zs}^{inc}(x,y)\,dy$$

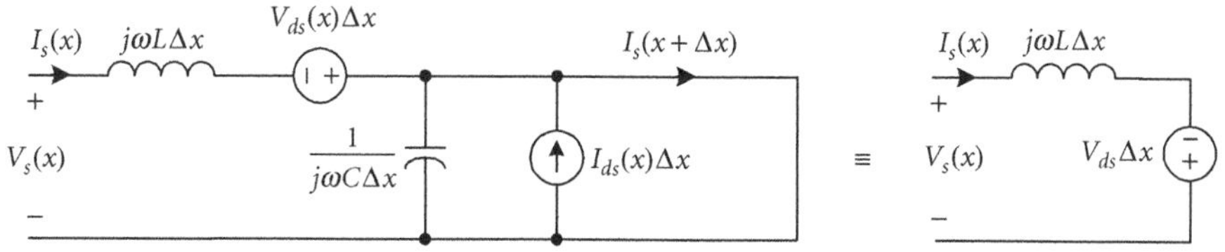

FIGURE 6.10 Susceptibility model for an electrically-short line with a short-circuited load.

[5]For a larger loop, both the inductance and capacitance are important since they determine the loop's first resonant frequency.

This integral relationship is essentially Faraday's law in the frequency domain. If the line is sufficiently short so that the incident magnetic field can be considered constant over its length and $H_{zs}^{inc}(x, y) \approx H_o$, then the open-circuit voltage is

$$V_{ds}\Delta x \approx j\omega\mu_o H_o d\Delta x = j\omega\mu_o H_o A \tag{6.32}$$

where $A = d\Delta x$ is the magnetic-field pickup area of the line. To reduce the susceptibility of the line to external noise, the dimensions of the line should be decreased. As indicated by (6.32), the induced noise voltage increases with frequency. The inductance should not be eliminated from this model. It models the counter-magnetic field generated by current in the loop whenever a finite impedance is connected across the line's input.

A Hertzian dipole consists of two electrically-short wires with open-circuited ends. If the electrically-small dipole is modeled as a twin-lead line with an open-circuited load, as shown in Figure 6.11, then it is clear that the voltage source and inductance are irrelevant: they are in series with the open-circuited load, located on the left side of the model. The inductance for an electrically-short open circuit is very small or negligible. The current source defined in (6.31) is a function of the capacitance and the electric field between the lines. If the distance between the wires, d, is sufficiently small so that the incident electric field can be considered constant and $E_{ys}^{inc}(x, y) \approx E_o$, then the short-circuit current is

$$I_{ds}\Delta x \approx -j\omega C E_o d\Delta x = j\omega(C\Delta x)V_s \tag{6.33}$$

where the voltage definition given in (6.19) was used. The equivalent time-domain expression is $(j\omega \Leftrightarrow d/dt)$

$$i_d(t)\Delta x = (C\Delta x)\frac{dv}{dt} \tag{6.34}$$

After dividing both sides by Δx, this equation is the familiar circuits definition for capacitance. For two wires with open-circuited ends, it is reasonable that the total capacitance, $C\Delta x$, is an important parameter.

As a simple use of the Hertzian susceptibility model, imagine that an electroexplosive device (or EED) with a resistance of 1.4 Ω is connected to two 0.5 m long parallel leads (with open-circuited ends) with 1.5 cm spacing. If the electric field in the vicinity of the leads is about 2 V/m at 39 MHz, then

$$I_{ds}\Delta x = -j2\pi(39\times10^6)(6.8\times10^{-12})(2)(0.015)(0.5) \approx -j25\,\mu\text{A}$$

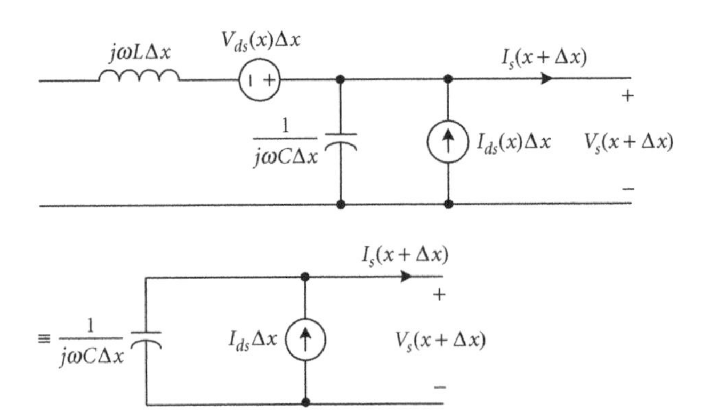

FIGURE 6.11 Susceptibility model for an electrically-short line with an open-circuited load.

where the capacitance per unit length is given in (6.41) as 6.8 pF/m. (The leads are electrically short at this frequency.) If the parasitic capacitance and inductance of the EED are neglected, the magnitude of the current through the resistance is

$$|I_{eeds}| = \left| I_{ds}\Delta x \frac{\dfrac{1}{j\omega C\Delta x}}{\dfrac{1}{j\omega C\Delta x}+1.4} \right| \approx 25\ \mu\text{A}$$

Whether this current amplitude is capable of firing the EED is dependent on many factors including the composition of the explosive materials, the temperature, and the shape and size of the excitation signal (e.g., sinusoid, pulse, or ramp). For one material presented in the reference, an average current level of 390 mA detonated an EED using a slowly varying ramp source. Using a repetitive rectangular pulse source, with a pulse width of 1 ms and spacing of 500 ms, the energy required to detonate the EED was 12 mJ.

The inquisitive mind might wonder why the "load" was taken on the right side of the model for the loop while on the left side of the model for the open-circuit wires. In the derivation of the model given in Figure 6.9, there was no mention of which side corresponds to the source or load. In many cases, the position of the elements in the model is not important, and, thus, the side corresponding to the source or load is not important. The locations for the load in the previous analysis were chosen to emphasize the real-life dominant circuit elements for the loop and open-circuited wire and to simplify the circuit. Without this insight, the model would still be valid if the "wrong" side were selected as the load. In Figure 6.12, for example, the left side is selected as the load side for the loop. The load is the short circuit. Now, the open-circuit voltage given by $V_s(x+\Delta x)$ is a function of both the voltage and current sources. Using superposition with voltage and current division,

$$\begin{aligned} V_s(x+\Delta x) &= V_{ds}(x)\Delta x \frac{\dfrac{1}{j\omega C\Delta x}}{\dfrac{1}{j\omega C\Delta x}+j\omega L\Delta x} + I_{ds}(x)\Delta x\left(j\omega L\Delta x \middle\| \frac{1}{j\omega C\Delta x}\right) \\ &= V_{ds}(x)\Delta x \frac{1}{1-\left(\dfrac{\omega}{\omega_c}\right)^2} + I_{ds}(x)\Delta x\left[\frac{j\omega L\Delta x}{1-\left(\dfrac{\omega}{\omega_c}\right)^2}\right] \end{aligned} \qquad (6.35)$$

where $\omega_c = 1/\sqrt{(L\Delta x)(C\Delta x)}$. Since the loop is electrically short, the frequency is much less than the resonant frequency, ω_c, of the loop and (6.35) simplifies to

$$V_s(x+\Delta x) \approx V_{ds}(x)\Delta x + I_{ds}(x)\Delta x(j\omega L\Delta x) \quad \text{if } \omega \ll \omega_c \qquad (6.36)$$

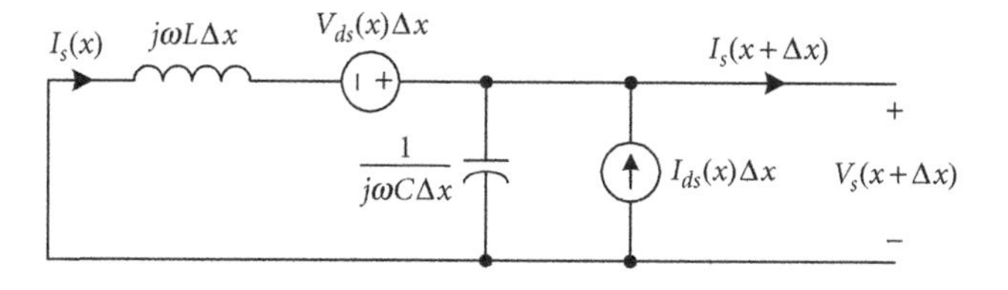

FIGURE 6.12 More complicated susceptibility model for an electrically-small loop.

Compared to (6.32), Equation (6.36) contains an additional term. This term is referred to as the electric-displacement correction factor since $I_{ds}(x)$ is a function of the capacitance of the loop. The capacitance in this correction factor can be obtained indirectly through a resonant frequency measurement.

6.10 Neglecting the Capacitance and Inductance

In reference to the susceptibility model repeated in Figure 6.13, it is stated that "Neglecting the line inductance and capacitance is typically valid so long as the termination impedances are not extreme values such as short and open circuits." Show that this statement is reasonable. [Paul, '92(b)]

The susceptibility model for an electrically-short line includes the induced voltage source, the induced current source, and the total inductance and capacitance of the cable. To demonstrate that the inductance and capacitance can be neglected in the model when the termination impedances are not extreme is straightforward. The voltage across the resistive load impedance will be determined to test the reasonableness of this statement. By superposition, the voltage across the right-side load termination is

$$V_{Lds} = V_{Is}\frac{\dfrac{R_{Ld}Z_C}{R_{Ld}+Z_C}}{\dfrac{R_{Ld}Z_C}{R_{Ld}+Z_C}+R_s+Z_L} + I_{Is}\frac{\dfrac{(R_s+Z_L)Z_C}{R_s+Z_L+Z_C}}{\dfrac{(R_s+Z_L)Z_C}{R_s+Z_L+Z_C}+R_{Ld}}R_{Ld} \tag{6.37}$$

The first term, obtained through voltage division, is due to the voltage source (with the current source turned off or replaced with an open circuit). The second term, obtained through current division, is due to the current source (with the voltage source turned off or replaced with a short circuit). This equation for the voltage across the load can be simplified if two assumptions are made:

$$R_{Ld} \ll Z_C = \frac{1}{j\omega(C\Delta x)} \quad \text{and} \quad R_s \gg Z_L = j\omega(L\Delta x) \tag{6.38}$$

where C and L are per unit length and Δx is the length of the electrically-short line. The first inequality states that the right-side load resistance must be much less than the impedance of the total line capacitance. Since this capacitive reactance is not normally small, at least at lower frequencies, the load resistance should not be large (or "extreme") to satisfy this inequality. The second inequality states that the left-side source resistance should be much greater than the impedance of the total line inductance. Since the reactance of the inductor is not normally large, at least at lower frequencies, the source resistor should not be small (or "extreme") to satisfy this inequality. With the assumptions that the source and load resistances are not extreme in value, the voltage across the load is approximately

$$V_{Lds} \approx V_{Is}\frac{R_{Ld}}{R_{Ld}+R_s} + I_{Is}\frac{\dfrac{R_sZ_C}{R_s+Z_C}}{\dfrac{R_sZ_C}{R_s+Z_C}+R_{Ld}}R_{Ld}$$

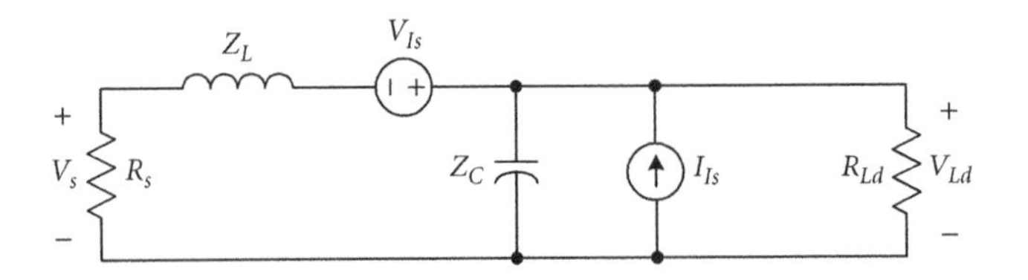

FIGURE 6.13 In some cases, Z_L and Z_C can be neglected.

If the source resistance is not too large,

$$R_s \ll Z_C = \frac{1}{j\omega(C\Delta x)}$$

the equation further reduces to

$$V_{Lds} \approx V_{Is}\frac{R_{Ld}}{R_{Ld}+R_s} + I_{Is}\frac{R_s}{R_{Ld}+R_s}R_{Ld} \quad (6.39)$$

This equation is independent of the capacitance and inductance of the cable.

If the entire process is repeated for the voltage across the source resistor and if the load resistor is not too small,

$$R_{Ld} \gg Z_L = j\omega(L\Delta x)$$

the voltage across the source resistor is also independent of the capacitance and inductance of the line:

$$V_s \approx -V_{Is}\frac{R_s}{R_{Ld}+R_s} + I_{Is}\frac{R_{Ld}}{R_{Ld}+R_s}R_s \quad (6.40)$$

To summarize, to neglect the capacitance and inductance in the model, the load resistance and source resistance should not be extreme relative to the impedances of the line capacitance and inductance. Clearly, if the load were an open circuit or the source were a short circuit, (6.38) would not be satisfied.

6.11 Probe Lead Pickup

A 0.5 W, 900 MHz cellular phone is transmitting about 1.5 meters from a scope that is measuring the voltage across a 1 kΩ resistor. The open-wire leads to the resistor are 5 cm in length. These probe leads are connected to the scope via a one meter long 50 Ω coaxial cable as shown in Figure 6.14. Assume that the 50 Ω cable capacitance is approximately 29.5 pF/ft. Determine the maximum possible voltage across the 1 kΩ resistor and scope due to the cellular phone.

Although other parts of the system, such as the 50 Ω transmission line, are also susceptible to the external radiation, only the 5 cm leads will be modeled in this discussion. As shown, the input to the scope is modeled as a 1 MΩ resistor in parallel with a 20 pF capacitor. Before the model given in Figure 6.9 is used, the electrical length of the leads must be determined. The free-space wavelength of the 900 MHz signal is

$$\lambda_o = \frac{c}{f} = \frac{3\times10^8}{900\times10^6} \approx 0.3 \text{ m}$$

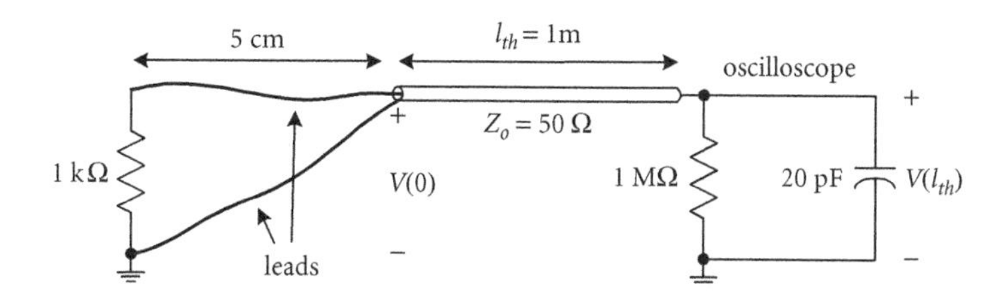

FIGURE 6.14 Long leads to the cable are also susceptible to field pickup.

For the leads to be electrically short, they should be less than about $\lambda_o/10$ or about 3 cm: the 5 cm leads can be considered electrically short. Although the spacing of the leads from the coax to the resistor is likely not uniform, to simplify the analysis, the spacing between the conductors, d, will be assumed constant at 1.5 cm and the wire gauge will be assumed to be #24 AWG, corresponding to a conductor radius, r_w, of 0.255 mm. The inductance and capacitance per meter for the leads is then

$$L = 4\times10^{-7}\ln\left\{\frac{d}{2r_w}\left[1+\sqrt{1-\left(\frac{2r_w}{d}\right)^2}\right]\right\} \approx 1.6\ \mu\text{H/m}$$

$$C = \frac{\pi\varepsilon_o}{\ln\left[\frac{d}{2r_w}+\sqrt{\left(\frac{d}{2r_w}\right)^2-1}\right]} \approx 6.8\ \text{pF/m} \qquad (6.41)$$

Although these expressions are most accurate when the length of the conductors is much greater than their separation, they will provide order-of-magnitude results. (Also, the calculated values are reasonable.)

To use the susceptibility model, the impedance of both the source and load are required. The source impedance is the 1 kΩ resistor, which will be assumed to have negligible lead or body inductance and a shunt parasitic capacitance of 2 pF. To determine the load impedance, the input impedance into the coaxial cable is required. Unfortunately, the input impedance of the scope is not matched to the cable. (Modern scopes also have a 50 Ω input-impedance setting.) This unmatched situation requires a simple understanding of transmission lines. The input impedance expression for a lossless transmission line of length l_{th} is

$$Z_{in} = Z_o\frac{Z_L + jZ_o\tan(\beta l_{th})}{Z_o + jZ_L\tan(\beta l_{th})}$$

where Z_L is the scope's impedance

$$Z_L = 10^6 \left\| \frac{1}{j2\pi(900\times10^6)(20\times10^{-12})} \right. \approx -j8.8\ \Omega$$

and β is the phase constant:

$$\beta = \omega\sqrt{LC} = \omega\sqrt{CZ_o^2C} = \omega CZ_o = 2\pi(900\times10^6)(29.5\times10^{-12})50 \approx 8.3\ \text{rad/m}$$

Using this simple RC model for the scope input, the load impedance presented to the cable is mainly capacitive at this high frequency. The input impedance for a 1 meter cable, seen by the leads, is also capacitive:

$$Z_{in} \approx 50\frac{-j8.8 + j50\tan(8.3\times1)}{50 + j(-j8.8)\tan(8.3\times1)} \approx -j160\ \Omega$$

At these high frequencies, it *cannot* be assumed that the scope's impedance is resistive or that it is high.

The impedance of the source or "input" of the leads at 900 MHz is far from 1 kΩ:

$$Z_s = 10^3 \left\| \frac{1}{j2\pi(900\times10^6)(2\times10^{-12})} \right. \approx 7.8 - j88\ \Omega$$

Again, this assumes that the resistor is operating well below its resonant frequency so that its inductance can be neglected.

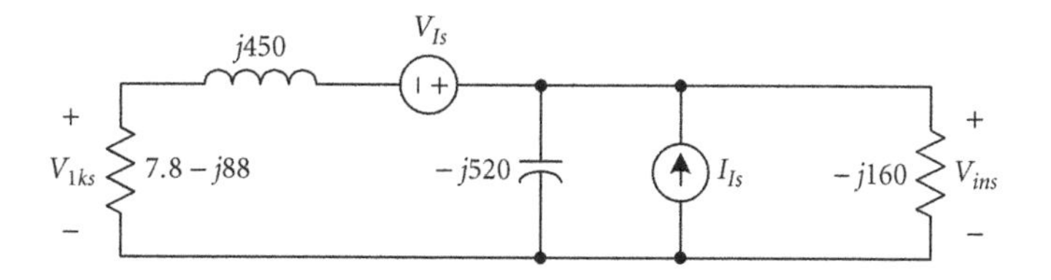

FIGURE 6.15 Susceptibility model of the probe leads.

Since the impedances at both terminations of the leads are known, they can be substituted into the electrically-short susceptibility model as shown in Figure 6.15. The total inductive and capacitive reactances of the 5 cm long leads are

$$\omega L\left(\frac{5}{100}\right) = 2\pi(900\times10^{6})(1.6\times10^{-6})(0.05) \approx 450\ \Omega$$

$$\frac{-1}{\omega C\left(\frac{5}{100}\right)} = \frac{-1}{2\pi(900\times10^{6})(6.8\times10^{-12})(0.05)} \approx -520\ \Omega$$

The inductance and capacitance of the leads cannot be neglected in this situation.

The voltage and current sources in this model are equal to

$$I_{Is} = \left[-j\omega C\int_0^d E_{ys}^{inc}(x,y)dy\right]\Delta x \tag{6.42}$$

$$V_{Is} = \left[j\omega\mu_o\int_0^d H_{zs}^{inc}(x,y)dy\right]\Delta x \tag{6.43}$$

where $\Delta x = 0.05$. If the variation in the electric and magnetic fields incident to the transmission line is small, (6.42) and (6.43) reduce to

$$I_{Is} \approx -j\omega C E_{ys}^{inc}(x,y)d\Delta x \tag{6.44}$$

$$V_{Is} \approx j\omega\mu_o H_{zs}^{inc}(x,y)d\Delta x \tag{6.45}$$

The electric and magnetic fields are required near the probe. To obtain a rough estimate of the power density at a distance of 1.5 m from the phone, an omnidirectional radiation pattern will be assumed:

$$P_d = \frac{P_{rad}}{4\pi r^2} = \frac{0.5}{4\pi(1.5)^2} \approx 18\ \text{mW/m}^2$$

Of course, if the gain of the phone's antenna (and the effects of the user on the radiation pattern) was known, then this power density estimate could be improved. Since $1.5 > \lambda_o$, the field at the leads can be considered in the far field of the phone and a plane wave. Thus, the relationship between the electric field and time-average power density in free space is

$$P_d = \frac{E^2}{2(377)} \quad \Rightarrow E = \sqrt{2(377)P_d} \approx 3.7\ \text{V/m}$$

(This number is the peak value of the electric field.) For plane waves, the electric and magnetic fields are related via the intrinsic impedance of the medium or 377 Ω for free space:

$$\frac{E}{H} = 377 \quad \Rightarrow H = \frac{3.7}{377} \approx 9.7 \text{ mA/m}$$

(This number is the peak value of the magnetic field.) Although the orientation of the leads relative to the phone was not provided, it will be assumed that the electric field is across the wires and the magnetic field is through the plane generated by the wires. This excitation is referred to as endfire. Substituting these fields into (6.44) and (6.45),

$$\begin{aligned} I_{Is} &\approx -j2\pi(900\times10^{6})(6.8\times10^{-12})(3.7)(0.015)(0.05) \\ &\approx -j0.1\times10^{-3} = .1\angle -90^{\circ} \text{ mA} \end{aligned} \tag{6.46}$$

$$\begin{aligned} V_{Is} &\approx j2\pi(900\times10^{6})(4\pi\times10^{-7})(9.7\times10^{-3})(0.015)(0.05) \\ &\approx j52\times10^{-3} = 52\angle 90^{\circ} \text{ mV} \end{aligned} \tag{6.47}$$

Finally, the voltages across the resistor and input of the cable can be determined using superposition and voltage and current division:

$$\begin{aligned} V_{1ks} &= -(j52\times10^{-3})\frac{7.8-j88}{7.8-j88+j450+\left(-j520\|-j160\right)} \\ &\quad +(-j0.1\times10^{-3})\frac{-j520\|-j160}{7.8-j88+j450+\left(-j520\|-j160\right)}(7.8-j88) \\ &\approx 20\angle 84^{\circ} \text{ mV} \end{aligned}$$

$$\begin{aligned} V_{ins} &= (j52\times10^{-3})\frac{-j520\|-j160}{7.8-j88+j450+\left(-j520\|-j160\right)} \\ &\quad +(-j0.1\times10^{-3})\frac{7.8-j88+j450}{7.8-j88+j450+\left(-j520\|-j160\right)}\left(-j520\|-j160\right) \\ &\approx 32\angle 240^{\circ} \text{ mV} \end{aligned}$$

The voltage measured by the scope is obtained using the following relationship relating the input and output voltages for a cable of length l_{th}:

$$\frac{V_{ins}}{V_{scope}} = \frac{1+\rho_L[\cos(2\beta l_{th}) - j\sin(2\beta l_{th})]}{(1+\rho_L)[\cos(\beta l_{th}) - j\sin(\beta l_{th})]}$$

where the reflection coefficient is

$$\rho_L = \frac{Z_L - Z_o}{Z_L + Z_o} = \frac{-j8.8-50}{-j8.8+50}$$

Solving for the voltage across the scope,

$$V_{scope} = V_{ins}\frac{(1+\rho_L)[\cos(\beta l_{th}) - j\sin(\beta l_{th})]}{1+\rho_L[\cos(2\beta l_{th}) - j\sin(2\beta l_{th})]} \approx 6\angle 60^\circ \text{ mV}$$

This voltage is small but measurable.

To minimize the induced signal on the short twin-lead line, the amplitude of the induced voltage and current sources in the model should be minimized. For a given frequency, reducing the lead line length, Δx, is probably the most effective method of reducing the susceptibility of the leads. According to (6.44), to reduce the susceptibility of the leads, the capacitance, C, between the lines should be small and the distance between the conductors, d, should be small. However, referring to (6.41), the capacitance per unit length between the two parallel conductors of radius r_w increases as d decreases. The product dC decreases with decreasing d until the minimum is reached. The minimum of the function dC occurs when $d \approx 3.6r_w$. The orientation of the leads and cable also affects the strength of the induced noise. This will be shown for electrically-long twin-lead lines in a later discussion.

6.12 Wave Equation

Verify that the differential equations previously derived for the susceptibility of twin-lead line

$$\frac{dI_s(x)}{dx} + j\omega CV_s(x) = -j\omega C\int_0^d E_{ys}^{inc}(x,y)dy = I_{ds}(x) \tag{6.48}$$

$$\frac{dV_s(x)}{dx} + j\omega LI_s(x) = j\omega\mu_o\int_0^d H_{zs}^{inc}(x,y)dy = V_{ds}(x) \tag{6.49}$$

correspond to the circuit given in Figure 6.9. Do they reduce to the standard lossless transmission line voltage equation when the induced sources are eliminated?

The correspondence between the transmission-line equation and the susceptibility model is easily seen. Equation (6.48) states the change in current from the output to the input of the line is due to the current "lost" through the capacitor and the induced current from the incident electric field. Equation (6.49) states that the voltage drop from the output to the input of the line is due to the voltage across the inductor and the induced voltage from the incident magnetic field. The correspondence between the circuit and these equations is rigorously obtained by using KCL at the top center node and KVL around the outer perimeter of the circuit given in Figure 6.9:

$$-I_s(x) + I_s(x+\Delta x) - I_{ds}(x)\Delta x + \frac{V_s(x+\Delta x)}{\dfrac{1}{j\omega C\Delta x}} = 0$$

$$-V_s(x) + j\omega L\Delta x I_s(x) - V_{ds}(x)\Delta x + V_s(x+\Delta x) = 0$$

Dividing both of these equations by Δx and rearranging,

$$\frac{I_s(x+\Delta x) - I_s(x)}{\Delta x} + j\omega CV_s(x+\Delta x) = I_{ds}(x) \tag{6.50}$$

$$\frac{V_s(x+\Delta x) - V_s(x)}{\Delta x} + j\omega LI_s(x) = V_{ds}(x) \tag{6.51}$$

Since $df(x)/dx = \lim_{\Delta x \to 0} \{[f(x+\Delta x) - f(x)]/\Delta x\}$, in the limit as $\Delta x \to 0$, equations (6.48) and (6.49) are obtained.

As a simple check on these results, when the induced voltage and current sources are set to zero, (6.48) and (6.49) reduce to

$$\frac{dI_s(x)}{dx} + j\omega C V_s(x) = 0 \tag{6.52}$$

$$\frac{dV_s(x)}{dx} + j\omega L I_s(x) = 0 \tag{6.53}$$

Differentiating (6.53) with respect to *x*, solving for the derivative of the current,

$$\frac{dI_s(x)}{dx} = -\frac{1}{j\omega L}\frac{d^2V_s(x)}{dx^2}$$

and substituting into (6.52), a famous second-order differential equation known as the wave equation is obtained:

$$-\frac{1}{j\omega L}\frac{d^2V_s(x)}{dx^2} + j\omega C V_s(x) = 0 \;\Rightarrow \frac{d^2V_s(x)}{dx^2} + \omega^2 LC V_s(x) = 0 \tag{6.54}$$

This equation is of the same form as that used to describe plane waves in lossless medium.

The solution for the voltage along a lossless transmission line is

$$V_s(x) = V^+ e^{-j\beta x}\left[1 + \rho_L e^{j2\beta(x - l_{th})}\right] \tag{6.55}$$

where $\beta = \omega\sqrt{LC}$. To check whether (6.55) is actually the solution for the voltage along a transmission line, it must be differentiated twice with respect to *x*:

$$\frac{dV(x)}{dx} = -j\beta V^+ e^{-j\beta x}\left[1 + \rho_L e^{j2\beta(x-l_{th})}\right] + V^+ e^{-j\beta x}\left[j2\beta\rho_L e^{j2\beta(x-l_{th})}\right]$$

$$\begin{aligned}\frac{d^2V(x)}{dx^2} &= -\beta^2 V^+ e^{-j\beta x}\left[1 + \rho_L e^{j2\beta(x-l_{th})}\right] - j\beta V^+ e^{-j\beta x}\left[j2\beta\rho_L e^{j2\beta(x-l_{th})}\right] \\ &\quad - j\beta V^+ e^{-j\beta x}\left[j2\beta\rho_L e^{j2\beta(x-l_{th})}\right] + V^+ e^{-j\beta x}\left[-4\beta^2\rho_L e^{j2\beta(x-l_{th})}\right] \\ &= -\beta^2 V^+ e^{-j\beta x}\left[1 + \rho_L e^{j2\beta(x-l_{th})}\right] = -\omega^2 LC V^+ e^{-j\beta x}\left[1 + \rho_L e^{j2\beta(x-l_{th})}\right]\end{aligned} \tag{6.56}$$

Substituting (6.55) and (6.56) into the wave equation, it is clear that (6.55) is a legitimate solution for the voltage along a transmission line:

$$-\omega^2 LC V^+ e^{-j\beta x}\left[1 + \rho_L e^{j2\beta(x-l_{th})}\right] + \omega^2 LC V^+ e^{-j\beta x}\left[1 + \rho_L e^{j2\beta(x-l_{th})}\right] = 0$$

In this case, the source of the voltage or current along the transmission line is not the incident plane wave, since these sources were set to zero, but voltage or current sources at the input or output of the transmission line.

The differential equation describing the voltage (or current) when the sources are not set to zero is obtained in a similar manner. First, (6.49) is differentiated with respect to x and rearranged:

$$\frac{dI_s(x)}{dx} = \frac{\mu_o}{L}\int_0^d \frac{\partial H_{zs}^{inc}(x,y)}{\partial x}dy - \frac{1}{j\omega L}\frac{d^2V_s(x)}{dx^2} \tag{6.57}$$

Second, (6.57) is substituted into (6.48) and rearranged:

$$\frac{d^2V_s(x)}{dx^2} + \omega^2 LCV_s(x) = j\omega\mu_o\int_0^d \frac{\partial H_{zs}^{inc}(x,y)}{\partial x}dy - \omega^2 LC\int_0^d E_{ys}^{inc}(x,y)dy \tag{6.58}$$

Equation (6.58) is the wave equation with two driving functions. Using the language of mathematics, it has both a homogeneous and particular solution. Its general solution will be left for the excessively curious. From Faraday's law, $\nabla\times\vec{E}_s = -j\omega\vec{B}_s$, the incident magnetic field under the integral can be written in terms of the incident electric field:

$$\left[\nabla\times\vec{E}_s^{inc}(x,y)\right]_z = \left[\frac{\partial E_{ys}^{inc}(x,y)}{\partial x} - \frac{\partial E_{xs}^{inc}(x,y)}{\partial y}\right] = -j\omega\mu_o H_{zs}^{inc}(x,y)$$

$$H_{zs}^{inc}(x,y) = \frac{1}{j\omega\mu_o}\left[\frac{\partial E_{xs}^{inc}(x,y)}{\partial y} - \frac{\partial E_{ys}^{inc}(x,y)}{\partial x}\right] \tag{6.59}$$

The solution to (6.58) will provide the value of the induced voltage along even electrically-long parallel conductors illuminated by a uniform plane wave along its entire length. With the use of (6.59), the solution can be written entirely in terms of the incident electric field.

6.13 Susceptibility of Electrically-Long Twin-Lead Line

Referring to Figure 6.16, when a lossy twin-lead line that is not necessarily electrically short is in the presence of electromagnetic fields, the currents at the source and load of the line are given by, respectively,[6]

$$\begin{aligned} I_s(z=0) &= \frac{1}{D}\int_0^{l_{th}} K_s(z)\{Z_o\cosh[\gamma(l_{th}-z)] + Z_L\sinh[\gamma(l_{th}-z)]\}dz \\ &+ \frac{1}{D}[Z_o\cosh(\gamma l_{th}) + Z_L\sinh(\gamma l_{th})]\int_0^d E_{xs}^{inc}(x,0)dx - \frac{Z_o}{D}\int_0^d E_{xs}^{inc}(x,l_{th})dx \end{aligned} \tag{6.60}$$

$$\begin{aligned} I_s(z=l_{th}) &= \frac{1}{D}\int_0^{l_{th}} K_s(z)[Z_o\cosh(\gamma z) + Z_s\sinh(\gamma z)]dz \\ &+ \frac{Z_o}{D}\int_0^d E_{xs}^{inc}(x,0)dx - \frac{1}{D}[Z_o\cosh(\gamma l_{th}) + Z_s\sinh(\gamma l_{th})]\int_0^d E_{xs}^{inc}(x,l_{th})dx \end{aligned} \tag{6.61}$$

[6] To be consistent with most of the related literature, the conductors are parallel to the z axis, instead of parallel to the x axis as in previous discussions.

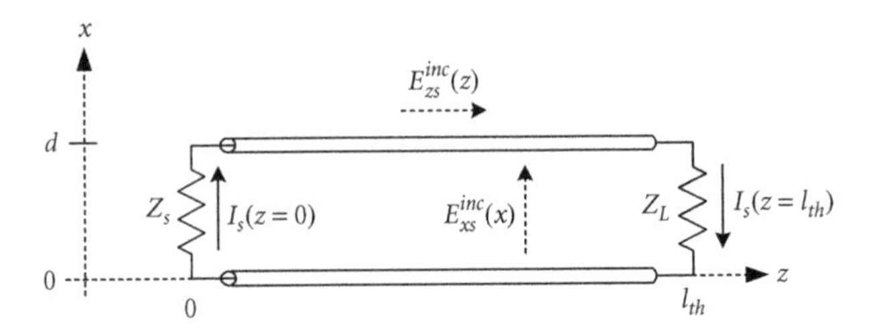

FIGURE 6.16 Twin-lead line illuminated by an electromagnetic field.

where

$$D = (Z_o Z_s + Z_o Z_L)\cosh(\gamma l_{th}) + (Z_o^2 + Z_s Z_L)\sinh(\gamma l_{th})$$
$$\gamma = \text{propagation constant of the lossy twin-lead line}$$
$$Z_o = \text{characteristic impedance of the lossy twin-lead line}$$
$$K_s(z) = E_{zs}^{inc}(d,z) - E_{zs}^{inc}(0,z)$$
$$E_{zs}^{inc}(d,z) = \text{incident electric field in the } z \text{ direction along the upper conductor}$$
$$E_{zs}^{inc}(0,z) = \text{incident electric field in the } z \text{ direction along the lower conductor}$$
$$E_{xs}^{inc}(x,0) = \text{incident electric field in the } x \text{ direction along the left termination}$$
$$E_{xs}^{inc}(x,l_{th}) = \text{incident electric field in the } x \text{ direction along the right termination}$$

To obtain some insight into how the current can vary at the two terminations, simplify these complicated expressions by sequentially assuming i) the line is lossless, ii) the line is electrically short and lossless, iii) the line is lossless and the incident field is a plane wave polarized in the x direction, and iv) the line is lossless and the incident field is a plane wave polarized in the z direction. [Smith, '77; Paul, '94; DeGauque]

Lossless Line

The characteristic impedance and propagation constant for a lossy line are defined as

$$Z_o = \sqrt{\frac{R + j\omega L}{G + j\omega C}}, \quad \gamma = \alpha + j\beta = \sqrt{(R + j\omega L)(G + j\omega C)} \tag{6.62}$$

where R, L, G, and C are the per-unit-length values for the resistance, inductance, conductance, and capacitance of the line, respectively. When the line is lossless, both the resistance of the conductors, R, and conductance between the conductors, G, are zero:

$$Z_o = \sqrt{\frac{L}{C}}, \quad \gamma = j\beta = j\omega\sqrt{LC} \quad \text{if } R = G = 0 \tag{6.63}$$

Then, the hyperbolic functions simplify to ordinary sinusoidal functions since

$$\sinh(jx) = j\sin(x), \quad \cosh(jx) = \cos(x)$$

Therefore, substituting these results into (6.60) and (6.61), the lossless expressions for the current at the source and load[7] are obtained:

$$I_s(z=0)=\frac{1}{D}\int_0^{l_{th}} K_s(z)\{Z_o\cos[\beta(l_{th}-z)]+jZ_L\sin[\beta(l_{th}-z)]\}dz$$
$$+\frac{1}{D}[Z_o\cos(\beta l_{th})+jZ_L\sin(\beta l_{th})]\int_0^d E_{xs}^{inc}(x,0)dx-\frac{Z_o}{D}\int_0^d E_{xs}^{inc}(x,l_{th})dx \tag{6.64}$$

$$I_s(z=l_{th})=\frac{1}{D}\int_0^{l_{th}} K_s(z)[Z_o\cos(\beta z)+jZ_s\sin(\beta z)]dz$$
$$+\frac{Z_o}{D}\int_0^d E_{xs}^{inc}(x,0)dx-\frac{1}{D}[Z_o\cos(\beta l_{th})+jZ_s\sin(\beta l_{th})]\int_0^d E_{xs}^{inc}(x,l_{th})dx \tag{6.65}$$

where $D=(Z_oZ_s+Z_oZ_L)\cos(\beta l_{th})+j(Z_o^2+Z_sZ_L)\sin(\beta l_{th})$ and for two parallel conductors of radius r_w,

$$Z_o=120\ln\left[\frac{d}{2r_w}+\sqrt{\left(\frac{d}{2r_w}\right)^2-1}\right]\approx 120\ln\left(\frac{d}{r_w}\right)\quad \text{if } d\gg r_w$$

Electrically-Short Lossless Line

When the line is electrically short, its length is small compared to a wavelength:

$$l_{th}\ll\lambda=\frac{v}{f}=2\pi\frac{\frac{1}{\sqrt{LC}}}{\omega}=\frac{2\pi}{\sqrt{LC}}\frac{1}{\frac{\beta}{\sqrt{LC}}}=\frac{2\pi}{\beta}\quad\Rightarrow l_{th}\beta\ll 2\pi \tag{6.66}$$

Since $\cos x\approx 1-(x^2/2)$ and $\sin x\approx x$ when x is much less than one, (6.64) and (6.65) are approximately equal to

$$I_s(z=0)\approx\frac{1}{D}\int_0^{l_{th}} K_s(z)\left\{Z_o\left[1-\frac{\beta^2(l_{th}-z)^2}{2}\right]+jZ_L\beta(l_{th}-z)\right\}dz$$
$$+\frac{1}{D}\left[Z_o\left(1-\frac{\beta^2 l_{th}^2}{2}\right)+jZ_L\beta l_{th}\right]\int_0^d E_{xs}^{inc}(x,0)dx-\frac{Z_o}{D}\int_0^d E_{xs}^{inc}(x,l_{th})dx \tag{6.67}$$

$$I_s(z=l_{th})\approx\frac{1}{D}\int_0^{l_{th}} K_s(z)\left[Z_o\left(1-\frac{\beta^2 z^2}{2}\right)+jZ_s\beta z\right]dz$$
$$+\frac{Z_o}{D}\int_0^d E_{xs}^{inc}(x,0)dx-\frac{1}{D}\left[Z_o\left(1-\frac{\beta^2 l_{th}^2}{2}\right)+jZ_s\beta l_{th}\right]\int_0^d E_{xs}^{inc}(x,l_{th})dx \tag{6.68}$$

[7]Actually, the labeling of the left-hand termination as the source and the right-hand termination as the load is arbitrary. The actual source could be on the right and the actual load on the left.

where $D \approx (Z_o Z_s + Z_o Z_L)[1-(\beta^2 l_{th}^2)/2] + j(Z_o^2 + Z_s Z_L)\beta l_{th}$. These expressions further simplify when the source and load impedances are both short circuits:

$$I_s(z=0) \approx \frac{1}{jZ_o\beta l_{th}} \int_0^{l_{th}} K_s(z)\left[1-\frac{\beta^2(l_{th}-z)^2}{2}\right]dz + \frac{1}{jZ_o\beta l_{th}}\left(1-\frac{\beta^2 l_{th}^2}{2}\right)\int_0^d E_{xs}^{inc}(x,0)dx - \frac{1}{jZ_o\beta l_{th}}\int_0^d E_{xs}^{inc}(x,l_{th})dx \quad \text{if } Z_s = Z_L = 0 \tag{6.69}$$

$$I_s(z=l_{th}) \approx \frac{1}{jZ_o\beta l_{th}} \int_0^{l_{th}} K_s(z)\left(1-\frac{\beta^2 z^2}{2}\right)dz + \frac{1}{jZ_o\beta l_{th}}\int_0^d E_{xs}^{inc}(x,0)dx - \frac{1}{jZ_o\beta l_{th}}\left(1-\frac{\beta^2 l_{th}^2}{2}\right)\int_0^d E_{xs}^{inc}(x,l_{th})dx \quad \text{if } Z_s = Z_L = 0 \tag{6.70}$$

When the source is a short circuit and load is an open circuit,

$$I_s(z=0) \approx \frac{1}{Z_o}\int_0^{l_{th}} K_s(z) j\beta(l_{th}-z)dz + \frac{j\beta l_{th}}{Z_o}\int_0^d E_{xs}^{inc}(x,0)dx \quad \text{if } Z_s = 0,\ Z_L = \infty \tag{6.71}$$

Of course, $I_s(z=l_{th})=0$. Short circuiting both terminations or open circuiting one termination does not force the current to be zero everywhere. If these expressions still seem too complicated, by assuming that both components of the electric field are constant but E_{zs}^{inc} varies in phase

$$E_{zs}^{inc}(0,z) \approx E_{zo}^{inc},\ E_{zs}^{inc}(d,z) \approx E_{zo}^{inc}(1-j\beta d),\ E_{xs}^{inc}(x,0) \approx E_{xo}^{inc},\ E_{xs}^{inc}(x,l_{th}) \approx E_{xo}^{inc} \tag{6.72}$$

they reduce even further:

$$I_s(z=0) = I_s(z=l_{th}) \approx -E_{zo}^{inc}\frac{d}{Z_o} + jE_{xo}^{inc}\frac{dl_{th}\beta}{2Z_o} \quad \text{if } Z_s = Z_L = 0 \tag{6.73}$$

$$I_s(z=0) \approx E_{zo}^{inc}\left(\frac{dl_{th}^2\beta^2}{2Z_o}\right) + jE_{xo}^{inc}\left(\frac{dl_{th}\beta}{Z_o}\right) \quad \text{if } Z_s = 0,\ Z_L = \infty \tag{6.74}$$

It is interesting that when both the source and load are short circuits and the E_{xo}^{inc} term is negligible, the current is not a function of the frequency. Since the line is electrically short, the current along the line is also constant. The current in this rectangular loop at low frequencies is the EMF induced (via Faraday's law) by the magnetic field passing through the loop divided by the impedance of the lossless loop:

$$|I| = \left|\frac{-j\omega\mu_o H_n d l_{th}}{j\omega L l_{th}}\right| = \left|\frac{\mu_o H_n d}{L}\right|$$

This expression is not a function of the frequency. From Faraday's law, $\nabla \times \vec{E}_s = -j\omega\vec{B}_s$, the incident magnetic field normal to the loop can be written in terms of the incident electric field parallel to the loop:

$$H_{ys}^{inc}(x,z) = \frac{1}{j\omega\mu_o}\left[\frac{\partial E_{zs}^{inc}(x,z)}{\partial x} - \frac{\partial E_{xs}^{inc}(x,z)}{\partial z}\right] \tag{6.75}$$

The normal component of the incident magnetic field is a function of E_{zs}^{inc} since it was assumed in (6.72) that E_{xs}^{inc} is not varying.

Lossless Line and Plane Wave Polarized in the x Direction

Before proceeding with the analysis, it is important to be able to visualize the direction of the $\vec{E}$ and $\vec{H}$ fields of the incident plane wave. The "direction" of the plane wave is normally taken as the direction of its power described by the Poynting vector:

$$\vec{P} = \vec{E} \times \vec{H}$$

The Poynting vector is perpendicular to both the electric and magnetic fields. Its direction is determined by the right-hand rule. An x-polarized wave implies that the *electric* field is entirely in the x direction. If the twin-lead conductors are oriented as shown in Figure 6.16, then an x-polarized wave would be parallel to both the source and load terminations. Referring to Figure 6.17, the angle between the Poynting vector and y axis is θ. Since the incident electromagnetic field is a plane wave parallel to both terminations, its electric field is constant (or uniform) in magnitude over the entire length of the line. However, the phase of the electric field can vary significantly over the length of the line. Taking the zero phase reference at $z = 0$, the electric field polarized in the x direction can be described as

$$E_{xs}^{inc}(x,z) = E_{xo}^{inc} e^{-j\beta z \sin\theta} \Rightarrow E_{xs}^{inc}(x,0) = E_{xo}^{inc},\ E_{xs}^{inc}(x,l_{th}) = E_{xo}^{inc} e^{-j\beta l_{th} \sin\theta} \tag{6.76}$$

where $\beta = \omega\sqrt{LC}$. Since the wave is polarized in the x direction, $E_{zs}^{inc}(x,z) = 0$. (These concepts are fully discussed in the later section on the susceptibility of a wire above a ground plane.) Once the angle of incidence, θ, and strength of the incident field are known, equations (6.60) and (6.61) can be used to determine the current at the source and load of any length line illuminated by an uniform plane wave polarized in the x direction. When $\theta = 0°$, the wave is traveling in the $-y$ direction, and

$$E_{xs}^{inc}(x,0) = E_{xo}^{inc},\ E_{xs}^{inc}(x,l_{th}) = E_{xo}^{inc} \tag{6.77}$$

This excitation is referred to as broadside, and both the phase and amplitude of the incident electric field are constant over the entire length of the line. When $\theta = 90°$, the wave is traveling in the $+z$ direction, and

$$E_{xs}^{inc}(x,0) = E_{xo}^{inc},\ E_{xs}^{inc}(x,l_{th}) = E_{xo}^{inc} e^{-j\beta l_{th}} \tag{6.78}$$

This excitation is referred to as endfire. The magnetic field is entirely normal to the plane generated by the line. The magnetic field will induce noise on the line in addition to the electric field. Although the

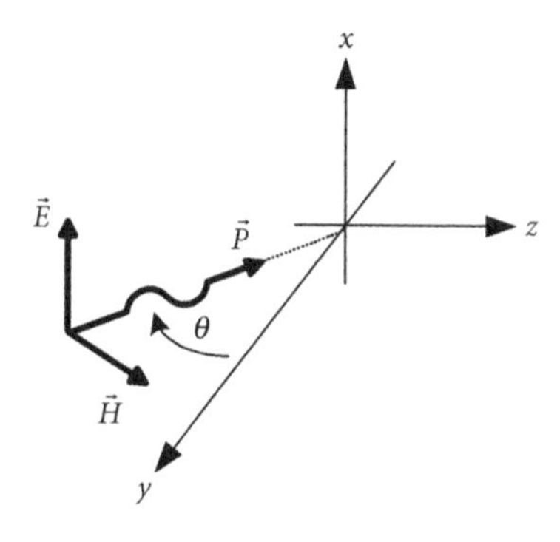

FIGURE 6.17 Electric field, magnetic field, and corresponding Poynting vector. The electric field is entirely in the x direction.

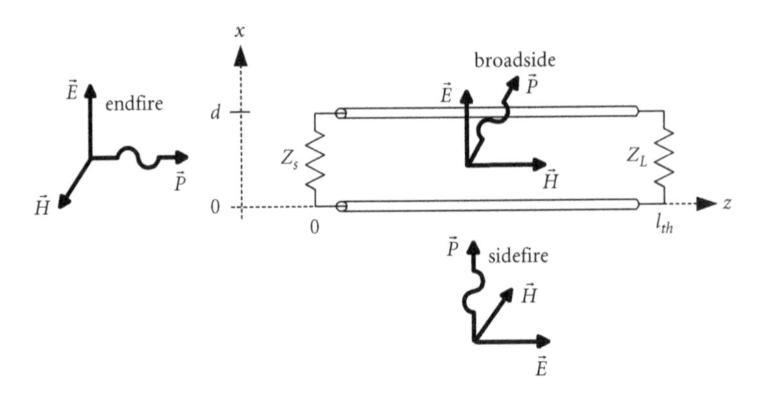

FIGURE 6.18 Endfire, broadside, and sidefire excitation of a transmission line.

magnetic field is not specifically shown in the equations for the induced current, its effect has been incorporated into the equations. The incident electric field across the source is not necessarily the same as the incident electric field across the load unless (using Euler's formula)

$$\beta l_{th} = \frac{2\pi}{\lambda} l_{th} = 2n\pi \quad (n = \text{integer}) \;\Rightarrow\; E_{xo}^{inc} e^{-j\beta l_{th}} = E_{xo}^{inc}$$

That is, if the line length is an integer multiple of a wavelength, then the electric field at the load will have the same phase and magnitude as at the source. Three special cases of excitation, including when the electric field is polarized in the z direction, are illustrated in Figure 6.18.

The expressions for the current at the source and at the load can be simplified if the distance between the conductors, d, is electrically small and the electric field is entirely in the x direction (and assuming a lossless line):

$$I_s(x=0) = \frac{d}{D}[Z_o \cos(\beta l_{th}) + jZ_L \sin(\beta l_{th})]E_{xo}^{inc} - \frac{Z_o d}{D} E_{xo}^{inc} e^{-j\beta l_{th} \sin\theta} \tag{6.79}$$

$$I_s(x=l_{th}) = -\frac{d}{D}[Z_o \cos(\beta l_{th}) + jZ_s \sin(\beta l_{th})]E_{xo}^{inc} e^{-j\beta l_{th} \sin\theta} + \frac{Z_o d}{D} E_{xo}^{inc} \tag{6.80}$$

Through some algebraic manipulations, when $\theta = 0°$ these equations reduce to

$$I_s(x=0) = \frac{E_{xo}^{inc} d}{D}\{Z_o[\cos(\beta l_{th}) - 1] + jZ_L \sin(\beta l_{th})\} \tag{6.81}$$

$$I_s(x=l_{th}) = -\frac{E_{xo}^{inc} d}{D}\{Z_o[\cos(\beta l_{th}) - 1] + jZ_s \sin(\beta l_{th})\} \tag{6.82}$$

and when $\theta = 90°$

$$I_s(x=0) = \frac{jE_{xo}^{inc}(Z_o + Z_L)d}{D}\sin(\beta l_{th}) \tag{6.83}$$

$$I_s(x=l_{th}) = \frac{E_{xo}^{inc}(Z_o - Z_s)d}{2D}\{[1 - \cos(2\beta l_{th})] + j\sin(2\beta l_{th})\} \tag{6.84}$$

Euler's formula and the trigonometric identities

$$\sin^2(\theta) = \frac{1-\cos(2\theta)}{2}, \quad \cos^2(\theta) = \frac{1+\cos(2\theta)}{2}, \quad \cos(\theta)\sin(\theta) = \frac{1}{2}\sin(2\theta) \tag{6.85}$$

were used to obtain these expressions.

Lossless Line and Plane Wave Polarized in the *z* Direction

For a plane wave polarized in the *z* direction, the incident electric field is entirely parallel to the conductors of the transmission line. Referring to Figure 6.19, $E_{xs}^{inc}(x,z) = 0$, and if the zero phase reference is taken at $\phi = 0°$,

$$E_{zs}^{inc}(x,z) = E_{zo}^{inc} e^{-j\beta\left(x-\frac{d}{2}\right)\sin\phi} \Rightarrow E_{zs}^{inc}(0,z) = E_{zo}^{inc} e^{j\frac{\beta d}{2}\sin\phi}, \quad E_{zs}^{inc}(d,z) = E_{zo}^{inc} e^{-j\frac{\beta d}{2}\sin\phi} \tag{6.86}$$

where $\beta = \omega\sqrt{LC}$. The angle of the Poynting vector is measured in the *xy* plane. Again, these expressions can be used in (6.64) and (6.65). When $\phi = 0°$, the Poynting vector is in the $-y$ direction. When $\phi = 90°$ the Poynting vector is in the *x* direction, and the excitation is referred to as sidefire. For this special case shown in Figure 6.18, the magnetic field is in the $-y$ direction.

For a *z*-polarized plane wave, the current expressions for a lossless line reduce to

$$\begin{aligned} I_s(z=0) &= \frac{-2jE_{zo}^{inc}}{D}\sin\left(\frac{\beta d}{2}\sin\phi\right)\int_0^{l_{th}}\{Z_o\cos[\beta(l_{th}-z)] + jZ_L\sin[\beta(l_{th}-z)]\}dz \\ &= \frac{-2jE_{zo}^{inc}}{\beta D}\sin\left(\frac{\beta d}{2}\sin\phi\right)\{Z_o\sin(\beta l_{th}) + jZ_L[1-\cos(\beta l_{th})]\} \end{aligned} \tag{6.87}$$

$$\begin{aligned} I_s(z=l_{th}) &= \frac{-2jE_{zo}^{inc}}{D}\sin\left(\frac{\beta d}{2}\sin\phi\right)\int_0^{l_{th}}[Z_o\cos(\beta z) + jZ_s\sin(\beta z)]dz \\ &= \frac{-2jE_{zo}^{inc}}{\beta D}\sin\left(\frac{\beta d}{2}\sin\phi\right)\{Z_o\sin(\beta l_{th}) + jZ_s[1-\cos(\beta l_{th})]\} \end{aligned} \tag{6.88}$$

where the sine identity, $\sin x = (e^{jx} - e^{-jx})/2j$, was used to rewrite $K_s(x)$. When $\phi = 0°$, or the difference in phase between the electric field at the two conductors is zero, both currents vanish.

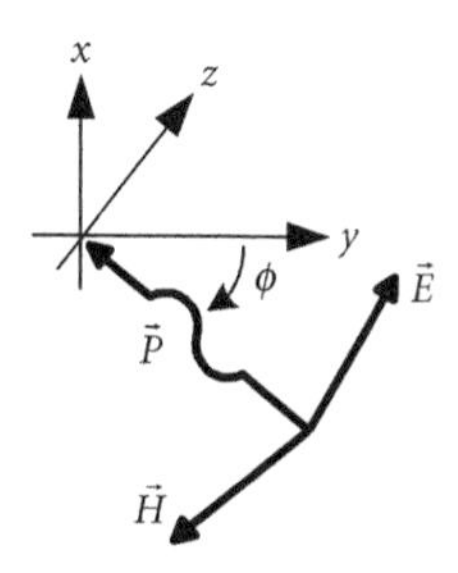

FIGURE 6.19 Electric field, magnetic field, and corresponding Poynting vector. The electric field is parallel to the *z* axis. Note the labeling of the axes.

$Z_s := 302 \quad Z_L := 302 \quad l_{th} := 10 \quad d := 0.0125 \quad r_w := 0.001 \quad \theta := 0 \quad \phi := \frac{\pi}{2} \quad E_{xo} := 0 \quad j := \sqrt{-1}$

$x := 40, 40.1..80 \qquad R := 0 \quad G := 0 \quad E_{zo} := 10^{-3}$

$$f(x) := \left(x + 1 - 10 \cdot \text{floor}\left(\frac{x}{10}\right)\right) \cdot 10^{\text{floor}\left(\frac{x}{10}\right)} \qquad L := 4 \cdot 10^{-7} \cdot \text{acosh}\left(\frac{d}{2 \cdot r_w}\right) \qquad C := \frac{\pi \cdot 8.854 \cdot 10^{-12}}{\text{acosh}\left(\frac{d}{2 \cdot r_w}\right)}$$

$$\beta(f) := 2 \cdot \pi \cdot f \cdot \sqrt{L \cdot C}$$

$$E_{zs}(x, z, f) := E_{zo} \cdot e^{-j \cdot \beta(f) \cdot \left(x - \frac{d}{2}\right) \cdot \sin(\phi)} \qquad E_{xs}(x, z, f) := E_{xo} \cdot e^{-j \cdot \beta(f) \cdot z \cdot \sin(\theta)}$$

$$Z_o(f) := \sqrt{\frac{(R + j \cdot 2 \cdot \pi \cdot f \cdot L)}{(G + j \cdot 2 \cdot \pi \cdot f \cdot C)}} \qquad \gamma(f) := \sqrt{(R + j \cdot 2 \cdot \pi \cdot f \cdot L) \cdot (G + j \cdot 2 \cdot \pi \cdot f \cdot C)} \qquad K_s(z, f) := E_{zs}(d, z, f) - E_{zs}(0, z, f)$$

$$D(f, Z_o, Z_s, Z_L) := (Z_o(f) \cdot Z_s + Z_o(f) \cdot Z_L) \cdot \cosh(\gamma(f) \cdot l_{th}) + (Z_o(f)^2 + Z_s \cdot Z_L) \cdot \sinh(\gamma(f) \cdot l_{th})$$

$$I_{sL}(f, Z_o, Z_s, Z_L) := 20 \cdot \log\left(\left|\frac{\int_0^{l_{th}} K_s(z, f) \cdot (Z_o(f) \cdot \cosh(\gamma(f) \cdot z) + Z_s \cdot \sinh(\gamma(f) \cdot z))\, dz \ldots + Z_o(f) \cdot \int_0^d E_{xs}(x, 0, f)\, dx \ldots + -(Z_o(f) \cdot \cosh(\gamma(f) \cdot l_{th}) + Z_s \cdot \sinh(\gamma(f) \cdot l_{th})) \cdot \int_0^d E_{xs}(x, l_{th}, f)\, dx}{D(f, Z_o, Z_s, Z_L) \cdot E_{zo}}\right|\right)$$

MATHCAD 6.1 Sidefire response of twin-lead line.

In Mathcad 6.1, the magnitude of the load current divided by the electric field is plotted in dB for a 10 m long lossless line with a characteristic impedance of 302 Ω. The excitation is sidefire to the line in Mathcad 6.1. Three cases are plotted: (i) the source and load impedances both equal to 302 Ω, (ii) the source and load impedances both equal to 1 Ω ≪ 302 Ω, and (iii) the source impedance equal to 10^5 Ω ≫ 302 Ω while the load impedance equal to 1 Ω ≪ 302 Ω.[8] As expected, when the line is electrically short,

[8]Although the source and load impedances are entirely real in this analysis, they can also be complex.

the current magnitude is the greatest when the source and load impedances are both small. A lossless line in free space that is 10 m in length is electrically short for frequencies less than 3 MHz:

$$f \leq \frac{1}{10}\frac{3\times 10^8}{10} = 3\,\text{MHz}$$

The first resonant frequency for any of the plots is greater than 3 MHz. Due to the complexity of even the simplified equations, it can be difficult to predict analytically the effect of the source impedance, load impedance, and incident wave angle on the current through the load. Notice that the first resonant frequency occurs around 8 MHz, corresponding to a wavelength of about 40 m. The frequency corresponding to one-quarter wavelength of the line is

$$f = \frac{v}{\lambda} = \frac{3\times 10^8}{4(10)} \approx 8\,\text{MHz}$$

At 8 MHz, the 10 m line has a length of one-quarter wavelength. With a high-resistance $10^5\ \Omega$ source connected to the transmission line, the other end presents a low resistance similar to the 1 Ω load.

When the source and load are matched to the transmission line, the currents at the source and load are not necessarily zero. There are a few cases where the current at the source and load are zero independent of the source and load impedance. According to (6.84), when $Z_s = Z_o$ with endfire excitation, the current at the load is always zero for lossless lines. Under the conditions specified in the derivation of (6.81) through (6.84) and (6.87) through (6.88), the load and source currents are also zero whenever $\beta l_{th} = 2n\pi$. Generally, matching does not eliminate the noise induced on the line. For a line that is not electrically long, the noise level is reduced by decreasing the length of the line (and reducing the spacing between the conductors of the line). This reduction in the noise level is illustrated in Mathcad 6.2, where the line length is varied from 1 to 10 to 100 m for matched source and load terminations. The induced current decreases as the line length decreases, assuming the line is electrically short. Of course, the shorter the line, the higher the first resonant frequency.

6.14 Susceptibility of Electrically-Long Wire Above a Ground Plane

Referring to Figure 6.20, a lossy conductor that is not necessarily electrically short is above a perfectly conducting large ground plane. When this transmission line is in the presence of electromagnetic fields, the currents at the source and load of the line are given by, respectively,

$$I_s(z=0) = \frac{2hE_{os}^{inc}}{D}\frac{\sin(\beta_x h)}{\beta_x h}\left\{\begin{array}{l} -j\beta_x e_z \int_0^{l_{th}} \{Z_o \cosh[\gamma(l_{th}-z)] + Z_L \sinh[\gamma(l_{th}-z)]\}\, e^{-j\beta_z z} dz \\ +e_x[Z_o\cosh(\gamma l_{th}) + Z_L \sinh(\gamma l_{th}) - Z_o e^{-j\beta_z l_{th}}] \end{array}\right\} \tag{6.89}$$

$$I_s(z=l_{th}) = \frac{2hE_{os}^{inc}}{D}\frac{\sin(\beta_x h)}{\beta_x h}\left[\begin{array}{l} -j\beta_x e_z \int_0^{l_{th}} [Z_o\cosh(\gamma z) + Z_s \sinh(\gamma z)] e^{-j\beta_z z} dz \\ +e_x\{Z_o - [Z_o\cosh(\gamma l_{th}) + Z_s\sinh(\gamma l_{th})] e^{-j\beta_z l_{th}}\} \end{array}\right] \tag{6.90}$$

$Z_s := 302 \quad Z_L := 302 \quad l_{th} := 10 \quad d := 0.0125 \quad r_w := 0.001 \quad \theta := 0 \quad \phi := 0 \quad E_{xo} := 10^{-3} \quad j := \sqrt{-1}$

$x := 40, 40.1 .. 80 \qquad R := 0 \quad G := 0 \quad E_{zo} := 0$

$$f(x) := \left(x + 1 - 10 \cdot \text{floor}\left(\frac{x}{10}\right)\right) \cdot 10^{\text{floor}\left(\frac{x}{10}\right)} \qquad L := 4 \cdot 10^{-7} \cdot \text{acosh}\left(\frac{d}{2 \cdot r_w}\right) \qquad C := \frac{\pi \cdot 8.854 \cdot 10^{-12}}{\text{acosh}\left(\frac{d}{2 \cdot r_w}\right)}$$

$$\beta(f) := 2 \cdot \pi \cdot f \cdot \sqrt{L \cdot C}$$

$$E_{zs}(x, z, f) := E_{zo} \cdot e^{-j \cdot \beta(f) \cdot \left(x - \frac{d}{2}\right) \cdot \sin(\phi)} \qquad E_{xs}(x, z, f) := E_{xo} \cdot e^{-j \cdot \beta(f) \cdot z \cdot \sin(\theta)}$$

$$Z_o(f) := \sqrt{\frac{(R + j \cdot 2 \cdot \pi \cdot f \cdot L)}{(G + j \cdot 2 \cdot \pi \cdot f \cdot C)}} \qquad \gamma(f) := \sqrt{(R + j \cdot 2 \cdot \pi \cdot f \cdot L) \cdot (G + j \cdot 2 \cdot \pi \cdot f \cdot C)} \qquad K_s(z, f) := E_{zs}(d, z, f) - E_{zs}(0, z, f)$$

$$D(f, Z_o, Z_s, Z_L) := (Z_o(f) \cdot Z_s + Z_o(f) \cdot Z_L) \cdot \cosh(\gamma(f) \cdot l_{th}) + (Z_o(f)^2 + Z_s \cdot Z_L) \cdot \sinh(\gamma(f) \cdot l_{th})$$

$$I_{sL}(f, Z_o, Z_s, Z_L, l_{th}) := 20 \cdot \log\left[\left|\frac{\displaystyle\int_0^{l_{th}} K_s(z, f) \cdot (Z_o(f) \cdot \cosh(\gamma(f) \cdot z) + Z_s \cdot \sinh(\gamma(f) \cdot z))\, dz \ldots + Z_o(f) \cdot \int_0^d E_{xs}(x, 0, f)\, dx \ldots + -(Z_o(f) \cdot \cosh(\gamma(f) \cdot l_{th}) + Z_s \cdot \sinh(\gamma(f) \cdot l_{th})) \cdot \int_0^d E_{xs}(x, l_{th}, f)\, dx}{D(f, Z_o, Z_s, Z_L) \cdot E_{xo}}\right|\right]$$

Broadside Excitation

dB: $I_{sL}\left(f(x), Z_o, Z_s, Z_L, \frac{l_{th}}{10}\right)$ (solid), $I_{sL}(f(x), Z_o, Z_s, Z_L, l_{th})$ (dotted), $I_{sL}(f(x), Z_o, Z_s, Z_L, 10 \cdot l_{th})$ (dashed)

Vertical axis: −80, −100, −120, −140, −160, −180

Horizontal axis: $1 \cdot 10^4$, $1 \cdot 10^5$, $1 \cdot 10^6$, $1 \cdot 10^7$, $1 \cdot 10^8$

f(x)
Hz

MATHCAD 6.2 Broadside response of twin-lead line.

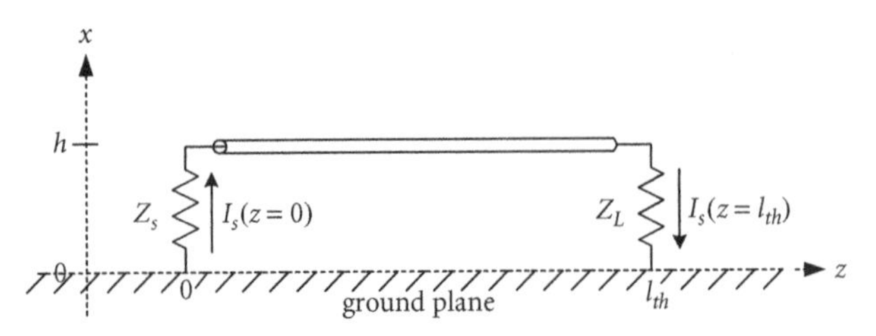

FIGURE 6.20 Conductor above a ground plane illuminated by an electromagnetic field.

where

$$D = (Z_o Z_s + Z_o Z_L)\cosh(\gamma l_{th}) + (Z_o^2 + Z_s Z_L)\sinh(\gamma l_{th})$$

γ = propagation constant of the lossy conductor

Z_o = characteristic impedance of the lossy conductor above a large perfectly-conducting ground plane

$$E_{zs}^{total}(x,z) = -j2E_{os}^{inc} e_z \sin(\beta_x x) e^{-j\beta_z z}$$

$$E_{xs}^{total}(x,z) = 2E_{os}^{inc} e_x \cos(\beta_x x) e^{-j\beta_z z}$$

To obtain some insight into how the current can vary at the two terminations, simplify these complicated expressions by assuming i) the line is lossless, ii) the line is lossless and the incident field is a plane wave polarized in the x direction, and iii) the line is lossless and the incident field is a plane wave polarized in the z direction. [Paul, '94; Smith, '77; DeGauque]

Lossless Line

Because the method of images can be used to derive (6.89) and (6.90), the expressions are similar in form to those for the twin-lead line. For this reason, many of the repetitive steps will not be shown. However, the high-frequency (and lossless) characteristic impedance specified in (6.89) and (6.90) and the corresponding inductance and capacitance per unit length are

$$Z_o = 60\ln\left[\frac{h}{r_w} + \sqrt{\left(\frac{h}{r_w}\right)^2 - 1}\right] \approx 60\ln\left(\frac{2h}{r_w}\right) \quad \text{if } h \gg r_w$$

$$L = 2\times10^{-7}\ln\left[\frac{h+r_w}{r_w} + \sqrt{\left(\frac{h+r_w}{r_w}\right)^2 - 1}\right] \approx 2\times10^{-7}\ln\left(\frac{2h}{r_w}\right) \quad \text{if } h \gg r_w$$

$$C = \frac{2\pi\varepsilon_o}{\ln\left[\frac{h+r_w}{r_w} + \sqrt{\left(\frac{h+r_w}{r_w}\right)^2 - 1}\right]} \approx \frac{2\pi\varepsilon_o}{\ln\left(\frac{2h}{r_w}\right)} \quad \text{if } h \gg r_w$$

where r_w is the radius of the conductor. The lossless versions of (6.89) and (6.90) are

$$I_s(z=0) = \frac{2hE_{os}^{inc}}{D}\frac{\sin(\beta_x h)}{\beta_x h}\left[\begin{array}{l} -j\beta_x e_z \int_0^{l_{th}} \{Z_o\cos[\beta(l_{th}-z)] + jZ_L\sin[\beta(l_{th}-z)]\}\, e^{-j\beta_z z}dz \\ +e_x\left[Z_o\cos(\beta l_{th}) + jZ_L\sin(\beta l_{th}) - Z_o e^{-j\beta_z l_{th}}\right] \end{array}\right] \tag{6.91}$$

$$I_s(z=l_{th}) = \frac{2hE_{os}^{inc}}{D}\frac{\sin(\beta_x h)}{\beta_x h}\left[\begin{array}{l} -j\beta_x e_z \int_0^{l_{th}} [Z_o\cos(\beta z) + jZ_s\sin(\beta z)]\, e^{-j\beta_z z}dz \\ +e_x\left\{Z_o - [Z_o\cos(\beta l_{th}) + jZ_s\sin(\beta l_{th})]e^{-j\beta_z l_{th}}\right\} \end{array}\right] \tag{6.92}$$

where $D = (Z_o Z_s + Z_o Z_L)\cos(\beta l_{th}) + j(Z_o^2 + Z_s Z_L)\sin(\beta l_{th})$.

Lossless Line and Plane Wave Polarized in the x Direction

Before the simplified current expressions are obtained for this special case, the expressions and relevant notation given for the total electric field above the ground plane

$$E_{zs}^{total}(x,z) = -j2E_{os}^{inc}e_z \sin(\beta_x x)e^{-j\beta_z z}, \; E_{xs}^{total}(x,z) = 2E_{os}^{inc}e_x \cos(\beta_x x)e^{-j\beta_z z} \tag{6.93}$$

must be explained. When an electric field is incident to a ground plane, a wave is reflected off the plane. The total electric field, E_s^{total}, is the sum of the incident and reflected fields. The current expressions (6.89) and (6.90) have assumed the total electric fields are of the form given in (6.93), which include both the incident and reflected terms. The strength of the incident electric field is $|E_{os}^{inc}|$. Also, in the derivation of these total-field expressions, it was assumed that the incident field was of the form

$$\vec{E}_{is} = E_{os}^{inc}\left(e_x^{inc}\hat{a}_x + e_y^{inc}\hat{a}_y + e_z^{inc}\hat{a}_z\right)e^{-j\beta_x x}e^{-j\beta_y y}e^{-j\beta_z z} \tag{6.94}$$

When an individual (including the author as a studious and underpaid graduate student) is first exposed to an equation of the form in (6.94), confusion is common. To clear this confusion, energy must be devoted to understanding this description of a general electromagnetic wave. An electromagnetic wave is described by i) the strength and phase of its electric field (or magnetic field), ii) the orientation or polarization of its electric field, and iii) the direction of its power, via the Poynting vector (or propagation vector). Two waves with the same electric field amplitude could both be traveling in the $+z$ direction. However, the electric field of one wave could be in the x direction while the electric field of the other wave could be in the y direction. The direction, polarization, and, of course, strength of an electromagnetic wave can dramatically affect the current induced on a transmission line.

First, the strength or magnitude of the electric field will be discussed. Using the previous notation, but dropping the *inc* superscript, note that since the magnitude of a complex exponential is one,

$$\left|e^{-j\beta_x x}e^{-j\beta_y y}e^{-j\beta_z z}\right| = \left|e^{-j(\beta_x x+\beta_y y+\beta_z z)}\right| = 1$$

$(e^{jx} = \cos x + j\sin x)$, it follows that

$$\left|\vec{E}_{is}\right| = \left|E_{os}e_x\hat{a}_x + E_{os}e_y\hat{a}_y + E_{os}e_z\hat{a}_z\right| = \sqrt{(E_{os}e_x)^2 + (E_{os}e_y)^2 + (E_{os}e_z)^2}$$

It should be clear that the products $|E_{os}e_x|$, $|E_{os}e_y|$, and $|E_{os}e_z|$ are the strengths of the electric field in the x, y, and z directions, respectively. The relative strengths in the three Cartesian coordinate directions are determined by the size of the e's (see Equation (6.96)). Furthermore, it will be soon possible to show that $e_x^2 + e_y^2 + e_z^2 = 1$ so

$$\left|\vec{E}_{is}\right| = |E_{os}|\sqrt{e_x^2 + e_y^2 + e_z^2} = |E_{os}| \tag{6.95}$$

Therefore, the strength of the electric field of the wave is given by $|E_{os}|$. The magnitude operation is required since E_{os} is a phasor with both an amplitude and a phase.

Second, the direction of polarization and its relationship to the e's will be discussed. In Figure 6.21, the Poynting vector (directed toward the origin) and electric field are shown. The direction of the Poynting vector can be described by two spherical coordinate-like angles given as θ_p and ϕ_p. The angle from the x axis to the Poynting vector is given by θ_p, while the angle from the y axis to the projection of the Poynting vector onto the yz plane is given by ϕ_p. The angle θ_E describes the polarization or orientation of the electric field. The angle the electric field vector makes with the unit vector $\hat{a}_\phi$ in the direction of the

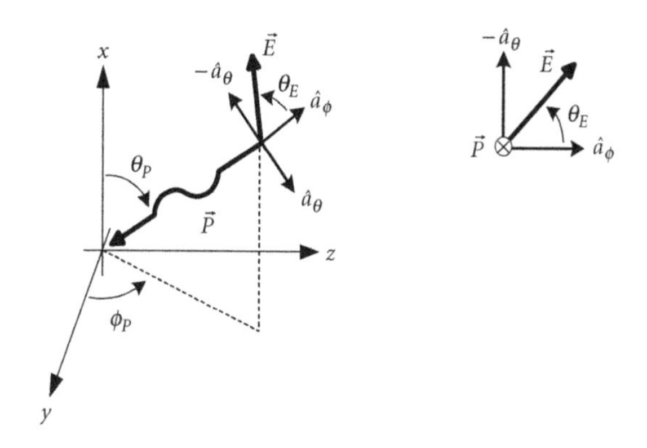

FIGURE 6.21 Angles used to describe both the direction and polarization of a wave directed toward the origin.

unit vector $-\hat{a}_\theta$ is equal to θ_E (these unit vectors are defined at the point of interest). It can be shown that the projections of the electric field vector, $\vec{E}$, onto the three axes are equal to

$$\begin{aligned} e_x &= \sin\theta_E \sin\theta_P \\ e_y &= -\sin\theta_E \cos\theta_P \cos\phi_P - \cos\theta_E \sin\phi_P \\ e_z &= -\sin\theta_E \cos\theta_P \sin\phi_P + \cos\theta_E \cos\phi_P \end{aligned} \tag{6.96}$$

As a simple check on these expressions, if $\theta_E = 0°$, $\theta_P = 90°$, and $\phi_P = 45°$, then

$$e_x = 0,\ e_y = -\frac{1}{\sqrt{2}},\ e_z = \frac{1}{\sqrt{2}}$$

The electric field is in the yz plane and directed equally in the $-y$ and $+z$ directions as shown in Figure 6.22. As another check, if $\theta_E = -90°$, $\theta_P = 45°$, and $\phi_P = 90°$, then

$$e_x = -\frac{1}{\sqrt{2}},\ e_y = 0,\ e_z = \frac{1}{\sqrt{2}}$$

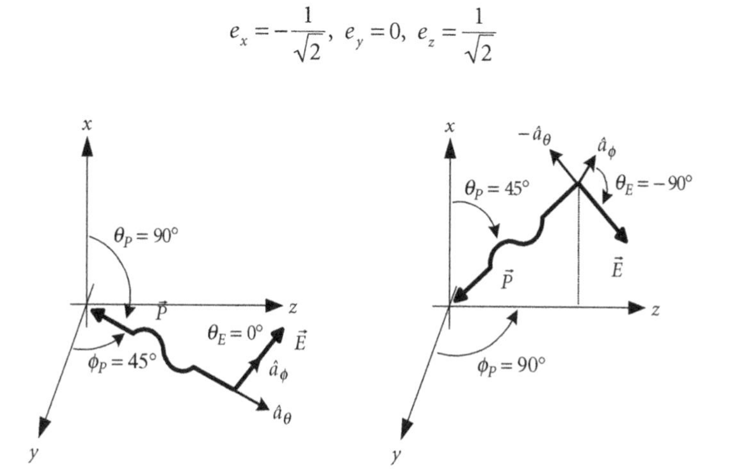

FIGURE 6.22 Two examples of the use of the Poynting vector and polarization angles.

The electric field is in the *xz* plane and directed equally in the $-x$ and $+z$ directions as shown in Figure 6.22.

Finally, the direction of the Poynting vector will be discussed. The relationships between the phase constants and $\beta = \omega\sqrt{\mu\varepsilon}$ for the homogeneous material between the conductor and ground plane are

$$\beta_x = -\beta\cos\theta_P,\ \beta_y = -\beta\sin\theta_P\cos\phi_P,\ \beta_z = -\beta\sin\theta_P\sin\phi_P \tag{6.97}$$

These three components of the phase constant are not a function of the θ_E since this angle describes the orientation of the electric field. Using the same checks, if $\theta_P = 90°$ and $\phi_P = 45°$,

$$\beta_x = 0,\ \beta_y = -\frac{\beta}{\sqrt{2}},\ \beta_z = -\frac{\beta}{\sqrt{2}}$$

The Poynting vector is in the *yz* plane and directed equally in the $-y$ and $-z$ directions. If $\theta_P = 45°$ and $\phi_P = 90°$, then

$$\beta_x = -\frac{\beta}{\sqrt{2}},\ \beta_y = 0,\ \beta_z = -\frac{\beta}{\sqrt{2}}$$

The Poynting vector is in the *xz* plane and directed equally in the $-x$ and $-z$ directions. The phase constants are negative in these examples since the wave is directed toward the origin. When they are placed into (6.94), the arguments of the complex exponentials are positive. A positive argument is reasonable since a wave traveling entirely in the $-z$ direction toward the origin would have $\beta_x = 0$, $\beta_y = 0$, and $\beta_z = -\beta$ and

$$\vec{E} = E_{os}(e_x\hat{a}_x + e_y\hat{a}_y)\, e^{j\beta z}$$

The positive $j\beta$ implies a wave traveling in the $-z$ direction. A wave traveling entirely in the $+y$ direction away from the origin would be described with a negative argument in the complex exponential (with $\beta > 0$):

$$\vec{E} = E_{os}(e_x\hat{a}_x + e_z\hat{a}_z)e^{-j\beta y}$$

To summarize, the strength of the electric field is determined by the magnitude of E_{os}, the orientation of the electric field is determined by the *e*'s, and the direction of the wave is determined by the complex exponentials.

Now with the proper background, the relationship between these angles and the angle θ shown in Figure 6.17 can be quickly determined. The angle θ is just a special case of that shown in Figure 6.21: $\theta_E = 90°$, $\theta_P = 90°$, and $\phi_P = -\theta$. Substituting into (6.96), (6.97), (6.94), and setting $y = 0$

$$e_x = 1,\ e_y = 0,\ e_z = 0,\ \beta_x = 0,\ \beta_y = -\beta\cos\theta,\ \beta_z = \beta\sin\theta$$

$$\vec{E}_{is} = E_{os}e^{j(\beta\cos\theta)y}e^{-j(\beta\sin\theta)z}\hat{a}_x\Big|_{y=0} = E_{os}e^{-j(\beta\sin\theta)z}\hat{a}_x \tag{6.98}$$

After setting $z = l_{th}$, (6.98) is equal to (6.76) The relationship between these angles and the angle ϕ shown in Figure 6.19 can also be quickly determined: $\theta_E = 0°$, $\theta_P + 90° = \phi$, and $\phi_P = 0°$. Substituting into (6.96),

(6.97), (6.94), and setting $y = 0$

$$e_x = 0,\ e_y = 0,\ e_z = 1$$

$$\beta_x = -\beta\cos(\phi - 90°) = -\beta\sin\phi,\ \beta_y = -\beta\sin(\phi - 90°) = \beta\cos\phi,\ \beta_z = 0$$

$$\vec{E}_{is} = E_{os}e^{j(\beta\sin\phi)x}e^{-j(\beta\cos\phi)y}\hat{a}_z\Big|_{y=0} = E_{os}e^{j(\beta\sin\phi)x}\hat{a}_z \tag{6.99}$$

With a shift in the zero-phase reference, (6.99) is equivalent to (6.86).

For a plane wave polarized in the *x* direction, $e_x = 1$, $e_y = 0$, $e_z = 0$, $\beta_x = 0$ and (6.91) and (6.92) reduce to

$$I_s(z=0) = \frac{2hE_{os}^{inc}}{D}\left[Z_o\cos(\beta l_{th}) + jZ_L\sin(\beta l_{th}) - Z_o e^{-j\beta_z l_{th}}\right] \tag{6.100}$$

$$I_s(z=l_{th}) = \frac{2hE_{os}^{inc}}{D}\left\{Z_o - [Z_o\cos(\beta l_{th}) + jZ_s\sin(\beta l_{th})]e^{-j\beta_z l_{th}}\right\} \tag{6.101}$$

The relationship $\lim_{x\to 0}[\sin(x)/x] = 1$ was used in this simplification. The broadside and endfire cases will now be given. For the broadside case, the *x*-polarized wave is traveling entirely in the –*y* direction and $\theta_E = 90°$, $\theta_P = 90°$, and $\phi_P = 0°$. Therefore,

$$e_x = 1,\ e_y = 0,\ e_z = 0,\ \beta_x = 0,\ \beta_y = -\beta,\ \beta_z = 0$$

and (6.100) and (6.101) reduce to

$$I_s(z=0) = \frac{2h}{D}E_{os}^{inc}\{Z_o[\cos(\beta l_{th}) - 1] + jZ_L\sin(\beta l_{th})\} \tag{6.102}$$

$$I_s(z=l_{th}) = \frac{2h}{D}E_{os}^{inc}\{Z_o[1 - \cos(\beta l_{th})] - jZ_s\sin(\beta l_{th})\} \tag{6.103}$$

Also, from (6.93)

$$E_{zs}^{total}(x,z) = 0,\ E_{xs}^{total}(x,z) = 2E_{os}^{inc} \tag{6.104}$$

The electric field, which is entirely in the *x* direction, is twice the incident and constant in magnitude and phase along the entire length of the conductor, which it should be for broadside excitation. When the conductor is electrically short, the current expressions further reduce to

$$I_s(z=0) \approx \frac{2h}{D}E_{os}^{inc}\left\{Z_o\left[\left(1 - \frac{\beta^2 l_{th}^2}{2}\right) - 1\right] + jZ_L\beta l_{th}\right\} = E_{os}^{inc}\frac{2h\beta l_{th}}{D}\left(-Z_o\frac{\beta l_{th}}{2} + jZ_L\right) \tag{6.105}$$

$$I_s(z=l_{th}) \approx \frac{2h}{D}E_{os}^{inc}\left\{Z_o\left[1 - \left(1 - \frac{\beta^2 l_{th}^2}{2}\right)\right] - jZ_s\beta l_{th}\right\} = E_{os}^{inc}\frac{2h\beta l_{th}}{D}\left(Z_o\frac{\beta l_{th}}{2} - jZ_s\right) \tag{6.106}$$

where $D \approx (Z_o Z_s + Z_o Z_L)[1-(\beta^2 l_{th}^2/2)] + j(Z_o^2 + Z_s Z_L)\beta l_{th}$. When both terminations are short circuits, sometimes referred to as a symmetrical connection,

$$I_s(z=0) = jE_{os}^{inc}\frac{h\beta l_{th}}{Z_o} \quad \text{if } Z_s = Z_L = 0 \tag{6.107}$$

$$I_s(z=l_{th}) = -jE_{os}^{inc}\frac{h\beta l_{th}}{Z_o} \quad \text{if } Z_s = Z_L = 0 \tag{6.108}$$

These results are sometimes useful since this scenario represents the short circuiting of the shield of a cable to ground on both ends of the cable. In this case, Z_o is not the characteristic impedance between the inner conductor and shield of the cable but the characteristic impedance between the outer shield and ground plane. Equations (6.107) and (6.108) are equal but of opposite sign. Referring to Figure 6.9, for an electrically-short line, the incident electric field partially determines the strength of the source of these currents:

$$I_{ds}\Delta x = -j\omega C h l_{th} E_{os}^{inc} = -j\left(\frac{\beta}{\sqrt{LC}}\right) C h l_{th} E_{os}^{inc} = -j\beta\sqrt{\frac{C}{L}} h l_{th} E_{os}^{inc} = -jE_{os}^{inc}\frac{h\beta l_{th}}{Z_o}$$

Since the magnetic field for broadside excitation is parallel to the surface generated by the conductors, V_{ds} in Figure 6.9 is zero. When the source is an open circuit and load is a short circuit, sometimes referred to as an asymmetrical connection,

$$I_s(z=0) = 0 \quad \text{if } Z_s = \infty,\ Z_L = 0 \tag{6.109}$$

$$I_s(z=l_{th}) = -jE_{os}^{inc}\frac{2h\beta l_{th}}{Z_o} \quad \text{if } Z_s = \infty,\ Z_L = 0 \tag{6.110}$$

(The magnitude of the current through the load is double that of the symmetrical connection.) This result is sometimes useful since this scenario represents the shorting of the shield of a cable to ground at only one end. At low frequencies, cables are often grounded at only one end to help prevent troublesome ground loops.

For the endfire case, the x-polarized wave is traveling in entirely the $+z$ direction and $\theta_E = 90°$, $\theta_P = 90°$, $\phi_P = -90°$ and

$$e_x = 1,\ e_y = 0,\ e_z = 0,\ \beta_x = 0,\ \beta_y = 0,\ \beta_z = \beta$$

Equations (6.100), (6.101), and (6.93) reduce to

$$\begin{aligned} I_s(z=0) &= \frac{2hE_{os}^{inc}}{D}\left[Z_o\cos(\beta l_{th}) + jZ_L\sin(\beta l_{th}) - Z_o e^{-j\beta l_{th}}\right] \\ &= j\frac{2hE_{os}^{inc}}{D}\sin(\beta l_{th})(Z_L - Z_o) \end{aligned} \tag{6.111}$$

$$\begin{aligned} I_s(z=l_{th}) &= \frac{2hE_{os}^{inc}}{D}\left\{Z_o - [Z_o\cos(\beta l_{th}) + jZ_s\sin(\beta l_{th})]e^{-j\beta l_{th}}\right\} \\ &= \frac{hE_{os}^{inc}}{D}(Z_s - Z_o)[\cos(2\beta l_{th}) - 1 - j\sin(2\beta l_{th})] \end{aligned} \tag{6.112}$$

$$E_{zs}^{total}(x,z) = 0,\ E_{xs}^{total}(x,z) = 2E_{os}^{inc}e^{-j\beta z} \tag{6.113}$$

where $e^{-j\beta l_{th}} = \cos(\beta l_{th}) - j\sin(\beta l_{th})$. The electric field, which is entirely in the x direction, is twice the incident but varies in phase along the length of the conductor. For endfire excitation, it takes time for the signal to travel along the length of the conductor. When the conductor is electrically short,

$$I_s(z=0) \approx j\frac{2hE_{os}^{inc}}{D}\beta l_{th}(Z_L - Z_o) \tag{6.114}$$

$$\begin{aligned} I_s(z=l_{th}) &\approx \frac{hE_{os}^{inc}}{D}(Z_s - Z_o)\left[\left(1-\frac{4\beta^2 l_{th}^2}{2}\right) - 1 - j2\beta l_{th}\right] \\ &= -E_{os}^{inc}\frac{2h\beta l_{th}}{D}(Z_s - Z_o)(\beta l_{th} + j) \end{aligned} \tag{6.115}$$

where $D \approx (Z_o Z_s + Z_o Z_L)[1-(\beta^2 l_{th}^2/2)] + j(Z_o^2 + Z_s Z_L)\beta l_{th}$. When both the source and load are short circuits,

$$I_s(z=0) = -E_{os}^{inc}\frac{2h}{Z_o} \quad \text{if } Z_s = Z_L = 0 \tag{6.116}$$

$$I_s(z=l_{th}) = E_{os}^{inc}\frac{2h}{Z_o}(1 - j\beta l_{th}) \quad \text{if } Z_s = Z_L = 0 \tag{6.117}$$

while when the source is an open circuit and load is a short circuit,

$$I_s(z=0) = 0 \quad \text{if } Z_s = \infty,\ Z_L = 0 \tag{6.118}$$

$$I_s(z=l_{th}) = -E_{os}^{inc}\frac{2h\beta l_{th}}{Z_o}(\beta l_{th} + j) \quad \text{if } Z_s = \infty,\ Z_L = 0 \tag{6.119}$$

Comparing the expressions for the induced current in the source or load for an electrically-short conductor above a ground plane when both the source and load are short circuits, it is clear that endfire excitation currents, (6.116) and (6.117), are much greater than the broadside excitations currents, (6.107) and (6.108):

$$\left|E_{os}^{inc}\right|\frac{2h}{Z_o} \gg \left|E_{os}^{inc}\right|\frac{h\beta l_{th}}{Z_o} \quad \text{if } Z_s = Z_L = 0$$

since $\beta l_{th} \ll 1$. With endfire excitation, there is both a phase difference of the electric field along the length of the conductor and a magnetic field normal to the plane generated by the conductor and ground plane directly below it. In "two shakes of a cow's tail," it will be shown that the induced current magnitude for sidefire excitation is similar to endfire excitation for electrically-short lines with short-circuit terminations.

Lossless Line and Plane Wave Polarized in the z Direction

For a plane wave polarized in the z direction, $e_x = 0$, $e_y = 0$, $e_z = 1$, $\beta_z = 0$ and (6.91) and (6.92) reduce to

$$I_s(z=0) = \frac{2hE_{os}^{inc}}{D}\frac{\sin(\beta_x h)}{\beta_x h}\left[-j\beta_x\int_0^{l_{th}}\{Z_o\cos[\beta(l_{th}-z)] + jZ_L\sin[\beta(l_{th}-z)]\}\,dz\right] \tag{6.120}$$

$$I_s(z=l_{th}) = \frac{2hE_{os}^{inc}}{D}\frac{\sin(\beta_x h)}{\beta_x h}\left\{-j\beta_x\int_0^{l_{th}}[Z_o\cos(\beta z) + jZ_s\sin(\beta z)]\,dz\right\} \tag{6.121}$$

For sidefire excitation, $\theta_E = 0°$, $\theta_P = 180°$, and $\phi_P = 0°$,

$$e_x = 0,\ e_y = 0,\ e_z = 1,\ \beta_x = \beta,\ \beta_y = 0,\ \beta_z = 0$$

and

$$I_s(z=0) = -\frac{2hE_{os}^{inc}}{D}\frac{\sin(\beta h)}{\beta h} j\beta \int_0^{l_{th}} \{Z_o \cos[\beta(l_{th}-z)] + jZ_L \sin[\beta(l_{th}-z)]\} dz \tag{6.122}$$

$$I_s(z=l_{th}) = -\frac{2hE_{os}^{inc}}{D}\frac{\sin(\beta h)}{\beta h} j\beta \int_0^{l_{th}} [Z_o \cos(\beta z) + jZ_s \sin(\beta z)] dz \tag{6.123}$$

When the conductor is electrically short,

$$I_s(z=0) \approx -\frac{2hE_{os}^{inc}}{D}\frac{\sin(\beta h)}{\beta h} j\beta \int_0^{l_{th}} \left\{ Z_o\left[1-\frac{\beta^2(l_{th}-z)^2}{2}\right] + jZ_L\beta(l_{th}-z)\right\} dz \tag{6.124}$$

$$I_s(z=l_{th}) \approx -\frac{2hE_{os}^{inc}}{D}\frac{\sin(\beta h)}{\beta h} j\beta \int_0^{l_{th}} \left[Z_o\left(1-\frac{\beta^2 z^2}{2}\right) + jZ_s\beta z\right] dz \tag{6.125}$$

where $D \approx (Z_o Z_s + Z_o Z_L)[1-(\beta^2 l_{th}^2/2)] + j(Z_o^2 + Z_s Z_L)\beta l_{th}$. When both the source and load are short circuits,

$$\begin{aligned} I_s(z=0) &= -\frac{2hE_{os}^{inc}}{Z_o l_{th}}\frac{\sin(\beta h)}{\beta h} \int_0^{l_{th}} \left[1-\frac{\beta^2(l_{th}-z)^2}{2}\right] dz \\ &\approx -E_{os}^{inc}\frac{2\sin(\beta h)}{\beta Z_o} \approx -E_{os}^{inc}\frac{2h}{Z_o} \quad \text{if } Z_s = Z_L = 0 \end{aligned} \tag{6.126}$$

$$\begin{aligned} I_s(z=l_{th}) &= -\frac{2hE_{os}^{inc}}{Z_o l_{th}}\frac{\sin(\beta h)}{\beta h} \int_0^{l_{th}} \left(1-\frac{\beta^2 z^2}{2}\right) dz \\ &\approx -E_{os}^{inc}\frac{2\sin(\beta h)}{\beta Z_o} \approx -E_{os}^{inc}\frac{2h}{Z_o} \quad \text{if } Z_s = Z_L = 0 \end{aligned} \tag{6.127}$$

while when the source is an open circuit and load is a short circuit,

$$I_s(z=0) \approx 0 \quad \text{if } Z_s = \infty,\ Z_L = 0 \tag{6.128}$$

$$I_s(z=l_{th}) \approx E_{os}^{inc}\frac{\beta l_{th}^2 \sin(\beta h)}{Z_o} \approx E_{os}^{inc}\frac{h\beta^2 l_{th}^2}{Z_o} \quad \text{if } Z_s = \infty,\ Z_L = 0 \tag{6.129}$$

6.15 Theory of Current Probes

Use Maxwell's equations to explain how current can be measured using a current probe. [Heller; Heumann; Radun; Anderson, '71; Marshall, '96; Klein; Destefan; Ramboz; Karrer]

It seems as though most electrical engineering graduates believe that to measure a current through an element the current must be measured either intrusively by inserting an ammeter in series with the

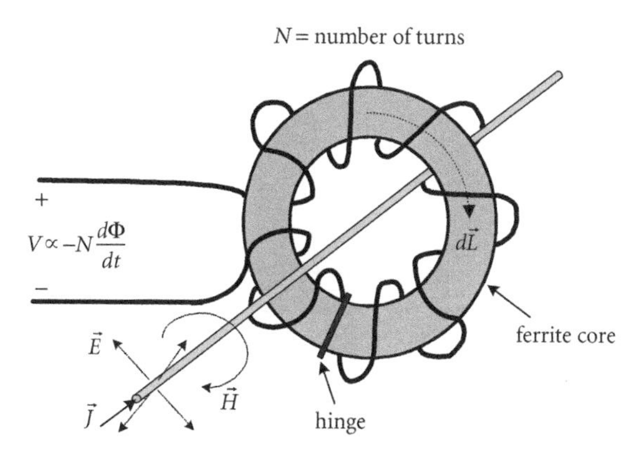

FIGURE 6.23 Induced voltage across the coil is a measure of the current passing through the window of the probe.

element or indirectly by measuring the voltage across a current-sensing resistor in series with the element. Actually, the current can be measured without any direct ohmic contact with the current-carrying wire by using a current probe, assuming the current is time varying.[9] Since it is not always convenient or possible to make direct electrical contact with the current-carrying wire, a current probe can be invaluable. A current probe can be used to determine the current magnitude in a wire or net current through a bundle of wires without breaking the circuit. Generally, the emissions level from these conductors increase with the strength of this net current. A current probe can also be employed to monitor fault, ripple, transformer, weld, and motor transient currents.

A current probe measures the time-varying current in a wire by measuring the voltage induced across a flux-collecting coil. This coil picks up the magnetic field generated by the time-varying current. One form of a current probe, shown in Figure 6.23, is a coil wrapped around a hinged toroidal core constructed of ferrite. The core is clamped around the conductor(s). Because the current-carrying wire is passed straight through the center or "window" of the probe, this probe is sometimes referred to as a window probe.

By examining the theoretical foundation of the current probe, the limitations of the commonly used expressions related to it can be clearly seen. Furthermore, Lenz's law can be introduced. Imagine that one or more wires carrying time-varying currents are surrounded by the probe. The net current and the corresponding time-varying fields they generate are related through one of Maxwell's equations (also referred to as Ampère's law):

$$\nabla \times \vec{H} = \vec{J} + \frac{d\vec{D}}{dt}$$

where $\vec{H}$ is the magnetic field, $\vec{J}$ is the current density, and $\vec{D}$ is the electric flux density. By using Stokes' theorem, this equation can be rewritten in its standard integral form:

$$\iint (\nabla \times \vec{H}) \cdot d\vec{s} = \iint \vec{J} \cdot d\vec{s} + \iint \frac{d\vec{D}}{dt} \cdot d\vec{s}$$

$$\oint \vec{H} \cdot d\vec{L} = \iint \vec{J} \cdot d\vec{s} + \frac{d}{dt}\varepsilon \iint \vec{E} \cdot d\vec{s} \tag{6.130}$$

[9]If the current is dc or not time varying, then a Hall effect sensor can be used to measure the current.

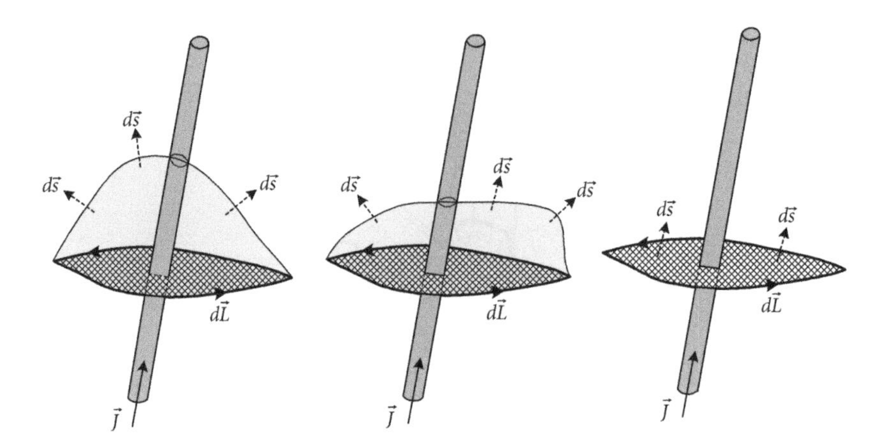

FIGURE 6.24 Identical closed line paths with different open surfaces. (Since these surfaces are open, the cross-hatched regions shown for the first two illustrations are not part of the surface.)

The line integral on the left side of (6.130) is the summation of the magnetic fields generated by the one or more wires along a closed path surrounding these wires. The direction and position of the closed path is given by $d\vec{L}$ and the corresponding limits of integration. This closed path generates many possible open surfaces.[10] The directions of these surfaces are given by the surface integration vector, $d\vec{s}$. When the surface is flat, then the direction of $d\vec{s}$ is in one direction and is easily obtained using the right-hand rule: if the fingers of the right hand are curled in the direction of $d\vec{L}$, then the thumb is the direction of the surface vector, $d\vec{s}$. In Figure 6.24, one closed path is shown completely enclosing a single current-carrying wire. Three different surfaces are shown with the same closed path.

Even though any open surface with the same closed path will yield the same result, by selecting the "right" closed path and corresponding open surface, the second term on the right side of (6.130) can sometimes be set to zero by inspection. For example, suppose the current-carrying wire is in the z direction, and the closed path surrounding this wire is entirely in the xy plane as shown in Figure 6.25. For the given direction of the closed path integration, the surface integration vector is entirely in the positive z direction. If the electric field is entirely in the x and y directions, then (6.130) reduces to

$$\oint \vec{H}\cdot(dx\hat{a}_x + dy\hat{a}_y) = \iint J_o\hat{a}_z\cdot dxdy\hat{a}_z + \frac{d}{dt}\varepsilon\iint (E_x\hat{a}_x + E_y\hat{a}_y)\cdot dxdy\hat{a}_z = \iint J_o dxdy + 0$$

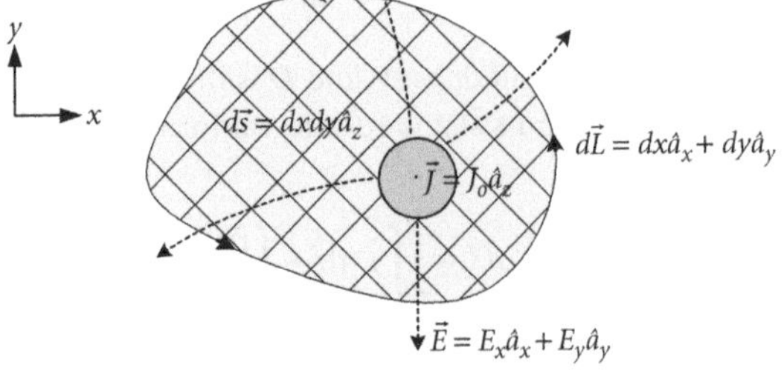

FIGURE 6.25 Current-carrying wire in the z direction with a closed path entirely in the xy plane.

[10]Imagine that the path is the fixed, closed wire loop of a bubble wand. If a forming bubble on the wand represents the open surface, it is clear that many possible bubbles have the same wand loop shape and dimensions.

since $\hat{a}_x \cdot \hat{a}_z = 0$ and $\hat{a}_y \cdot \hat{a}_z = 0$. When evaluating (6.130), the closed integration path can sometimes be selected so that its surface vector, $d\vec{s}$ is entirely or mostly perpendicular to the electric field. The fields for many low-loss transmission lines are mostly TEM (both the electric and magnetic fields are transverse to the direction of propagation).

For some situations, (6.130) reduces to a simplified form of Ampère's law

$$\oint \vec{H} \cdot d\vec{L} \approx \iint \vec{J} \cdot d\vec{s} = I_{encl} \quad \text{if} \iint \vec{J} \cdot d\vec{s} \gg \frac{d}{dt} \varepsilon \iint \vec{E} \cdot d\vec{s} \tag{6.131}$$

where I_{encl} is the total current enclosed by the closed contour. The time-derivative integral term, which is neglected in this expression, is referred to as the displacement current. It is not necessarily zero unless the total time-varying electric field summed over the surface is zero. It is approximately zero for magnetoquasistatic (MQS) fields.[11] The current probe is partially based on (6.131). If the magnetic field can be measured around any closed path enclosing the current to be measured, the net current can be determined. The total magnetic field can be measured by using a coil with one or more turns.

If the current in the wire is varying with time, then the magnetic field surrounding the wire is also varying with time. Another of Maxwell's equations, referred to as Faraday's law,

$$\nabla \times \vec{E} = -\frac{d\vec{B}}{dt} \tag{6.132}$$

can be used, in its integral form, to show how to convert this time-varying magnetic field to a voltage. Integrating both sides of (6.132) with respect to an open surface, and then using Stokes' theorem to convert one of the integrals to a closed line integral,

$$\iint (\nabla \times \vec{E}) \cdot d\vec{s} = -\iint \frac{d\vec{B}}{dt} \cdot d\vec{s}$$

$$\oint \vec{E} \cdot d\vec{L} = -\frac{d}{dt} \iint \vec{B} \cdot d\vec{s} \tag{6.133}$$

By taking *d*/*dt* past the integral, it is assumed that the surface of interest (i.e., the pickup coil) does not change with time or is not moving. Integral expression (6.133) can be applied to a pickup loop to show how a magnetic field can be converted to a voltage. To simplify matters, the pickup loop consists of a single-turn loop connected to two leads as shown in Figure 6.26. The time-varying magnetic flux density, $\vec{B}$, passes through both the pickup loop and leads connecting to an ideal voltage measurement instrument.

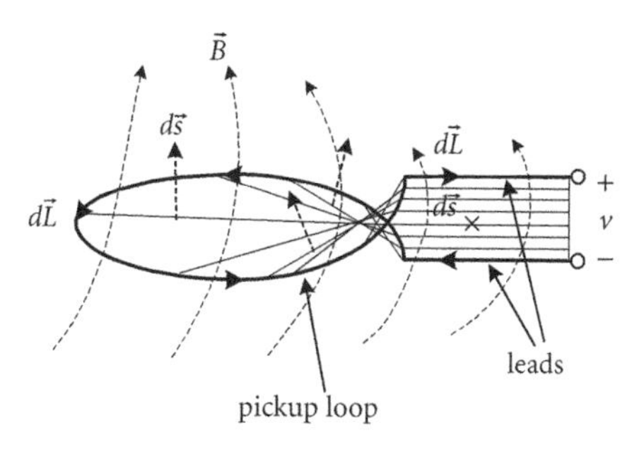

FIGURE 6.26 Single-turn pickup loop used to convert a time-varying magnetic field to a voltage, *v*.

[11]The current probe is sometimes enclosed in a metallic shield to reduce this electric field term. This electric field shield is not continuous (i.e., it is broken in one or more places). If it were continuous, the induced current in this shorted-turn shield could generate a significant counter magnetic field.

The closed line integral corresponding to $d\vec{L}$ is taken along the path of the loop and test leads as shown. The resistance of the loop and leads is assumed negligible so that $\vec{E}$ along this portion of the path along the conducting wire is zero. Hence, the only nonzero value for $\vec{E} \cdot d\vec{L}$ is at the open circuit where the voltage, v, is measured:

$$v = -\frac{d}{dt}\iint \vec{B} \cdot d\vec{s} \tag{6.134}$$

This induced voltage, v, is a measure of the total magnetic flux density passing through both the pickup loop and leads connecting the loop to the instrument:

$$v = -\frac{d}{dt}\iint_{\text{loop}} \vec{B} \cdot d\vec{s} - \frac{d}{dt}\iint_{\text{leads}} \vec{B} \cdot d\vec{s} \tag{6.135}$$

The open surface has the perimeter of both the loop and lead conductors. For the situation shown in Figure 6.26, $d\vec{s}$ is not in one direction. It is also easy to forget the leads portion of the surface. In some cases, the field is mostly parallel to the surface between the leads and then the second integral in (6.135) is small. Often, the lead length, orientation, and sometimes even spacing are not fixed. Also, the plane or surface generated by the leads may not be parallel to the magnetic field. If the relationship between the output voltage and total "input" flux is to be predictable, the flux collected by the leads should be known or at least small. By reducing the area between the leads (i.e., decreasing the spacing between or length of the leads) or twisting the leads, the fraction of the flux collected by the leads compared to that collected by the intended pickup loop can be made negligible. In the remaining analysis, the magnetic flux picked up by the leads will be neglected. Nevertheless, the total flux density integrated or summed over any open surface is referred to as the total flux, Φ. Therefore,

$$v = -\frac{d\Phi}{dt} \tag{6.136}$$

This Φ is the same flux used to determine inductance. Equation (6.136) is referred to as Faraday's law. It is the basis for motors, generators, transformers, and current probes.

Before an N-turn loop is analyzed, Faraday's law will be further studied. As stated, the major consequence of this relationship is that a net nonzero, time-varying magnetic flux passing through a loop will generate a voltage (also referred to as an electromotive force or emf) across the loop. This voltage generates or induces a current in a loop that is not open circuited. This induced current is due to the forces on the electrons in the conducting loop from this external field. The induced voltage is not at a fixed single location in the loop, which is a strange concept for those readers with only a circuits background. However, by opening the loop anywhere and placing a voltmeter across this opening, this emf can be measured. Even when the loop is perfectly conducting, this emf is present. This voltage is accounted for in circuits through inductive elements. (The single loop is a simple single-turn inductor.) Even when an inductor is perfectly conducting with no ohmic losses, a voltage exists across the inductor. However, the current through the inductor must be time varying:

$$v = L\frac{di}{dt}$$

If the current is not varying with time, then the voltage across the inductor is zero. Faraday's law also indicates that the flux must also be time varying to generate a nonzero voltage.

The negative sign in (6.136) indicates that the induced magnetic field generated by the current in the loop tends to oppose any *change* in the field contained within the loop. (This is referred to as Lenz's law.) This is consistent with the circuit's concept that inductors resist change in current. The key to determining the actual direction of the current induced in a loop at a particular instant in time or the sign of the voltage

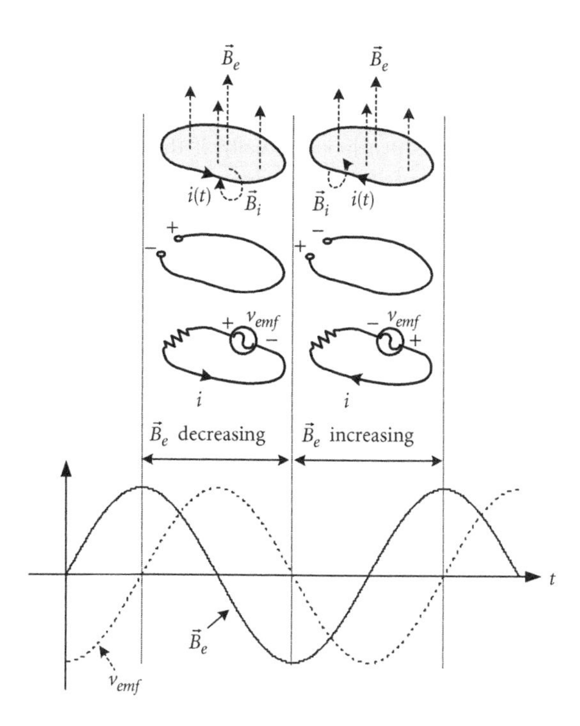

FIGURE 6.27 Direction of the induced current in a closed loop and the sign of the induced voltage in an opened loop as a function of the magnetic flux density.

across an open in the loop at a particular instant in time, is realizing that the induced magnetic field opposes *change*. As illustrated in Figure 6.27, when the external or applied magnetic flux density, $\vec{B}_e$, is decreasing with time (i.e., a negative slope), then the current direction shown will generate a magnetic field directed upward through the loop. This induced magnetic flux density attempts to keep the total flux within the loop constant. When the external magnetic field is increasing with time, however, the induced current is in the opposite direction. This induced magnetic field generates a magnetic field directed downward through the loop again attempting to keep the total flux within the loop constant. The sign of the induced voltage when the loop is open circuited is also shown for these two situations. Since current is the rate of positive charge flow, the charge builds up at these end points with the sign shown. In the more realistic situation where the voltmeter or loop has a finite resistance, the current is still in the direction shown. A voltage source can be inserted in the loop as shown to model this induced emf. The sign or polarity of this voltage source during the given interval will generate the direction of the current shown.

Now, a 2-turn pickup loop will be examined. A 2-turn coil is shown in Figure 6.28. Although the area of each loop is a little difficult to determine for this helical winding, each loop will be assumed to have

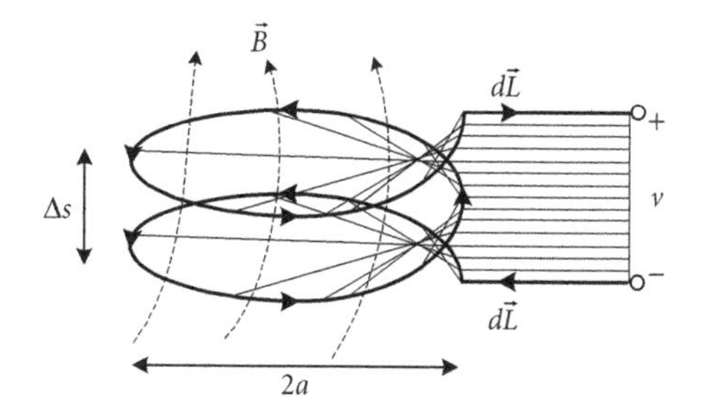

FIGURE 6.28 Tightly wound 2-turn coil will pick up more flux than a single-turn loop (of the same radius).

an area of πa^2. As with the single-turn loop, the closed contour or line integration is taken tangential to the conductor's surface. The corresponding open surface involves two loops rather than just one loop. Neglecting the lead surface, it should be clear for the magnetic flux density shown that this 2-turn loop is collecting more flux than the single-turn loop. Actually, if the turns are tightly wound so that the variation in $\vec{B}$ is small, in the direction of $d\vec{s}$, over the loop spacing, Δs, then the flux collected is about twice that of a single turn. With N tightly wound turns, and with the variation in $\vec{B}$ over the entire length of the coil small, then

$$v = -N\frac{d\Phi}{dt} \tag{6.137}$$

where Φ is the total flux collected by one of the loops. Again, this expression assumes that the loops are similar so they have approximately the same area and that the loops are closely spaced so the field variation along the N loops is small.

Finally, the relationship between the total current enclosed by a current probe and the resultant output voltage can be determined. Assume that a pickup coil consists of N closely spaced, uniformly distributed, equal-area turns. The shape and length of the helical coil can be adjusted so that it can follow any path.[12] The area of each turn is equal to $A = \pi a^2$. The axis of the coil is not necessarily straight or circular. Actually, so that its relationship to $\oint \vec{H}\cdot d\vec{L}$ is clearly seen, the axis of the coil will follow the closed path of integration. Initially, the area of each loop is assumed very small compared to the variation of the magnetic field so that an integration of $\vec{B}$ is not required. In this case, the induced voltage generated by each m turn of the coil is given by

$$\begin{aligned}\Delta v(m,t) &= -\frac{d}{dt}\iint \vec{B}(m,t)\cdot d\vec{s} = -\frac{d}{dt}[B_n(m,t)A] \\ &= -A\frac{dB_n(m,t)}{dt} = -A\mu\frac{dH_n(m,t)}{dt}\end{aligned} \tag{6.138}$$

where $B_n(m,t) = \mu H_n(m,t)$ is the component of the flux density vector normal to each loop's surface or the component in the direction of $d\vec{s}$ for each loop. This direction to the plane of each loop is not necessarily constant unless the coil is straight. Although $B_n(m,t)$ is assumed constant over each loop's surface, it can vary from loop to loop. This possible space variation is via m. For a given m, $B_n(m,t)$ is constant with respect to position. To determine the total voltage induced across the entire coil, $\Delta v(m,t)$ is summed over all N turns of the coil:

$$v(t) = \sum_{m=1}^{N}\Delta v(m,t) = -A\mu\sum_{m=1}^{N}\frac{dH_n(m,t)}{dt} = -A\mu\frac{d}{dt}\sum_{m=1}^{N}H_n(m,t) \tag{6.139}$$

This summation of the normal magnetic field components should be converted to an integration if it is to be related to Ampère's law, (6.131). If the total length of the integration path is L, then the number of turns over an incremental length ΔL is L/N. Therefore, assuming N is sufficiently large, (6.139) can be written as

$$\begin{aligned}v(t) &= -A\mu\frac{d}{dt}\sum_{m=1}^{N}H_n(m,t)\left(\Delta L\frac{N}{L}\right) = -\frac{A\mu N}{L}\frac{d}{dt}\sum_{m=1}^{N}H_n(m,t)\Delta L \\ &\approx -\frac{A\mu N}{L}\frac{d}{dt}\oint H_n(r,t)dL\end{aligned} \tag{6.140}$$

[12]Imagine the coil consists of many turns on a flexible plastic tube.

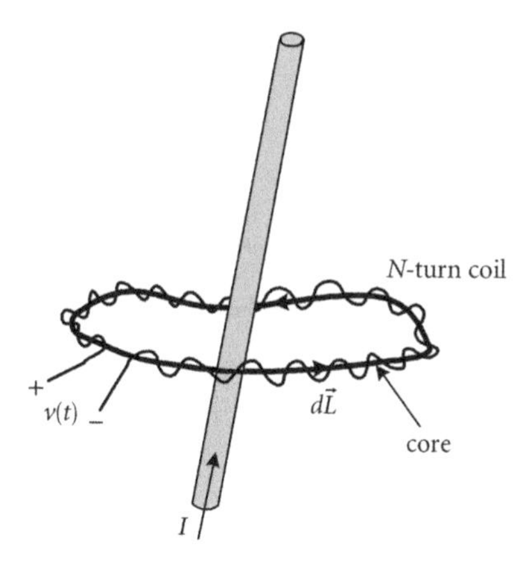

FIGURE 6.29 Induced voltage across the N-turn, tightly wrapped coil[13] is a measure of the total current enclosed.

since $\Delta L(N/L) = 1$. The variable r is the spatial position for each loop m of the coil. As illustrated in Figure 6.29, the tightly wound coil is oriented in the direction of $d\vec{L}$ so that the axis of the coil is in the direction of $d\vec{L}$. Therefore,

$$v(t) = -\frac{A\mu N}{L}\frac{d}{dt}\oint \vec{H}(r,t)\cdot d\vec{L}$$

Using Ampère's law,

$$v(t) = -\frac{A\mu N}{L}\frac{dI_{encl}(t)}{dt} \tag{6.141}$$

where $I_{encl}(t)$ is the total current enclosed by the coil. For sinusoidal steady-state conditions, the ratio of the output voltage to the "input" current is, in the frequency domain,

$$V_s = -\frac{A\mu Nj\omega I_{encls}}{L} \Rightarrow \left|\frac{V_s}{I_{encls}}\right| = \frac{A\mu N\omega}{L} \tag{6.142}$$

This fascinating result indicates that under certain conditions the induced voltage across the N-turn, tightly wound coil is not a function of the position of the enclosed current (or currents) or the shape of the coil!

Before further discussing this probe, a numeric example will be presented. A quickly constructed probe for detecting a 250 mA 10 kHz sinusoidal signal consists of 50 turns on a thin, flexible plastic tubing. This tubing or coil form has a diameter of 1 inch and a total length of 2 ft. The induced open-circuit voltage across this coil is expected to be (since it is not simple to measure) about 0.8 mV:

$$|V_s| = \frac{A\mu_o N\omega |I_{encls}|}{L} = \frac{\pi\left(\frac{0.0254}{2}\right)^2 (4\pi\times 10^{-7})50(2\pi\times 10^4)(0.25)}{2(12\times 0.0254)} \approx 0.8 \text{ mV}$$

This quickly constructed current probe would be more appropriate for detecting larger currents. More reasonable output voltages are obtained with many more turns in conjunction with a high-permeability core form.

Imagine that the current-carrying conductor shown in Figure 6.29 is shifted near one small segment of the coil. Although the induced voltage across the turns of the nearby small segment will increase, the induced voltage across the larger remaining segments will compensate resulting in the same output

[13]If the reader squints, or special glasses are used, this coil will appear tightly wrapped.

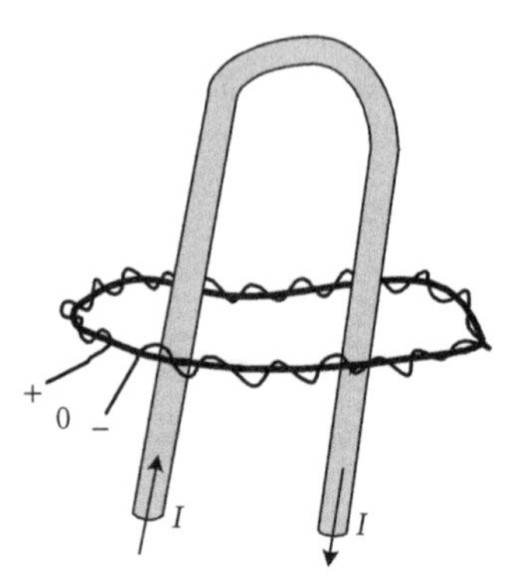

FIGURE 6.30 Net enclosed current by the current probe is zero and hence the induced voltage is zero.

voltage. The current probe measures the net current enclosed within its window. Multiple line currents can also exist within its window. The current probe will indicate the net current at a specific instant of time. For the situation shown in Figure 6.30, the net enclosed current is zero; therefore, the voltage across the output of the coil is zero.

There are several restrictions associated with (6.142). First, the current sources enclosed should not be too close to the coil so that the constant field assumption is not violated. Second, the current sources should not be too close to the coil relative to the spacing between the turns so that the magnetic field does not change too rapidly between neighboring loops.[14] This was assumed in the conversion of the summation to the integral in (6.140). Third, the permeability of the core that the coil is wrapped on should have a constant value. Depending on the core material selected and position of the coil, sometimes large currents (e.g., 1 kA) can saturate part or all of the core. If the current source is close to the coil, then the core may only be saturated near the source. Even if the core is inadvertently magnetized by a magnet in one or more locations, the induced voltage can also be affected. Fourth, both the magnetic field from sources not enclosed by the probe and from induced currents in the coil windings should be small. For tightly wound, uniformly spaced windings, external fields should have little effect on the induced voltage.

It is interesting to note that even when the area of each turn of a tightly wound coil is not small, such as with certain clamp-on current probes, the output voltage is *relatively* insensitive to the position of the current-carrying conductor contained within its window. In these situations, the mutual inductance between the current-carrying conductor and probe coil is insensitive to conductor position. When high precision is required, the window opening within the probe can be restricted by using plastic or nonmetallic nonmagnetic inserts. This will limit any positioning error. There is another source of error when using the simple current probe that is a function of the nonzero pitch of the helical winding. As shown in Figure 6.28, real multiturn coils are not generally composed of perfectly flat loops. When performing the integration in (6.138), the contribution of the magnetic field to the surfaces between the loops was neglected. Although only shown for a 2-turn straight coil, the model shown in Figure 6.31 applies to an

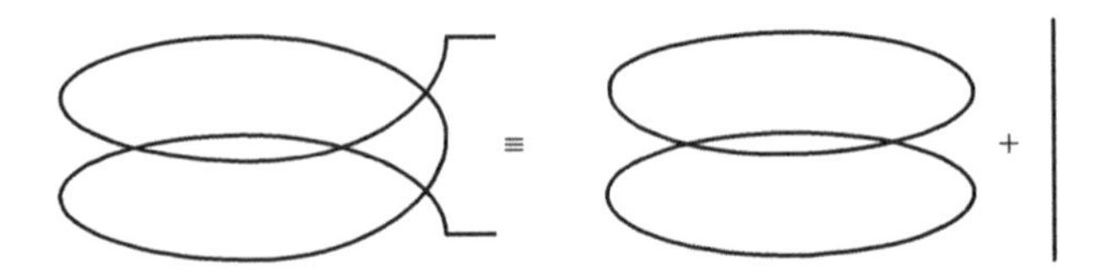

FIGURE 6.31 Helical winding with a nonzero pitch angle modeled as the sum of flat loops and a conductor parallel to the axis of the coil.

[14]One guideline seen is that the distance between the current-carrying conductor in the window of the probe and the coil should be large compared to the distance between the turns.

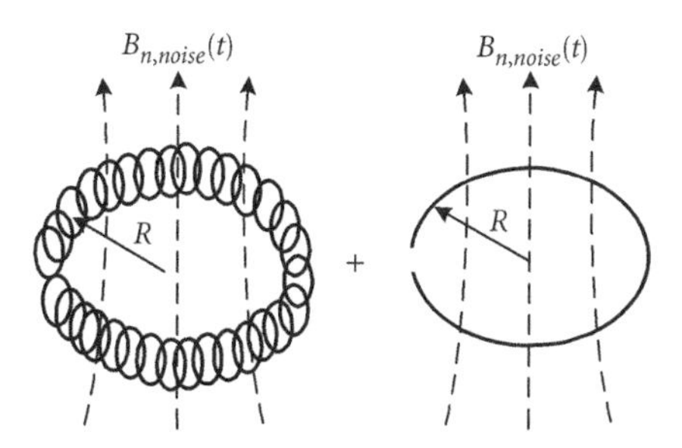

FIGURE 6.32 Magnetic fields perpendicular to the window opening can be a source of current probe error.

N-turn coil with curvature. The nonzero pitch angle of the winding is modeled using a conductor parallel to the axis of the winding. Both the circumferential component and axial (or longitudinal) component of the field contributions can be modeled as shown. The probe can be susceptible to external fields such as strong AM stations or other external current-carrying conductors because of this axial component. A helical circular-coil current probe is equivalent to the zero-pitch solenoid summed with the circular loop as shown in Figure 6.32. The effect of the nonzero pitch of the windings is modeled using a single-turn circular loop of radius R. The coil's major loop radius is not equal to the radius of the individual loops, previously given as a. Ideally, a magnetic field perpendicular to the area of the probe window would not be detected since the field is parallel to the perfectly flat loops of the coil. In reality, this field induces an error voltage equal to

$$v_{error} = -N\frac{d\Phi}{dt} = -(1)\frac{d}{dt}\left[\pi R^2 B_{n,noise}(t)\right] = -\pi R^2 \frac{dB_{n,noise}(t)}{dt} \tag{6.143}$$

or in the frequency domain for sinusoidal signals,

$$|V_{errors}| = \pi R^2 \omega |B_{n,noises}| \tag{6.144}$$

This pitch effect can be reduced or partially compensated for by feeding one end of the helical winding back through the coil. Two versions of feeding a wire through the coil are shown in Figure 6.33. The voltage induced in this extra loop through the coil's axis is about the same magnitude but opposite polarity[15] to the axial term given in (6.143). A delay cable, which contains a wound inner conductor about a ferromagnetic core, can be modified to generate this probe. (The outer jacket is stripped, the outer conductor is removed, and the inner nylon string through the center of the ferromagnetic core is pulled out. The return conductor is then passed through where the nylon string once existed.)

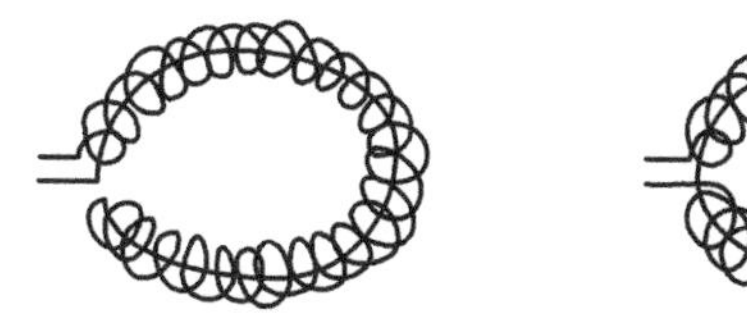

FIGURE 6.33 Two methods of feeding a wire back through the coil.

[15]To understand why this extra loop has the opposite polarity, imagine injecting a current into this coil. Using the right-hand rule, the field generated by the axial component of the helical coil is in the opposite direction to that generated by the extra loop passed through the coil.

Another inexpensive and novel method discussed in the literature is to wind insulated wire around larger diameter iron wire. Since the iron wire is both magnetic and conductive it can be used as both the magnetic core for the insulated wire windings and the return lead for the extra loop fed back through the windings.

If the enclosed current is not varying with time, then the change in the flux with respect to time is zero, and the induced voltage across the coil is zero. In other words, this probe does not work for dc currents.[16] This probe is *similar* to a 1:N transformer. When the input impedance of the voltmeter used to measure the output voltage is large, the current in the secondary of this transformer is small, and the loading effects on the primary wire will be extremely small. Although a low-resistance current-sensing resistor in series with the circuit is very inexpensive, it generally has much more effect on a circuit then a noncontacting current probe. (The parasitic inductance of the current-sensing resistor should not be forgotten.)

When the core of the current probe is nonmagnetic and nonconducting (e.g., insulating plastic), sometimes referred to as an air core, the probe is named a Rogowski coil or Maxwell Worm. Both rigid and flexible versions of these coils are commercially available. Probably the principal advantage of using an air core is that the response is very linear. That is, the output voltage is linearly proportional to the net current of the conductors passing through the center of the probe. This permits easy calibration of the probe since the relationship between the output voltage and input current can be determined at a convenient current level. The dynamic range of the Rogowski coil is large, limited by the voltage measurement circuitry, noise, and arcing or breakdown. When a magnetic core is used, the relationship is not linear and saturation can occur based on the core material selected. When the core saturates, the output voltage will cease to increase at the same rate as the input current. The turns of the coil around the nonmagnetic toroidal form should be evenly distributed. This reduces the susceptibility of the coil to magnetic fields from other sources besides the current-carrying conductor(s) passing through the window of the probe. For high currents and frequencies, the voltage generated can be large, even with a nonmagnetic core. Arcing between the turns can also occur. Another advantage of the air core is the lower weight and cost compared to a magnetic core. Obvious disadvantages of using a nonmagnetic core are the lower output voltage or gain. Hence, for a given output voltage, the number of turns for an air core is much greater than for a magnetic core.

As with all inductors, the parasitic capacitance of the coil itself will limit the upper, useful operating frequency of the current probe. This capacitance will limit the bandwidth of the probe whether the core is magnetic or nonmagnetic. The performance of both air core and magnetic core probes at these higher frequencies is more difficult to model because of this parasitic capacitance (and inductance). Generally, the parasitic capacitance of the coil will increase with the number of turns. Hence, for a given output voltage, the operating frequency and first resonant frequency of the air core are less than for a magnetic core.[17] Air-core probes have been constructed with resonant frequencies up to 180 MHz. (Upper frequency ranges of about 1 MHz are more typical.) In addition to the parasitic capacitance of the coil, there is a capacitance between the coil and conductor under test. This capacitance, which is quite small (e.g., pF's), also affects the probe's response at high frequencies. As with ordinary transformers, one or more shields can be used to reduce, or at least control, this electric field coupling.

Current probes can easily measure the common-mode current for two bundled wires by clamping around both wires. To measure the differential-mode current in each wire, however, the clamp must be placed around each individual wire. For many cables, it is not reasonable to pull apart the conductors to allow for this measurement. Current probes are great at measuring common-mode currents.

---

[16]A Hall effect sensor can be used for dc. However, these sensors require external driving circuitry and temperature compensation.

[17]These resonances can be dampened by uniformly tapping the coil with resistors. The other end of the resistors are tied to a common annular ring. These resistors are essentially reducing the Q of the coil.

6.16 Loaded Current Probe

Determine the expression for the output voltage for a resistively loaded current probe if the current carrying test wire is centered in the window of the probe. As shown in Figure 6.34, the probe consists of a rectangular cross-section, air-core N-turn coil. When is this probe acting like a 1:N ideal transformer? Determine the output if the load consists of a series RC integrator circuit. [Zahn; Karrer]

To determine the output voltage across this resistively loaded current probe, the mutual inductance between the wire and coil is required, as well as the self inductance of the coil. The core will be assumed nonmagnetic and insulating. For this type of core, the analysis is easy since the core will not disturb or distort the magnetic field from the current-carrying wire. For such cores, the mutual inductance between the coil and straight wire and the self inductance of the coil can be quickly determined. (When a coil on a toroidal ferrite core is placed around the wire, the fields are disturbed around the wire: the medium surrounding the wire is no longer homogeneous and the magnetic field is not circumferential everywhere around the wire.)[18]

The derivation of the mutual inductance expression will be given to illustrate a few basic concepts. The magnetic field around the current-carrying wire including through the air-core coil is

$$H_\phi = \frac{I}{2\pi\rho}$$

where ρ is the radial distance from the center of the wire. The total flux collected by the N-turn coil is then

$$\Phi = \iint \vec{B}\cdot d\vec{s} = N\iint \mu_o H_\phi \hat{a}_\phi \cdot d\rho dz \hat{a}_\phi = N\int_0^s \int_b^a \mu_o \frac{I}{2\pi r} d\rho dz = \frac{NI\mu_o s}{2\pi}\ln\left(\frac{a}{b}\right) \tag{6.145}$$

where a is the outer radius and b is the inner radius of the rectangular cross-section coil. This expression assumes that the turns are tightly wound since the surface of each loop was assumed perpendicular to the magnetic field. The mutual inductance between the wire and coil is the ratio of the flux collected by the coil to the current in the wire:

$$M = \frac{\Phi}{I} = \frac{N\mu_o s}{2\pi}\ln\left(\frac{a}{b}\right) \tag{6.146}$$

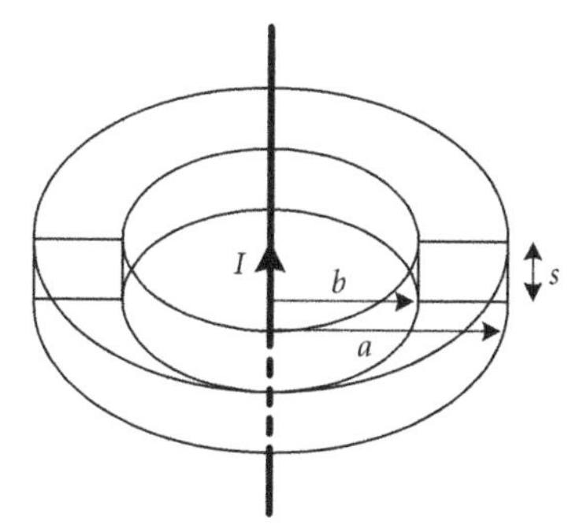

FIGURE 6.34 Current-carrying conductor passing through the center of an air-core coil. The cross-section of the core is rectangular. (The windings are not shown.)

[18]If the ferrite surrounding the wire were in the shape of a very long concentric cylindrical shell, then the field distortion would be small away from the ends of the shell. The tangential components of the magnetic field are continuous across a boundary unless surface current is present along the boundary.

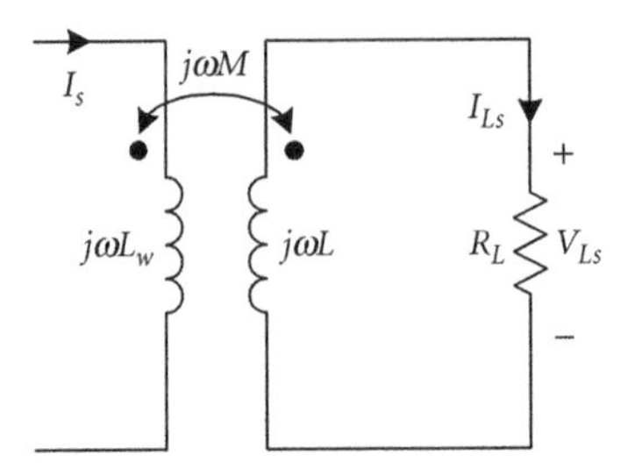

FIGURE 6.35 Model of a resistively loaded current probe using a linear transformer.

The expression for the self inductance of a tightly wound, rectangular cross-section, toroid-shaped coil is given by

$$L = \frac{N^2 \mu_o s}{2\pi} \ln\left(\frac{a}{b}\right) \tag{6.147}$$

This self inductance is defined as the ratio of the flux through the coil to the current I in the coil (not in the center conductor) generating this flux. Notice that it varies with the square of the number of turns.

With the mutual and self inductances known, the current probe can be modeled using a linear transformer as shown in Figure 6.35. The self inductance of the wire is L_w.[19] The expression for ratio of the current through the resistive load to the current in the wire is obtained by writing a single KVL expression around the secondary loop:

$$R_L I_{Ls} + j\omega L I_{Ls} - j\omega M I_s = 0$$

Therefore, the expression for the current ratio is

$$\frac{I_{Ls}}{I_s} = \frac{j\omega M}{j\omega L + R_L} = \frac{j\omega \frac{N\mu_o s}{2\pi} \ln\left(\frac{a}{b}\right)}{j\omega \frac{N^2 \mu_o s}{2\pi} \ln\left(\frac{a}{b}\right) + R_L} \tag{6.148}$$

When $R_L = 0$, corresponding to heavy-loading of the probe, or the load resistance is small compared to the coil's inductive reactance, $R_L \ll \omega L$, this ratio reduces to

$$\frac{I_{Ls}}{I_s} \approx \frac{j\omega \frac{N\mu_o s}{2\pi} \ln\left(\frac{a}{b}\right)}{j\omega \frac{N^2 \mu_o s}{2\pi} \ln\left(\frac{a}{b}\right)} = \frac{1}{N} \tag{6.149}$$

This result is identical to the current expression for ideal transformers. For open-circuit conditions, the current through the load is small as expected:

$$\frac{I_{Ls}}{I_s} = \frac{j\omega \frac{N\mu_o s}{2\pi} \ln\left(\frac{a}{b}\right)}{j\omega \frac{N^2 \mu_o s}{2\pi} \ln\left(\frac{a}{b}\right) + R_L} \approx 0 \quad \text{if } R_L \gg \omega L \text{ and } R_L \gg \omega M$$

[19]Actually, this wire inductance is a self partial inductance since it is only part of a complete path.

In general, the ratio of the load current to the input current is not equal to $1/N$. If the load resistance is small compared to the inductive reactance of the coil, the expression for the current ratio can be rewritten as a function of the coefficient of coupling, k, between the wire and coil:

$$\frac{I_{Ls}}{I_s} = \frac{j\omega M}{j\omega L + R_L} = \frac{j\omega k\sqrt{L_w L}}{j\omega L + R_L}$$
$$\approx \frac{j\omega k\sqrt{L_w L}}{j\omega L} = k\sqrt{\frac{L_w}{L}} \propto k\sqrt{\frac{L_w}{N^2}} = k\frac{\sqrt{L_w}}{N} \quad \text{if } \omega L \gg R_L \tag{6.150}$$

For good ordinary transformers where the permeability of the core is large, most of the flux of the primary is collected by the secondary, and the coupling is usually strong ($k \approx 1$). For current probes, the coupling is not necessarily strong, and the leakage flux can be greater than the coupled flux.

The voltage across the load is

$$V_{Ls} = R_L I_{Ls} = R_L \frac{j\omega M}{j\omega L + R_L} I_s = R_L \frac{j\omega \frac{N\mu_o s}{2\pi} \ln\left(\frac{a}{b}\right)}{j\omega \frac{N^2 \mu_o s}{2\pi} \ln\left(\frac{a}{b}\right) + R_L} I_s \tag{6.151}$$

The open-circuit voltage, or the voltage across a load resistance that is much greater than the reactance of the coil, is

$$V_{Ls} \approx R_L \frac{j\omega \frac{N\mu_o s}{2\pi} \ln\left(\frac{a}{b}\right)}{R_L} I_s = j\omega \frac{N\mu_o s}{2\pi} \ln\left(\frac{a}{b}\right) I_s = j\omega M I_s \quad \text{if } R_L \gg \omega L \tag{6.152}$$

In agreement with (6.142), the open-circuit voltage increases with the number of turns, area of each turn, and frequency of operation (or change in the current with respect to time).

Without substituting the expressions for the mutual and self inductances, the output voltage across a resistively load probe can be approximated as

$$V_{Ls} = R_L \frac{j\omega M}{j\omega L + R_L} I_s = \frac{j\omega M}{\frac{j\omega}{\frac{R_L}{L}} + 1} I_s \approx \begin{cases} j\omega M I_s & \text{if } \omega \ll \frac{R_L}{L} \\ \frac{M R_L}{L} I_s & \text{if } \omega \gg \frac{R_L}{L} \end{cases} \tag{6.153}$$

For $\omega \ll R_L/L$, the output voltage is a function of the frequency. Since $j\omega$ in the frequency domain corresponds to d/dt in the time domain, for frequencies much less than R_L/L, the output voltage is proportional to the derivative of the input current. In theory, this voltage could be integrated to obtain the current. For frequencies much greater than R_L/L, the voltage is directly proportional to the input current.[20] It is, however, a function of the load resistance. To extend the lower operating frequency of the probe in this directly proportional region, R_L/L should be decreased. Of course, if the load were a short, then this would be easily accomplished. However, instrumentation has nonzero input resistance. For a fixed R_L, a large L would be required for a low operating frequency. For air-core Rogowski coils, a large L may not be possible. Furthermore, a large L implies a large parasitic capacitance and low resonant

[20]In this region, the coil is sometimes referred to as "self integrating."

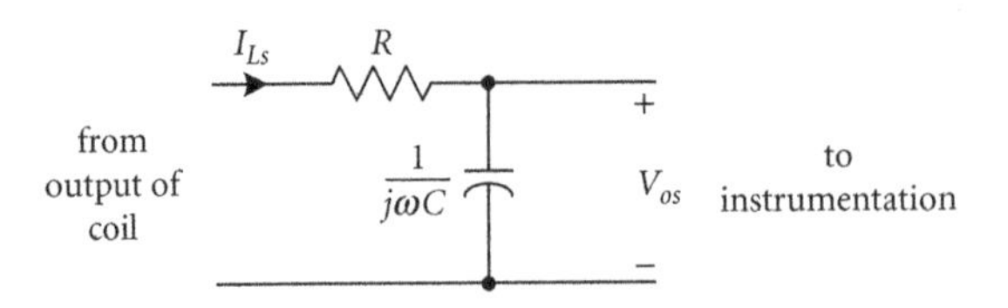

FIGURE 6.36 Simple *RC* integrator circuit.

(and upper operating) frequency. Rogowski coils, with the addition of an integrator circuit, are used for frequencies less than R_L/L. Other current transformers are often operated above R_L/L.

As stated, it is common to use an integrator circuit across the output of a Rogowski coil. A simple nonactive *RC* version is shown in Figure 6.36. The expression for the output voltage is obtained by replacing R_L in (6.148) with $R+(1/j\omega C)$:

$$V_{os}=\frac{1}{j\omega C}I_{Ls}=\frac{1}{j\omega C}\frac{j\omega M}{j\omega L+R_L}I_s=\frac{1}{C}\frac{M}{j\omega L+R+\frac{1}{j\omega C}}I_s=\frac{j\omega M}{-\omega^2LC+j\omega RC+1}I_s$$

$$=\frac{j\omega M}{-\left(\frac{\omega}{\frac{1}{\sqrt{LC}}}\right)^2+\frac{j\omega}{\frac{1}{RC}}+1}I_s \tag{6.154}$$

The instrumentation loading is assumed negligible, which can be accomplished using an active *RC* filter. For frequencies much less than the resonant frequency $1/\sqrt{LC}$, this expression reduces to (assuming $1/(RC)\ll 1/\sqrt{LC}$)

$$V_{os}\approx\frac{j\omega M}{\frac{j\omega}{\frac{1}{RC}}+1}I_s \quad \text{if } \omega\ll\frac{1}{\sqrt{LC}} \tag{6.155}$$

Furthermore,

$$V_{os}\approx\begin{cases} j\omega MI_s & \text{if } \omega\ll\frac{1}{RC},\ \omega\ll\frac{1}{\sqrt{LC}} \\ \frac{M}{RC}I_s & \text{if } \frac{1}{RC}\ll\omega\ll\frac{1}{\sqrt{LC}} \end{cases} \tag{6.156}$$

For frequencies much less than $1/(RC)$, the output voltage is proportional to the derivative of the input current ($j\omega\Leftrightarrow d/dt$). However, for frequencies much greater than $1/(RC)$, the output voltage is proportional to the input current and not a function of the frequency. Unlike the resistively load probe, the output voltage in this region is not a function of L (but still a function of M). To extend the lower operating frequency of the probe where the voltage is proportional to the input current, the *RC* product should be large. Unfortunately, according to (6.156), the output voltage or probe gain is $M/(RC)$. A large C can also reduce the upper operating frequency of the probe via $1/\sqrt{LC}$.

A numerical example is always helpful. If $R = 47$ kΩ, $C = 0.016$ μF, and $L = 0.8$ μH, then

$$\frac{1}{2\pi RC} \approx 210\text{ Hz}, \quad \frac{1}{2\pi\sqrt{LC}} \approx 1.4\text{ MHz}, \quad \frac{R}{2\pi L} \approx 9.4\text{ GHz}$$

Although $R/(2\pi L)$ is extremely large, the first resonant frequency of the system is about 1.4 MHz. This limits the upper operating frequency of the probe. Well above 210 Hz, the output of the integrator is proportional to the input current. Although the mutual inductance between the circuit conductor under test and the coil is not specified, it must be less than L since $L_w < L$ and $k < 1$:

$$M = k\sqrt{LL_w} \leq \sqrt{LL_w} < L \tag{6.157}$$

Therefore, the gain of the system in this region is less than $M/(RC) < L/(RC) \approx 10^{-3}$. The output of the integrator to a 100 mA sinusoidal input signal would be less than about 0.1 mV!

6.17 Transfer Impedance of Current Probes

The transfer impedance of a current probe typically increases with frequency and then flattens out at some cutoff frequency. The lower frequency region is called the "voltage region" and the higher frequency region is called the "current region." Working in the frequency domain, assume a cable of impedance Z_o matched to its load is connected to the probe. The mutual inductance between the circuit wire and probe coil is M, and the self inductance of the probe coil is L. Determine the transfer impedance for this system, defined as the ratio of output voltage across the probe to the current through the circuit wire. Using this expression, verify the response of the standard current probe. [Smith, '93; Fischer; Paul, '92(b)]

Current probes (and current transformers) are commercially manufactured for a wide variety of frequencies and current amplitudes. Current probes are available to measure high level 60 Hz currents and low level microwave currents. They are designed for a maximum rated current level (before core saturation occurs) and maximum operating frequency (before resonance occurs). Generally, because of the finite bandwidth of any probe, a probe designed to measure 60 Hz power currents would not be suitable for RF current measurements. When a noncontact current measurement is required at very low frequencies and dc, Hall-effect devices are used instead of current probes (since a transformer cannot couple dc currents).

Current probe data sheets provide the relationship between the input current enclosed by the probe and output voltage of the probe via the probe's transfer impedance. This transfer impedance is a calibration curve, experimentally determined, for the probe. The transfer impedance, in ohms, is defined as

$$Z_T = \frac{V_s}{I_s} \tag{6.158}$$

where V_s is the phasor voltage across the output of the coil into a specified load and I_s is the net phasor current enclosed by the probe. A transfer impedance of 1 Ω, for example, implies that a 1 A sinusoidal input current would produce a 1 V sinusoidal output voltage. The magnitude of the transfer impedance in dB (relative to 1 Ω) is

$$Z_{TdB} = 20\log\left|\frac{Z_T}{1}\right| = 20\log\left|\frac{V_s}{I_s}\right| = 20\log|V_s| - 20\log|V_s| \text{ dB}\Omega \tag{6.159}$$

Calibration curves for probes are frequently given in Z_{TdB}. Current probes can be purchased with operating frequencies from about 10 Hz to 3 GHz. The transfer impedance can vary from about –35 dB to 25 dB (or dBΩ).

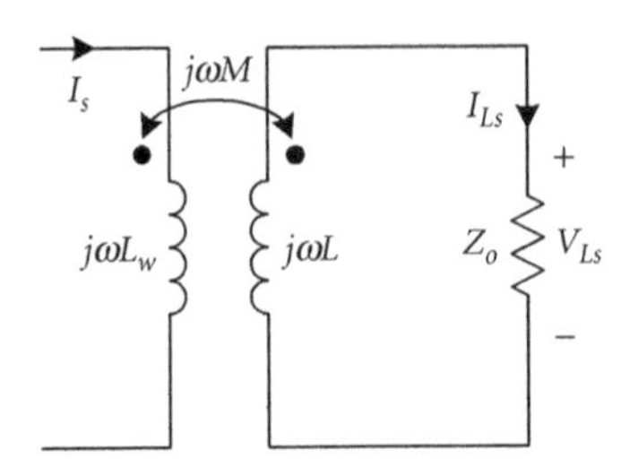

FIGURE 6.37 Transformer model of the current probe connected to a matched transmission line of impedance Z_o.

The output voltage of the probe and its calibration curve is a function of the probe's load impedance. Commercial filter specifications are similar in that they are valid for a given load (and source) impedance. If a cable of characteristic impedance Z_o connects the probe to a spectrum analyzer of input impedance Z_o, then the cable is matched to its load. Therefore, the input impedance seen by the probe looking into the cable is also Z_o. A typical value used for a load impedance is 50 Ω.

The self partial inductance of the wire under test is given as L_w in Figure 6.37. It is much less than the inductance of the multiturn ferrite-core coil of the probe. Although this load current expression was previously derived, it will now be obtained using the T-equivalent model for the transformer. Referring to Figure 6.38, the load current is equal to (via current division)

$$I_{Ls} = \frac{j\omega M}{j\omega M + j\omega(L-M) + Z_o} I_s = \frac{j\omega M}{j\omega L + Z_o} I_s \tag{6.160}$$

The ratio of the load voltage to the input current, or the transfer impedance, is equal to

$$Z_T = \frac{V_{Ls}}{I_s} = \frac{Z_o I_{Ls}}{I_s} = \frac{j\omega M Z_o}{j\omega L + Z_o} = \frac{j\omega M}{1 + \dfrac{j\omega}{\dfrac{Z_o}{L}}} \tag{6.161}$$

For real Z_o, the Bode magnitude plot for (6.161) is immediately obtained. It is given in Figure 6.39. The standard, current-probe calibration curve has this same basic shape. The probe is normally used above the cutoff frequency:

$$\omega_c = \frac{Z_o}{L} \text{ or } f_c = \frac{Z_o}{2\pi L} \tag{6.162}$$

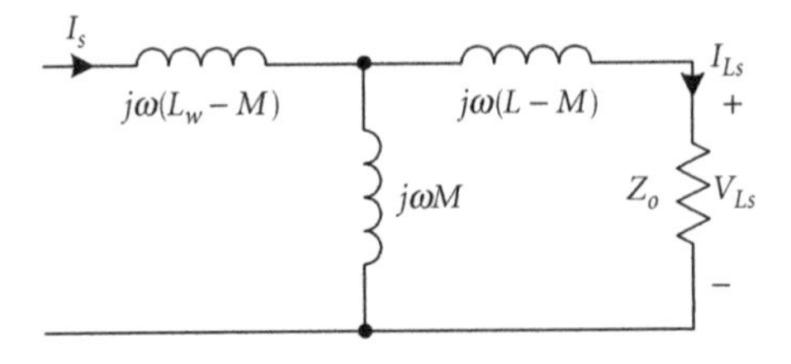

FIGURE 6.38 T-equivalent model of the circuit in Figure 6.37. This equivalent model does not contain any magnetic coupling between the inductors.

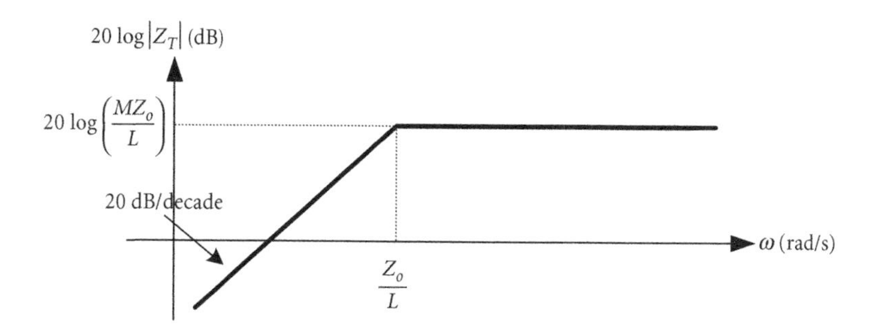

FIGURE 6.39 Bode magnitude plot of (6.161).

The region above the cutoff frequency is referred to as the current region, and its response is constant with frequency (below the resonant frequencies). The region below the cutoff frequency is referred to as the voltage region. Of course, at some frequency greater than the cutoff frequency, the probe characteristics begin to change due to various nonidealities (e.g., parasitic capacitance). As Z_o decreases, the cutoff frequency and gain (or probe sensitivity) decreases. Since the flat region of the response is the normal operating range of the *current* probe, decreasing Z_o increases the operating range or bandwidth of the probe at the expense of reduced gain. Since Z_o is the input impedance of the line (and characteristic impedance of the line since the load presented by the measuring instrument is matched to the line), placing a resistor, R_p, in parallel or shunt with the input of the line at the current probe will decrease the cutoff frequency to

$$\omega_c = \frac{Z_o \| R_p}{L}$$

This will decrease the probe's gain but increase its bandwidth.

As the number of turns on the coil increase, L increases and the cutoff frequency decreases. The output voltage or "gain" of the probe increases with Z_T:

$$V_{dB} = Z_{TdB} + I_{dB} \tag{6.163}$$

In the voltage region, well below the cutoff frequency, the output voltage does indeed increase with M:

$$V_{Ls} = \frac{j\omega M}{1+\dfrac{j\omega}{\dfrac{Z_o}{L}}} I_s \approx \frac{j\omega M}{1} I_s = j\omega M I_s \quad \text{if } \omega \ll \frac{Z_o}{L} \tag{6.164}$$

From (6.146), the output voltage increases with the number of turns, N. Since both M and L increase with the number of turns (but not at the same rate), it is clear that a greater output voltage in the voltage region is obtained at the expense of a lower cutoff frequency, bandwidth, or operating frequency range. This is the common gain-bandwidth relationship. When the secondary is open circuited, or $Z_o \to \infty$, the cutoff frequency is also at infinity (for finite L) and there is no "counter" current in the secondary.

Below the cutoff frequency, the transfer impedance increases with frequency at a rate of 20 dB/decade:

$$20\log|Z_T| \approx 20\log|j\omega M| = 20\log(\omega M) \tag{6.165}$$

The voltage across an open-circuited square loop sniffer is equal to $M\,di/dt$, where i is the current through the circuit wire and M is the mutual inductance between the circuit wire and sniffer loop. This transfer impedance also increases at 20 dB/decade:

$$V_s = j\omega M I_s \quad \Rightarrow \frac{V_s}{I_s} = j\omega M \quad \Rightarrow 20\log\left|\frac{V_s}{I_s}\right| = 20\log(\omega M) \tag{6.166}$$

This increase in voltage with frequency is why this range is referred to as the voltage region. The voltage across the probe increases at 20 dB/decade for a constant input current. The constant "gain" region above the cutoff frequency is referred to as the current region since this is where a current probe is normally used. In this constant gain region, the output voltage and input current are related by Z_T Using (6.146) and (6.147), in the current region

$$Z_T \approx \frac{MZ_o}{L} = \frac{Z_o}{N}$$

In this region, for a given input current, the output voltage decreases with the number of turns. However, in this situation, as N increases, the coefficient of coupling, k, would not necessarily remain constant as implied by equations (6.146) and (6.147).

References

Abramowitz, Milton, and Irene A. Stegun, *Handbook of Mathematical Functions with Formulas, Graphs, and Mathematical Tables*, National Bureau of Standards, 1964.

Abrie, Pieter L.D., *The Design of Impedance-Matching Networks for Radio-Frequency and Microwave Amplifiers*, Artech House, 1985.

Adamczewski, Ignacy, *Ionization, Conductivity, and Breakdown in Dielectric Liquids*, Barnes & Noble, 1969.

Adams, John W. and Dennis S. Friday, *Measurement Procedures for Electromagnetic Compatibility Assessment of Electroexplosive Devices*, IEEE Transactions on Electromagnetic Compatibility, Vol. 30, No. 4, November 1988.

Adler, Richard B., Lan Jen Chu, and Robert M. Fano, *Electromagnetic Energy Transmission and Radiation*, John Wiley & Sons, 1960.

Agarwal, A., C.J. Reddy, and M.Y. Joshi, *Coupled Bars in Rectangular Coaxial*, Electronics Letters, Vol. 25, No. 1, January 1989.

AIEE Committee Report, *Voltage Gradients Through the Ground Under Fault Conditions*, Transactions of AIEE, Vol. 77, Part III, October 1958.

Amin, M.B. and I.A. Benson, *Nonlinear Effects in Coaxial Cables at Microwave Frequencies*, Electronics Letters, Vol. 13, No. 25, December 1977.

Analog Devices, *Instrumentation Amplifier Data Sheets*, Analog Devices, 1999.

Anderson, Dan, *Electrostatic Discharge*, Richmond Technology, Videocassette, 1994.

Anderson, Edwin M., *Electric Transmission Line Fundamentals*, Reston, 1985.

Anderson, John M., *Wide Frequency Range Current Transformers*, The Review of Scientific Instruments, Vol. 42, No. 7, July 1971.

Anderson, Leonard R. and Jack Macneill, *Electric Machines & Transformers*, Prentice-Hall, 1988.

Andrew Corporation, *Andrew Catalog 37*, 1995.

ANSI/IEEE Grounding of Industrial and Commercial Power Systems, Standard 142, *Green Book*, 1982.

Arnold, A.H.M., *The Alternating-Current Resistance of Parallel Conductors of Circular Cross-Section*, Journal of IEE, Vol. 77, No. 463, July 1935.

ARRL Antenna Compendium: Vol. 4, American Radio Relay League, 1995.

ARRL Antenna Handbook, American Radio Relay League, 1997.

ARRL Handbook for Radio Amateurs, American Radio Relay League, 1996.

Asano, Kazutoshi, *On the Electrostatic Problem Consisting of a Charged Nonconducting Disk and a Metal Sphere*, Journal of Applied Physics, Vol. 48, No. 3, March 1977.

ASM International, *Electronic Materials Handbook, Volume 1: Packaging*, 1989.

ASM International, *Metals Handbook Vol. 2 Properties and Selection: Nonferrous Alloys and Special-Purpose Materials*, 1990.

ASM International, *Metals Handbook*, 1998.

Attwood, Stephen S., *Electric and Magnetic Fields*, Dover, 1967.

Auger, Raymond W., *The Relay Guide*, Reinhold, 1960.

Baboian, Robert, ed., *Corrosion Tests and Standards: Application and Interpretation*, ASTM, 1995.

Ballou, Glen M., ed., *Handbook for Sound Engineers*, Howard W. Sams, 1991.

Banks, Robert S. and Tieu Vinh, *An Assessment of the 5-mA 60 Hz Contact Current Safety Level*, IEEE Transactions on Power Apparatus and Systems, Vol. PAS-103, No. 12, December 1984.

Barbano, Normand, *The Aerodiscone Antenna*, The Microwave Journal, November 1966.

Bardell, P.R., *Magnetic Materials in the Electrical Industry*, Purnell and Sons, 1955.

Barke, Erich, *Line-to-Ground Capacitance Calculation for VLSI: A Comparison*, IEEE Transactions on Computer-Aided Design, Vol. 7, No. 2, February 1988.

Barnes, John R., *Electronic System Design: Interference and Noise Control Techniques*, Prentice-Hall, 1987.

Bazelyan, E.M. and Yu P. Raizer, *Spark Discharge*, CRC Press, 1998.

Beeman, Donald, ed., *Industrial Power Systems Handbook*, McGraw-Hill, 1955.

Belden Wire and Cable Company, *Master Catalog*, 1995.

Bell Telephone Laboratories, *Integrated Device and Connection Technology*, Vol. 3, Prentice-Hall, 1971.

Bell Telephone Laboratories, *Physical Design of Electronic Systems*, Vol. 1, Prentice-Hall, 1970.

Bendjamin, J., R. Thottappillil, and V. Scuka, *Time Varying Magnetic Fields Generated by Human Metal (ESD) Electrostatic Discharges*, Journal of Electrostatics, Vol. 46, 1999.

Benner, Linda S., T. Suzuki, K. Meguro, and S. Tanaka, ed., *Precious Metals: Science and Technology*, International Precious Metals Institute, 1991.

Berghe, S. Van den, F. Olyslager, D. De Zutter, J. De Moerloose, and W. Temmerman, *Power Plane Resonances as a Source of Delta-I Noise and the Influence of Decoupling Capacitors*, IEEE International Symposium on Electromagnetic Compatibility, 1997.

Bernstein, Theodore, *Electrical Shock Hazards and Safety Standards*, IEEE Transactions on Education, Vol. 34, No. 3, August 1991.

Bhartia, P., K.V.S. Rao, and R.S. Tomar, *Millimeter-Wave Microstrip and Printed Circuit Antennas*, Artech House, 1991.

Biddle Instruments, *"Getting Down to Earth…" A Manual on Earth-Resistance Testing for the Practical Man*, Biddle Instruments, 1981.

Bird, Dick, *The G4ZU "Bird Cage" Aerial*, CQ, April 1960.

Blackburn, John F., ed., *Components Handbook*, McGraw-Hill, 1949.

Blinchikoff, Herman J. and Anatol I. Zverev, *Filtering in the Time & Frequency Domains*, John Wiley & Sons, 1976.

Blood, William R., Jr., *MECL System Design Handbook*, Motorola Semiconductor Products, Mesa, Arizona, 1972.

Blume, L.F., A. Boyajian, G. Camilli, T.C. Lennox, S. Minneci, and V.M. Montsinger, *Transformer Engineering*, John Wiley & Sons, 1951.

Boast, Warren B., *Vector Fields*, Harper & Row, 1964.

Bodson, Dennis, *Electromagnetic Pulse and the Radio Amateur* — Part 4, QST, November 1986.

Bogle, A.G., *The Effective Inductance and Resistance of Screened Coils*, Journal of IEE, Vol. 87, No. 525, September 1940.

Bonsen, Robert, *Building a Semi-anechoic Chamber: An Overview*, ITEM 1999.

Bosch, *Interference Suppression* (Technical Instruction), Delta Press, Great Britain, 1979.

Bosich, Joseph F., *Corrosion Prevention for Practicing Engineers*, Barnes & Noble, 1970.

Bouwers, A. and P.G. Cath, *The Maximum Electric Field Strength For Several Simple Electrode Configurations*, Philips Technical Review, Vol. 6, No. 9, September 1941.

Boxleitner, Warren, *Electrostatic discharge and electronic equipment: a practical guide for designing to prevent ESD problems*, IEEE Press, 1989.

Bozorth, Richard M., *Ferromagnetism*, D. Van Nostrand, 1951.

Bradley, Fennimore N., *Materials for Magnetic Functions*, Hayden, 1971.

Brailsford, F., *Physical Principles of Magnetism*, D. Van Nostrand, 1966.

Brandes, E.A. and G.B. Brook, ed., *Smithells Metals Reference Book*, Butterworths-Heinemann, 1988.

Bridges, Jack E., *An Update on the Circuit Approach to Calculate Shielding Effectiveness*, IEEE Transactions on Electromagnetic Compatibility, Vol. 30, No. 3, August 1988.

Bridges, Jack E., *New Development in Electrical Shock Safety*, IEEE International Symposium on Electromagnetic Compatibility, 1994.

Britt, Scott D., David M. Hockanson, Fei Sha, James L. Drewniak, Todd H. Hubing, and Thomas P. Van Doren, *Effects of Gapped Groundplanes and Guard Traces on Radiated EMI*, IEEE International Symposium on Electromagnetic Compatibility, 1997.

Bronaugh, Edwin L. and William S. Lambdin, *Electromagnetic Interference Test Methodology and Procedures,* Interference Control Technologies, Vol. 6, 1988.

Brotherton, M., *Capacitors: Their Use in Electronic Circuits,* D. Van Nostrand, 1946.

Brown, Robert Grover, Robert A. Sharpe, and William Lewis Hughes, *Lines, Waves, and Antennas: The Transmission of Electric Energy,* Ronald Press, 1961.

Browne, Thomas E., Jr., ed., *Circuit Interruption: Theory and Techniques,* Marcel Dekker, 1984.

Burberry, R.A., *VHF and UHF Antennas,* The Redwood Press, 1992.

Burgett, Richard R., Richard E. Massoll, and Donald R. Van Uum, *Relationship Between Spark Plugs and Engine-Radiated Electromagnetic Interference,* IEEE Transactions on Electromagnetic Compatibility, Vol. EMC-16, No. 3, August 1974.

Butler, Chalmers M., *Capacitance of a Finite-Length Conducting Cylindrical Tube,* Journal of Applied Physics, Vol. 51, No. 11, November 1980.

Butler, Chalmers M., Yahya Rahmat-Samii, and Raj Mittra, *Electromagnetic Penetration Through Apertures in Conducting Surfaces,* IEEE Transactions on Antennas and Propagation, Vol. AP-26, No. 1, January 1978.

Cadick, John, Mary Capelli-Schellpfeffer, and Dennis Neitzel, *Electrical Safety Handbook,* McGraw-Hill, 2000.

Cahill, L.W., *Approximate Formulae for Microstrip Transmission Lines,* Proceedings of the I.R.E., Australia 35, No. 10, October 1974.

Campbell, George A. and Ronald M. Foster, *Fourier Integrals for Practical Applications,* D. Van Nostrand, 1948.

Campbell, Peter, *Comments on Energy Stored in Permanent Magnets,* IEEE Transactions on Magnetics, Vol. 36, No. 1, January 2000.

Caola, R.J., D.W. Deno, and V.S.W. Dymek, *Measurements of Electric and Magnetic Fields In and Around Homes Near a 500 kV Transmission Line,* IEEE Transactions on Power Apparatus and Systems, Vol. PAS-102, No. 10, October 1983.

Carlisle, Ben H., *A New Way to Prevent Electrocution,* Machine Design, Vol. 60, No. 7, April 7, 1988.

Carr, Clifford C., *Electric Machinery,* John Wiley & Sons, 1958.

Carr, Joseph J., *Automobile Electronics Servicing Guide,* Howard W. Sams, 1973.

Carr, Joseph J., *Practical Antenna Handbook,* McGraw-Hill, 1998.

Carson, John R., *A Generalization of the Reciprocal Theorem,* The Bell System Technical Journal, Vol. 3, July 1924.

Carson, John R., *Reciprocal Theorems in Radio Communication,* Proceedings of the IRE, Vol. 17, No. 6, June 1929.

Cebik, L.B., W4RNL, www.cebik.com, 2001.

Ceramic Source '86, The American Ceramic Society, 1985.

Chapman, Stephen J., *Electric Machinery Fundamentals,* McGraw-Hill, 1991.

Chatterjee, R., *Elements of Microwave Engineering,* Ellis Horwood Limited, 1986.

Chatterton, P.A. and M.A. Houlden, *EMC: Electromagnetic Theory to Practical Design,* John Wiley & Sons, 1992.

Chen, Wai-Kai and Run-Sheng Liang, *A General n-Port Network Reciprocity Theorem,* IEEE Transactions on Education, Vol. 33, No. 4, November 1980.

Cheng, David K., *Field and Wave Electromagnetics,* Addison-Wesley, 1989.

Chikazumi, Soshin and Stanley H. Charap, *Physics of Magnetism,* John Wiley & Sons, 1964.

Chipman, Robert A., *Theory and Problems of Transmission Lines,* Schaum's Outline Series, McGraw-Hill, 1968.

Chou, Chung-Kwang, Arthur W. Guy, and John A. McDougall, *Shielding Effectiveness of Improved Microwave-Protective Suits,* IEEE Transactions on Microwave Theory and Techniques, Vol. MTT-35, No. 11, November 1987.

Chow, Y.L. and M.M. Yovanovich, *The Shape Factor of the Capacitance of a Conductor,* Journal of Applied Physics, Vol. 53, No. 12, December 1982.

Christiansen, Donald, ed., *Electronic Engineers' Handbook*, McGraw-Hill, 1996.

Chubb, J.N. and G.J. Butterworth, *Charge Transfer and Current Flow Measurements in Electrostatic Discharges*, Journal of Electrostatics, Vol. 13, 1982.

Cleveland, Robert F., Jr. and Jerry L. Ulcek, *Questions and Answers about Biological Effects and Potential Hazards of Radiofrequency Electromagnetic Fields*, Federal Communications Commission, OET Bulletin 56, August 1999.

Cleveland, Robert F., Jr., David M. Sylvar, and Jerry L. Ulcek, *Evaluating Compliance with FCC Guidelines for Human Exposure to Radiofrequency Electromagnetic Fields*, Federal Communications Commission, OET Bulletin 65, August 1997.

Cogger, Steve and Hugh Hemphill, *Filter Selections Simplified with 0.1/100-Ohm Data*, ITEM 1997.

Cohn, Seymour B., *Characteristic Impedances of Broadside-Coupled Strip Transmission Lines*, IRE Transactions on Microwave Theory and Techniques, Vol. MTT-8, No. 6, November 1960.

Cohn, Seymour B., *The Electric Polarizability of Apertures of Arbitrary Shape*, Proceedings of the IRE, Vol. 40, No. 9, September 1952.

Collin, Robert E., *Antennas and Radiowave Propagation*, McGraw-Hill, 1985.

Collin, Robert E. and Francis J. Zucker, *Antenna Theory Part 1 and 2*, McGraw-Hill, 1969.

Collis, Mike, *Omni-Gain Vertical Collinear for VHF and UHF*, 73, August 1990.

COMAR, *Possible Health Hazards From Exposure to Power-Frequency Electric and Magnetic Fields — A COMAR Technical Information Statement*, IEEE Engineering in Medicine and Biology Magazine, Vol. 19, No. 1, January/February 2000.

Cooley, William W., *Low-Frequency Shielding Effectiveness of Nonuniform Enclosures*, IEEE Transactions on Electromagnetic Compatibility, Vol. EMC-10, No. 1, March 1968.

Coombs, Clyde F., ed., *Printed Circuits Handbook*, McGraw-Hill, 1988.

Cooper, W. Fordham and D.A. Dolbey Jones, *Electrical Safety Engineering*, 1993.

Cooperman, Gene, *A New Current-Voltage Relation for Duct Precipitators Valid for Low and High Current Densities*, IEEE Transactions on Industry Applications, Vol. 1A-17, No. 2, March/April 1981.

Cooperman, P., *A Theory for Space-Charge-Limited Currents with Application to Electrical Precipitation*, Transactions of the AIEE, Vol. 79, Part 1, 1960.

Copson, David A., *Microwave Heating*, AVI, 1975.

Corona, Paolo, Gaetano Latmiral, and Enrico Paolini, *Performance and Analysis of a Reverberating Enclosure with Variable Geometry*, IEEE Transactions on Electromagnetic Compatibility, Vol. EMC-22, No. 1, February 1980.

Corona, Paolo, John Ladbury, and Gaetano Latmiral, *Reverberation-Chamber Research — Then and Now: A Review of Early Work and Comparison With Current Understanding*, IEEE Transactions on Electromagnetic Compatibility, Vol. 44, No. 1, February 2002.

Corp, M. Bruce, *ZZAAP Taming ESD, RFI, and EMI*, Academic Press, 1990.

Corson, Dale R. and Paul Lorrain, *Introduction to Electromagnetic Fields and Waves*, W.H. Freeman, 1962.

Cote, Authur E., ed., *Fire Protection Handbook*, 1991.

Crawford, Myron L., *Generation of Standard EM Fields Using TEM Transmission Cells*, IEEE Transactions on Electromagnetic Compatibility, Vol. EMC-16, No. 4, November 1974.

Crawford, Myron L., John L. Workman, and Curtis L. Thomas, *Expanding the Bandwidth of TEM Cells for EMC Measurements*, IEEE Transactions on Electromagnetic Compatibility, Vol. EMC-20, No. 3, August 1978.

Creed, Frank C., *The Generation and Measurement of High Voltage Impulses*, Center Book Publishers, 1989.

Cross, Jean, *Electrostatics: Principles, Problems and Applications*, Adam Hilger, 1987.

Crowley, Joseph M., *Fundamentals of Applied Electrostatics*, John Wiley & Sons, 1986.

Crowley, Joseph M., *The Electrostatics of Static-Dissipative Worksurfaces*, Journal of Electrostatics, Vol. 24, 1990.

Cruft Laboratory Electronics Training Staff, *Electronic Circuits and Tubes*, McGraw-Hill, 1947.

Cullity, B.D., *Introduction to Magnetic Materials*, Addison-Wesley, 1972.

Cuming Microwave Corporation, *High Performance Broadbanded Pyramidal RF Absorber Data Sheet,* Technical Bulletin 390–1.

Cumming, W.A., *A Nonresonant Endfire Array for VHF and UHF,* IRE Transactions-Antennas and Propagation, Vol. AP-3, No. 2, April 1955.

Dalziel, C.F. and F.P. Massoglia, *Let-Go Currents and Voltages,* Transactions of AIEE, Vol. 75, Part II, 1956.

Dalziel, Charles F., *A Study of Hazards of Impulse Currents,* Transactions of AIEE, Vol. 72, Part III, 1953.

Dalziel, Charles F., *Electric shock hazard,* IEEE Spectrum, Vol. 9, No. 2, February 1972.

Dalziel, Charles F. and W.R. Lee, *Lethal Electric Currents,* IEEE Spectrum, Vol. 6, No. 2, February 1969.

Das, B. N., S.B. Chakrabarty, and K. Siva Ramo Rao, *Capacitance of Transmission Line of Parallel Cylinders in the Presence of Dielectric Coating,* IEEE Transactions on Electromagnetic Compatibility, Vol. 37, No. 1, 1995.

Das, Sisir K. and B.K. Sinha, *Numerical Solution of Higher Order Mode Cut-Off Frequencies in Symmetric TEM Cells Using Finite Element Method,* IEEE Transactions on Electromagnetic Compatibility, Vol. 32, No. 4, November 1990.

Dascalescu, Lucian, Patrick Ribardière, Claude Duvanaud, and Jean-Marie Paillot, *Electrostatic Discharges from Charged Spheres Approaching a Grounded Surface,* Journal of Electrostatics, Vol. 47, 1999.

Daugherty, Kevin M., *Analog-to-Digital Conversion: A Practical Approach,* McGraw-Hill, 1995.

Davidson, J.L., T.J. Williams, A.G. Bailey, and G.L. Hearn, *Characterisation of Electrostatic Discharges from insulating surfaces,* Journal of Electrostatics, Vol. 51–52, 2001.

Davis, Fred E., *Speaker Testing for Audibility Cables,* Audio, July 1993.

DeGauque, Pierre and Joel Hamelin, ed., *Electromagnetic Compatibility,* Oxford University Press, 1993.

Demarest, Kenneth R., *Engineering Electromagnetics,* Prentice-Hall, 1998.

DeMaw, M.F., *Ferromagnetic Core Design & Applications Handbook,* Prentice-Hall, 1981.

Deno, Don W., *Currents Induced in the Human Body by High Voltage Transmission Line Electric Field — Measurement and Calculation of Distribution and Dose,* IEEE Transactions on Power Apparatus and Systems, Vol. PAS-96, No. 5, September/October 1977.

Deno, Don W., *Electrostatic Effect Induction Formulae,* IEEE Transactions on Power Apparatus and Systems, Vol. PAS-94, No. 5, September/October 1975.

Destefan, Dennis E., *Calibration and Testing Facility for Resistance Welding Current Monitors,* IEEE Transactions on Instrumentation and Measurement, Vol. 45, No. 2, April 1996.

Devoldere, John, *Antennas and Techniques for Low-Band DXing,* ARRL, 1994.

Doetsch, Gustav, *Guide to the Applications of the Laplace and Z-Transforms,* Van Nostrand Reinhold, 1971.

Domininghaus, Hans, *Plastics for Engineers: Materials, Properties, Applications,* Hanser, 1988.

Donovan, John E., *Triboelectric Noise Generation in Some Cables Commonly Used with Underwater Electroacoustic Transducers,* The Journal of the Acoustical Society of America, Vol. 48, No. 3 (part 2), 1970.

Dorf, Richard D., ed., *The Electrical Engineering Handbook,* CRC Press, 1993.

Dorsett, Henry G., Jr. and John Nagy, *Dust Explosibility of Chemicals, Drugs, Dyes, and Pesticides,* Bureau of Mines, Report of Investigations 7132, 1968.

Drabowitch, S., A. Papiernik, H. Griffiths, and J. Encinas, *Modern Antennas,* Chapman & Hall, 1998.

Dummer, G.W.A., *Materials for Conductive and Resistive Functions,* Hayden, 1970.

Dunn, John M., *Local, High-Frequency Analysis of the Fields in a Mode-Stirred Chamber,* IEEE Transactions on Electromagnetic Compatibility, Vol. 32, No. 1, February 1990.

Durham, Marcus O., Karen D. Durham, and Robert A. Durham, *TVSS Design,* IEEE Industry Applications Magazine, Vol. 8, No. 5, September/October 2002.

Dwight, H.B., *Calculation of Resistances to Ground,* AIEE Transactions, Vol. 55, December 1936.

Dwight, Herbert B., *Effective Resistance of Isolated NonMagnetic Rectangular Conductors,* AIEE Transactions, Vol. 66, 1947.

Dworsky, Lawrence N., *Modern Transmission Line Theory and Applications,* John Wiley & Sons, 1979.

Eaton, Robert J., *Electric Power Transmission Systems,* Prentice-Hall, 1972.

Eggers, Bruce A., *An Analysis of the Balun,* QST, April 1980.

Eilers, Keith W., Mark Wingate, and Eric Pham, *Application and Safety Issues for Transient Voltage Surge Suppressors*, IEEE Transactions on Industry Applications, Vol. 36, No. 6, November/December 2000.

Electronic Materials Handbook: Volume 1 Packaging, ASM International, 1989.

Elliot, Robert S., *An Introduction to Guided Waves and Microwave Circuits*, Prentice-Hall, 1993.

Elliot, Robert S., *Antenna Theory and Design*, Prentice-Hall, 1981.

Elliott, S.R., *The Physics and Chemistry of Solids*, John Wiley & Sons, 1998.

Elmore, W.C., *The Transient Response of Damped Linear Networks with Particular Regard to Wideband Amplifiers*, Journal of Applied Physics, Vol. 19, January 1948.

Emerson, William H., *Electromagnetic Wave Absorbers and Anechoic Chambers Through the Years*, IEEE Transactions on Antennas and Propagation, Vol. AP-21, No. 4, July 1973.

Engdahl, Tomi, *Telephone Line Audio Interface Circuits*, www.hut.fi/Misc/Electronics/circuits/teleinterface.html, 1999.

Enthone Electronic Materials, *A Comparison of Conductive Coatings for EMI Shielding Applications*, 2001.

Everitt, W.L. and G.E. Anner, *Communication Engineering*, McGraw-Hill, 1956.

Fair-Rite Products Corp., *Fair-Rite Soft Ferrites*, 1994.

Fano, Robert M., Lan Jen Chu, and Richard B. Adler, *Electromagnetic Fields, Energy, and Forces*, John Wiley & Sons, 1960.

Felici, N.J., *Electrostatics and Electrostatic Engineering*, Proceedings of the Static Electrification Conference, Institute of Physics and the Physical Society, 1967.

Feng, James Q., *An analysis of corona currents between two concentric cylindrical electrodes*, Journal of Electrostatics, Vol. 46, 1999.

Fenn, John B., *Lean Flammability Limit and Minimum Spark Ignition Energy*, Industrial and Engineering Chemistry, Vol. 43, No. 12, December 1951.

Ferronics Incorporated, *Ferrite Cores Design Guide and Product Catalog*, 1991.

Ficchi, Rocco F., ed., *Practical Design for Electromagnetic Compatability*, Hayden, 1971.

Fink, Donald G. and H. Wayne Beaty, *Standard Handbook for Electrical Engineers*, McGraw-Hill, 1978.

Fischer Custom Communications, Inc., *Current Probes Catalog*, Torrance, CA.

Flanagan, William M., *Handbook of Transformer Applications*, McGraw-Hill, 1986.

Flügge, S., ed., *Electric Fields and Waves, Encyclopedia of Physics*, Vol. 16, Springer-Verlag, 1958.

Fowler, E.P., *Microphony of Coaxial Cables*, Proceedings of the IEE, Vol. 123, No. 10, October 1976.

Frankel, Sidney, *Characteristic Impedance of Parallel Wires in Rectangular Troughs*, Proceedings of the I.R.E., Vol. 30, No. 4, April 1942.

Freeman, Roger L., *Practical Data Communications*, John Wiley & Sons, 1995.

Freeman, Roger L., *Reference Manual for Telecommunication Engineering*, John Wiley & Sons, 1994.

Frerking, Marvin E., *Crystal Oscillator Design and Temperature Compensation*, Van Nostrand Reinhold, 1978.

Gaertner, Reinhold, Karl-Heinz Helling, Gerhard Biermann, Erich Brazda, Roland Haberhauer, Wilfried Koehl, Richard Mueller, Werner Niggemeier, and Bernhard Soder, *Grounding Personnel via the Floor/Footwear System*, Electrostatic Overstress/Electrostatic Discharge Symposium Proceedings, 1997.

Gandhi, O.P. and I. Chatterjee, *Radio-Frequency Hazards in the VLF and MF Band*, Proceedings of the IEEE, Vol. 70, No. 12, December 1982.

Gandhi, Om P. and Abbas Riazi, *Absorption of Millimeter Waves by Human Beings and Its Biological Implications*, IEEE Transactions on Microwave Theory and Techniques, Vol. MTT-34, No. 2, February 1986.

Gandhi, Om P., ed., *Biological Effects and Medical Applications of Electromagnetic Energy*, Prentice-Hall, 1990.

Gardner, Murray F. and John L. Barnes, *Transients in Linear Systems*, Vol. 1, John Wiley & Sons, 1942.

Gauger, J.R., *Household Appliance Magnetic Field Survey*, IEEE Transactions on Power Apparatus and Systems, Vol. PAS-104, No. 9, September 1985.

Geddes, Leslie A., ed., *Handbook of Electrical Hazards and Accidents*, CRC Press, 1995.

General Motors Corporation, *Automotive Component EMC Specification Overview of Test Requirements: GM9100P*, 1993.

Gerber, Eduard A. and Arthur Ballato, *Precision Frequency Control, Volume 2: Oscillators and Standards*, Academic Press, 1985.

German, Robert F., Henry W. Ott, and Clayton R. Paul, *Effect of an Image Plane on Printed Circuit Board Radiation*, IEEE International Symposium on Electromagnetic Compatibility, 1990.

Gibson, Norbert, *Static Electricity — An Industrial Hazard Under Control?*, Journal of Electrostatics, Vol. 40–41, 1997.

Glasband, Martin, *"Lifting" the Grounding Enigma*, MIX Magazine, November 1994.

Glasford, Glenn M., *Fundamentals of Television Engineering*, McGraw-Hill, 1955.

Glor, Martin, *Electrostatic Hazards in Powder Handling*, Research Studies Press, 1988.

Goedbloed, Jasper, *Electromagnetic Compatibility*, Prentice Hall, 1992.

Golde, R.H., *Lightning: Lightning Protection*, Vol. 2, Academic Press, 1977.

Goldman, Alex, *Modern Ferrite Technology*, Van Nostrand Reinhold, 1990.

Gray, Dwight E., ed., *American Institute of Physics Handbook*, McGraw-Hill, 1972.

Greason, William D., *Electrostatic Damage in Electronics: Devices and Systems*, Research Studies Press, 1987.

Greason, William D., *Generalized Model of Electrostatic Discharge (ESD) for Bodies in Approach: Analyses of Multiple Discharges and Speed of Approach*, Journal of Electrostatics, Vol. 54, 2002.

Greason, William D., *Quasi-static Analysis of Electrostatic Discharge (ESD) and the Human Body Using a Capacitance Model*, Journal of Electrostatics, Vol. 35, 1995.

Green, Harry E., *Determination of the Cutoff of the First Higher Order Mode in a Coaxial Line by the Transverse Resonance Technique*, IEEE Transactions on Microwave Theory and Techniques, Vol. 37, No. 10, October 1989.

Greene, Frank M., *NBS Field-Strength Standards and Measurements (30 Hz to 1000 MHz)*, Proceedings of the IEEE, Vol. 55, No. 6, June 1967.

Greenwald, E.K., ed., *Electrical Hazards and Accidents: Their Cause and Prevention*, Van Nostrand Reinhold, 1991.

Grigsby, L.L., ed., *The Electric Power Engineering Handbook*, CRC Press LLC, 2001.

Grossner, Nathan R., *Transformers for Electronic Circuits*, McGraw-Hill, 1983.

Grover, Frederick W., *Inductance Calculations*, Dover Publications, 1973.

Gruber, Mike, *Learn how the ARRL Laboratory Evaluates New Products — and What Those Numbers Mean to You!*, QST, October 1994.

Gunn, Ross, *The Electrification of Precipitation and Thunderstorms*, Proceedings of the IRE, Vol. 45, No. 10, 1957.

Guofu, Zhou and Gong Lian, *An Improved Analytical Model for Braided Cable Shields*, IEEE Transactions on Electromagnetic Compatibility, Vol. 32, No. 2, May 1990.

Gupta, K.C., Ramesh Garg, and I.J. Bahl, *Microstrip Lines and Slotlines*, Artech House, 1979.

Gupta, K.C., Ramesh Garg, and Rakesh Chadha, *Computer-Aided Design of Microwave Circuits*, Artech House, 1981.

Gupta, Someshwar C., *Transform and State Variable Methods in Linear Systems*, John Wiley & Sons, 1966.

Guru, Bhag Singh and Hüseyin R. Hiziroğlu, *Electromagnetic Field Theory Fundamentals*, PWS, 1998.

Gustafson, W.G., *Magnetic Shielding of Transformers at Audio Frequencies*, The Bell System Technical Journal, Vol. 17, No. 3, July 1938.

Guy, Arthur W., Chung-Kwang Chou, John A. McDougall, and Carrol Sorensen, *Measurement of Shielding Effectiveness of Microwave-Protective Suits*, IEEE Transactions on Microwave Theory and Techniques, Vol. MTT-35, No. 11, November 1987.

Haase, Heinz, *Electrostatic Hazards: Their Evaluation and Control*, Veriag Chemie-Weinheim, 1977.

Hagen, Jon B., *Radio-Frequency Electronics: Circuits and Applications*, Cambridge University Press, 1996.

Hall, Stephen H., Garret W. Hall, and James A. McCall, *High-Speed Digital System Design: A Handbook of Interconnect Theory and Design Practices*, John Wiley & Sons, 2000.

Hansen, R.C., *Shielding Formulas for Near Fields*, Microwave Journal, Vol. 43, No. 11, November 2000.

Hansson, Kjell, Monica Sandström, and Anders Johnsson, *Measured 50 Hz Electric and Magnetic Fields in Swedish and Norwegian Residential Buildings*, IEEE Transactions on Instrumentation and Measurement, Vol. 45, No. 3, June 1996.

Hara, Masanori and Masanori Akazaki, *A Method for Prediction of Gaseous Discharge Threshold Voltage in the Presence of a Conducting Particle*, Journal of Electrostatics, Vol. 2, 1976/1977.

Hare, Ed and Robert Schetgen, ed., *Radio Frequency Interference: How to Find It and Fix It*, Publication No. 149, American Radio Relay League, 1991.

Harper, Charles A., ed., *Handbook of Wiring, Cabling, and Interconnecting for Electronics*, McGraw-Hill, 1972.

Harper, Charles A., ed., *High Performance Printed Circuit Boards*, McGraw-Hill, 2000.

Harper, Charles A., ed., *Passive Electronic Component Handbook*, McGraw-Hill, 1997.

Harper, W.R., *Contact and Frictional Electrification*, Oxford University Press, 1967.

Harrington, Roger F., *Time-Harmonic Electromagnetic Fields*, McGraw-Hill, 1961.

Harris Corporation, *Applications Notes 9304, 9307, 9310, 9768, and 9771*, 1998–1999.

Harris, John W., and Horst Stocker, *Handbook of Mathematics and Computational Science*, Springer, 1998.

Hartmann, Irving, John Nagy, and Murray Jacobson, *Explosive Characteristics of Titanium, Zirconium, Thorium, Uranium and Their Hyrides*, Bureau of Mines, Report of Investigations 4835, 1951.

Hartmann, Irving, *Recent Research on the Explosibility of Dust Dispersion*, Industrial and Engineering Chemistry, Vol. 40, 1948.

Hasselgren, Lennart and Jorma Luomi, *Geometrical Aspects of Magnetic Shielding at Extremely Low Frequencies*, IEEE Transactions on Electromagnetic Compatibility, Vol. 37, No. 3, August 1995.

Hatsuda, Takeshi and Tadashi Matsumoto, *Computation of Impedance of Partially Filled and Slotted Coaxial Line*, IEEE Transactions on Microwave Theory and Techniques, Vol. MTT-15, No. 11, November 1967.

Haus, Herman A. and James R. Melcher, *Electromagnetic Fields and Energy*, Prentice-Hall, 1989.

Haussmann, Gary, Marty Matthews, and Franz Gisin, *Impact on Radiated Emissions on Printed Circuit Board Stitching*, IEEE International Symposium on Electromagnetic Compatibility, 1999.

Haykin, S.S., *Active Network Theory*, Addison-Wesley, 1970.

Hayt, William H., Jr. and John A. Buck, *Engineering Electromagnetics*, McGraw-Hill, 2001.

Hayward, W.H., *Introduction to Radio Frequency Design*, Prentice-Hall, 1982.

Heck, Carl, *Magnetic Materials and Their Applications*, Butterworth, 1974.

Heidelberg, E., *Generation of Igniting Brush Discharges by Charged Layers on Earthed Conductors*, Proceedings of the Static Electrification Conference, Institute of Physics and the Physical Society, 1967.

Heidler, F., J.M. Cvetić, and B.V. Stanić, *Calculation of Lightning Current Parameters*, IEEE Transactions on Power Delivery, Vol. 14, No. 2, April 1999.

Heiland, C.A., *Geophysical Exploration*, Prentice-Hall, 1946.

Helfrick, A.O., *Modern Aviation Electronics*, Prentice-Hall, 1994.

Heller, Peter, *Analog Demonstrations of Ampere's Law and Magnetic Flux*, American Journal of Physics, Vol. 60, No. 1, January 1992 (figure corrections in March 1992).

Helszajn, J., *Microwave Planar Passive Circuits and Filters*, John Wiley & Sons, 1994.

Hemming, LeLand H., *Architectural Electromagnetic Shielding Handbook: A Design and Specifications Guide*, IEEE Press, 1992.

Hemming, LeLand H., *Electromagnetic Anechoic Chambers: A Fundamental Design and Specifications Guide*, IEEE Press, 2002.

Heumann, Klemens, *Magnetic Potentiometer of High Precision*, IEEE Transactions on Instrumentation and Measurement, Vol. IM-15, No. 4, December 1966.

Heydt, G.T., *Electric Power Quality*, Stars in a Circle Publications, 1991.

Hilberg, Wolfgang, *Electrical Characteristics of Transmission Lines*, Artech House, 1979.

Hill, David A., *Electromagnetic Theory of Reverberation Chambers*, National Institute of Standards and Technology, Technical Note 1506, 1998.

Hill, David A., *Electronic Mode Stirring for Reverberation Chambers*, IEEE Transactions on Electromagnetic Compatibility, Vol. 36, No. 4, November 1994.

Hingorani, Narain G., *Introducing Custom Power*, IEEE Spectrum, Vol. 32, No. 6, June 1995.

Hippel, Arthur von, ed., *Dielectric Materials and Applications*, Artech House, 1995.

Hiziroğlu, Hüseyin R., Unpublished Results, January 1986.

Hoburg, J.F., *A Computational Methodology and Results for Quasistatic Multilayered Magnetic Shielding*, IEEE Transactions on Electromagnetic Compatibility, Vol. 38, No. 1, February 1996.

Hoburg, J.F., *Principles of Quasistatic Magnetic Shielding with Cylindrical and Spherical Shields*, IEEE Transactions on Electromagnetic Compatibility, Vol. 37, No. 4, November 1995.

Hoeft, Lothar O. and Joseph S. Hofstra, *Measured Electromagnetic Shielding Performance of Commonly Used Cables and Connectors*, IEEE Transactions on Electromagnetic Compatibility, Vol. 30, No. 3, August 1988.

Hoer, Cletus and Carl Love, *Exact Inductance Equations for Rectangular Conductors With Applications to More Complicated Geometries*, Journal of Research of the National Bureau of Standards–C. Engineering and Instrumentation, Vol. 69C, No. 2, April–June 1965.

Hoffmann, Reinmut K., *Handbook of Microwave Integrated Circuits*, Artech House, 1987.

Holloway, Christopher L. and Edward F. Kuester, *Net and Partial Inductance of a Microstrip Ground Plane*, IEEE Transactions on Electromagnetic Compatibility, Vol. 40, No. 1, February 1998.

Holloway, Christopher L., Paul M. McKenna, Roger A. Dalke, Rodney A. Perala, and Charles L. Devor, Jr., *Time-Domain Modeling, Characterization, and Measurements of Anechoic and Semi-Anechoic Electromagnetic Test Chambers*, IEEE Transactions on Electromagnetic Compatibility, Vol. 44, No. 1, February 2002.

Holloway, Christopher L., Ronald R. DeLyser, Robert F. German, Paul McKenna, and Motohisa Kanda, *Comparison of Electromagnetic Absorbers Used in Anechoic and Semi-Anechoic Chambers for Emissions and Immunity Testing of Digital Devices*, IEEE Transactions on Electromagnetic Compatibility, Vol. 39, No. 1, February 1997.

Holm, Ragnar, *Electric Contacts: Theory and Application*, Springer-Verlag, 1967.

Horenstein, Mark N., *Microelectronic Circuits and Devices*, Prentice-Hall, 1996.

Horowitz, Paul and Winfield Hill, *The Art of Electronics*, Cambridge University Press, 1989.

Horváth, T. and I. Berta, *Static Elimination*, Research Studies Press, 1982.

Hotte, P.W., G. Gela, J.D. Mitchell, Jr., and P.F. Lyons, *Electrical Performance of Conductive Suits*, IEEE Transactions on Power Delivery, Vol. 12, No. 3, July 1997.

Howard W. Sams, *Reference Data for Radio Engineers*, 1975.

Huang, Yi, *Conducting Triangular Chambers for EMC Measurements*, Measurement Science Technology, Vol. 10, 1999.

Hubing, Todd H., James L. Drewniak, Thomas P. Van Doren, and David M. Hockanson, *Power Bus Decoupling on Multilayer Printed Circuit Boards*, IEEE Transactions on Electromagnetic Compatibility, Vol. 37, No. 2, 1995.

Hubing, Todd H., Thomas P. Van Doren, and James L. Drewniak, *Identifying and Quantifying Printed Circuit Board Inductance*, IEEE International Symposium on Electromagnetic Compatibility, 1994.

Hughes, J.F., *Electrostatic Powder Coating*, Research Studies Press, 1984.

Hunt, William T., Jr. and Robert Stein, *Static Electromagnetic Devices*, Allyn and Bacon, 1963.

Hymers, W., *Star-delta Method of Earth Electrode Measurements*, Electrical Review, Vol. 196, No. 1, January 1975.

IEC Report Publication, *Effects of Current Passing Through the Human Body, Part 1: General Aspects*, Publication 479-1, Geneva: International Electrotechnical Commission, 1984.

Inan, Umran S. and Aziz S. Inan, *Engineering Electromagnetics*, Addison-Wesley, 1999.

IPCS, Environmental Health Criteria 137-Electromagnetic Fields (300 Hz to 300 GHz), International Programme on Chemical Safety-World Health Association, 1992.

IPCS, Environmental Health Criteria 69, International Programme on Chemical Safety-World Health Association, 1987.

Jackson, J.D., *Classical Electrodynamics*, John Wiley & Sons, 1975.

Jacobson, Murray, Austin R. Cooper, and John Nagy, *Explosibility of Metal Powders*, Bureau of Mines, Report of Investigations 6516, 1964.

Jacobson, Murray, John Nagy, and Austin R. Cooper, *Explosibililty of Dusts Used in the Plastics Industry*, Bureau of Mines, Report of Investigations 5971, 1962.

Jacobson, Murray, John Nagy, Austin R. Cooper, and Frank J. Ball, *Explosibililty of Agricultural Dusts*, Bureau of Mines, Report of Investigations 5753, 1961.

Janssen W. and F. Nilber, *High-Frequency Circuit Engineering*, IEE, 1996.

Jefimenko, Oleg D., *Electricity and Magnetism*, Electret Scientific, 1966.

Johnson, Howard and Martin Graham, *High-Speed Signal Propagation: Advanced Black Magic*, Prentice-Hall, 2003.

Johnson, Howard W. and Martin Graham, *High-Speed Digital Design: A Handbook of Black Magic*, Prentice-Hall, 1993.

Johnson, Richard C. and Henry Jasik, ed., *Antenna Engineering Handbook*, McGraw-Hill, 1961.

Johnson, Walter C., *Transmission Lines and Networks*, McGraw-Hill, 1950.

Jonassen, Niels, *Electrostatics*, Chapman & Hall, 1998.

Jones, T.B. and S. Chan, *Charge Relaxation in Partially Filled Vessels*, Journal of Electrostatics, Vol. 22, 1989.

Jones, Thomas B. and Jack L. King, *Powder Handling and Electrostatics: Understanding and Preventing Hazards*, Lewis, 1991.

Jones, Thomas B. and Kit-Ming Tang, *Charge Relaxation in Powder Beds*, Journal of Electrostatics, Vol. 19, 1987.

Jordon, Edward C. and Keith G. Balmain, *Electromagnetic Waves and Radiating Systems*, Prentice-Hall, 1968.

Joyner, Kenneth H., Paul R. Copeland, and Ian P. MacFarlane, *An Evaluation of a Radiofrequency Protective Suit and Electrically Conductive Fabrics*, IEEE Transactions on Electromagnetic Compatibility, Vol. 31, No. 2, May 1989.

Kaiser, Kenneth L., *A Study of Free-Surface Electrohydrodynamics*, Ph.D. Thesis, Purdue University, 1989.

Kallman, Raymond, *Realities of Wrist Strap Monitoring Systems*, Electrostatic Overstress/Electrostatic Discharge Symposium Proceedings, 1994.

Kam, Dong Gun, Heeseok Lee, Seungyong Baek, Bongcheol Park, and Joungho Kim, *Enhanced Immunity against Crosstalk and EMI Using GHz Twisted Differential Line Structure on PCB*, IEEE International Symposium on Electromagnetic Compatibility, 2002.

Kanda, Motohisa, *Electromagnetic-Field Distortion Due to a Conducting Rectangular Cylinder in a Transverse Electromagnetic Cell*, IEEE Transactions on Electromagnetic Compatibility, Vol. EMC-24, No. 3, August 1982.

Kaplan, Wilfred, *Operational Methods for Linear Systems*, Addision-Wesley, 1962.

Karrer, Nicolas and Patrick Hofer-Noser, *A New Current Measuring Principle for Power Electronic Applications*, Proceedings of ISPSD, 1999.

Keiser, Bernhard, *Principles of Electromagnetic Compatibility*, Artech House, 1987.

Keithley, *Low Level Measurements*, Keithley Test Instruments, 1993.

Kerr, Donald E., ed., *Propagation of Short Radio Waves*, McGraw-Hill, 1951.

Kessler, LeAnn and W. Keith Fisher, *A Study of the Electrostatic Behavior of Carpets Containing Conductive Yarns*, Journal of Electrostatics, Vol. 39, 1997.

Khalifa, M., ed., *High-Voltage Engineering: Theory and Practice*, Marcel Dekker, 1990.

Kibble, B.P. and G.H. Raymer, *Coaxial AC Bridges*, Adam Hilger, 1984.

Kimbark, Edward Wilson, *Electrical Transmission of Power and Signals*, John Wiley & Sons, 1949.

Kimmel, William D. and Daryl D. Gerke, *Electromagnetic Compatibility in Medical Equipment: A Guide for Designers and Installers*, IEEE, 1995.

Kimmel, William D., *Wide Frequency Impedance Modeling of EMI Ferrites*, IEEE International Symposium on Electromagnetic Compatibility, 1994.

King, Ronald W. P., *Transmission Line Theory*, McGraw-Hill, 1955.

Kinsman, Robert G., *Crystal Filters: Design, Manufacture, and Application*, John Wiley & Sons, 1987.
Klein, A.G., *Demonstration of Ampere's Circuital Law Using a Rogowski Coil*, American Journal of Physics, Vol. 43, No. 4, April 1975.
Kobb, Bennett Z., *Spectrum Guide: Radio Frequency Allocations in the United States 30 MHz-300 GHz*, New Signals Press, 1995.
Korn, Granino A., and Theresa M. Korn, *Mathematical Handbook for Scientists and Engineers*, McGraw-Hill, 1968.
Koucherng, Roger Lee, John E. Bennett, William H. Pinkston, and J.E. Bryant, *New Method for Assessing EED Susceptibility to Electromagnetic Radiation*, IEEE Transactions on Electromagnetic Compatibility, Vol. 33, No. 4, November 1991.
Koyler, John M. and Donald E. Watson, *ESD from A to Z: Electrostatic Discharge Control for Electronics*, Van Nostrand Reinhold, 1990.
Kraus, John D., *Antennas*, McGraw-Hill, 1988.
Kraus, John D., *Electromagnetics*, McGraw-Hill, 1992.
Krauss, Herbert L. and Charles W. Allen, *Designing Toroidal Transformers to Optimize Wideband Performance*, Electronics, August 16, 1973.
Krevelen, D.W. Van, *Properties of Polymers*, Elsevier, 1990.
Ku, Y.H., *Transient Circuit Analysis*, D. Van Nostrand, 1961.
Kumar, A., *Fixed and Mobile Terminal Antennas*, Artech House, 1991.
Kussy, Frank W. and Jack L. Warren, *Design Fundamentals for Low-Voltage Distribution and Control*, Marcel Dekker, 1987.
Lacanette, Kerry, *A Basic Introduction to Filters-Active, Passive, and Switched Capacitor*, National Semiconductor Application Note 779, April 1991.
Ladbury, John M. and Galen H. Koepke, *Reverberation Chamber Relationships: Corrections and Improvements or Three Wrongs Can (almost) Make a Right*, IEEE International Symposium on Electromagnetic Compatibility, 1999.
Ladbury, John M. and Kevin Goldsmith, *Reverberation Chamber Verification Procedures, or, How to Check if Your Chamber Ain't Broke and Suggestions on How to Fix It if It Is*, IEEE International Symposium on Electromagnetic Compatibility, 2000.
Ladbury, John M., Galen H. Koepke, and Dennis G. Camell, *Improvements in the CW Evaluation of Mode-Stirred Chambers*, IEEE International Symposium on Electromagnetic Compatibility, 1997.
Lago, Gladwyn V. and Donald L. Waidelich, *Transients in Electrical Circuits*, Ronald Press, 1958.
Lai, Fang-Shi and J. Kenneth Watson, *A Sectional Spheroid Model for Cylindrical Permanent Magnets*, IEEE Transactions on Magnetics, Vol. MAG-16, No. 2, 1980.
Laport, Edmund A., *Radio Antenna Engineering*, McGraw-Hill, 1952.
LaRocca, Robert L., *Personnel Protection Devices for Use on Appliances, IEEE Transactions on Industry Applications*, Vol. 28, No. 1, January/February 1992.
Larsen, Øystein, Janicke H. Hagen, and Kees van Wingerden, *Ignition of Dust Clouds by Brush Discharges in Oxygen Enriched Atmospheres*, Journal of Loss Prevention in the process industries, Vol. 14, 2001.
Lasitter, Homer A., *Low-Frequency Shielding Effectiveness of Conductive Glass*, IEEE Transactions on Electromagnetic Compatibility, Vol. EMC-6, No. 2, July 1964.
Lawrence, Ralph R., *Principles of Alternating-Current Machinery*, McGraw-Hill, 1953.
Lee, Capt. Paul H., *The Amateur Radio Vertical Antenna Handbook*, CQ Communications, 1984(a).
Lee, K.S.H and F.C. Yang, *A Wire Passing by a Circular Aperture in an Infinite Ground Plane*, EMP Interaction Notes, IN 317, Air Force Weapons Laboratory, Kirtland AFB, NM, February 1977.
Lee, K.S.H., *EMP Interaction: Principles, Techniques, and Reference Data, A Handbook of Technology from the EMP Interaction Notes*, Hemisphere, 1986.
Lee, Kai Fong, *Principles of Antenna Theory*, John Wiley & Sons, 1984(b).

Lee, Thomas H., *The Design of CMOS Radio-Frequency Integrated Circuits*, Cambridge University Press, 1998.

Leferink, Frank B.J. and Marcel J.C.M. van Doorn, *Inductance of Printed Circuit Board Planes*, IEEE International Symposium on Electromagnetic Compatibility, 1993.

Leferink, Frank B.J., *High Field Strength in a Large Volume: The Balanced Stripline TEM Antenna*, IEEE International Symposium on Electromagnetic Compatibility, 1998.

Leferink, Frank B.J., *Inductance Calculations; Methods and Equations*, IEEE International Symposium on Electromagnetic Compatibility, 1995.

Lefferson, Peter, *Twisted Magnetic Wire Transmission Line*, IEEE Transactions on Parts, Hybrids, and Packaging, Vol. PHP-7, No. 4, December 1971.

Lenk, John D., *Handbook of Electronic Circuit Designs*, Prentice-Hall, 1976.

Lerner, C.M., *Problems and Solutions in Electromagnetic Theory*, John-Wiley & Sons, 1985.

LeVasseur, Dave, *Midcom's Tips for Transformer Modeling*, Midcom Technical Note #82, 1998.

Liao, Samuel Y., *Microwave Devices and Circuits*, Prentice-Hall, 1990.

Licari, James, J., *Plastic Coatings for Electronics*, McGraw-Hill, 1970.

Lindgren, Erik A., *Contemporary RF Enclosures*, 1967.

Lo, Y.T. and S.W. Lee, ed., *Antenna Handbook*, Van Nostrand Reinhold, 1988.

Looms, J.S.T., *Insulators for High Voltages*, Peter Peregrinus, 1988.

Lopez, Michael, John L. Prince, and Andreas C. Cangellaris, *Influence of a Floating Plane on Effective Ground Plane Inductance in Multilayer and Coplanar Packages*, IEEE Transactions on Advanced Packaging, Vol. 22, No. 2, May 1999.

Lovatt, Howard C. and Peter A. Watterson, *Energy Stored in Permanent Magnets*, IEEE Transactions on Magnetics, Vol. 35, No. 1, January 1999.

Lubkin, Yale Jay, *Filter Systems and Design: Electrical, Microwave, and Digital*, Addison-Wesley, 1970.

Lüttgens, Günter and Norman Wilson, *Electrostatic Hazards*, Butterworth-Heinemann, 1977.

MIT Staff, *Magnetic Circuits and Transformers*, The MIT Press, 1943.

Ma, Mark T., Motohisa Kanda, Myron L. Crawford, and Ezra B. Larsen, *A Review of Electromagnetic Compatibility/Interference Measurement Methodologies*, Proceedings of the IEEE, Vol. 73, No. 3, March 1985.

Macalpine, W.W. and R.O. Schildknecht, *Coaxial Resonators with Helical Inner Conductor*, Proceedings of the IRE, Vol. 47, No. 12, December 1959.

Macatee, Stephen R., *Considerations in Grounding and Shielding Audio Devices*, Journal of Audio Engineering Society, Vol. 43, No. 6, June 1995.

Mager, A., *Magnetic Shielding Efficiencies of Cylindrical Shells with Axis Parallel to the Field*, Journal of Applied Physics, Vol. 39, No. 3, February 1968.

Mager, Albrecht J., *Magnetic Shields*, IEEE Transactions on Magnetics, Vol. MAG-6, No. 1, March 1970.

Magnusson, Philip C., *Transmission Lines and Wave Propagation*, Allyn and Bacon, 1970.

Magrab, Edward B. and Donald S. Blomquist, *The Measurement of Time-Varying Phenomena*, John Wiley & Sons, 1971.

Maissel, Leon I. and Reinhard Glang, *Handbook of Thin Film Technology*, McGraw-Hill, 1970.

Mardiguian, Michel, *Controlling Radiated Emissions by Design*, Van Nostrand Reinhold, 1992.

Mardiguian, Michel, *Electromagnetic Control in Components and Devices, Vol. 5*, Interference Control Technologies, 1988(a).

Mardiguian, Michel, *EMI Troubleshooting Techniques*, McGraw-Hill, 2000.

Mardiguian, Michel, *Grounding and Bonding, Vol. 2*, Interference Control Technologies, 1988(b).

Mardiguian, Michel, *Interference Control in Computers and Microprocessor-Based Equipment*, Don White Consultants, 1984.

Marshall, J.L., *Lightning Protection*, John Wiley & Sons, 1973.

Marshall, S.V., Richard E. DuBroff, and G.G. Skitek, *Electromagnetic Concepts and Applications*, Prentice-Hall, 1996.

Martin, R.E., *Electrical Interference in Electronic Systems: Its Avoidance within High-Voltage Substations and Elsewhere*, Research Studies Press, 1979.
Maruvada, P. Sarma, A. Turgeon, D.L. Goulet, and C. Cardinal, *An Experimental Study of Residential Magnetic Fields in the Vicinity of Transmission Lines*, IEEE Transactions on Power Delivery, Vol. 13, No. 4, October 1998.
Matisoff, Bernard S., *Wiring and Cable Designer's Handbook*, TAB, 1987.
Maurer, B., M. Glor, G. Lüttgens, and L. Post, *Hazards Associated with Propagating Brush Discharges on Flexible Intermediate Bulk Containers, Compounds and Coated Materials*, Electrostatics 1987, Institute of Physics Conference Series, No. 85, 1987.
McCaig, Malcolm, *Permanent Magnets in Theory and Practice*, John Wiley & Sons, 1977.
McCleery, D.K., *Introduction to Transients*, John Wiley & Sons, 1961.
McDonald, Jack R., Wallace B. Smith, and Herbert W. Spencer, III, *A Mathematical Model for Calculating Electrical Conditions in Wire-Duct Electrostatic Precipitation Devices*, Journal of Applied Physics, Vol. 48, No. 6, June 1977.
McElroy, P.K., *Designing Resistive Attenuating Networks*, Proceedings of the IRE, Vol. 23, No. 3, March 1935.
McGranaghan, Mark F., David R. Mueller, and Marek J. Samotyj, *Voltage Sags in Industrial Systems*, IEEE Transactions on Industrial Applications, Vol. 29, No. 2, March/April 1993.
McLachlan, N.W., *Bessel Functions for Engineers*, Oxford University Press, 1934.
Measures, Richard L., *A Balanced Antenna Tuner*, QST, February 1990.
Meeldijk, Victor, *Electronic Components*, John Wiley & Sons, 1995.
Melcher, James R., *Continuum Electromechanics*, MIT Press, 1981.
Meliopoulos, A.P. Sakis, *Power System Grounding and Transients: An Introduction*, Marcel Dekker, 1988.
Metaxas, A.C. and R.J. Meredith, *Industrial Microwave Heating*, Peter Peregrinus, 1983.
Meulenaere, Filip De and Jean Van Bladel, *Polarizability of Some Small Apertures*, IEEE Transactions on Antennas and Propagation, Vol. AP-25, No. 2, March 1977.
Mild, Kjell Hansson, *Occupational Exposure to Radio Frequency Electromagnetic Fields*, Proceedings of the IEEE, Vol. 68, No. 1, January 1980.
Mills, Jeffrey P., *Electro-Magnetic Interference Reduction in Electronic Systems*, Prentice-Hall, 1993.
Miner, Gayle F., *Lines and Electromagnetic Fields for Engineers*, Oxford University Press, 1996.
Mitchell, M.J., *Analyzing the Double Exponential Pulse*, EMC Test & Design, Vol. 5, No. 3, March 1994.
Mizuno, A., *Electrostatic Precipitation*, IEEE Transactions on Dielectrics and Electrical Insulation, Vol. 7, No. 5, October 2000.
Montrose, Mark I., *EMC and the Printed Circuit Board*, IEEE Press, 1999.
Montrose, Mark I., *Printed Circuit Board Techniques For EMC Compliance*, IEEE Press, 2000.
Moon, Parry and Domina Eberle Spencer, *Field Theory for Engineers*, D. Van Nostrand, 1961.
Moon, Parry and Domina Eberle Spencer, *Foundations of Electrodynamics*, D. Van Nostrand, 1960.
Moore, A.D., ed., *Electrostatics and Its Applications*, John Wiley & Sons, 1973.
Moore, Richard K., *Traveling-Wave Engineering*, McGraw-Hill, 1960.
Morrish, Allan H., *The Physical Principles of Magnetism*, John Wiley & Sons, 1965.
Morrison, Ralph, *Grounding and Shielding Techniques*, John Wiley & Sons, 1998.
Morrison, Ralph, *Solving Interference Problems in Electronics*, John Wiley & Sons, 1995.
Morse, Philip M. and Herman Feshbach, *Methods of Theoretical Physics*, McGraw-Hill, 1953.
Moser, J. Ronald and Ralph F. Spencer, Jr., *Predicting the Magnetic Fields from a Twisted-Pair Cable*, IEEE Transactions on Electromagnetic Compatibility, Vol. EMC-10, No. 3, September 1968(a).
Moser, J. Ronald, *An Empirical Study of ELF and VLF Shield Cans*, IEEE Transactions on Electromagnetic Compatibility, Vol. EMC-10, No. 1, March 1968(b).
Moser, J. Ronald, *Low-Frequency Shielding of a Circular Loop Electromagnetic Field Source*, IEEE Transactions on Electromagnetic Compatibility, Vol. EMC-9, No. 1, March 1967.
Mullin, William F., *abc's of Capacitors*, Howard W. Sams, 1967.
MuRata, *EMI Suppression Filter*, Cat. No. C30E, Murata.

Nagy, John and Harry C. Verakis, *Development and Control of Dust Explosions*, Marcel Dekker, 1983.

Nagy, John, Austin R. Cooper, and Henry G. Dorsett, Jr., *Explosibility of Miscellaneous Dusts*, Bureau of Mines, Report of Investigation 7208, 1968.

Nagy, John, Henry G. Dorsett, Jr., and Austin R. Cooper, *Explosibility of Carbonaceous Dusts*, Bureau of Mines, Report of Investigations 6597, 1965.

Nahman, Norris S. and Donald R. Holt, *Transient Analysis of Coaxial Cables Using the Skin Effect Approximation* $A+B\sqrt{s}$, IEEE Transactions on Circuit Theory, Vol. 19, No. 5, September 1972.

Nanevicz, J.E., *Some Techniques for the Elimination of Corona Discharge Noise in Aircraft Antennas*, Proceedings of the IEEE, Vol. 52, No. 1, January 1964.

Nanevicz, Joseph E., *Static Charging and its Effects on Avionic Systems*, IEEE Transactions on Electromagnetic Compatibility, Vol. EMC-24, No. 2, May 1982.

National Association of Relay Manufacturers, *Engineer's Relay Handbook*, Hayden, 1960.

Nave, Mark J., *Power Line Filter Design for Switched-Mode Power Supplies*, Van Nostrand Reinhold, 1991.

Nelson, W.R., *Interference Handbook*, Radio Publications, 1981.

Nicholson, J.R. and J.A. Malack, *RF Impedance of Power Lines and Line Impedance Stabilization Networks in Conducted Interference Measurements*, IEEE Transactions on Electromagnetic Compatibility, EMC-15, No. 2, May 1973.

Nilsson, James W. and Susan A. Riedel, *Electric Circuits*, Prentice-Hall, 2001.

Nishiyama, Hitoshi and Mitsunobu Nakamura, *Capacitance of Disk Capacitors*, IEEE Transactions on Components, Hybrids, and Manufacturing Technology, Vol. 16, No. 3, May 1993.

Nixon, Floyd E., *Handbook of Laplace Transformation: Fundamentals, Applications, Tables, and Examples*, Prentice-Hall, 1965.

Novak, Istvan, *Reducing Simultaneous Switching Noise and EMI on Ground/Power Planes by Dissipative Edge Termination*, IEEE Transactions on Advanced Packaging, Vol. 22, No. 3, August 1999.

Nussbaum, Allen, *Electromagnetic Theory for Engineers and Scientists*, Prentice-Hall, 1965.

O'Riley, Ronald P., *Electrical Grounding: Bringing Grounding Back to Earth*, Delmar, 1999.

Oliver, Frank J., *Practical Instrumentation Transducers*, Hayden, 1971.

Olsen, E., *Applied Magnetism: A Study in Quantities*, Springer-Verlag, 1966.

Olsen, R.G. and T.A. Pankaskie, *On the Exact, Carson and Image Theories for Wires at or Above the Earth's Interface*, IEEE Transactions on Power Apparatus and Systems, Vol. PAS-102, No. 4, April 1983.

Olsen, Robert G., *Some Observations About Shielding Extremely Low-Frequency Magnetic Shields by Finite Width Shields*, IEEE Transactions on Electromagnetic Compatibility, Vol. 38, No. 3, August 1996.

Olsher, Dick, *Linday-Geyer Highly Magnetic Cable*, Stereophile, February 1991.

Oppenheim, Alan V., Alan S. Willsky, and S. Hamid Nawab, *Signals & Systems*, Prentice-Hall, 1997.

Ott, Henry W., *Noise Reduction Techniques in Electronic Systems*, John Wiley & Sons, 1988.

Ozenbaugh, Richard Lee, *EMI Filter Design*, Dekker, 1996.

Page, H., *Principles of Aerial Design*, D. Van Nostrand, 1966.

Papakanellos, Panagiotis J., Dimitra I. Kaklamani, and Christos N. Capsalis, *Analysis of an Infinite Current Source Above a Semi-Infinite Lossy Ground Using Fictitious Current Auxiliary Sources in Conjunction With Complex Image Theory Techniques*, IEEE Transactions on Antennas and Propagation, No. 49, No. 10, October 2001.

Parker, Carole U., *Choosing a Ferrite for the Suppression of EMI*, Fair-Rite Technical Paper, Fair-rite Products Corp., 1991.

Parker, Rollin J. and Robert J. Studders, *Permanent Magnets and Their Application*, John Wiley & Sons, 1962.

Paul, Clayton R. and Jack W. McKnight, *Prediction of Crosstalk Involving Twisted Pairs of Wires-Part I: A Transmission-Line Model for Twisted-Wire Pairs*, IEEE Transactions on Electromagnetic Compatibility, Vol. EMC-21, No. 2, May 1979.

Paul, Clayton R. and Jack W. McKnight, *Prediction of Crosstalk Involving Twisted Pairs of Wires-Part II: A Simplified Low-Frequency Prediction Model*, IEEE Transactions on Electromagnetic Compatibility, Vol. EMC-21, No. 2, May 1979.

Paul, Clayton R. and Keith B. Hardin, *Diagnosis and Reduction of Conducted Noise Emissions*, IEEE Transactions on Electromagnetic Compatibility, Vol. 30, No. 4, November 1988.

Paul, Clayton R. and Syed A. Nasar, *Introduction to Electromagnetic Fields*, McGraw-Hill, 1987.

Paul, Clayton R., *Analysis of Multiconductor Transmission Lines*, John Wiley & Sons, 1994.

Paul, Clayton R., *Effectiveness of Multiple Decoupling Capacitors*, IEEE Transactions on Electromagnetic Compatibility, Vol. EMC-34, No. 2, May 1992(a).

Paul, Clayton R., *Frequency Response of Multiconductor Transmission Lines Illuminated by an Electromagnetic Field*, IEEE Transactions on Electromagnetic Compatibility, Vol. EMC-18, No. 4, November 1976.

Paul, Clayton R., *Introduction to Electromagnetic Compatibility*, John Wiley & Sons, 1992(b).

Paul, Clayton R., *Prediction of Crosstalk in Ribbon Cables*, IEEE International Symposium on Electromagnetic Compatibility, 1978.

Peek, F.W., *Dielectric Phenomena in High Voltage Engineering*, McGraw-Hill, 1929.

Perez, Reinaldo, ed., *Handbook of Electromagnetic Compatibility*, Academic Press, 1995.

Perls, Thomas A., *Electrical Noise from Instrument Cables Subjected to Shock and Vibration*, Journal of Applied Physics, Vol. 23, No. 6, June 1952.

Perry, Michael P., *Low Frequency Electromagnetic Design*, Marcel Dekker, 1985.

Perry, Tekla S., *Today's View of Magnetic Fields*, IEEE Spectrum, Vol. 31, No. 12, December 1994.

Perz, M.C. and M.R. Raghuveer, *Transmission Lines — Effects of Earth Resistivity on Magnetic Field, Images and Equivalent Circuit*, IEEE Transactions on Power Apparatus and Systems, Vol. PAS-98, No. 6, November/December 1979.

Pettersson, Per, *Image Representation of Wave Propagation on Wires Above, On, and Under Ground*, IEEE Transactions on Power Delivery, Vol. 9, No. 2, April 1994.

Pidoll, Ulrich von, Helmut Krämer, and Heino Bothe, *Avoidance of Electrostatic Hazards during Refueling of Motorcars*, Journal of Electrostatics, Vol. 40–41, 1997.

Plastics for Electronics: Desk-top Data Bank, The International Plastics Selector, 1979.

Plonsey, Robert and Robert E. Collin, *Principles and Applications of Electromagnetic Fields*, McGraw-Hill, 1961.

Pratt, Thomas H., *Electrostatic Ignition Hazards*, Burgoynes, 1995.

Prentiss, Stan, *Modern Television: Service and Repair*, Prentice-Hall, 1989.

Pucel, Robert A., Daniel J. Massé, and Curtis P. Hartwig, *Losses in Microstrip*, IEEE Transactions on Microwave Theory and Techniques, Vol. MTT-16, No. 6, June 1968.

Pugh, Emerson M. and Emerson W. Pugh, *Principles of Electricity and Magnetism*, Addison-Wesley, 1970.

Püschner, H., *Heating With Microwaves: Fundamentals, Components and Circuit Technique*, Springer-Verlag, 1966.

Queen, Ralph H., *Common or Differential-Mode Noise? It Makes a Difference in Your Filter*, EMC Technology, Vol. 4, No. 3, July/September 1985.

Räde, Lennart, and Bertil Westergren, *Beta Mathematics Handbook*, Studentlitteratur, 1992.

Radun, Arthur, *An Alternative Low-Cost Current-Sensing Scheme for High-Current Power Electronics Circuits*, IEEE Transactions on Industrial Electronics, Vol. 42, No. 1, February 1995.

Rakov, Vladimir A. and Martin A. Uman, *Review and Evaluation of Lightning Return Stroke Models Including Some Aspects of Their Application*, IEEE Transactions on Electromagnetic Compatibility, Vol. EMC-40, No. 4, November 1998.

Raloff, Janet, *EMFs Run Aground*, Science News, Vol. 144, August 1993.

Ramboz, John D., *Machinable Rogowski Coil, Design, and Calibration*, IEEE Transactions on Instrumentation and Measurement, Vol. 45, No. 2, April 1996.

Ramo, Simon, John R. Whinnery, and Theodore Van Duzer, *Fields and Waves in Communication Electronics*, John Wiley & Sons, 1965.

Rao, Nannapaneni Narayana, *Elements of Engineering Electromagnetics*, Prentice-Hall, 1994.

Ratz, Alfred G., *Triboelectric Noise*, Instrument Society of America Transactions, Vol. 9, No. 2, 1970.

Rauch, Gregory B., Gary Johnson, Paul Johnson, Andreas Stamm, Seietsu Tomita, and John Swanson, *A Comparison of International Residential Grounding Practices and Associated Magnetic Fields*, IEEE Transactions on Power Delivery, Vol. 7, No. 2, April 1992.

Raychem Wire and Cable, *Electrical Screening*, Tyco Electronics Corporation.

Redl, Richard, Paolo Tenti, and J. Daan Van Wyk, *Power Electronic's Polluting Effects*, IEEE Spectrum, Vol. 34, No. 5, May 1997.

Reed, Michael and Ron Rohrer, *Applied Introductory Circuit Analysis for Electrical and Computer Engineers*, Prentice-Hall, 1999.

Reilly, J. Patrick, *Electrical Stimulation and Electropathology*, Cambridge University Press, 1992.

Riggs, Olen L., Jr. and Carl E. Locke, *Anodic Protection*, Plenum Press, 1981.

Rikitake, T., *Magnetic and Electromagnetic Shielding*, D. Reidel, 1987.

Riley, Karl, *Tracing EMFs in Building Wiring and Grounding: A Practical Guide for Reducing AC Magnetic Fields Produced by Building Wiring and Grounding Practices*, Magnetic Sciences International, 1995.

Rizk, F.A.M., *Low-Frequency Shielding Effectiveness of a Double Cylinder Enclosure*, IEEE Transactions on Electromagnetic Compatibility, Vol. EMC-19, No. 1, February 1977.

Rizk, Farouk A.M., *Effect of Floating Conducting Objects on Critical Switching Impulse Breakdown of Air Insulation*, IEEE Transactions on Power Delivery, Vol. 10, No. 3, July 1995.

Roberts, Willmar K., *A New Wide-Band Balun*, Proceedings of the IRE, Vol. 45, No. 12, December 1957.

Rochester, Louis, *Coaxial Cables Application Data*, Boston Technical, 1969.

Rogers, Walter E., *Introduction to Electric Fields*, McGraw-Hill, 1954.

Roseberry, B.E. and R.B. Schulz, *A Parallel-Strip Line for Testing RF Susceptibility*, IEEE Transactions on Electromagnetic Compatibility, EMC-7, No. 2, June 1965.

Rosenstark, Sol, *Transmission Lines in Computer Engineering*, McGraw-Hill, 1994.

Roters, Herbert C., *Electromagnetic Devices*, John Wiley & Sons, 1941.

Rotman, Walter and Nicholas Karas, *Printed Circuit Radiators: The Sandwich Wire Antenna*, The Microwave Journal, August 1959.

Rowland, Phillip W., *Industrial System Grounding for Power, Static, Lightning, and Instrumentation, Practical Applications*, IEEE Transactions on Industry Applications, Vol. 31, No. 6, November/December 1995.

Rudge, A.W., K. Milne, A.D. Olver, and P. Knight, *The Handbook of Antenna Design, Volume 1 and 2*, Peter Peregrinus Ltd, 1982–83.

Ruehli, A.E., *Inductance Calculations in a Complex Integrated Circuit Environment*, IBM J. Res. Develop., September 1972.

Ruehli, Albert, Clayton Paul, and Jan Garrett, *Inductance Calculations using Partial Inductances and Macromodels*, IEEE International Symposium on Electromagnetic Compatibility, 1995.

Rulf, Benjamin and Gregory A. Robertshaw, *Understand Antennas for Radar, Communications, and Avionics*, Van Nostrand Reinhold, 1987.

Ruston, Henry and Joseph Bordogna, *Electric Networks: Functions, Filters, and Analysis*, McGraw-Hill, 1966.

Ruthroff, C.L., *Some Broad-Band Transformers*, Proceedings of the IRE, Vol. 47, No. 8, August 1959.

Rutledge, David B., *The Electronics of Radio*, Cambridge University Press, 1999.

Saums, Harry L. and Wesley W. Pendleton, *Materials for Electrical Insulating and Dielectric Functions*, Hayden, 1973.

Schelkunoff, S.A., *The Electromagnetic Theory of Coaxial Transmission Lines and Cylindrical Shields*, The Bell System Technical Journal, Vol. 13, No. 4, October 1934.

Schlabbach, J., D. Blume, and T. Stephanblome, *Voltage Quality in Electrical Power Systems*, The Institution of Electrical Engineers, 2001.

Schneider, M.V., Bernard Glance, and W.F. Bodtmann, *Microwave and Millimeter Wave Hybrid Integrated Circuits for Radio Systems*, The Bell System Technical Journal, Vol. 48, No. 6, July–August 1969.

Schultz, John J., *The G4ZU X Beam for 20*, CQ, June 1965.

Schulz, Richard B., George C. Huang, and Walter L. Williams, *RF Shielding Design*, IEEE Transactions on Electromagnetic Compatibility, Vol. EMC-10, No. 1, March 1968.

Schulz, Richard B., *Shielding Theory and Practice*, IEEE Transactions on Electromagnetic Compatibility, Vol. EMC-30, No. 3, August 1988.

Schwab, Adolf J., *High-Voltage Measurement Techniques*, The MIT. Press, 1972.

Schwarz, S.J., *Analytical Expressions for the Resistance of Grounding Systems*, Transactions of AIEE, Part III-B, August 1954.

Schweitzer, Philip A., ed., *Corrosion and Corrosion Protection Handbook*, Marcel Dekker, 1988.

Sclater, Neil, *Electrostatic Discharge Protection for Electronics*, TAB, 1990.

Secker, P.E. and J.N. Chubb, *Instrumentation for Electrostatic Measurements*, Journal of Electrostatics, Vol. 16, 1984.

Sedra, Adel S. and Kenneth C. Smith, *Microelectronic Circuits*, Saunders, 1991.

Seshadri, S.R., *Fundamentals of Transmission Lines and Electromagnetic Fields*, Addison-Wesley, 1971.

Setian, Leo, *Practical Communication Antennas with Wireless Applications*, Prentice-Hall, 1998.

Sevick, Jerry, *Building Baluns and Ununs: Practical Designs for the Experimenter*, CQ Communications, 1994.

Sevick, Jerry, *Transmission Line Transformers*, The American Radio Relay League, 1987.

Shen, L.C., S.A. Long, M.R. Allerding, and M.D. Walton, *Resonant Frequency of a Circular Disc, Printed-Circuit Antenna*, IEEE Transactions on Antennas and Propagation, Vol. AP-25, No. 4, July 1977.

Shenfeld, Saul, *Magnetic Fields of Twisted-Wire Pairs*, IEEE Transactions on Electromagnetic Compatibility, Vol. EMC-11, No. 4, November 1969.

Shenfeld, Saul, *Shielding of Cylindrical Tubes*, IEEE Transactions on Electromagnetic Compatibility, Vol. EMC-10, No. 1, March 1968.

Siebert, William McC., *Circuits, Signals, and Systems*, McGraw-Hill, 1986.

Silver, Samuel, ed., *Microwave Antenna Theory and Design*, Peter Peregrinus, 1984.

Silvester, P., *Modern Electromagnetic Fields*, Prentice-Hill, 1968.

Sim, A.C., *New H.F. Proximity-Effect Formula*, Wireless Engineer, Vol. 30, No. 7, August 1953.

Simpson, Ted L., *Effect of a Conducting Shield on the Inductance of an Air-Core Solenoid*, IEEE Transactions on Magnetics, Vol. 35, No. 1, January 1999.

Single Side Band, Collins Radio Company, 1959.

Sjögren, C.T.H., *Application of the Wave Impedance Concept when Calculating Shielding Effectiveness*, Ninth International Conference on Electromagnetic Compatibility, No. 396, September 1994.

Skilling, Hugh Hildreth, *Electric Transmission Lines*, McGraw-Hill, 1951.

Slade, Paul G., ed., *Electrical Contacts: Principles and Applications*, Marcel Dekker, 1999.

Smith, Albert A., Jr., *Coupling of External Electromagnetic Fields to Transmission Lines*, John Wiley & Sons, 1977.

Smith, Douglas C., *High Frequency Measurements and Noise in Electronic Circuits*, Van Nostrand Reinhold, 1993.

Smith, Douglas C., *Unusual Forms of ESD and Their Effects*, Electrostatic Overstress/Electrostatic Discharge Symposium Proceedings, 1999.

Smith, Glenn S., *Radiation Efficiency of Electrically Small Multiturn Loop Antennas*, IEEE Transactions on Antennas and Propagation, Vol. AP-20, No. 5, September 1972.

Smits, F.M., *Measurement of Sheet Resistivities with the Four-Point Probe*, The Bell System Technical Journal, No. 3, May 1958.

Smythe, William R., *Static and Dynamic Electricity*, McGraw-Hill, 1968.

Snelling, E.C., *Soft Ferrites: Properties and Applications*, CRC Press, 1969.

Soliman, Samir S. and Mandyam D. Srinath, *Continuous and Discrete Signals and Systems*, Prentice-Hall, 1998.

Somerville, J.M., *The Electric Arc*, John Wiley & Sons, 1959.

Southwick, Roger, *EMI Signal Measurements at Open Area Test Sites*, EMC Test & Design, Vol. 3, No. 4, July/August 1992.

Spangenberg, Karl R., *Vacuum Tubes*, McGraw-Hill, 1948.

Spencer, Ned A., *An Antenna for 30-50 MC Service Having Substantial Freedom from Noise Caused by Precipitation Static and Corona*, IRE Transactions on Vehicular Communications, VC-9, No. 2, August 1960.

Spiegel, Murray R., and John Liu, *Mathematical Handbook of Formulas and Tables*, Schaum's Outline Series, 1999.

Standler, Ronald B., *Protection of Electronic Circuits from Overvoltages*, John Wiley & Sons, 1989.

Static Control Components, Inc., *Choosing the Right Static Bag*, Technical Bulletin P1, 1996.

Stevenson, William D., Jr., *Elements of Power System Analysis*, McGraw-Hill, 1975.

Streetman, Ben G. and Sanjay Banerjee, *Solid State Electronic Devices*, Prentice-Hall, 2000.

Stremler, Ferrel G., *Introduction to Communication Systems*, Addison-Wesley, 1982.

Strojny, Jan A., *Some Factors Influencing Electrostatic Discharge from a Human Body*, Journal of Electrostatics, Vol. 40–41, 1997.

Stuchly, M.A. and S.S. Stuchly, *Dielectric Properties of Biological Substances — Tabulated*, Journal of Microwave Power, Vol. 15, No. 1, 1980.

Stutzman, Warren L. and Gary A. Thiele, *Antenna Theory and Design*, John Wiley & Sons, 1998.

Su, Kendall L., *Time-Domain Synthesis of Linear Networks*, Prentice-Hall, 1971.

Sze, S.M., *Physics of Semiconductor Devices*, John Wiley & Sons, 1981.

Tagg, G.F., *Earth Resistances*, Pitman, 1964.

Takuma, T., M. Yashima, and T. Kawamoto, *Principle of Surface Charge Measurement for Thick Insulating Specimens*, IEEE Transactions on Dielectrics and Electrical Insulation, Vol. 5, No. 4, August 1998.

Taylor, Clayborne D., Earl Harper, Charles W. Harrison, Jr., and William A. Davis, *On Bounding the Excitation of a Terminated Wire Behind an Aperture in a Shield*, IEEE Transactions on Antennas and Propagation, Vol. AP-34, No. 2, February 1986.

Taylor, D.M. and P.E. Secker, *Industrial Electrostatics: Fundamentals and Measurements*, Research Studies Press, 1994.

Taylor, D.M., *Measuring Techniques for Electrostatics*, Journal of Electrostatics, Vol. 51–52, 2001.

Tektronix, *ABCs of Probes*, Tektronix, 1997.

Tell, Richard A. and Edwin D. Mantiply, *Population Exposure to VHF and UHF Broadcast Radiation in the United States*, Proceedings of the IEEE, Vol. 68, No. 1, January 1980.

Thomas, Alan K., *Magnetic Shielded Enclosure Design in the DC and VLF Region*, IEEE Transactions on Electromagnetic Compatibility, Vol. EMC-10, No. 1, March 1968.

Thomson, J.J. and G.P. Thomson, *Conduction of Electricity Through Gases*, Vol. II, Dover, 1969.

Tippet, John C. and David C. Chang, *Characteristic Impedance of a Rectangular Coaxial Line with Offset Inner Conductor*, IEEE Transactions on Microwave Theory and Techniques, Vol. MTT-26, No. 11, November 1978.

Tolson, P., *Assessing the Safety of Electrically Powered Static Eliminators for Use in Flammable Atmospheres*, Journal of Electrostatics, Vol. 11, 1981.

Tomboulian, D.H., *Electric and Magnetic Fields*, Harcourt, Brace, & World, 1965.

Tremaine, Howard M., *Audio Cyclopedia*, Howard W. Sams, 1969.

Tsaliovich, Anatoly, *Cable Shielding for Electromagnetic Compatibility*, Van Nostrand Reinhold, 1995.

Uitert, LeGrand G. Van, *Dielectric Properties of and Conductivity in Ferrites*, Proceedings of the IRE, Vol. 44, No. 10, October 1956.

Ulaby, Fawwaz T., *Applied Electromagnetics*, Prentice-Hall, 1999.

Uman, Martin A., *Understanding Lightning*, Bek Technical Publications, 1971.

Upton, Miles E.G., and Andrew C. Marvin, *The Fields Due to a Small Loaded Loop in Free Space*, IEEE Transactions on Electromagnetic Compatibility, Vol. EMC-36, No. 1, February 1994.

Valdes, L.B., *Resistivity Measurements on Germanium for Transistors*, Proceedings of the IRE, February 1954.

Valley, George E., Jr. and Henry Wallman, *Vacuum Tube Amplifiers*, McGraw-Hill, 1948.

Vance, Edward F., *Coupling to Shielded Cables*, John Wiley & Sons, 1978.

Vendelin, George D., Anthony M. Pavio, and Ulrich L. Rohde, *Microwave Circuit Design Using Linear and Nonlinear Techniques*, John Wiley & Sons, 1990.

Vines, Roger M., H. Joel Trussell, Kenneth C. Shuey, and J.B. O'Neal, Jr., *Impedance of the Residential Power-Distribution Circuit*, IEEE Transactions on Electromagnetic Compatibility, Vol. EMC-27, No. 1, February 1985.

Vines, Roger M., H. Joel Trussell, Louis J. Gale, and J. Ben O'Neal, Jr., *Noise on Residential Power Distribution Circuits*, IEEE Transactions on Electromagnetic Compatibility, Vol. EMC-26, No. 4, November 1984.

Violette, J.L. Norman, Donald R.J. White, and Michael F. Violette, *Electromagnetic Compatibility Handbook*, Van Nostrand Reinhold, 1987.

Vo, Joe, *A Comparison of Differential Termination Techniques*, National Semiconductor Application Note 903, August 1993.

Vosteen, William E., *A Review of Current Electrostatic Measurement Techniques and Their Limitations*, Monroe Electronics, 1984.

Wadell, Brian C., *Transmission Line Design Handbook*, Artech House, 1991.

Wait, James R. and David A. Hill, *Electromagnetic Shielding of Sources Within a Metal-Cased Bore Hole*, IEEE Transactions on Geoscience Electronics, Vol. GE-15, No. 2, April 1977.

Wait, James R., *Electromagnetic Induction Technique for Locating a Buried Source*, IEEE Transactions on Geoscience Electronics, Vol. GE-9, No. 2, April 1971.

Wait, James R., *The False Image of a Line Current Within a Conducting Half-Space*, IEEE Transactions on Electromagnetic Compatibility, Vol. 39, No. 3, August 1997.

Wakerly, John F., *Digital Design: Principles and Practices*, Prentice-Hall, 2001.

Walker, Charles S., *Capacitance, Inductance, and Crosstalk Analysis*, Artech House, 1990.

Warne, L.K. and K.S.H. Lee, *Some Remarks on Antenna Response in a Reverberation Chamber*, IEEE Transactions on Electromagnetic Compatibility, Vol. 43, No. 2, May 2001.

Watkins, J., *Circular Resonant Structures in Microstrip*, Electronics Letters, Vol. 5, No. 21, October 1969.

Watson, J.K., *Applications of Magnetism*, self published, 1985.

Watt, Arthur D., *VLF Radio Engineering*, Pergamon Press, 1967.

Watterson, Peter A., *Energy Calculation of a Permanent Magnetic System by Surface and Flux Integrals (the Flux-mmf Method)*, IEEE Transactions on Magnets, Vol. 36, No. 2, March 2000.

Weber, Ernst, *Electromagnetic Theory: Static Fields and Their Mapping*, Dover, 1965.

Weeks, W.L., *Antenna Engineering*, McGraw-Hill, 1968.

Weeks, W.L., *Electromagnetic Theory for Engineering Applications*, John Wiley & Sons, 1964.

Welsby, V.G., *The Theory and Design of Inductance Coils*, MacDonald, 1960.

Wenner, Frank, *A Method of Measuring Earth Resistivity*, Bulletin of the Bureau of Standards, Vol. 12, No. 4, October 11, 1915.

Weston, David A., *Electromagnetic Compatibility: Principles and Applications*, Marcel Dekker, 1991.

Wheeler, Harold A., *Formulas for the Skin Effect*, Proceedings of the IRE, Vol. 30, No. 9, September 1942.

Wheeler, Harold A., *Inductance Formula for Circular and Square Coils*, Proceedings of the IEEE, Vol. 70, No. 12, December 1982.

Wheeler, Harold A., *The Spherical Coil as an Inductor, Shield, or Antenna*, Proceedings of the IRE, Vol. 46, No. 9, September 1958.

Wheeler, Harold A., *Transmission-Line Properties of a Round Wire in a Polygon Shield*, IEEE Transactions on Microwave Theory and Technique, Vol. MTT-27, No. 8, August 1979.

Whitaker, Jerry C., *AC Power Systems Handbook*, CRC Press LLC, 1999.

White, Donald R.J., *EMI Test Instrumentation and Systems, Vol. 4*, Don White Consultants, 1971.

White, Joseph F., *Microwave Semiconductor Engineering*, Van Nostrand Reinhold, 1982.

White, R.J. and Michel Mardiguian, *Electromagnetic Shielding, Vol. 3*, Interference Control Technologies, 1988.

Whitehouse, A.C.D., *Screening: New Wave Impedance for the Transmission-Line Analogy*, Proceedings of the IEE, Vol. 116, No. 7, July 1969.

Whiteside, H. and R. King, *The Loop Antenna as a Probe*, IEEE Transactions on Antennas and Propagation, Vol. AP-12, No. 3, May 1964.

Whitlock, Bill, *Answers to Common Questions About Audio Transformers*, Jensen Application Note AN-002.

Whitlock, Bill, *Interconnection of Balanced and Unbalanced Equipment*, Jensen Application Note AN-003.

Whitlock, Bill, *Subtleties Count in Wide-Dynamic-Range Analog Interfaces*, EDN, June 4, 1998.

Williams, George M. and Thomas H. Pratt, *Characteristics and hazards of electrostatic discharges in air*, Journal of Loss Prevention in the Process Industries, Vol. 3, October 1990.

Wilson, Perry F. and Mark T. Ma, *Simple Approximate Expressions for Higher Order Mode Cutoff and Resonant Frequencies in TEM Cells*, IEEE Transactions on Electromagnetic Compatibility, Vol. EMC-28, No. 3, August 1986.

Wilson, Robert, *The Offset Multiband Trapless Antenna (OMTA)*, QST, October 1995.

Winburn, D.C., *Practical Electrical Safety*, Marcel Dekker, 1988.

Winder, Steve, *Filter Design*, Newnes, 1997.

Wolff, Edward A., *Antenna Analysis*, John Wiley & Sons, 1967.

Wolff, Ingo and Norbert Knoppik, *Rectangular and Circular Microstrip Disk Capacitors and Resonators*, IEEE Transactions on Microwave Theory and Techniques, Vol. MTT-22, No. 10, October 1974.

Wood, Jody W., Personal Communications, New England Wire Technologies, 2003.

Xu, Xiao-Bang and Xiao-Mei Yang, *A Simple Computational Method for Predicting Magnetic Field in the Vicinity of a Three-Phase Underground Cable with a Fluid-Filled Steel-Pipe Enclosure*, IEEE Transactions on Power Delivery, Vol. 10, No. 1, January 1995.

Xu, Xiao-Bang and Xiao-Mei Yang, *Computation of the Magnetic Field Generated by an Underground Pipe-Type Cable*, IEEE Transactions on Power Delivery, Vol. 11, No. 2, April 1996.

Young, F.J., D.J. Boomgaard, and D.A. Colling, *Transfer Impedance of Soft Magnetic Materials*, Journal of Applied Physics, Vol. 42, No. 4, March 1971.

Yu, Luke, Mark Chow, and Jay Bowen, *Safety and Ground Fault Protection in Electrical Systems*, IEEE Industry Applications Magazine, March/April 1998.

Zaborszky, John and Joseph W. Rittenhouse, *Electric Power Transmission*, Ronald Press, 1954.

Zaffanella, L.E., *Pilot Study of Residential Power Frequency Magnetic Fields*, Electric Power Research Institute, EL-6509, Research Project 2942, September 1989.

Zahn, Markus, *Electromagnetic Field Theory: a problem solving approach*, Robert E. Krieger, 1979.

Zemanian, A.H., *Distribution Theory and Transform Analysis: An Introduction to Generalized Functions, with Applications*, Dover, 1965.

Zverev, Anatol I., *Handbook of Filter Synthesis*, John Wiley & Sons, 1967.

Index

D

E

F

M

N

S

T

U

V

W

X

Y

Z

Printed and bound by CPI Group (UK) Ltd, Croydon, CR0 4YY
17/10/2024
01775700-0006